Novel Developments in Aquaculture and Fisheries Science

Novel Developments in Aquaculture and Fisheries Science

Editor: Geoffrey Gilbert

www.callistoreference.com

Callisto Reference,
118-35 Queens Blvd., Suite 400,
Forest Hills, NY 11375, USA

Visit us on the World Wide Web at:
www.callistoreference.com

ISBN: 978-1-64116-067-4 (Hardback)

Cataloging-in-Publication Data

Novel developments in aquaculture and fisheries science / edited by Geoffrey Gilbert.
p. cm.
Includes bibliographical references and index.
ISBN 978-1-64116-067-4
1. Aquaculture. 2. Fishery sciences. 3. Fisheries. I. Gilbert, Geoffrey.
SH135 .N68 2019
639.8--dc21

Table of Contents

Preface

Aquaculture and fisheries science are two distinct yet interrelated fields of study. Aquaculture is the farming of aquatic animals such as fishes, molluscs, algae and aquatic plants. Fisheries science is concerned with the management of fisheries by integrating the principles of oceanography, marine biology, ecology, limnology, etc. A common concern in both these fields is the sustainable exploitation of aquatic organisms in order to prevent extinction. This book aims to shed light on some of the unexplored aspects of aquaculture and fisheries science. It strives to provide significant in-depth knowledge of these disciplines and develop a comprehensive understanding. This book is appropriate for students seeking detailed information in this area as well as for experts.

This book unites the global concepts and researches in an organized manner for a comprehensive understanding of the subject. It is a ripe text for all researchers, students, scientists or anyone else who is interested in acquiring a better knowledge of this dynamic field.

I extend my sincere thanks to the contributors for such eloquent research chapters. Finally, I thank my family for being a source of support and help.

Editor

1

A Common Concept of Population Dynamics Applicable to Both *Thrips imaginis* (*Thysanoptera*) and the Pacific Stock of the Japanese Sardine (*Sardinops melanostictus*)

Kazumi Sakuramoto*

Tokyo University of Marine Science and Technology, Konan, Minato-ku, Tokyo, Japan

Abstract

The aim of this study was to discuss a common concept of population dynamics applicable to both *Thrips imaginis* (*Thysanoptera*) and the Pacific stock of the Japanese sardine (*Sardinops melanostictus*). First, I elucidate the mechanism that produces the false density-dependent effect on population changes of *Thrips imaginis*, I conducted simple deterministic simulations to discuss the qualitative viewpoints. Second, I conducted Monte Carlo simulations by using the average population size and standard deviation of the thrip data used by Davidson and Andrewartha for the quantitative discussion. In simple deterministic simulations, the resultant plots of population change against population size showed a decreasing trend for which the slope was statistically significant even though the true relationship between the population change and population size had no density-dependent effect. The results of the Monte Carlo simulations indicated that nearly 70% of the trials showed false density-dependent effects. The provability of the false decision, which was to recognize the existence of density-dependent effects, increased as the standard deviation in population size in a month *i*-1 increased. When the number of samples increased, the probability of the false decision greatly increased. The conclusion from these simulations was that the density-dependent effect that emerged in the population change of *T. imaginis* was artificially produced and invalid. Further, the thrip population size in month *i* was determined in proportion to that in month *i*-1; and environmental conditions in month *i*. This mechanism was completely same of that shown in the Japanese sardine (*Sardinops melanostictus*). Therefore, the fluctuations in populations of *T. imaginis* and Japanese sardine could be explained with the same concept of population dynamics shown here.

Keywords: *Thrips imagines*; Density-Dependent Effect; Population Change; Reproductive Success; Japanese Sardine

Introduction

When discussing a fluctuation mechanism in biotic resources, one of the key issues is whether a density-dependent effect exists. In fisheries science, nearly all scientists, at one time, have believed in the existence of a density-dependent effect and have discussed management schemes under the assumption that a density-dependent effect plays an important role in controlling population fluctuations. In other words, if a density-dependent effect truly exists in the stock-recruitment relationship, which is the relationship between recruitment and spawning-stock biomass, or is a key factor in controlling population fluctuations, then an optimal population size that achieves the maximum sustainable yield (MSY) can be defined, which is one of the most important concepts in fisheries management. The basic idea of fisheries management is essentially constructed by this concept, which has been widely used all over the world [1,2]. However, if a density-dependent effect does not exist or is nearly negligible in explaining population fluctuations, then an MSY cannot be defined, and the optimal population size could never be determined biologically. Therefore, depending on whether a density-dependent effect exits or is negligible, the management procedure will differ. The concept of "regime shift," which is large, abrupt, and persistent changes in the ocean ecosystem, has recently become popular [3]. By using the concept of regime shift, Wada and Jacobson [4] analyzed Japanese sardines and concluded that density-dependent effects existed in the favorable and unfavorable regimes, and the carrying capacities during the favorable regime were 75 times greater than those during the unfavorable regime. Wada and Jacobson's [4] work has been cited in many paper [5-9] and is truly a key paper in the analysis of *Sardinips* population dynamics. However, Sakuramoto [10] reanalyzed the data used by Wada and Jacobson [4] and showed diametrically opposed results. That is, a false decreasing trend was produced in the regression line of the natural logarithm for reproductive success (recruitment divided by egg production) against the natural logarithm for egg production in response to the observation error, and thus, the density-dependent effect detected was artificial and invalid. Sakuramoto [10] concluded that the recruitment for the Japanese sardine was basically determined in proportion to the spawning stock biomass, then a proportional model was reasonable to accept as the optimal stock-recruitment relationship model.

Sakuramoto and Suzuki [11] showed the mechanism that the density-dependent effect was erroneously detected in response to the observed and/or process errors in the variables. In particular, the effect was large when the errors in the independent variables increased. Further, Sakuramoto and Suzuki [11] specifically proposed that an opposite result was not possible. That is, when the true model showed a density-dependent effect, such as that expressed by the Ricker [12] or Beverton and Holt model [13], the proportional model was seldom selected in response to process and/or observation errors. This suggests that if a proportional model is identified for the actual data, then it

***Corresponding author:** Kazumi Sakuramoto, Tokyo University of Marine Science and Technology, Konan, Minato-ku, Tokyo, Japan, E-mail: sakurak@kaiyodai.ac.jp

does not erroneously produce results in response to process and/or observation errors.

I was surprised to recently learn of the existence of a similar polemic on the density-dependent effect in the field of entomology, more than 50 years ago. The apple blossom thrips (*Thrips imaginis Bagnall*) is widely distributed in Australia, where is an indigenous species [1]. The females do not lay eggs in batches, but lay several eggs each day continuously throughout their adult life; and an adult may live for several months at the temperatures experienced during the spring [14]. The population at any time includes individuals of various stages and ages; the generation effects is therefore not likely to be pronounces [15]. During the years 1932-8 the number of *T. imaginis* in 20 roses in the garden at the White Institute was recorded throughout the year. Since 1938 samples have been taken only during the spring and early summer (September to December) [16].

By using these data, Davidson and Andrewartha [1,16] showed that the population size of *T. imaginis* can be predicted accurately based on the data regarding the temperature and precipitation, and they concluded that the population size could be explained by a succession of good and bad seasons and by the heterogeneity of places where they live [17]. That is, they denied the conventional understanding that a natural population can be regulated only by density-dependent factors; however, Frederick Smith [18] reanalyzed Davidson and Andrewartha's data to test whether changes in the thrip population size also showed signs of density-dependent effects. That is, whether the changes in the thrip population size showed a negative trend against the population size. I believe that the polemic expressed regarding the thrip population size is essentially the same problem discussed in fisheries researches. The issue discussed in the field of entomology will be resolved using the same logic as that proposed by Sakuramoto [10]. The aim of this study is to elucidate the mechanism that produces the false density-dependent effect in population changes of *T. imaginis* and to show the mechanisms in population fluctuations of *T. imaginis* can be explained with essentially the same mechanisms in population fluctuations of Japanese sardine.

Materials and Methods

Simple deterministic simulations

First, in order to show the mechanism that produces the false density-dependent effect in population changes, simple deterministic simulations were conducted according to Sakuramoto [10] as follows: N_1 and N_2 denote the average thrip population size in months t_1 and t_2, respectively. For simplicity, I considered the case when N_1 and N_2 independently fluctuated in response to environmental conditions in month t_1 and t_2. Only 3 environmental conditions were assumed: bad, moderate, and good. When the condition was bad, the population deceased at half the average rate, and when the condition was good, the population increased by 2 times the average rate. When the condition was moderate, the population equaled the average size. When the number of thrips in t_1 and t_2 were independently determined by environmental conditions in the months studied, 9 pairs of population sizes were obtained ($n_{1,i}$, $n_{2,i}$) (i=1, 2,...9), where $n_{1,i}$, $n_{2,i}$ denote the observed thrip population size in months t_1 and t_2, respectively. The population changes during these 2 months were defined by using the following logarithm: $\log(dn_i) = \log(n_{2,i}) - \log(n_{1,i})$. That is, 9 pairs of $\log(n_{1,i})$ and $\log(dn_i)$ (i=1, 2,...9) were obtained. Table 1 shows the 3 cases of (N_1, N_2). That is, N_1 was set at 1000 for all 3 cases; however, N_2 was set at 2000 (increasing pattern), 1000 (no different pattern) and 500 (decreasing pattern) respectively.

Monte Carlo simulations

Second, in order to check the probability that produces the false density-dependent effect in population changes of *T. imaginis*, Monte Carlo simulations were conducted by using the data for thrip populations listed in Table 3 of Davidson and Andrewartha [1] (Table 2). In accordance with Smith [18], I first focused on the data in October and November. According to the period of data from 1932 to 1938, I randomly generated 7 pairs of artificial data, $\log(n_{1,i})$ and $\log(n_{2,i})$ from the normal distribution $N\left(m_{Oct}, sd^2_{Oct}\right)$ and $N\left(m_{Nov}, sd^2_{Nov}\right)$, respectively, where m_{Oct}, sd_{Oct} and m_{Nov}, sd_{Nov} were the means and standard deviations of the observed logarithm transformed thrip data by October and November,

Deterministic simulation 1 Increasing pattern			N_2=2000		
			$n_{2.b}$	$n_{2.m}$	$n_{2.g}$
			1000	2000	4000
n_1=1000	$n_{1.b}$	500	(n_1,dn_1)=(500,2)	(n_1,dn_1)=(500,4)	(n_1, dn_1)=(500,8)
	$n_{1.m}$	1000	(n_1,dn_1)=(1000,1)	(n_1,dn_1)=(1000,2)	(n_1, dn_1)=(1000,4)
	$n_{1.g}$	2000	(n_1,dn_1)=(2000,0.5)	(n_1,dn_1)=(2000,1)	(n_1, dn_1)=(2000,2)

Deterministic simulation 2 No different pattern			N_2=1000		
			$n_{2.b}$	$n_{2.m}$	$n_{2.g}$
			500	1000	2000
n_1=1000	$n_{1.b}$	500	(n_1, dn_1)=(500,1)	(n_1, dn_1)=(500,2)	(n_1, dn_1)=(500,4)
	$n_{1.m}$	1000	(n_1, dn_1)=(1000,0.5)	(n_1, dn_1)=(1000,1)	(n_1, dn_1)=(1000,2)
	$n_{1.g}$	2000	(n_1, dn_1)=(2000,0.25)	(n_1, dn_1)=(2000,0.5)	(n_1, dn_1)=(2000,1)

Deterministic simulation 3 Decreasing pattern			N_2=500		
			$n_{2.b}$	$n_{2.m}$	$n_{2.g}$
			250	500	1000
n_1=1000	$n_{1.b}$	500	(n_1,dn_1)=(500,0.5)	(n_1,dn_1)=(500,1)	(n_1,dn_1)=(500,2)
	$n_{1.m}$	1000	(n_1,dn_1)=(1000,0.25)	(n_1,dn_1)=(1000,0.5)	(n_1,dn_1)=(1000,1)
	$n_{1.g}$	2000	(n_1,dn_1)=(2000,0.125)	(n_1,dn_1)=(2000,0.25)	(n_1,dn_1)=(2000,0.5)

Table 1: Population sizes in month t_1 and t_2 assumed in the simple deterministic simulations. (n_1, dn_1) denotes the observed pairs of population size and population change determined by each environmental condition. $n_{i,b}$, $n_{i,m}$, and $n_{i,g}$, (i=1,2) denote bad, moderate and good environmental conditions, respectively.

	1932-33	1933-34	1934-35	1935-36	1936-37	1937-38	1938-	Mean	Standard deviation
Apr.	0.65	0.38	0.79	0.41	0.52	0.30	0.54	0.51	0.17
May	1.37	0.59	1.16	0.79	0.91	0.48	1.04	0.91	0.31
Jun.	1.25	0.72	1.43	1.22	1.11	0.68	0.75	1.02	0.30
Jul.	0.64	0.74	0.83	-0.10	0.96	0.40	0.90	0.62	0.37
Aug	0.52	0.23	0.57	0.23	0.26	0.23	0.73	0.40	0.21
Sep.	1.53	0.45	0.77	0.74	0.63	0.66	1.71	0.93	0.49
Oct.	1.14	1.00	0.20	1.34	1.05	1.39	2.20	1.19	0.59
Nov.	2.13	1.89	1.56	2.13	1.61	2.49	2.76	2.08	0.44
Dec.	2.43	1.85	1.88	2.14	1.84	2.11	2.14	2.05	0.21
Jan.	1.58	1.19	1.16	1.20	0.77	1.19		1.18	0.26
Feb.	0.99	0.71	0.89	0.76	0.46	0.79		0.77	0.18
Mar	0.45	0.86	0.57	0.80	0.57	0.60		0.64	0.16

Table 2: The data used by Davidson and Andrewartha [1] transformed by logarithm.

Number of samples = 7		s.d. (Nov.)		
		0.75	1.00	1.25
s.d. (Oct.)	0.75	687	493	346
	1.00	849	687	529
	1.25	932	817	687

Table 3: Results of sensitivity tests when the standard deviation was changed.s.d denotes the standard deviation. Figures show the number of trials in which the slope was significantly negative.

respectively. Monte Carlo simulations were conducted 1000 times and the slopes of the regression lines of $\log(dn_i)$ were calculated against $\log(n_{1,i})$ (i = 1, 2,...,7). Then, I counted the number of trials in which the slope was significantly negative.

Sensitivity tests

The sensitivity tests for the number of trials in which the slope was significantly negative were conducted when the standard deviations changed in $\log(n_{1,i})$ and/or $\log(n_{2,i})$. In addition, the tests were conducted when the number of observed samples changed from 7 to 14.

Reanalysis of the data by Davidson and Andrewartha [1]

In accordance with Smith [18], I first reanalyzed the October and November data by Davidson and Andrewartha [1]. Second, I analyzed the relationships between the population size in month *i* and that in month *i*-1 for 7 years (*i*=May, Jun,...,Sep., and Dec.) and 6 years (*i*=Dec,..., Feb.) respectively. I also analyzed the relationship using all the data from April 1932 to December 1938. I used 2 regression methods for plotting $\log(n_{2,i})$ against $\log(n_{1,i})$ (i.e., single regression analysis and Deming regression analysis [19]. A single regression analysis is problematic in that it assumes no observation error of the independent variable; therefore, parameter estimates derived from a single regression analysis exhibit bias [20]. When both independent and dependent variables have observational errors, the Deming regression analysis [19] is effective to remove the bias derived from a single regression analysis for the case when the sample size is enough large. The estimation using the Deming regression was conducted using the program developed by Aoki [21].

Comparison of the model proposed by Davidson and Andrewartha [1] and that proposed in this paper

If the population size in month *i* is determined in proportion to that in month *i*-1 and the effect of the environmental conditions in month *i* multiplicatively add to the population size in month *i*, the relationship of population size between the 2 months can be expressed as follows:

$$n_i = a \cdot n_{i-1} \bullet f(\mathbf{x}_i) \quad (1)$$

Where a and $f(\cdot)$ denote a proportional constant and functions that determine the effects in response to environmental conditions $\mathbf{x}_i=[x_{i,1},...,x_{i,k}]$ (*k* denotes the number of environmental factors), respectively. When n_0 denotes the first month or reference month, n_i can be written as follows:

$$n_i=a_i \cdot n_0 \bullet a^{i-1} \cdot f(\mathbf{x}1) \bullet a^{i-2} \cdot f(\mathbf{x}_2) \bullet .... \bullet af(\mathbf{x}_{i-1}) \bullet f(\mathbf{x}_i) \quad (2)$$

This indicates that the population size in month *i* can be expressed using the population size in reference to month 0 and the environmental conditions of the reference month over that of month *i*. That is, the difference of population size in month *i* from the reference month 0 can be explained only by the environmental factors, $\mathbf{x}_1$,...,$\mathbf{x}_i$. On the other hand, if equation (2) is correct, then the following relationship between month *i* and month *i*–1 can be obtained:

$$n_i=a\{a^{i-1} \cdot n_0 \bullet a^{i-2} \cdot f(\mathbf{x}_1) \bullet a^{i-3} \cdot f(\mathbf{x}_2) \bullet .... \bullet f(\mathbf{x}^{i-1})\} \bullet f(\mathbf{x}_i)=a \cdot n_{i-1} \bullet f(\mathbf{x}_i) \quad (3)$$

Davidson and Andrewartha [16] used the partial regression analysis to explain the population fluctuation in *T. imaginis* and noted in the summary of their paper as follows: The method of partial regression was used to measure the degree of association between the number of thrips present during the spring and the weather experienced during the preceding months. Therefore, their partial regression models can be summarized as follows:

$$\log(n_i) = \log(n) + \log h(\mathbf{x}_1) + \log h(\mathbf{x}_2) + ... + \log h(\mathbf{x}_i), \quad (4)$$

Where, $\log(n)$ and $h(\cdot)$ denote the logarithm of the geometric mean of population size in spring and functions that determine the effects in response to environmental conditions $\mathbf{x}_i=[x_{i,1},..., x_{i,q}]$ (*q* denotes the number of environmental factors).

Results

Results of simple deterministic simulations

Figure 1 shows the results of the simple deterministic simulations. In these Figure 1 simulations, although the values on the y-axis were different, all the patterns were the same. That is, I did not assume a density-dependent effect in the relationship between $\log(dn_i)$ and $\log(n_{1,i})$ in these simulations, even though the resultant plots showed decreasing trends in all cases. All the slopes of the regression lines of

$\log(dn_i)$ against $\log(n_{1,i})$ were -1, and all were statistically significant. In Figures 1a, b, and c, the significant provability (p values) was 0.033 (<0.05); in Figure 1d, it was 0.022, which was smaller than those in Figures 1a, b, and c, because the number of data sets increased from 9 to 18.

Results of Monte Carlo simulations

Table 2 shows the logarithm transformed thrip data from Table 3 in Davidson and Andrewartha [1]. The averages and standard deviations, respectively, were 1.19 and 0.59 in October and 2.08 and 0.44 in November. By using these averages and standard deviations, Monte Carlo simulations were conducted. Figure 2 shows the histogram of the slopes for 1000 trails (Figure 2a) and that of the slopes that were significantly negative by a 5% significance level in the 1000 trails (Figure 2b). In this case, 687 out of 1000 trials showed significantly negative slopes with a 5% significance level. That is, 68.7% of the trials erroneously indicated the existence of density-dependent effects in the relationship between population change and population size.

Results of sensitivity tests

Table 3 shows the results of the sensitivity tests for the number of trials in which the slope was significantly negative when the standard deviation in the population size in October and/or November was changed. When the standard deviation in October (independent variable) increased by 1.25 times than that observed, the percentage of misjudges increased. On the other hand, when the standard deviation in November (dependent variable) decreased by 0.75 times than that observed, the percentage of misjudges increased. When the levels of decreasing and increasing standard deviations were the same in both October and November (e.g., 0.75 times in October and 0.75 times in November or 1.25 times in October and 1.25 times in November), the percentage of misjudges was the same at 68.7%. The standard deviation in October was large and that in November was small, the percentage of misjudges was the highest, and vice versa.

Table 4 shows the results of sensitivity tests for the number of trials in which the slope was significantly negative when the number of samples changed from 7 to 14. This effect was extremely large. In nearly all cases, the slopes of the regression lines indicated the false decreasing trends.

Re-analysis results of the data from Davidson and Andrewartha [1]

Figure 3 shows the reanalysis results of the data from Davidson and Andrewartha [1]. Figure 3 (Top left) shows the trajectories of the log transformed thrip population size in October and November. The patterns of the trajectories were similar to each other. Figure 3 (Top right) shows the relationship between population change log(dn) and population size in October. As mentioned previously, the population change $\log(dn)$ was defined using $\log(n_{Nov})$ - $\log(n_{Oct})$. The regression line indicated the negative slope at -0.342 and showed a 10% statistical significance. Figure 3 (Bottom left) shows the regression line of the population size in November against that in October. The slope of the regression line estimated using a simple regression method was 0.656 (p=0.008), and the 95% confidence interval of the slope was [0.260, 1.051]. The slope of the regression line estimated using the Deming regression method was 0.713, and the 95% confidence interval of the slope was [0.529, 1.718].

Figure 4 shows the trajectories of the log transformed thrip population size in month i-1 and month i (i=May, Jun, ..., Mar., except Oct.), respectively. Figure 5 shows the relationship between population size in month i against that in month i-1, (i=May, Jun, ..., Mar., except Oct.), and shows the regression lines estimated by simple method (blue line) and Deming regression (red line), respectively. The slopes estimated are also shown in Table 5. In the simple regression, in only 3 cases (Oct.-Nov., Dec.- Jan., and Jan.-Feb.), the slopes were not significantly different from unity. For other 2 cases (Apr.-May, and Aug.-Sep.), the 95% confidence interval of the slopes were extremely

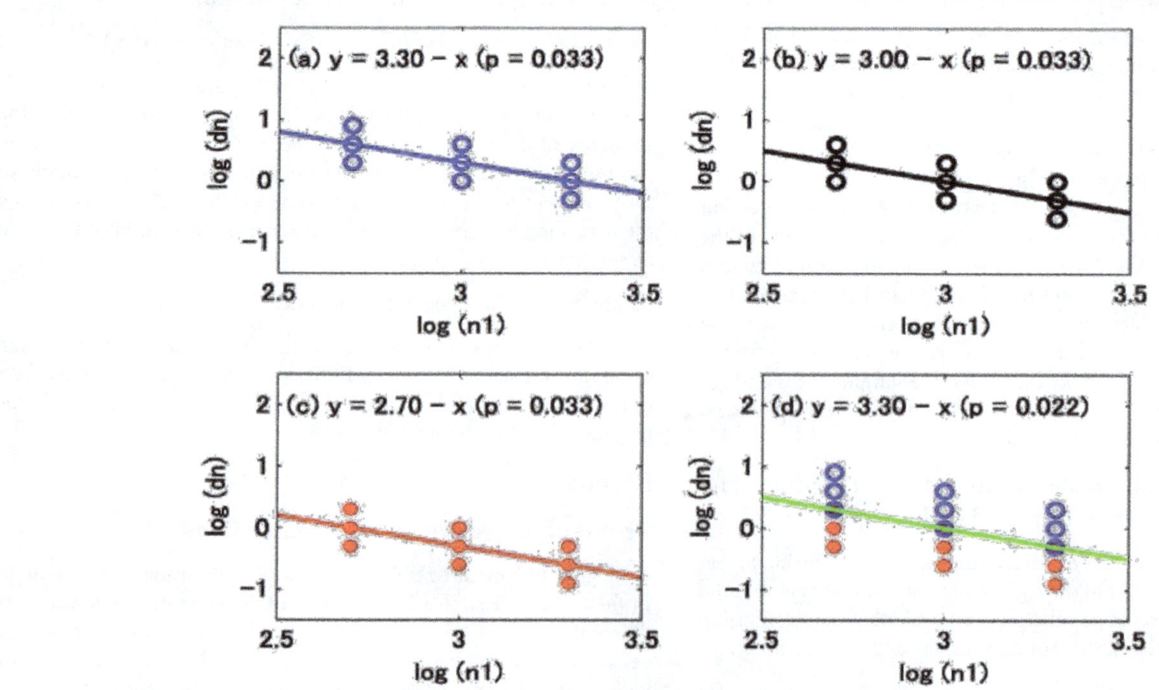

Figure 1: Results of simple deterministic simulations. a: Increasing pattern; b: constant pattern; c: decreasing pattern; and d: a (black) and c (red) combined.

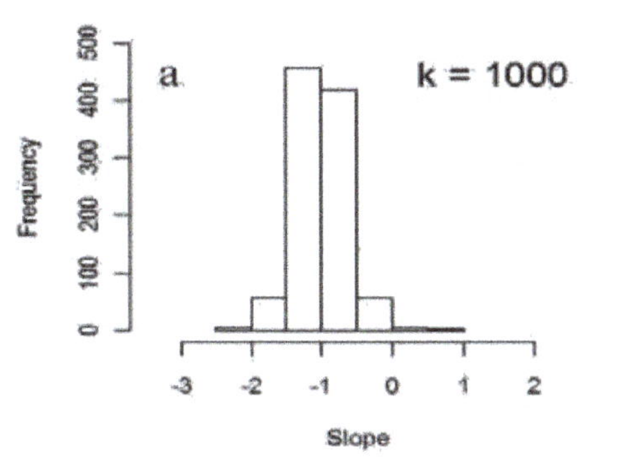

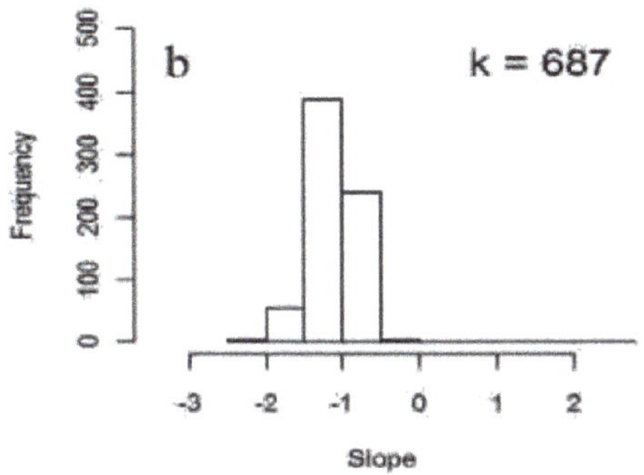

Figure 2: a: Histogram of the slope of the regression lines obtained in 1000 Monte Carlo simulations; and b: histogram when the slope was significantly negative at a 5% significance level. *k* denotes the total number of trials shown in the histogram.

large, however they were not significantly less than unity. Figure 6 shows the results when all data from April 1932 to December 1938 were combined. Then the number of samples was 74. In this case the slope of the regression line estimated using the simple regression method was 0.597 (p=1.897 × 10^{-8}), and the 95% confidence interval of the slope was [0.409, 0.786]; The slope estimated by the Deming regression was 1.002 and the 95% confidence interval was [0.814, 1.338], then the slope was not significantly different from unity.

Discussion

The results of the simulations showed that even when a density-dependent effect did not exist, an apparent density-dependent effect was erroneously detected from the regression line plotted for population change against population size. Therefore, when such negative slopes of the regression lines are obtained, we should not judge that a density-dependent effect truly exists.

In this analysis, I defined population change by $\log(n_2)$ - $\log(n_1)$ instead of $\log[(n_2-n_1)/n_1]$, because I wanted to analyze the change when n_2 is smaller than n_1. The population size in November was greater than that in October; therefore, when the population size in October is used as n_1 and that in November is used as n_2, the definition $\log[(n_2 - n_1)/n_1]$ could be used for the population change. However, the results obtained were nearly same as that by using the former definition. The slope of the regression line was -0.393, and the significant probability was 0.0808. The analyses for each month, because the sample sizes were limited, then only in 5 cases, the slope of the regression lines between population size in month i against that in month i-1 were significant with 5% significant level. In there, only for 3cases (Oct.-Nov., Dec.- Jan., and Jan.-Feb.), the slopes were not significantly different from unity. For other 2 cases (Apr.-May and Aug.-Sep.), the 95% confidence interval of

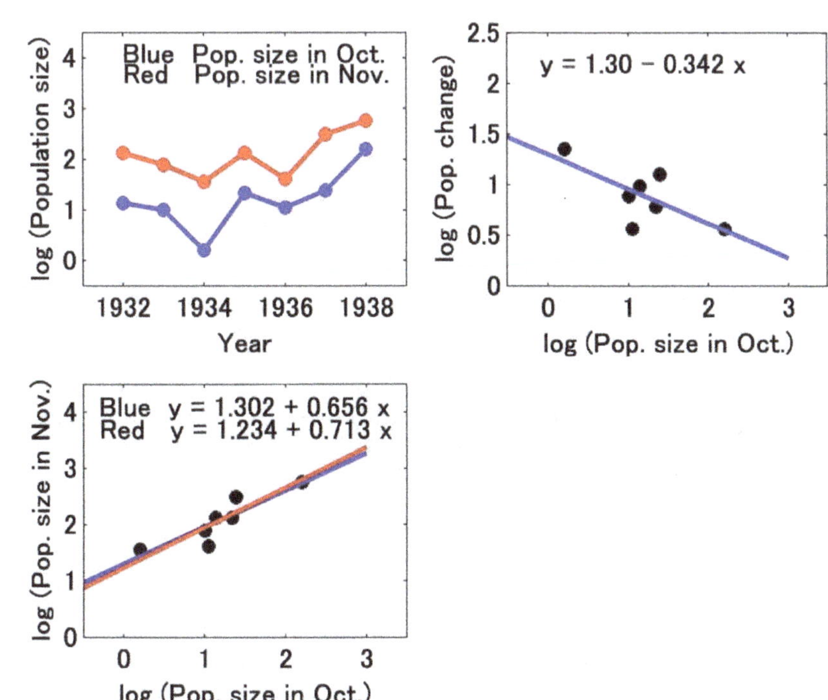

Figure 3: Reanalysis results of the October and November data used by Davidson and Andrewartha [1]. Top left: Trajectories of the population sizes in October $\log(n_1)$ and November $\log(n_2)$; Top right: regression line for population change $\log(dn)$ against $\log(n_1)$; and Bottom left: regression line for $\log(n_2)$ against $\log(n_1)$. Blue and red lines indicate the regression lines estimated by simple and Deming regression methods, respectively.

Number of samples = 14		s.d. (Nov.)		
		0.75	1.00	1.25
s.d. (Oct.)	0.75	964	883	737
	1.00	999	964	914
	1.25	1000	995	964

Table 4: Results of sensitivity tests when the standard deviation was changed. Figures show the number of trials in which the slope was significantly negative.

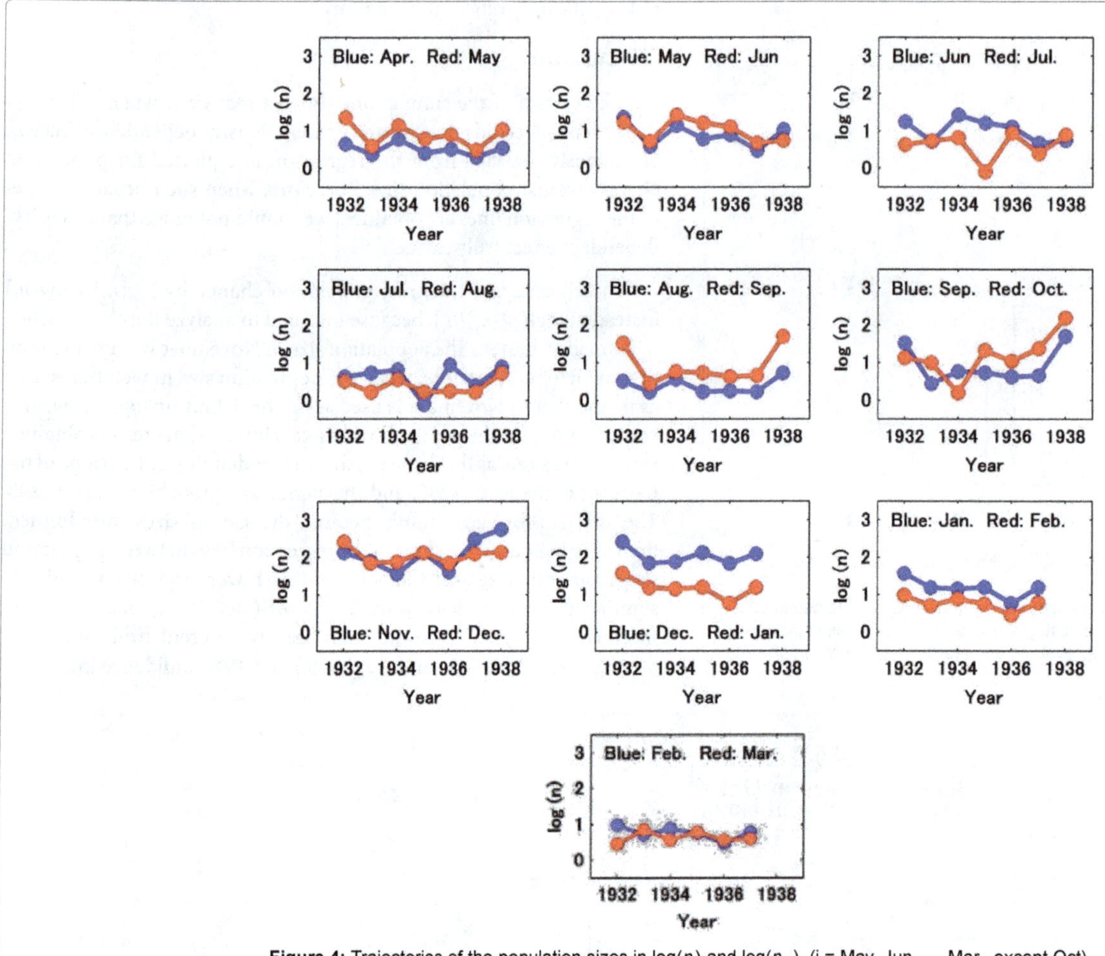

Figure 4: Trajectories of the population sizes in log(n_i) and log(n_{i-1}), (i = May, Jun, ..., Mar., except Oct).

the slopes were extremely large, however, the interval did not show less than unity. This means that there was no case that showed the negative density-dependent effect between the population sizes in the successive two months. When the regression line of log(n_i) against log(n_{i-1}) for all data was estimated using the simple regression method, the slope was less than unity. However, as Sakuramoto and Suzuki [11] showed, the regression line less than unity was considered to be produced by process and/or observed errors. When the Deming regression was applied to the data, the slope was not statistically different from unity. Then, the relationship can be expressed using log(n_{Nov})=log(a) + log(n_{Oct}) or n_{Nov} = $a{\bullet}n_{Oct}$. This indicates that the population size in month i could be determined in proportion to the population size in month i-1, i.e., the relationship between the populations sizes in month i and that in month i-1 can be expressed using a simple proportional model. The validity of a proportional model is not explicitly noted, however, the fluctuations of the catches in the species such as snow crab [22], sandfish [23,24] and walleye pollock [25] are already explained without assuming a density-dependent effect in the fluctuation mechanism in catches. The population fluctuation in the Peruvian anchoveta [26] is explained well using the same concept of the fluctuation mechanism proposed in this paper. That is, my concept for the mechanism in population fluctuation coincides well with that by Davidson and Andrewartha [16] and Andrewartha and Birch [27] in that the density-dependent effect is not a key factor or a negligible factor in controlling the population size.

However, an important difference in the concept between Andrewartha and Birch [27] and that proposed by me is that they believed that the essentially important factors in controlling population size were the succession of good and bad seasons and the heterogeneity of places where they live [17], and they did not refer to the proportional relationships of population size between each month. On the contrary, my proposed concept is that the population size in month i, is proportional to that in month i-1 and environmental conditions in month i multiplicatively add to the population size in month i. This simple mechanism can also explain the concept of the fluctuation mechanism proposed by Davidson and Andrewartha [16] and Andrewartha and Birch [27] as shown in equations (2) and

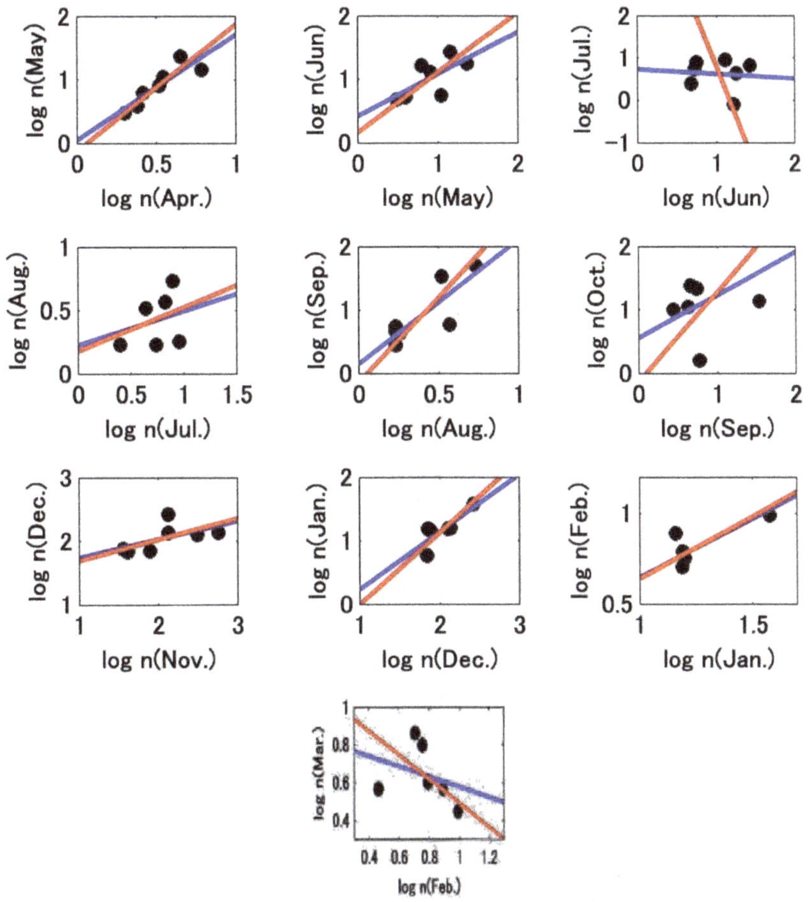

Figure 5: Trajectories of the population sizes for $\log(n_i)$ against $\log(n_{i-1})$, (i = May, Jun, ..., Mar. , except Oct).

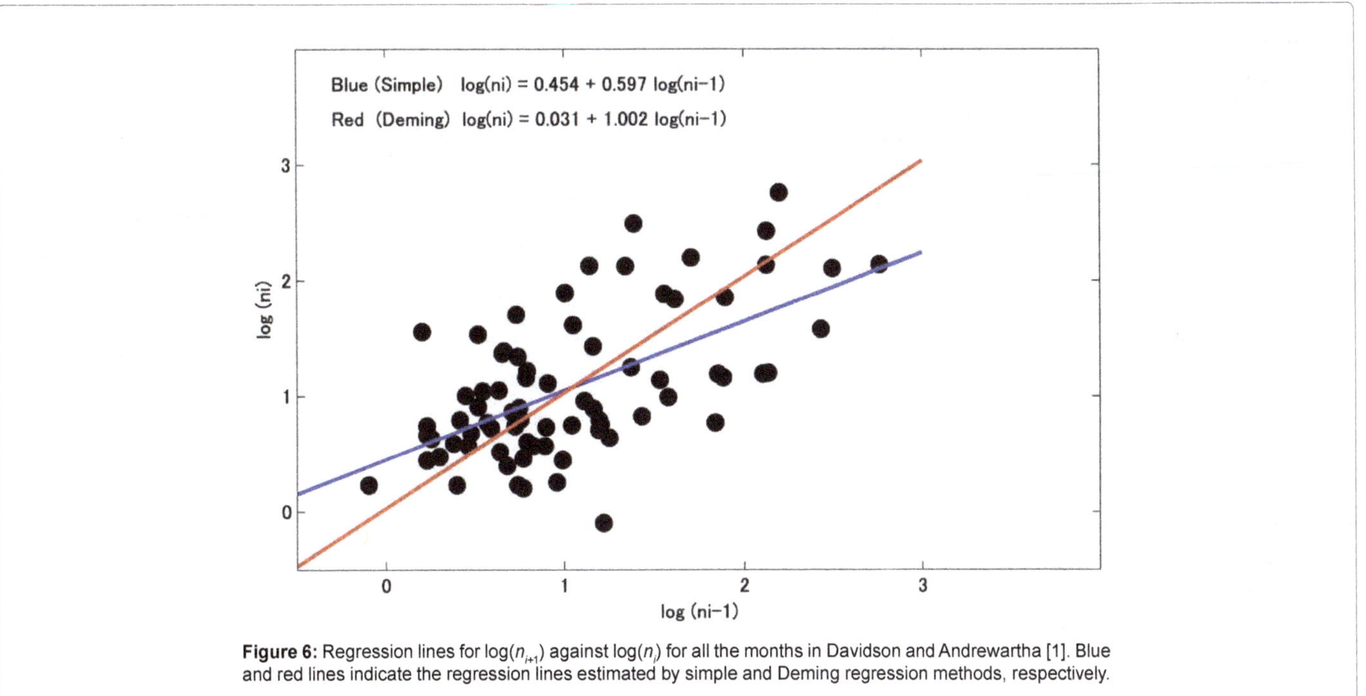

Figure 6: Regression lines for $\log(n_{i+1})$ against $\log(n_i)$ for all the months in Davidson and Andrewartha [1]. Blue and red lines indicate the regression lines estimated by simple and Deming regression methods, respectively.

Simple regression				
n_{i-1}	n_i	**Slope**	**95% confidence interval**	**p-value**
Apr.	May	1.661	(0.680, 2.64)	0.00733
May	Jun	0.661	(-0.142, 1.463)	0.0880
Jun	Jul.	-0.109	(-1.510, 1.292)	0.850
Jul.	Aug.	0.271	(-0.294, 0.836)	0.272
Aug.	Sep.	1.950	(0.449,3.451)	0.0206
Sep.	Oct.	0.679	(-0.487, 1.845)	0.195
Oct.	Nov.	0.656	(0.260, 1.051)	0.00802
Nov.	Dec.	0.288	(-0.166, 0.741)	0.164
Dec.	Jan.	0.901	(0.00776, 1.793)	0.0488
Jan.	Feb.	0.645	(0.253, 1.036)	0.0103
Feb.	Mar.	-0.271	(-1.409, 0.868)	0.546
All data		0.597	(0.409, 0.786)	1.897(10^{-8})
Deming regression				
n_{i-1}	n_i	**Slope**	**95% confidence interval**	
Apr.	May	1.998	(1.715, 2.735)	
May	Jun	0.944	(-2.114, 1.686)	
Jun	Jul.	-4.775	-5.581, 0.603	
Jul.	Aug.	0.348	(-, -)	
Aug.	Sep.	2.684	(2.137, -)	
Sep.	Oct.	1.424	(0.664, 11.978)	
Oct.	Nov.	0.713	(0.528, 1.718)	
Nov.	Dec.	0.335	(0.133, 0.841)	
Dec.	Jan.	1.133	(0.279, 1.356)	
Jan.	Feb.	0.682	(-2.450, 0.738)	
Feb.	Mar.	-0.635	(-2.004, 0.600)	
All data		1.002	(0.814, 1.338)	

Table 5: Intervals and p values estimated using a simple or the Deming regression.

(4). This seems to be the same mechanism that Sakuramoto [10,28] proposed for the stock-recruitment relationship in Japanese sardine. That is, recruitment R_t is essentially determined in proportion to spawning stock biomass S_t, and then environmental factors in year *t* further change the recruitments. The subscripts showing years are all *t*, however, there are time lags. S_t, and R_t denote the biomass in January and the number of recruitment over March to May, respectively. The environmental factors are sea surface temperature over the southern area of the Kuroshio Extension in February and Arctic oscillation in February. Therefore, the fluctuations mechanism for Japanese sardine, which lives in the sea, and that for *T. imaginis*, which lives on the land, could be explained by the same concept of fluctuation mechanism; population size in time *i* is essentially determined in proportion to that in time *i*-1, which is further changed in response to environmental conditions in time *i* [29].

Acknowledgements

I thank to the great efforts expended by Davidson, Andrewartha and their groups in collecting the detailed data. I also respect their contributed works. I thank to Drs A. Ohno, N. Suzuki, H. Sugiyama and S. Hasegawa for their useful comments. I thank two anonymous reviewers for their many useful suggestions that led to improvements in the manuscript. Editage, a division of Cactus Communications, for their assistance with translating and editing this manuscript.

References

1. Davidson J, Andrewartha HG (1948) Annual trends in a natural population of *Thrips imaginis* (Thysanoptera). Journal Animal Ecology 17: 193-199.
2. Caddy JF, Mahon R (1995) Reference points for fisheries management. FAO Fisheries Technical Paper 347. Rome. FAO: 1-83.
3. Kawasaki T (1994) A decade of the regime shift of small pelagic–from the FAO expert Consultation (1983) to the PICES III. Bull. Japanese Society of Fisheries Oceanography 58, 321–333.
4. Wada T, Jacobson LD (1998) Regimes and stock-recruitment relationships in Japanese sardine (*Sardinops melanostictus*) 1951-1995. Canadian Journal of Fisheries and Aquatic Sciences 55: 2455-2463.
5. Cergole MC, Saccardo SA, Rossi-Wongtschowski (2002) Fluctuations in the spawning stock biomass and recruitment of the brazilian sardine (*Sardinella brasiliensis*) 1977-1997. Rev Bras Oceanogr 50:13-26.
6. Tanaka E (2003) A method for estimating dynamics of carrying capacity using time series of stock and recruitment. Fish Sci 69:677-686.
7. Yatsu A,Watanabe T, Ishida M, Sugisaki H, Jacobson LD (2005) Environmental effects on recruitment and productivity of Japanese sardine *Sardinops melanostictus* and chub mackerel Scomber japonicus with recommendations for management. Fish Oceanogr 14:263-278.
8. Munch SB, Kottas A (2009) A Bayesian modeling approach for determining productivity regimes and their characteristics. Ecological Applications 19:527–537.
9. McClatchie S, Goericke R, Auad G, Hill K (2010) Re-assessment of the stock-recruit and temperature-recruit relationships for Pacific sardine (Sardinops sagax). Can J Aquat Sci 67: 1782-1790.
10. Sakuramoto K (2012) A new concept of the stock-recruitment relationship for the Japanese sardine, *Sardinops melanostictus*. The open Fish Science Journal 5: 60-69.
11. Sakuramoto K, Suzuki N (2011) Effects of process and/or observation errors on the stock–recruitment curve and the validity of the proportional model as a stock–recruitment relationship. Fish Sci 78: 41-45.
12. Ricker WE (1954) Stock and recruitment. Journal of the Fisheries Research Board of Canada 11: 559-623.
13. Beverton RJH, Holt SJ (1957) On the dynamics of exploited fish populations. Fisheries Investigations Series II 19: 1-533.
14. Andrewartha HG (1935) On the effect of temperature and food upon egg production and the length of adult life of *Thrips imaginis* Begnall. J Counc Scient and Indust Research Australia 8: 281-288.
15. Davidson J (1944) On the growth of insect populations with successive generations. Aust. J. Exp. Biol. Med. Sci. 22: 95-103.

16. Davidson J, Andrewartha HG (1948) The influence of rainfall, evaporation and atmospheric temperature on fluctuations in the size of a natural population of *Thrips imaginis* (Thysanoptera). Journal of Animal Ecology, 17: 200-222.

17. Birch LC, Browning TO (1993) Historical Records of Australian Science. Australian Academy of Science 9.

18. Smith FE (1961) Density dependence in the Australian Thrips. Ecological Society of America 42 (2): 403-407.

19. Deming WE (1943) Statistical adjustment of data, vol 40 Wiley, New York.

20. Stockl D, Dewitte K, Thienpont LM (1998) Validity of linear regression in method comparison studies: is it limited by the statistical model or the quality of the analytical input data? Clinical Chemistry 44: 2340-2346.

21. Aoki S (2009) Parameter estimation of a regression line by the Deming regression method.

22. Yamanaka D, Sakuramoto K, Suzuki S, Nagasawa T (2007) Catch forecasting and relationship between water temperature and catch fluctuations in snow crab Chionoecetes opilio in the western Sea of Japan. Fish Sci73: 873-844.

23. Sakuramoto K, Sugiyama H, Suzuki N (2001) Models for forecasting sandfish catch in the coastal waters off Akita Prefecture and the evaluation of the effect of a 3-year fisheries closure. Fisheries Sciences 67: 203-213.

24. Watanabe K, Sakuramoto K, Sugiyama H, Suzuki N (2005) Collapse of the arctoscopus japonicus catch in the Sea of Japan - Environmental factors or overfishing. Global Environmental Research 2: 131-137.

25. Oh T, Sakuramoto K, Hasegawa S, Suzuki N (2005) Relationship between sea-surface temperature and catch fluctuations in the Pacific stock of walleye pollock in Japan. Fish Sci 71: 855-861.

26. Singh AA, Sakuramoto K, Suzuki N (2014) Model for stock-recruitment dynamics of the Peruvian anchoveta (Eugraulis ringens) off Peru. Agricultural Sciences.

27. Andrewartha HG, Birch LC (1954) The distribution and abundance of animals. The university of Chicago press, USA.

28. Sakuramoto K (2013) A recruitment forecasting model for the Pacific stock of the Japanese sardine (*Sardinops melanostictus*) that does not assume density-dependent effects. Agricultural Sciences, 4, 6A: 1-8.

29. Sakuramoto K (2013) The common population dynamics for *Thrips imaginis* and the Pacific stock of the Japanese sardine. Conference Abstract in The Japanese Society of Fisheries Oceanography 2013.

Influence of Stocking Density on Growth and Survival of Post Fry of the African Mud Catfish, *Clarias gariepinus*

Nwipie GN, Erondu ES* and Zabbey N

Department of Fisheries, Faculty of Agriculture, University of Port Harcourt, East-West Road, PMB 5323, Choba, Rivers State, Nigeria

Abstract

The effect of stocking density on growth and survival of post fry of the African mud catfish, *Clarias gariepinus,* was investigated. *C. gariepinus* post fry were stocked at the rate of 5, 10, 15, 20, and 25 post fry/litre of water. The post fry were fed to satiation four times daily with crumbles of a commercial catfish feed. The survival, mean total length, mean body weight, condition factor, specific growth rate and performance index were found to be density dependent. Survival rate ranged from 29.7 ± 7.4 to 56.6 ± 33.3%, while specific growth rate was between 0.00143 ± 0.0014 to 0.00702 ± 0.0044. Optimum growth and survival rates were recorded at stocking densities of 5, 10, 15 post fry/litre of water. However, the determined optimum stocking density for rearing of *C. gariepinus* post fry in tanks is 15 post fry/litre of water. It is concluded that increased density impacts on growth and survival of the fish, a consequence of increased activity (frequent surfacing, feeding and swimming) and aggressiveness.

Keywords: Aquaculture; Fish consumption; Demand-supply; Retardation; Food security

Introduction

The production of fast-growing fry is vital to the development of a viable aquaculture venture for enhanced protein security [1]. In Nigeria, aquaculture production has increased tremendously due to increasing demand for fish protein. The Federal Department of Fisheries [2] estimated that over 1.6 million tonnes of both fin and shell fish were required to meet the animal protein need of the country's population in 2004. It was also noted [3] that about 50% of the fish supply in Nigeria comes from importation. The current national fish consumption figure has been put at about 2.66 million tonnes per annum and with an annual fish import figure of about 750,000 metric tons [4]. Nigeria is thus the highest importer of fish in Africa. This means that a huge sum of money is expended on fish importation annually. It is estimated that about 100 billion is spent on fish importation annually in Nigeria [5].

Aquaculture is considered key for bridging the national fish demand-supply gap. No doubt, the increasing demand for fish protein can be met when capture fisheries is supplemented by aquaculture. The Nigerian aquaculture industry has grown considerably, contributing to the production of about 20,475 metric tonnes of fish per year in the 1990s (Olaniyi, 2005) to about 85,087 metric tonnes per year in 2007 [5].

African walking mud catfish, *Clarias gariepinus,* is the most successful aquaculture species in Nigeria. The fish is widely cultured by both trained and untrained smallholder, medium and large-scale farmers. Other attributes that make *C. gariepinus* a first 'choice' farming fish and highly relished by consumers include biological (high food conversion ratio, fast growth rate, readily accepting artificial feeds, ease of artificial propagation, disease resistance), social (good market price, good table food quality) and ecological reasons (tolerance to wide range of environmental conditions) [6-8].

Rearing of the early life history stages of fish (fry and post fry) is the most critical aspect of aquaculture [9,10]. This is essentially because fish at these stages are very sensitive to the various factors or determinants of production. High mortality during fish seed production is common and is a major challenge confronting fish farmers. The import of this scenario is aptly captured by Brain and Amy [11] who noted that economically productive aquaculture is greatly dependent on adequate supply and management of fish seed with which to stock rearing enclosures. Appropriate management practices are, therefore, considered strategic in order to stem this trend. Stocking density has been highlighted as an important aspect of aquaculture practice, generally associated with problems such as reduction in feed conversion efficiency, condition factor and growth [12,13]. Information on stocking density, defined as the weight of fish per unit volume or per unit volume in unit time of water flowing through or contained in the holding environment [14] and its effect on the early life stages of *C. gariepinus* is scarcely available. The available data are skewed to fingerlings [15,16].

The effect of post fry stocking density on survival, growth and performance index of *C. gariepinus* was, therefore, investigated. It is envisaged that the result would provide information capable of aiding the design of appropriate protocols for fish seed production, especially early life history stages of *Clarias gariepinus*.

Materials and Method

Experimental procedure

C. gariepinus post fry were obtained from the University of Port Harcourt Faculty of Agriculture demonstration farm. The fry were acclimated for one week prior to the experiment and fed with commercial catfish diet (*Coppens*). Four thousand and fifty post fry (mean weight 0.68 ± 0.10 g; mean total length 4.03 ± 0.10 cm) *Clarias gariepinus* were counted and transferred to fifteen 20-litre rectangular plastic containers (30 cm width × 40 cm length × 26 cm depth). The fish were stocked at five densities: (5 fry/litre T1, 10 fry/litre T2, 15 fry/litre T3, 20 fry/litre T4, 25 fry/litre T5). Three replicate tanks were used for

***Corresponding author:** Erondu ES, Department of Fisheries, Faculty of Agriculture, University of Port Harcourt, East-West Road, PMB 5323, Choba, Rivers State, Nigeria, E-mail: ebere.erondu@uniport.edu.ng

each stocking density. Using a water flow through fish culture system, water depth was maintained at 20 cm (i.e. 18 litres) level daily. The post fry were fed to satiation 4 times (7 am, 10 am, 1 pm and 4 pm) daily with the commercial catfish feed.

During the experiment, leftover feed and wastes were siphoned out twice daily, in the morning (6.30 am) and in the evening (3.30 pm). Dead fish in the tanks were recorded. Water temperature, pH and dissolved oxygen (DO) were measured daily at 6.30 am with mercury in-glass thermometer, pH meter model WTW Ph 330 and DO meter (Model MW600), respectively. At weekly intervals, 10 larvae were randomly sampled from each tank, weighed individually and total length measured. After 21 days of rearing, all the surviving larvae were collected from each tank. For each individual, total length and body weight were measured and recorded. The survival rate (SR), specific growth rate (SGR), mean daily weight gain (MDWG) and Fulton's condition factor (K) were calculated as follows:

$$\text{SR (\%)} = 100 \times \frac{\textit{Final number of larvae}}{\textit{Initial number of larvae}}$$

$$\text{SGR (\% day)} = 100 \times \frac{\textit{In(Final body weight) - In(initial body weight)}}{\textit{Duration of rearing period (days)}}$$

$$K = \frac{\textit{Final mean body weight (mg)}}{L^3 (cm)} \times 100$$

$$\text{MDWG (mg/day)} = \frac{\textit{Final mean body weight (mg) - initialbody weight(mg)}}{\textit{duration of rearing (days)}} \times 100$$

To evaluate the effect of stocking density on production performance with more precision, the performance index (PI) was calculated [17,18]. This index was calculated by combining two responses such as growth and survival.

$$\text{PI} = \textit{Survival rate} \times \frac{\textit{Final mean body weight (mg) - initialbody weight(mg)}}{\textit{duration of rearing (days)}}$$

Statistical analysis

The experiment adopted the completely randomized design. Analysis of variance (ANOVA) and least significant difference (LSD) were determined to find out the effect of stocking densities on the growth and survival rate of *C. gariepinus* post fry, at 5% level of significance.

Results

Physico-chemical quality of the test media monitored throughout the study period is summarized in Table 1. Water temperature ranged between 30.7°C and 30.9°C.

The pH mean values in all tanks ranged from 6.4 to 7.1 and decreased with increasing stocking density. Values of dissolved oxygen were comparatively lower in higher density media (T_4 and T_5).

The results of Specific Growth Rate (SGR) and performance index PI show higher stocking densities resulted in lower SGR and PI; the best values were obtained in T_1, T_2 and T_3 (Table 2 and Figure 1). At the end of the experiment, the condition factor (K) of post fry was lower in T_4 and T_5 ($p>0.05$). Length and weight followed the same pattern (Table 2, Figures 2 and 3). But K was similar for post fry reared at stocking densities T_1, T_2 and T_3.

Survival rate was also significantly ($p<0.05$) affected by the stocking density. This decreased with increasing stocking density. The mean survival rates are summarized in Table 2. In all the treatments, post fry did not exhibit discernible cannibalism.

Discussion

Post fry stocking density influenced environmental conditions in culture tanks, which, in turn undermined growth and survival of the fish. High accumulation of faeces or excreta and metabolic wastes arising from higher density of post fry led to significant concentrations of nitrogen compounds and simultaneously lowered oxygen in T_4 and T_5. Low levels of dissolved oxygen and pH are considered limiting factors in intensive fish culture. For example, low levels of dissolved oxygen have been attributed to decreasing growth in channel catfish (*Ictalurus punctatus*) larvae reared in tanks [19] (Brazil and Wolters). In general, poor growth performance of cultured species takes place at pH <6.5 [20].

Growth is the manifestation of the net outcome of energy gains and losses within a framework of abiotic and biotic conditions. The effect of stocking density on growth (SGR and MDWG) and Performance Index (PI) was highly significant at higher stocking densities (T_4 and T_5), while there was no significant variation among SGR, MDWG and PI, at T_1,T_2, and T_3 . This implies that yield would be higher at T_1, T_2, and T_3 stocking densities than at T_4 and T_5. Under crowded conditions, fish suffer stress as a result of aggressive feeding interaction and eat less, resulting in growth retardation [21]. The results indicate that post fry *C. gariepinus* stocking densities above 15 individuals/litre will compromise growth. In a study on the effect of stocking density on growth and survival of *C. batrachus* larvae reared in tanks, Sahoo et al. [22] reported similar effects of high stocking densities on growth and SGR.

Parameters	T_1	T_2	T_3	T_4	T_5
Temperature °C	30.70	30.80	30.9	30.7	30.8
pH	7.1	7.0	6.6	6.4	6.6
DO (mg/l)	6.20	5.90	5.66	4.82	4.37

Table 1: Mean values of water quality parameters measured in the rearing tanks.

Parameters	T_1	T_2	T_3	T_4	T_5
Length	4.08 ± 0.2^a	4.05 ± 0.13^a	3.98 ± 0.23^a	3.64 ± 0.49^b	3.58 ± 0.46^b
Weight	0.82 ± 0.23^a	0.73 ± 0.24^a	0.70 ± 0.03^b	0.66 ± 0.12^b	0.61 ± 0.10^c
Specific Growth Rate (SGR)%	0.0070 ± 0.004^a	0.0052 ± 0.002^a	0.0026 ± 0.002^b	0.0017 ± 0.0009^b	0.0014 ± 0.00^c
Condition Factor (k)	1.93 ± 0.24	1.81 ± 0.48	1.72 ± 0.23	1.65 ± 0.13	1.54 ± 0.06
Mean Daily Weight Gain (MDWG)	0.06 ± 0.007^a	0.059 ± 0.02^a	0.051 ± 0.002^a	0.049 ± 0.005^b	0.044 ± 0.008^c
Survival Rate (SR)	56.66 ± 33.30^a	44.44 ± 10.30^a	36.04 ± 5.50^b	31.85 ± 3.80^c	29.70 ± 7.40^c
Performance index (PI)	2.80 ± 1.13^a	2.50 ± 0.51^a	2.19 ± 0.09^b	1.46 ± 0.32^c	1.42 ± 0.15^c

Table 2: Growth, survival and performance index of *C. gariepinus* post fry reared at different stocking densities.

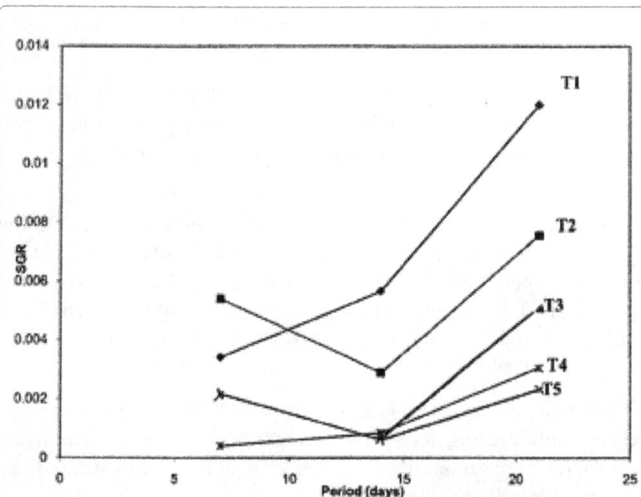

Figure 1: Specific growth rate of *C. gariepinus* at five stocking densities in tanks.

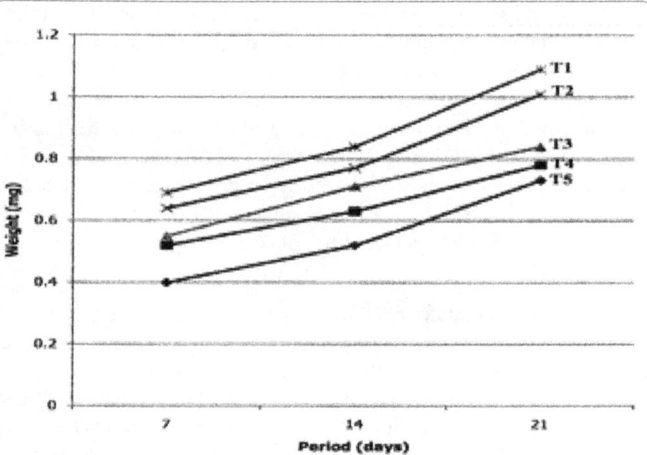

Figure 2: Total body weight of *C. gariepinus* fry cultured at five densities in tanks for 21 days.

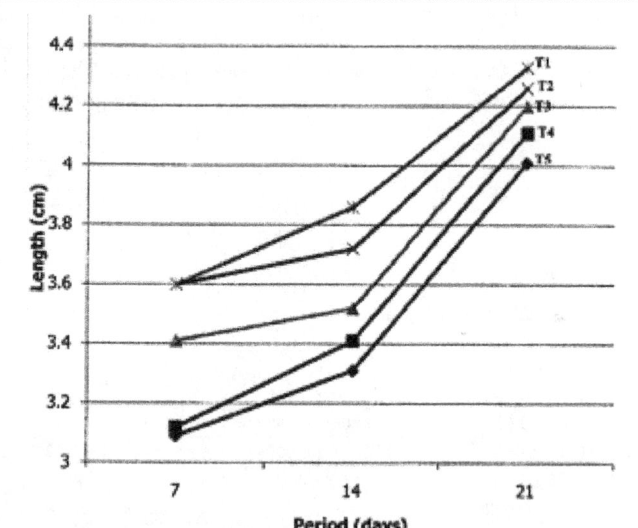

Figure 3: Total body length of *C. gariepinus* fry cultured at five densities in tanks for 21 days.

The condition factor of post fry *C. gariepinus* also decreased at higher stocking densities. At lower stocking densities, the post fry ate well and maximally converted ingested feeds to biomass. Mortality of the post fry was higher at high stocking densities and generally, dead individuals removed from the tanks were relatively of smaller sizes with flattened abdomen; a sign of empty stomach. It is conceivable that the smaller individuals were denied access to food, as they could not have competed favourably with relatively bigger individuals. The effect of stocking density on fry survival depends on the species. For the fry of *Rachycentron canadum* [22-24] reared in tanks, survival rate decreased as stocking density increased. On the other hand, Haylor [25] reported that stocking density did not affect survival of *C. gariepinus* fry reared in floating cages. However, Jamabo et al. [15] reported that survival rate, mean body weight, mean total length and specific growth rate were stocking density dependent for *C. gariepinus* fingerlings, which corroborates the findings of the present study.

For optimum productivity, profitability and good biological conditions (growth and survival), *C. gariepinus* post fry stocking density in tanks should be 15 individuals/litre.

References

1. Young-Sulem S, Brummett RE (2006) Intensity and profitability of *Clarias gariepinus* nursing systems in periurban Yaounde, Cameroon. Aquac Res: 601-605.
2. FDF (2003) Report of the presidential forum on aquaculture development in Nigeria.
3. FDF (2002) Fisheries policy in Nigeria. Federal department of fisheries Abuja: 1-20.
4. Oota L (2012) Is Nigeria Committed to Fish Production
5. Food and Agricultural Organization of the United Nations (2007). The State of the World Fisheries and Aquaculture. Fisheries Department, Rome, Italy: 30.
6. Ayinla OA, Nwadukwe FO (1988). Preliminary studies on the early rearing of *Clarias gariepinus*. N.I.O.M.R. Technical paper, N0.30:15.
7. Zabbey N, Nwadukwe FO, Deekae SN (2007) Effect of crude petroleum oil (Bonny light) on the survival of the fry of African catfish, *Clarias gariepinus*, Fry. Azazeb 9:20-25.
8. Nwadukwe FO, Ayinla OA (2004) Growth and survival of hybrid catfish fingerlings under three dietary treatments in concrete tanks. Azazeb 6: 102-106.
9. Madu CT, Udodike EBC, Ita EO (1990) Food and feeding habits of hatchlings of mudfish *Clarias anguillaris* (L). Afr J Aquat Sci 5: 27-31.
10. Ibrahim MSA, Mona HA, Mohammed A (2008) Zooplankton as live food for fry and fingerlings of Nile Tilapia (*Oreochromis niloticus*) and Catfish *Clarias gariepinus* in Concrete ponds. Central Laboratory for Aquaculture Research (CLAR), Abbassa, Sharkia, Egypt. 8th International Symposium on Tilapia in Aquaculture: 757-769.
11. Brain FD, Amy C (1980) Induced fish breeding in South East Asia. Working report, Singapore. 25th-28th November 1980. 10RC: 178.
12. Lymbery P (1992) The welfare of farmed fish. A comparison in world farming report, Petersfield Hampshire.
13. Ellis T, North B, Scott AP, Bromage NR, Porter M et al. (2002) The relationship between stocking density and welfare in farmed rainbow trout. J Fish Biol 61: 493-531.
14. Ashley PJ (2007) Fish welfare: Current issues in aquaculture. *Applied Animal Behavior Science*. 104: 199-235.
15. Jamabo NA, Keremah RI (2009) Effect of stocking density on the growth and survival of the fingerlings of *Clarias gariepinus*. Journal of fisheries international 4: 55-57.
16. Dada AA (2006) Effect of stocking density on the growth rates of *Claria gariepinus* in a floating Hapa system. FISON Conference proceeding: 276-282

17. Zacharia S, Kakati VS (2002) Growth and survival of Penaeus merguiensis post larvae at different salinities. Isr J Aquacult Bamidgeh 54:157-162.

18. Mohanty RK (2004) Density-dependent growth performance of Indian major carps in rainwater reservoirs. J Appl Ichthyology 20: 123-127.

19. Brazil BL, Wolters WR (2002) Hatching success and fingerling growth of channel catfish cultured in ozonated hatchery water. N Am J Aquac 64: 144-149.

20. Mount DI (1973) Chronic exposure of low pH on fathead minnow's survival, growth and reproduction. Water Res 7: 987-993.

21. Bjoernsson B (1994) Effect of stocking density on growth and survival of halibut (*Hippoglossus hippoglossus*) reared in large circular tanks for three years. Aquaculture 123: 259-271.

22. Sahoo SK, Giri SS, Sahu AK (2004) Effect of stocking density on growth and survival of *Clarias batrachus* (linn) larvae and fry during hatchery rearing. J Appl Ichthyol 20: 302-305.

23. Hitzfelder GM, Holt G, Fox JM, Mckee DA (2006) The effect of rearing density on growth and survival of Cobia, Rachycentron canadum larvae in a closed recirculating aquaculture system. J World Aquacult Soc 37: 204-218.

24. Schram E, Van der Heul JW, Kamstra A, Verdegem MCJ (2006) Stocking density dependent growth of dover (*Solea solea*). Aquaculture 252: 239 -247.

25. Haylor GS (1992) controlled hatchery production of *Clarias gariepinus* (Burchell, 1822): growth and survival of fry at higher stocking density. FAO 22: 405-422.

3

Dynamics of Infection of Juvenile Chinook Salmon with *Ceratomyxa shasta*

Masami Fujiwara*

Department of Wildlife and Fisheries Sciences, Texas A&M University College Station, TX 77843-2258, USA

Abstract

Mathematical models for the infection of juvenile Chinook salmon (*Oncorhynchus tshawytscha*) with *Ceratomyxa shasta* (a myxozoan parasite) in the Klamath River, California, were developed and parameterized with existing data. These models were then used to investigate the effect of three of the environmental conditions thought to be important to the parasite-induced mortality of juvenile Chinook salmon: the stream discharge during the exposure to parasite, the water temperature after infection, and the duration of the exposure to the parasites. The results of this study show that the sensitivity of parasite-induced fish mortality to environmental conditions is higher in spring than fall. Furthermore, the rate of parasite-induced mortality in fish increases with temperature within its range in the spring and summer, when a large number of juvenile fish migrate through the zone where the parasites are prevalent. These results suggest that temperature may strongly affect the parasite-induced mortality of salmon in this stream. Observed seasonal difference in actinospore concentration could not be explained by the dilution effect due to changes in the stream discharge. This suggests the potential importance of other processes such as seasonal fluctuations in the release and natural mortality rates of actinospores. Finally, a sensitivity analysis was used to compare the effects of various environmental conditions. Under the conditions experienced in June 2008, increasing the discharge by 1 m^3/sec would have an effect equivalent to decreasing the exposure duration by 0.26 hours or decreasing the temperature by 0.053°C. This type of analysis is expected to facilitate efforts to restore or mitigate salmon habitats.

Introduction

Pacific salmon frequently suffer from infectious diseases. For example, the mortality rate of Chinook salmon (*Oncorhynchus tshawytscha*) is known to be increased by infection from bacteria [1-5], viruses [5,6], and metazoan parasites, [2,4,7-12]. However, understanding the relationship between parasite infection and associated mortality is complicated because the infection and subsequent mortality of fish are affected by environmental factors [13-17].

Ceratomyxa shasta is a myxozoan parasite and known to cause mortality of juvenile Chinook salmon. Recently, an elevated density of *C. Shasta* has been reported in a section of the Klamath River, California [18-21]. Juvenile Chinook salmon passing through the section, which is herein called the "infectious zone" have shown very high mortality rates in some years (62% in 2005 [18]). The infectious zone extends approximately from the *Shasta* River confluence (284 River km) to just below the Scott River confluence (229 River km). Because of the commercial, cultural, and other importance of this fish stock, the infection of juvenile salmon with *C. Shasta* has been studied intensively in the field and laboratory. The infection rate is currently thought to be affected by discharge [17], water velocity, and the density of *C. Shasta* in the water [15]. In addition, the rate of death of fish after infection appears to increase with temperature [14]. As these environmental factors are likely to be affected by the planned removal of dams [22], their relationship with parasite-induced salmon mortality needs to be elucidated for the Klamath River system.

The life cycle of *C. Shasta* requires two obligate hosts. The first is the polychaete *Manayunkia speciosa* from which actinospores of the parasite are released into the water. The infectious zone has a much greater density of infected polychaetes compared to other locations within the river [20,23]. The released actinospores infect juvenile Chinook salmon. After the infection, the actinospores may propagate within infected salmon and eventually kill the juvenile salmon [14,15]; however, the death of juvenile salmon is often delayed relative to the time of infection [15], complicating field measurements of parasite-induced mortality.

The main objective of this study was to understand how the parasite-induced mortality rate of juvenile Chinook salmon in the Klamath River is affected by the prevalence of the parasite, the stream discharge, and the temperature. This was achieved by building models and parameterizing them with existing data. The primary sources of information were field and laboratory experiments conducted by Ray et al. [14,15]. They caged juvenile salmon *in situ* in the infectious zone for a defined period and then maintained the fish in the laboratory to investigate parasite-induced mortality [15]. In a second set of experiments, they varied the duration of field caging, and altered the laboratory water temperature to investigate the effect of temperature on the post-infection development of actinospores. Finally, they measured the *in situ* actinospore density and water discharge to determine the amount of exposure to parasites in the field [14].

Here, three models were developed to examine the infection process: a model for the stochastic infection of salmon with actinospores; a deterministic model for approximating the infection process; and a model representing the concentration of actinospores in the water. These models were parameterized with the data from Ray et al. [14,15] and existing stream-discharge information obtained from the USGS National Water Information System (http://waterdata.usgs.gov).

Materials and Methods

Stochastic infection process

Actinospores encounter and infect juvenile Chinook salmon,

***Corresponding author:** Masami Fujiwara, Department of Wildlife and Fisheries Sciences, Texas A&M University College Station, TX 77843-2258, USA
E-mail: fujiwara@tamu.edu

which die from the infection with some probability. Here, infection is considered "successful" if it can eventually lead to death from the infection; otherwise, it is considered "unsuccessful." Therefore, infection is considered successful based on its potential to kill rather than whether it actually killed its host or not. Because parasite-induced death is delayed, an individual salmon can experience multiple successful infections in addition to multiple unsuccessful infections.

In this analysis, both successful and unsuccessful infections were assumed to be stochastic and independent of each other; all actinospores were assumed to be identical in terms of the infection process; and infection was assumed not to influence subsequent infections. Under these assumptions (independent, homogeneous, and memoryless), the number of individual-specific successful infections χ has the Poisson distribution, see [24]:

$$\chi \sim \text{Poiss}(\mu) \tag{1}$$

Where μ is the mean rate of successful infection. This is used for estimating the parameter.

Deterministic model: Approximation of the infection process

To model the dynamics of juvenile salmon abundance, a system of differential equations was developed, as follows:

$$\frac{dS}{dt} = -\alpha cS - dS \tag{2}$$

$$\frac{dI}{dt} = \alpha cS - dI \tag{3}$$

Where S is the density of susceptible salmon without successful infection; I is the density of salmon successfully infected by *C. Shasta*; α is the infection rate; c is the concentration of actinospores; d is the instantaneous per-capita mortality of salmon from causes other than infection during the exposure period; and t is the duration of exposure. Ray et al. [15] used control groups to estimate the non-disease related mortality rate, and found that d was 0 during their exposure and laboratory-holding period. This largely reflected the lack of predators both *in situ* and in the laboratory and the use of antibiotics in the laboratory.

The density of susceptible salmon after exposure for time t is given by solving equation (2) after setting d=0:

$$S = S_0 \, e^{-act} \tag{4}$$

Where S_0 is the initial density of susceptible salmon. Finally, the proportion of fish that eventually die from infection (m) is given by:

$$m = 1 - e^{-act} = 1 - e^{-aE} \tag{5}$$

where E is exposure (calculated as the product of the actinospore concentration c and the exposure duration t). The unit for parameter E is the number of actinospores per liter of water. Consequently, the unit for the infection rate α, which also includes the filtration of water by fish, is the number of dead fish per liter of water filtered per unit time. Because the filtration rate of water is unknown in this study, it is included in the unknown parameter α to be estimated rather than estimating it separately.

Model for the concentration of actinospores in the water

The concentration of actinospores in the water was determined by their release from polychaetes, their reduction due to the infection of salmon, and their natural mortality, as:

$$\frac{dc}{d\tau} = \kappa - \Upsilon cN - \delta c \tag{6}$$

Where κ is the instantaneous release rate of actinospores from polychaetes, γ is the infection rate (including both successful and unsuccessful infections), N is the total density of salmon (N=S+I), and δ is the instantaneous per capita mortality rate of actinospores in the water. For a given environmental condition, therefore $\gamma \propto \alpha$. The difference between γ and α exists because not all infected salmon die after parasite infection. Therefore, $\gamma > \alpha$. The uptake of actinospores by other organisms was assumed to be included in the natural mortality term δ , and was expected to be small due to the host specificity of *C. Shasta*.

The effect of the stream discharge (V:m^3/sec) on the actinospore concentration (c) was incorporated. Assuming that the polychaete density and actinospore release rate per polychaete remain constant, κ can be expressed as:

$$\kappa \propto \frac{1}{V} \tag{7}$$

In other words, a high water discharge dilutes the concentration of actinospores in the water.

Because the processes governing the dynamics of actinospores are expected to be much faster than those governing salmon density, such that $\frac{dc}{d\tau} = 0$, the expression for the actinospore concentration c is obtained as the following:

$$c^* = \frac{\kappa}{\gamma N + \delta} \tag{8}$$

where the star indicates that the concentration is at equilibrium. Therefore, if infection rate $\gamma$and the per capita instantaneous mortality rate of actinospores (δ) is not affected by the discharge, the concentration is inversely proportional to the discharge, and may be written as:

$$c^* \propto \frac{1}{V} \tag{9}$$

It should be noted that the assumption $\frac{dc}{d\tau} = 0$ implies that salmon and actinospores in the water equilibrate quickly in the time scale of the model, but it does not imply that the equilibrium concentration c^* is constant in a longer time-scale. For example, c*can change because of slower changes in release rate κ and salmon density N. Parameter κ is discussed further in Discussion.

Sensitivity of mortality rate

In general, a sensitivity of mortality rate m with respect to any arbitrary parameter X is given by $\frac{\partial a}{\partial X}$. Thus, the sensitivities of the mortality rate to the duration of exposure t, the discharge V, or the temperature h were obtained by taking the derivative of the mortality function (5) with respect to the corresponding parameter. The sensitivity of the mortality rate to the duration of exposure is:

$$\frac{\partial m}{\partial t} = \alpha c e^{-\alpha ct} \tag{10}$$

the sensitivity of the mortality to the discharge is:

$$\frac{\partial m}{\partial V} = -\frac{\alpha bt}{V^2} e^{\frac{-\alpha bt}{V}} \tag{11}$$

and the sensitivity of the mortality to the temperature is:

$$\frac{\partial m}{\partial h} = \frac{\partial m}{\partial \alpha}\frac{\partial \alpha}{\partial h} = cte^{-\alpha ct}\frac{\partial \alpha}{\partial h} \tag{12}$$

Where, $\frac{\partial \alpha}{\partial h}$ is given by the slope of the function relating α and h.

Parameters for the sensitivity analysis were estimated as outlined in the following section. For the actinospore concentration c, the average number of actinospores per hour per liter in June and September 2008 were calculated from the values in Ray et al. [15]. For the sensitivity calculations, the temperature h was assumed to be 18°C, and the duration of exposure *t* was assumed to be 24 hours. Furthermore, average discharge over a month was obtained from the USGS National Water Information System for June and September 2008 was used for *V*.

Parameter estimation and data

The two basic parameters to be estimated were the infection rate (α) and the actinospore concentration in the water (c). Infection rate can be a function of temperature, while the concentration can be a function of the stream discharge. These parameters were estimated from the results in Ray et al. [14,15] and existing stream discharge data.

The probability of a fish being successfully infected is $\Pr(\chi>0)$. Therefore, the probability of mortality from infection, according to the Poisson Infection Process (1), is:

$\Pr(\chi>0)=1-e^{-\mu}$

By letting, $\mu = \alpha E$ probability (13) can be approximated by the deterministic model (5). In other words, by taking the mean probability over a large number of fish, this probability was interpreted as the finite mortality rate of salmon $\Pr(\chi>0)=m$, and the mean infection rate is given by αE. The relationship between the exposure *E* and the mortality rate was taken from Ray et al. [15] (Table 1), and α was obtained by fitting equation (13) to the data using the least squared-error method. The Jackknife method [25] was then used to estimate its variance.

Next, the rate of successful infection α was modeled as a function of the laboratory water temperature. The relationship between temperature and the mortality rate among fish exposed to actinospores for 72 hours was taken from Ray et al. [14] (Table 2). The data reported by Ray et al. [15], which were obtained from fish maintained at 18°C, were used to estimate α as described above. This estimated rate of successful infection is denoted hereafter by $\hat{\alpha}_{18}$. The infection rate at another temperature, h, was then derived from equation (13) by solving for E twice after substituting $\mu=\alpha_h E$ and $\mu=\alpha_{18}E$ and, equating them, and solving for α_h:

$$\alpha_h = \frac{\hat{\alpha}_{18}\ln(1-m_h)}{\ln(1-m_{18})} \qquad (14)$$

where $\hat{\alpha}_{18}$ is the value estimated from the data in Table 1, and m_h (including m_{18}) is the mortality rate at temperature h (Table 2). At low temperatures, the infection rate is thought to remain relatively constant [14]. Furthermore, it was assumed that 13°C is below the inflection point, based on a plot of the infection rate α_h as a function of *h*.

Finally, from the relationship between the actinospore concentration and the discharge (9), the actinospore concentration (c) was modeled as a function of the stream discharge as follows:

$$c = \frac{b}{V} \qquad (15)$$

Where b is the constant related to the actinospore release rate. Because the experiments by Ray et al. [15] were conducted in June and September of 2008, the average concentrations in June and September were calculated separately. Then, b_i (parameter *b* for month *i*) was estimated as:

$$\hat{b}_i = \bar{V}_{i,2008}\bar{c}_i \qquad (16)$$

where $\bar{V}$ is the mean discharge in month i of year 2008 and $\bar{c}$ is the average actinospore concentration over the experiments performed in month i of year 2008. The latter concentration was obtained by dividing the exposure by the number of hours the fish were exposed. For discharge, the mean stream discharge over one month (*V*) at Seiad Valley, California (USGS Station 11520500) and Scott River at Fort Jones, California (USGS Station 11519500) were obtained from the USGS National Water Information System (http://waterdata.usgs.gov). The latter discharge was subtracted from the former because the location of the experimental study was above the confluence with the Scott River whereas the Seiad Valley station was below the confluence. Here, the concentration in the experiment was assumed representative of the concentration for that month. If additional data become available, this assumption can be relaxed. For example, daily mean actinospore concentration and daily mean discharge can be used. The concentrations for years 2002 to 2012 were extrapolated as:

$$c_{i,y} = \frac{\hat{b}_i}{\bar{V}_{i,y}} \qquad (17)$$

Results

The observed and fitted relationships of the finite mortality rate (*m*) and actinospore exposure (*E*) are shown in Figure 1. They agree well, suggesting that it was reasonable to use the Poisson process to model the infection. According to the fitted relationship, 90% mortality is expected upon exposure to 6.0×10^8 actinospores; 50% mortality is expected upon exposure to1.8×10^8 actinospores; and 10% mortality is expected upon exposure to 2.7×10^7 actinospores over 24 hours.

The estimated per-fish infection rate per concentration of actinospores is shown as a function of temperature (Figure 2). First, the infection rate at 13°C was estimated and plotted as a horizontal line at the lower part of temperature. Then, a quadratic function was fitted to the infection rates at 15°C, 18°C, and 21°C.

The effect of stream discharge on the mortality rate is shown in Figure 3. In this analysis, it was assumed that fish were exposed to actinospores over 24 hours at 18°C and the discharge alone varied according to the estimated discharge. Results show that mortality rate fluctuation was larger in June than in September, but these fluctuations were smaller than the difference between the mean rates in June and September. The estimated coefficients, b_6 and b_9, were 1.1×10^9 (σ^2:v 2.9×10^{11}) and 2.4×10^7 (σ^2: 9×10^8) respectively.

Finally, Figure 4 shows the finite mortality rate of fish exposed to actinospores over a 24-hour period as a function of temperature and

	June				September		
Actinospore exposure E (×10⁶) per liter of water	153.2	535.4	594.5	612	4.4	6.6	13.2
Mortality rate (% death)	34.9	84.7	98.5	94.2	2.5	16.7	17.7

Table 1: Mortality rate (% death) and associated actinospore exposure (*E*) obtained from Ray et al. [15].

Temperature (°C)	13	15	18	21
Mortality rate (% death)	68.8	70.6	86.9	97.7

Table 2: Mortality rate (% death) resulting from C. shasta infection and associated post-exposure temperature, as obtained from Figure 3 in Ray et al. [14]

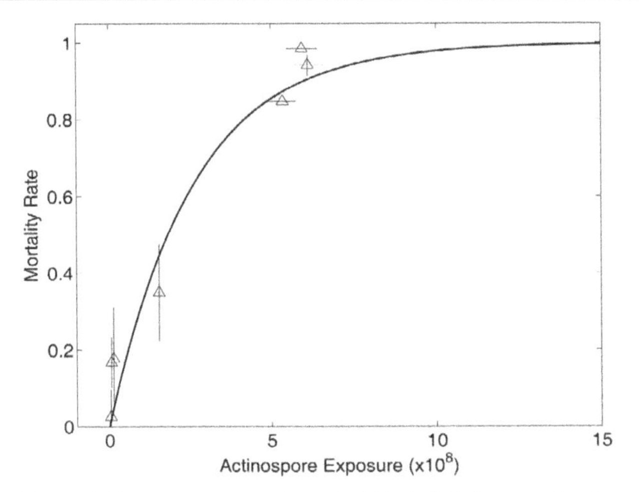

Figure 1: Mortality rate (proportion of individuals that died) as a function of actinospore exposure. Triangles represent the data from Ray et al. [15]. The curve is a fitted model; $\hat{\alpha} = 3.85 \times 10^{-9}$ (σ^2:1.2 x 10^{-15})

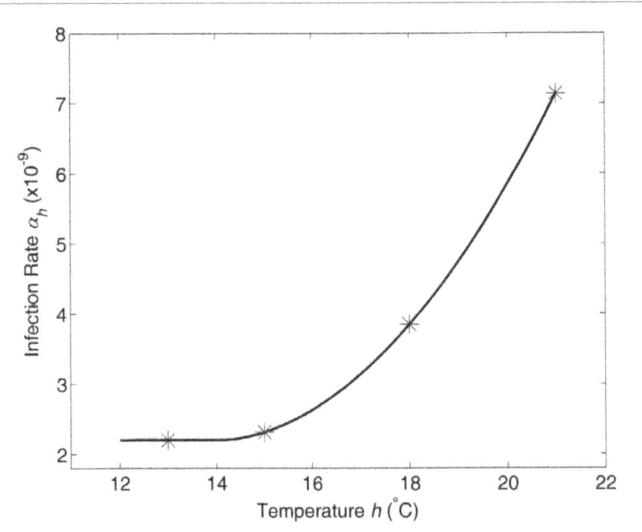

Figure 2: Instantaneous infection rate (α) as a function of temperature. Stars indicate the estimated instantaneous infection rate $\hat{\alpha}$, and the curve is a fitted quadratic function extrapolated to the infection rate at temperature 13°C

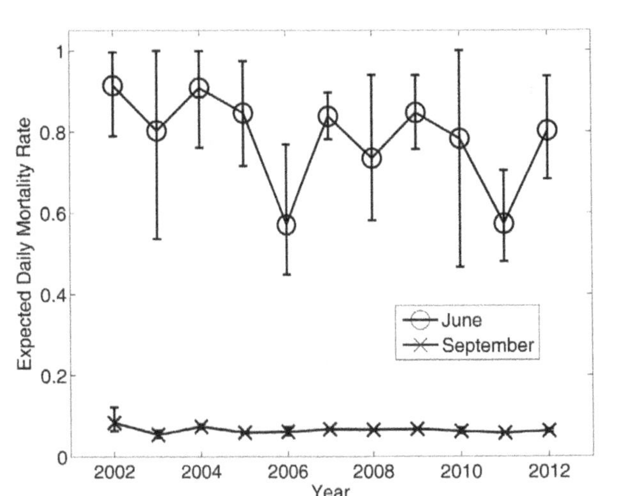

Figure 3: Expected daily mortality rate (proportion of individuals that died) in June (o) and September (x) from 2002 to 2012. The values were estimated using equations (5) and (17) along with average discharge over a month in the Klamath River, California, obtained from the USGS National Water Information System. Error bars show 95% confidence intervals reflecting variations in the daily discharge. It was assumed fish was exposed to *C. shasta* over 24 hours at 18°C.

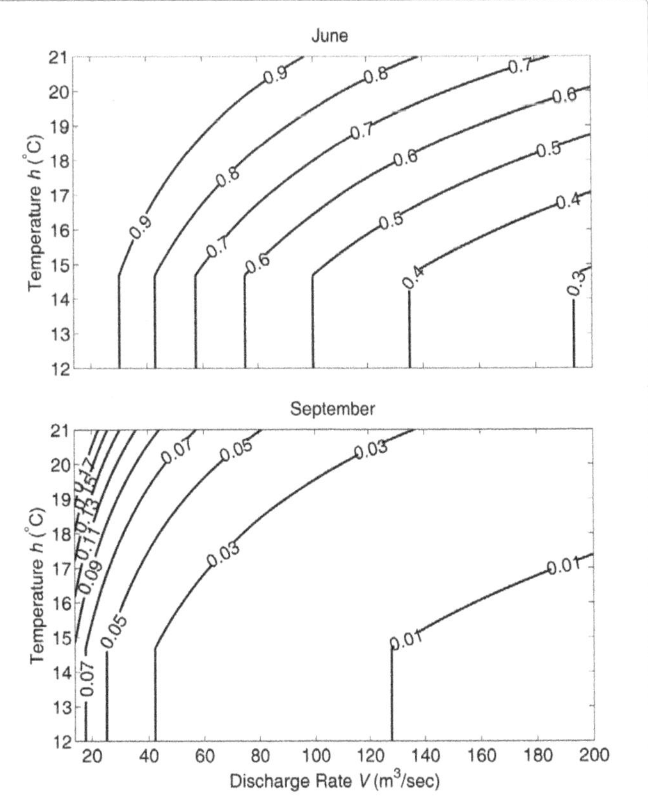

Figure 4: Expected daily mortality rate (shown in contours) as a function of the mean discharge and temperature in June (a) and September (b).

discharge. For a given discharge and temperature, the mortality rate in June was much higher than that in September.

The sensitivities of the mortality rate to the duration of exposure, discharge, and temperature were calculated around the estimated values obtained for June and September 2008. Table 3 shows the sensitivity values and their inverses, the latter of which reveal the relative amount of change in a given condition needed to increase the mortality by a unit amount. Their relative magnitudes suggest that increasing the discharge by 1 m^3/sec in June had an effect equivalent to that of decreasing the exposure duration by 0.26 hours or decreasing the temperature by 0.053°C. In September, increasing the discharge by 1 m^3/sec had an effect equivalent to that of decreasing the duration by 0.73 hours or decreasing the temperature by 0.15°C.

Discussion

The novelty of the study was its development of mathematical models based on existing *C. Shasta* data and extracting information useful for the management of Chinook salmon. Data for the study were primarily obtained from Ray et al. [14,15]. Each model was developed separately for different environmental factors, but could be integrated into equation (5) to model the overall mortality rate of

Parameters	Sensitivity $\left(\frac{\partial m}{\partial X}\right)$	Inverse of sensitivity $\left(\frac{\partial m}{\partial X}\right)^{-1}$	Sensitivity $\left(\frac{\partial m}{\partial X}\right)$	Inverse of sensitivity $\left(\frac{\partial m}{\partial X}\right)^{-1}$
	June		September	
Exposure Duration (I)	0.0147	68.2	0.0027	377.0
Discharge (*V*)	-0.0039	-258	-0.0019	-517
Temperature (*h*)	0.0735	13	0.0133	75

Table 3: Sensitivities of the mortality rate (% death) to exposure duration, discharge, and temperature. Sensitivities and the inverse of the sensitivities are shown. These values were evaluated with the estimated parameters for June and September 2008. See text for detail.

Chinook salmon. This, in turn, could be incorporated into a population model for Chinook salmon [17]. To my knowledge, a development of the model that allows the integration of various environmental factors into a single unifying model for the parasite-induced mortality of fish is new. Finally, the models can be used to analyze the sensitivity of salmon mortality rates to environmental conditions, which is expected to be especially useful for future salmon management.

The persistence of *C. Shasta* requires two hosts: salmon and polychaete. Both juvenile salmon migrating downstream in spring and summer and adult salmon migrating upstream in fall are infected with actinospores of *C. Shasta* through their gills [13]. If the host dies after infection, myxospores of *C. Shasta* are released into the water. The myxospores later infect polychaetes. Because of delay in the death of salmon after infection and the directions of fish migration, I suspect that adult salmon are more important in releasing the myxospores, which infect polychaete, but further research is needed to confirm this hypothesis. The concentration of actinospores in the water is a very important determinant of the infection rate of salmon by the parasite [16]; this is consistent with the fact that the infection takes place through the gills. Unfortunately, the life cycle of polychaetes is not well understood [20]. If further information becomes available, however, it can be incorporated into the release rate of actinospores (κ) in equation (6). Ultimately, the release rate κ can be a function of the rate of release of actinospores per polychaete, the density of polychaete, and the discharge of water. The first two rates can also be a function of environmental conditions.

The rate of successful infection was found to increase with temperature [14]. In the present study, this effect was quantified in terms of the instantaneous rate of successful infection α_h (Figure 2). The water temperature in the upper Klamath River currently ranges from below 5°C in winter to above 20°C in summer months [22]. According to the comparison of the estimated infection rate versus temperature (Figure 2), the increase in infection rate with increasing temperatures appears to accelerate within the higher temperature range. River temperatures in April and May, when numerous juvenile salmon typically pass through the infectious zone, is within this range. Furthermore, the sensitivity analysis suggests that the same amount of temperature changes has a greater effect on salmon mortality rate in June versus September (Table 3). This suggests that spring temperatures may contribute to determining the number of salmon surviving the infectious zone.

Ray and Bartholomew [16] showed that the total infection rate increases with temperature from 11°C to 18°C but declines from 18°C to 22°C. This appears to contradict with the results shown in Ray et al. [14]. However, the difference between the results from Ray et al. [14], the results presented herein, and the results from Ray and Bartholomew [16] is that the former quantifies the effect of temperature on salmon mortality, while the latter quantifies the effect of temperature on the number of transmissions of *C. Shasta* to salmon. Not all infections of salmon with parasites lead to the subsequent death of salmon [14]; therefore these results cannot be compared directly. Both studies, however, suggest the mortality rate per parasite after infection increases substantially with temperature although the total infection rate may decline. At temperatures above 18°C, the rate of successful infection (α) is expected to accelerate with the total infection rate (γ), and the results in Ray and Bartholomew [16] can be incorporated into changes in the total infection rate.

Bjork and Bartholomew [13] reported that the infection rate (γ) is affected by the velocity of water. It should be noted that the infection rate (γ) is the instantaneous rate at which actinospores enter salmon per actinospore concentration and salmon density. In the current study, this quantity was assumed to be constant regardless of the temperature and discharge. However, the effect of water velocity on infection rate is difficult to investigate. For example, the velocity could affect the amount of water filtered by salmon, which would affect the total exposure of salmon to the parasites. The velocity could also affect physical infection processes in fish gills. These processes, however, are confounded with the behavior of fish. For example, in a caging experiment, fish remained in a cage against water flow, but naturally swimming salmon might actually be carried with the flow or actively swim downstream [26]. If more detailed information on the effect of velocity on infection rate (γ) becomes available in the future, it would be interesting to incorporate it into the model.

There are clear differences in the mortality rate between June and September (Figure 4). The water temperature of the Klamath River is generally higher in September versus June, while the discharge is lower in September versus June. Both of these observations suggest that the parasite-induced mortality rate should be higher in September than in June. However, the estimated mortality was much higher in June versus September, regardless of the temperature and discharge. Therefore, the between-month difference in mortality cannot be explained by the effect of temperature on the development of *C. Shasta* after infection or the effect of discharge on the concentration of actinospores in the water. Instead, the difference is explained by the difference in the concentration of actinospores in the water [15]. This suggests that the natural mortality rate of actinospores (δ), their release rate from polychaetes (κ), and/or the infection rate (γ) differed substantially between the two months according to equation (8). This, in turn, suggests the importance of studying polychaete life cycle and actinospore mortality rate in order to understand the observed seasonal difference.

An important assumption in the model is that all juvenile salmon have the same susceptibility to actinospore infection. It is plausible that there is heterogeneity in susceptibility. If so, one way to incorporate such heterogeneity would be to make the infection rate, α, include individual heterogeneity. It is also plausible that hatchery-released and naturally spawned juveniles may have different susceptibilities. Such heterogeneity could be incorporated into mathematical models by dividing the juveniles into two groups and separately estimating their parameters.

The sensitivities of the mortality rate to the temperature, discharge, and exposure duration were calculated, but direct comparison was difficult because the environmental variables had different units. Therefore, the inverse of each sensitivity was taken in order to calculate the unit change in a given environmental condition that was required to change the mortality by the same unit amount. Although the changes in salmon mortality may appear to be very small, the abundance of juvenile salmon is high. Therefore, a small change in the mortality rate can translate into a large difference in the number of salmon surviving the infectious zone. The sensitivity results can be used in planning for the restoration or mitigation of salmon habitats. Discussions are currently underway regarding the removal of dams from the Klamath River, which would be expected to change the flow pattern and water temperature. This sensitivity analysis can be used to predict the effects of these changes on the mortality of Chinook salmon from *C. Shasta* infection.

The primary goal of this study was to build models and parameterize them with existing data, in order to understand the mortality of juvenile Chinook salmon resulting from the infection with *C. Shasta*. The results are constrained by the limited availability of data, but the basic model structures should be very useful for managing environmental conditions in the Klamath River and predicting the future conditions that will be seen following the removal of dams. The utility of these novel models should improve in the future, as more data are collected, allowing us to predict responses resulting from future environmental conditions.

Acknowledgement

I thank three anonymous reviewers and editor, who provided very constructive comments and suggestions on the previous version of the manuscript. I also thank Cyrenea Piper for assisting the preparation of this paper. This research was made possible through support provided by the U.S. National Oceanic and Atmospheric Administration, National Marine Fisheries Service (DOC Contract-NFFR7500-10-18114). The opinions expressed herein are those of the author and do not necessarily reflect the views of the funding agency.

References

1. Fryer JL, Sanders JE (1981) Bacterial kidney-disease of salmonid fish. Annual Review of Microbiology 35: 273-298.
2. Foott JS, Walker RL (1992) Disease survey of trinity river salmonid smolt populations 1991 report, US Fish and Wildlife Service, California-Nevada Fish Health Center, Anderson, California.
3. Fryer JL, Lannan CN (1993) The history and current status of *Renibacterium-salmoninarum*, the causative agent of bacterial kidney-disease in Pacific salmon. Fish Res 17: 15-33.
4. Walker RL, Foott JS (1993) Disease survey of Klamath River salmonid smolt populations: 1992 report, US Fish and Wildlife Service, California-Nevada Fish Health Center, Anderson, California.
5. Arkush KD, Giese AR, Mendonca HL, McBride AM, Marty GD, et al. (2002) Resistance to three pathogens in the endangered winter-run Chinook salmon (*Oncorhynchus tshawytscha*): Effects of inbreeding and major histocompatibility complex genotypes. Canadian Journal of Fisheries and Aquatic Sciences 59: 966-975.
6. Arkoosh MR, Casillas E, Huffman P, Clemons E, Evered J, et al. (1998) Increased susceptibility of juvenile Chinook salmon from a contaminated estuary to *Vibrio anguillarum*. Trans Am Fish Soc 127: 360-374.
7. Jacobson KC, Arkoosh MR, Kagley AN, Clemons ER, Collier TK, et al. (2003) Cumulative effects of natural and anthropogenic stress on immune function and disease resistance in juvenile Chinook salmon. Journal of Aquatic Animal Health 15: 1-12.
8. Nichols K, Foott JS (2005) Investigational report: Health monitoring of juvenile Klamath river Chinook salmon, US Fish and Wildlife Service, California-Nevada Fish Health Center, Anderson, California.
9. Stocking RW, Holt RA, Foott JS, Bartholomew JL (2006) Spatial and temporal occurrence of the salmonid parasite *Ceratomyxa shasta* in the Oregon-California Klamath River basin. Journal of Aquatic Animal Health 18: 194-202.
10. Nichols K, True K (2007) Investigational report: Monitoring incidence and severity of *Ceratomyxa shasta* and *Parvicapsula minibicornis* in juvenile Chinook salmon (*Oncorhynchus tshawytscha*) in the Klamath river, US Fish and Wildlife Service, California-Nevada Fish Health Center, Anderson, California.
11. Bjork SJ, Bartholomew JL (2009) Effects of *Ceratomyxa shasta* dose on a susceptible strain of rainbow trout and comparatively resistant Chinook and coho salmon. Diseases of Aquatic Organisms 86: 29-37.
12. Jacobson KC, Teel D, Van Doornik DM, Casillas E (2008) Parasite-associated mortality of juvenile Pacific salmon caused by the trematode *Nanophyetus salmincola* during early marine residence. Marine Ecology-Progress Series 354: 235-244.
13. Bjork SJ, Bartholomew JL (2009) The effects of water velocity on the *Ceratomyxa shasta* infectious cycle. Journal of Fish Diseases 32: 131-142.
14. Ray RA, Holt RA, Bartholomew JL (2012) Relationship between temperature and *Ceratomyxa shasta*-induced mortality in klamath river salmonids. Journal of Parasitology 98: 520-526.
15. Ray RA, Rossignol PA, Bartholomew JL (2010) Mortality threshold for juvenile Chinook salmon *Oncorhynchus tshawytscha* in an epidemiological model of *Ceratomyxa shasta*. Diseases of Aquatic Organisms 93: 63-70.
16. Ray RA, Bartholomew JL (2013) Estimation of transmission dynamics of the *Ceratomyxa shasta* actinospore to the salmonid host. Parasitology 140: 907-916.
17. Fujiwara M, Mohr MS, Greenberg A, Foott JS, Bartholomew JL (2011) Effects of ceratomyxosis on population dynamics of Klamath fall-run Chinook salmon. Trans Am Fish Soc 140: 1380-1391.
18. Nichols K, True K, Wiseman E, Foott S (2007) Investigational report: Incidence of *Ceratomyxa shasta* and *Parvicapsula minibicornis* infections by qpcr and histology in juvenile Klamath river Chinook salmon, US Fish and Wildlife Service, California-Nevada Fish Health Center, Anderson, California.
19. Nichols K, Foott JS (2006) Investigational report: Health monitoring of juvenile Klamath River Chinook salmon, US Fish and Wildlife Service, California-Nevada Fish Health Center, Anderson, California.
20. Stocking RW, Bartholomew JL (2007) Distribution and habitat characteristics of *Manayunkia speciosa* and infection prevalence with the parasite *Ceratomyxa shasta* in the Klamath river, Oregon-California. Journal of Parasitology 93: 78-88.
21. Nichols K, True K, Fogerty R, Ratcliff L (2008) Fy 2007 investigational report: Klamath river juvenile salmonid health monitoring, april-august 2007 US Fish and Wildlife Service, California-Nevada Fish Health Center, Anderson, California.
22. USDOI and CDFG (2012) Klamath facilities removal final environmental impact statement/ environmental impact report, U.S. Department of the Interior and California Department of Fish & Game, Siskiyou County, California and Klamath County, Oregon.
23. Bartholomew JL, Atkinson SD, Hallett SL, Zielinski CM, Foott JS (2007) Distribution and abundance of the salmonid parasite *Parvicapsula minibicornis* (myxozoa) in the Klamath River basin (Oregon-California, USA). Diseases of Aquatic Organisms 78: 137-146.
24. Gallager RG (1996) Discrete stochastic processes, Kluwer Academic Publishers, Nowell, Massachusetts, USA.
25. Efron B and Tibshirani BJ (1994) An introduction to the bootstrap Chapman and Hall/CRC, Boca Raton, Florida, USA.
26. Quinn TP (2005) The behavior and ecology of Pacific salmon and trout, University of Washington Press, Seattle, WA.

4

Evaluation of Maggot Meal (Muscadomestica) and Single Cell Protein (Mushroom) in the Diet of *Clarias gariepinus* Fingerlings (Burchell, 1822)

Michael KG and Sogbesan OA*

Department of Fisheries, ModibboAdama University of Technology, Yola, Nigeria

Abstract

The growth performance of Clarias gariepinus fingerlings (mean weight 0.9 g) fed diets of maggot meal and single cell protein supplemented diet were investigated for 56 days. Nine diet of 40% FM and 30% FM, 10%MM and 20% FM, 10%MM, 10%SCP and 10%FM, 20%MM, 10%SCP and 30%MM, 10%SCP and 40%MM and 10MM, 30% SCP and 10%FM, 10%MM, 20%SCP and 40%SCP. The result of the experiment showed that fingerlings on the diet of 30%FM and 10% MM plus dry pelleted feed (T2) has the best specific growth rate (1.22%), food conversion (2.14) indicating that feed with maggot in combination with other supplemented diet formed better balance diet for the fingerlings. Feed with 30% fish meal and 10% maggot meal is recommended by this study.

Keywords: *Clarias gariepinus*; Maggot; Single cell protein; Feed utilization; Cost benefits

Introduction

In fisheries feed constitute one of the major inputs in intensive and semi intensive fish farming and can reduce the economic viability of a farm if suitable feeds are not used. The use of commercial pelleted fish feeds is so expensive that it accounts for about 40% to 60% of the recurrent cost of fish farming venture [1-3].

Fish feed is highly affected by the cost of feed ingredients used in formulation of the feed [1]. The major component in fish feed formulation is the fish meal (that constitute about 50-70% by weight), which use to be limited by high cost, non-availability and competition from poultry and livestock sectors. The high price of fish meal and other ingredients such as soybean, groundnut cake and maize has led to the need for investigating into other alternate cheap sources of fish feed ingredients that will provide the requited for the fish at cheaper cost so as to increase production from aquaculture sector and bridge the gap between fish demand and supply in Nigeria.

Poultry manure and abattoir wastes impose threat of disposal to the poultry and cattle slaughtering industries and as well as a serious pollution problem to the incumbent environment and man's health, hence an efficient and effective means of poultry waste disposal becomes imperative.

Maggot, the larval form of housefly (Muscadomestica) is not being competed for as animal protein source by man. The production of Maggot from waste materials either from plants or animal origin dung and food waste where it digests then to odour free "scum" with high nutrient value.

Maggot is readily available and has been accredited for its quality protein with amino acids profile showing its biological value to be superior to soybean and groundnut cake [4,5]. Maggot and single cell protein (mould) could be a cheap source of protein ingredients in fish diet.

Single cell proteins from organisms such as bacteria and fungi have been considered as possible substitutes for fish meal in diets for Clarias. The use of petroleum yeasts in combination with other proteins (SCP), especially alkane grown yeasts and methanol fermenting bacteria provided that they are supplemented by the limiting amino acids, can serve as a high quality source [6]. They are rich in crude protein content (55-80%) as well as amino acid profile as potential feedstuffs in *Clarias* nutrition.

Hike in the price of fishmeal and consequently fish feed has led to the need for investigating into other alternative cheap sources of fish at cheaper cost if production from aquaculture sector and bridge the gap between fish demand and supply in Nigeria. Hence, partial or total replacement of fish meal protein with alternative source of protein could be considerable economic advantage especially of the ingredients in association with moderate reduction in feed efficiency.

The aim of this study is to evaluate the use of organic matter to produce usable protein for fish feed so as to prepare a cost efficient feed for fish, in search for a better disposal means and conversion of wastes to useful nutrients for fish. And also to evaluate the efficiency of maggot harvesting techniques as well as the economics and production capacity of feeding *clarias gariepinus* with maggots and supplemented diets.

Materials and Methods

Culture of maggot and single cells

Maggots used for this experiment were cultured from chicken manure using sack method as described by Madu and Ufodike [7]. The collection was done as described by Adejinmi [4,5] and Sogbesan et al. using screens. The maggots are photonegative, so in attempts to escape from the traces of sunlight they passed through the 3 mm mesh size net and is collected in a basin under the net. Maggots collected were weighted, oven dried and grounded into powdery form using blending machine. The Single cells were harvested from the wild.

***Corresponding author:** Sogbesan OA, Department of Fisheries, ModibboAdama University of Technology, Yola, Nigeria, E-mail: keccybaby1258@gmail.com

Proximate analysis of the experimental diets

The single cell protein, maggot and all the formulated diets were analysed for the proximate composition following Association of Analytical Chemist Methods AOAC, 2000.

Fish feed ingredients

Fish meal, maize, soybean, groundnut cake, salt, vitamins/mineral premix, palm oil, starch, dicalcium phosphate. The fish meals are substituted with single cell protein/maggot meal in the diet (Table 1).

Experimental site

The experiment was conducted at the fisheries experimental farm of the ModibboAdama University of Technology, Yola, Adamawa State.

Experimental fish

The experimental fish *C. gariepinus* fingerlings total 100 were randomly sorted, weighted, stocked at 10 fingerlings per each plastic bow and starved overnight before the commencement of the feeding trial. The fish *C. gariepinus* fingerling were fed at 5% body weight, twice a day between 8 am and 4 pm. The fish were monitored for mortality daily. Dead fish were removed, counted and recorded for determination of survival rate (Table 2).

Sampling

The initial length and weight of fish (fingerlings) before stocking and the final length and weight of the fish were taken. Weekly sampling of the fish was done, i.e. 5 fingerlings were sampled. The weekly weight-

Experimental diet	I	II	III	IV	V	VI	VII	VIII	IX
Fish meal %	40.00	30.00	20.00	10.00	-	-	-	10.00	-
Maggot meal %	-	10.00	10.00	20.00	30.00	40.00	10.00	10.00	-
Single cell protein	-	-	10.00	10.00	10.00	-	30.00	20.00	40.00
Maize	37.03	40.00	44.50	44.50	40.00	37.03	40.00	44.50	37.03
Groundnut cake	17.97	15.00	10.50	10.50	15.00	17.97	15.00	10.50	17.97
Palm oil	2.00	2.00	2.00	2.00	2.00	2.00	2.00	2.00	2.00
Cassava starch	0.50	0.50	0.50	0.50	0.50	0.50	0.50	0.50	0.50
DicalciumPhosphate salt	0.50	0.50	0.50	0.50	0.50	0.50	0.50	0.50	0.50
Vitamin/Mineral	0.50	0.50	0.50	0.50	0.50	0.50	0.50	0.50	0.50
Premix	1.50	1.50	1.50	1.50	1.50	1.50	1.50	1.50	1.50
Total	**100**	**100**	**100**	**100**	**100**	**100**	**100**	**100**	**100**

Table 1: Dry matter composition of experimental diets at 40% crude protein.

Parameters	T_1	T_2	T_3	T_4	T_5	T_6	T_7	T_8	T_9
Initial mean weight (g)	0.9	0.98	1.04	1.0	0.96	1.0	1.0	0.96	1.10
Final mean weight (g)	4.15	4.75	4.33	4.25	3.69	4.50	4.37	3.95	3.83
Weight gain (g)	3.25b	3.77d	3.29bc	3.25b	2.71a	3.50c	3.37c	2.99ab	2.93ab
Initial length (cm)	4.60	4.56	4.10	3.50	4.10	4.40	3.86	3.28	3.80
Final length(cm	7.80	7.55	7.65	7.43	7.25	8.13	7.0	7.15	7.15
Relative weight Gain (%)	361.a	384.6a	316.d	325.c	282.3	350.b	337.0b	311.e	248.2f
Specific Growth Rate (%/day)	1.185b	1.224a	1.106	1.122	1.039	1.166	1.143	1.09	0.967
Feed intake (g)	8.07	8.07	7.93	7.59	6.44	7.64	7.57	8.13	7.36
Feed conversion Ratio	2.48b	2.14d	2.41b	2.34b	2.38b	2.18d	2.25c	2.27c	2.70a
K_I	0.924	1.075	1.508	2.332	1.392	1.173	1.738	2.72	2.004
K_2	0.874	1.103	0.967	1.036	0.963	0.837	1.274	1.08	1.048
Protein Efficiency Rate	0.081	0.0943	0.082	0.081	0678	0.087	0.843	0.74	0.068

Data with dissimilar alphabets are significantly different ($p<0.05$).

Table 2: Growth performances of *Clarias gariepinus* fed maggot and single cell supplemented diets.

T_1		T_2	T_3	T_4	T_5	T_6	T_7	T_8	T_9
Cost of feed (₦)	218.1	170.6	96.2	126.9	119.4	128.6	113.0	124.5	115.8
Cost of Feeding(₦)	17.50	13.77	7.63	9.63	7.69	9.83	8.55	10.12	8.52
Cost of fingerlings (₦ Fish)	40	40	40	40	40	40	40	40	40
Miscellaneous	100	100	100	100	100	100	100	100	100
Expenditure (₦)	157.50	153.77	147.63	149.63	147.69	149.83	148.55	150.12	148.52
Value of fish (₦)	184.00	193.87	166.54	170.00	152.92	180.00	174.80	164.58	139.27
Net profit (₦)	26.50	40.10	18.91	2037	5.23	30.17	26.25	14.46	9.25
Incidence of cost	67.10	45.25	29.24	39.04	44.05	36.74	33.53	41.63	42.41
Profit index	10.51	14.07	21.83	17.65	19.88	18.31	17.78	16.26	16.34
Cost Benefit ratio	1.168	1.260	1.128	1.136	1.035	1.201	1.176	1.096	0.937

Table 3: Cost benefits analysis of *Clarias gariepinus* fed maggot and single cell supplemented diets.

length of fish recorded was used to determine the growth performance of the fish. The feed supplied were used to determine the feed utilization or nutrient parameter following the methods of Burelet [8].

Water sampling

Water temperature was taken with graduated mercury-in-glass thermometer while dissolved oxygen, ammonia and pH were determined using the methods described by Boyd [8].

Growth, feed and cost-benefits parameters

At the end of the culture and feeding trials, the growth rates, condition factor, survival rate and nutrient utilization were computed and analysed according to zaid and sogbesan, [9] (Table 3).

The production cost in naira of the experimental diets was calculated following the method of Faturoti and Lawal [10] based on the current market price of the ingredients used for formulating the diets. Economic evaluation was determined according to New Faturoti and Lawal and Mazid et al. [9,11,12] based on the following.

(i) Weight gain of fish fed each experimental diet.

(ii) Similar number of days of the experiment for all the treatments.

(iii) Survival of the experimental fish stocked.

(iv) Cost of ingredients processing and of formulated feeds using the non-conventional animal feedstuffs.

(v) Cost of stocked and cropped fish before and after the feeding trials respectively.

Statistical analysis

Data generated from the experiment were subjected to Analysis Of Variance (ANOVA), correlation, and graphical representation. Duncan Analysis (Duncan, 1984) was used to compare the mean differences.

Results

The total highest weight gain of 3.77 g was recorded in fish fed (30%) fishmeal and 10% maggot meal) followed by 3.50 g from 40% maggot, 2.71 g was computed from fish fed 30% maggot and 10% single cell protein supplemented diet.

The best specific growth rate of 1.22% /day was recorded from fish fed 30% fish meal, 10% maggot meal and 20% single cell 2.14 from diet containing 30% fish meal and 10% maggot meal supplemented diets.

The highest feed intake recorded goes to fish fed 10% fishmeal, 10% maggot meal and 20% single cell protein with total feed intake of 8.13 g.

When you compare k1, and k2 in Figure 1, k1, has 2.7 as the highest condition factor when fed 10% fishmeal, 10% maggot meal and 20% single cell protein and lowest condition factor 0.92 when fed 40% fishmeal while k2 has highest condition factor 1.27 when fed 10% maggot meal and 30% single cell protein and lowest condition factor 0.84 when fed 40% maggot meal supplement.

The results on the survival rate indicated that the feeding of *C. gariepinus* fingerlings on maggot diets resulted into high survival rate. This can't be connected to the high acceptability of this meal which was observed during the study and also in accordance to the earlier report of Babatude [13] (Tables 4 and 5).

The best net profit of N40.10 and Cost - Benefit Ratio of 1.260 was recorded from fish fed 30% fish meal and 10% maggot meal supplemented diets followed by diet supplemented 40% maggot meal with N36.74 incidence of cost, N26.25 Net profit and N1.176 Cost - Benefit Ratio and the least 40% single cell protein supplemented diet with N42, 41 incidence of cost, 9.25 Net profit and 0.937 Cost –Benefit Ratio (Figure 2).

Discussion

The high growth performance of fingerlings fed maggot in combination with other supplemented diet in this experiment have formed a better balance diet for the fingerling catfish. A similar observation was made by Ugwumba and Abumoye [14] who obtained the best growth performance, food conversion and survival of *C. gariepinus* fingerlings (1-3 g body weight) when maggot was fed as supplemented food (maggot artificial feed).

The best feed conversion ratio with diet Viii (10% FM, 10% MM, 20% SCP) followed by diet ix (single cell alone) goes to suggest that the diets containing maggot were better utilized by the fingerlings. According to Jhimgram maggots are easily digested by fish.

The control diet would have been expected to show the best growth performance especially in terms of weight gain since it contains fish which has high level of protein that has been known as the best feed ingredient for fish [15,16] but this was not so, However, Lovell [15]

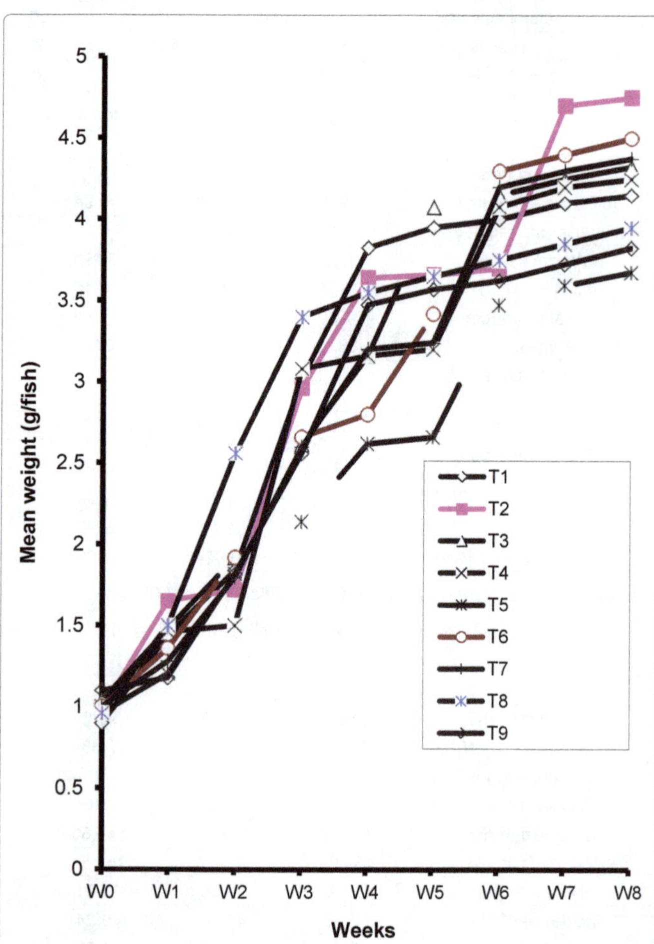

Figure 1: Weekly changes in mean weight of *Clarias gariepinus* fed maggot and single cell supplemented diets.

	Cost of feed (N)	Cost of Feeding(N)	Value of fish (N)	Net profit (N)	Incidence of cost	Profit index	Cost Benefit Ratio
Cost of feed (N)	1						
Cost of Feeding(N)	**0.983737**	1					
Value of fish (N)	0.592602	0.654475	1				
Net profit (N)	-0.07186	-0.07492	0.02516	1			
Incidence of cost	0.915516	0.858921	0.237713	-0.10587	1		
Profit index	-0.9115	-0.91817	-0.43038	0.072316	-0.86746	1	
Cost Benefit ratio	0.420056	0.487747	0.979202	0.05278	0.044476	-0.25036	1

Table 4: Correlation of the Cost benefits indices of using maggot and single cell supplemented diets.

	T1	T2	T3	T4	T5	T6	T7	T8	T9
T1	1								
T2	0.984943	1							
T3	0.972739	0.990114	1						
T4	-0.09986	-0.00029	-0.09661	1					
T5	0.985324	0.975228	0.98697	-0.20547	1				
T6	0.980632	0.997826	0.997131	-0.03968	0.982206	1			
T7	0.979054	0.996493	0.998318	-0.05437	0.98364	0.999811	1		
T8	0.989882	0.989922	0.994515	-0.13741	0.996432	0.993851	0.994482	1	
T9	0.991275	0.981061	0.9872	-0.17581	**0.999188**	0.985458	0.986151	0.997838	1

Table 5: Correlation of the experimental diets economy.

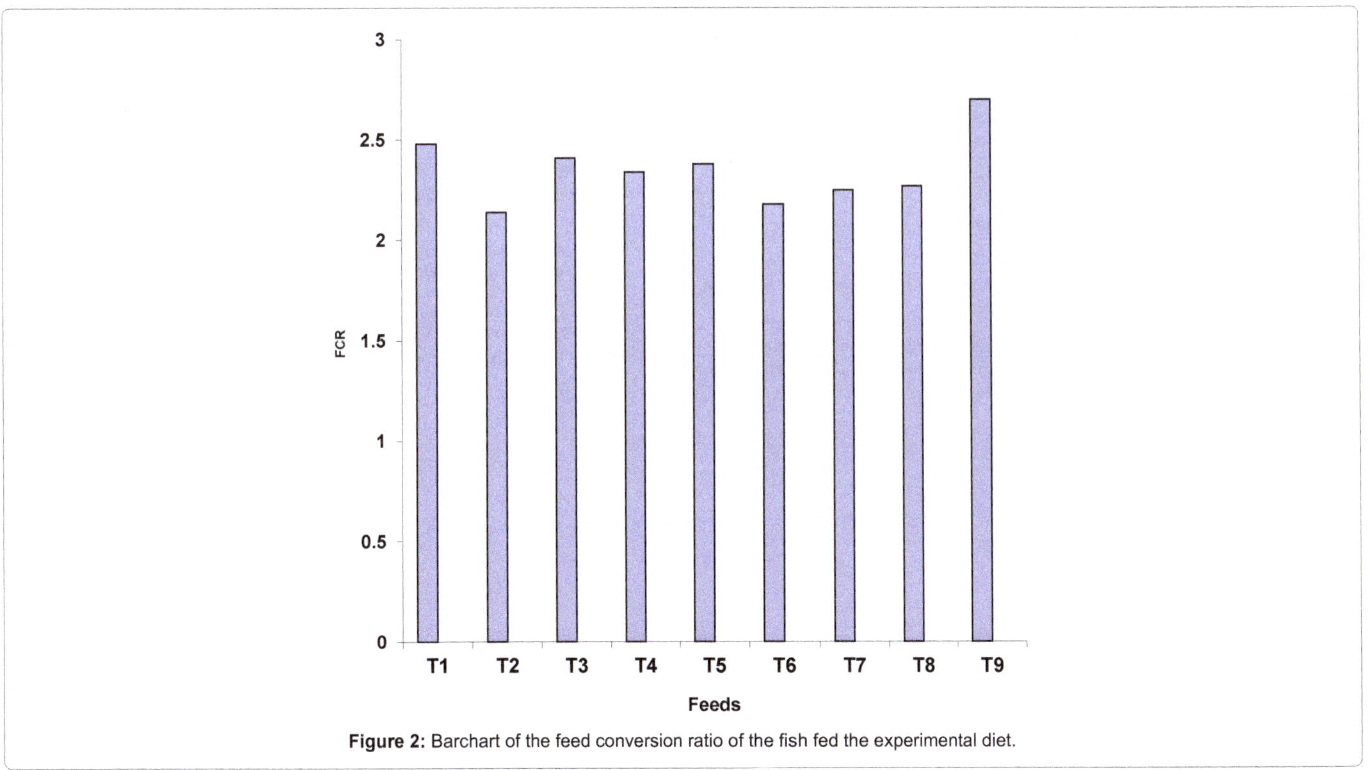

Figure 2: Barchart of the feed conversion ratio of the fish fed the experimental diet.

reported that the biological value of protein source does not only depend on its amino acid profile but also on its digestibility as indicated by digestibility energy which increased with maggot meal inclusion. Fibre content of feed has been documented to enhance growth performance in fish [17].

The optimum aim of every agricultural investor is to make profit at the end of the cultural season. This same phenomenon is as well applicable to fisheries, Since cost of feed has been one of the major constrain to the development of aquaculture sector, provision of an alternative ingredient that will be able to reduce certain percentage of the incurred overhead cost as a result of feeding should be embraced.

The actual feed product cost and harvest of maggot [5] is confounded by and associated benefit to livestock - poultry producers gain from manure management since fishmeal production requires labour, Fuel and equipment, one could assume that the equipment used to collect poultry manure, culture and harvest live maggot and process dried maggot meal might cost the same amount as reported by Newton et al. [18] but the cost of feed production did not agree with their report. The cost benefit report in this study also justifies the growth performance finding. Based on these results, the use of maggot to substitute the costly fishmeal to about 75% inclusion level is recommended to fish farmers and feed industry though there is a need to appraise large scale production of maggot. And the higher growth performance observed in combined feeding can explain by the synergetic effect of combining two biological compounds to have a single and superior effect than when individually applied. This observation is in agreement with suggestions by previous authors, that combined protein source is better than single protein source for fish diets [19-22].

The water quality parameters were within tolerates ranges throughout the period of the experiment. Temperature ranged between 28-32°C, Dissolved oxygen was between 4.5-5.8 ml while pH fluctuates between 7-8. The values of physico-chemical parameters observed in the culture system were within the range recommended for fresh water fish [19,20].

References

1. Olomola A (1990) Capture Fisheries and Aquaculture in Nigeria; a Comparative Economic Analysis. African Rural Social Sciences Series Report: 32.
2. Falaye AE (1992) Utilisation of Agro-Industrial Wastes as Fish Feedstuffs in Nigeria. Proceeding of the 10th Annual Conference of FISON: 47-57.
3. De Silva SS, Anderson TA (1995) Fish Nutrition in Aquaculture. Chapmann and Hall Aquaculture Series, Tokyo: 319.
4. Adejinmi OO (2000) The chemical composition and Nutritional potential of soldier fly larvae (Hermetiaelucens) in poultry rations. University of Ibadan, Ph.D.Thesis: 292.
5. Sogbesan OA, Ugwumba AAA (2006) Effect of different substrates on growth and productivity of Nigeria semi- arid zone earthworm (Hyperiodrilus Gropp J, Beck H & Erbersdobler H 1975). Tierphysiol Tiererniihr Futtm'tte Ur: 34-141.
6. Madu CT, Ufodike EBC (2003) Growth and survival of catfish (Clarias anguillaris) juveniles fed live tilapia and maggot as unconventional diets. 18: 47-51.
7. Boyd CE, Lickotkoper F (1990) Water quality in pond for aquaculture Alabama Agric. Experimental Station Auburn University, Alabama: 30.
8. Zaid AA, Sogbesan OA (2010) Evaluation and potential of cocoyam as carbohydrate source in catfish (C. gariepinus [Burchell,1822]) juvenile diet. African journal of Agriculture Research 5: 453-457.
9. Faturoti EO, Lawal LA (1986) Performance of Supplementary Feeding and organic manuring on the production of O. niloticu. Journal of West African Fisheries 1: 25-3.
10. New MB (1989) Formulated aquaculture feeds in Asia. Some thoughts on comparative economics, industrial potential, problems and research needs in relation to small scale farmers. In: Report of the workshop on Shrimps and fin fish feed development.
11. Mazid MA, Zaher M, Begun NN, Aliu MZ, Nahar F (1997) Formulation of Cost Effective Feeds from locally available ingredients for Carp culture system for increase production. Aquaculture 151: 7-78.
12. Babatunde AA (1990) Growth and survival of hatchery produced hybrid of Heterobranchusbidorsalis x Clarias gariepinus fry fed on varying protein diets and harvested plankton. M.Sc.thesis, University of Ibadan.
13. Ugwumba AAA, Abumoye OO (1998) Growth response of Clarias gariepinus fed live maggots from poultry droppings. In: sustainable utilization of Aquatic/ Wetland Resources: 182-188.
14. Lovell RT (1994) Compensatory gain in Fish. Aquaculture Management 20: 91-93
15. Massumotu T, Ruchmat T, Ito Y (1996) Amino acid availability values for several protein sources for yellow tail (Seriolaquinqueradiate). Aquaculture 146: 109-119.
16. Steffens W (1989) Principles of Fish Nutrition. Ellis Horwood Limited, UK: 38.
17. Newton GL, Sheppard DC, Watson DW, Burtle GJ, et al. (2005) The Black Soldier Fly, Hermetiaillucens, as a Manure Management/Resource Recovery Tool. State of the Science, Animal Manure and Waste Managmenent: 5-7.
18. Adekoya RA, Awojobi HA, Taiwo BBA (2004) The effect of partial replacement of maize with fall fat palm kernel on the performance of laying hens. Journal Agricultural Forest, Social Science 2: 89-94.
19. Adigun BA (2005) Water Quality Management in Aquaculture and Fresh water Zooplankton production for use in fish hatchery.
20. Washington DC (2000) Association of Official Analytical Chemists,Official methods of chemical Analysis. USA.
21. Burel C, Boujard T, Corraze G, Kaushik SJ, et al. (1998) Incorporation of high levels of extruded lupin in diets for rainbow trout (Oncorhynchusmykiss): nutritional value and effect on thyroid status. Aquaculture 163: 325-345.
22. Erondu ES, Nubian C, Nwadukwe O (1993) Haematological Studies on four catfish species raised in freshwater ponds in Nigeria. Journal of Pure and Applied Ichyology 9: 250-256.

5

Induced Breeding of Grass Carp (*Ctenopharyngodon idella*) and Silver Carp (*Hypophthalmichthys molitrix*) Using Ovatide as Synthetic Hormone at National Fish Seed Farm (Nfsf) Manasbal, Kashmir, J&K

Mudasir Rashid[1], Masood ul Hassan Balkhi[1], Gulzar Ah.Naiko[1] and Tamim Ahamad[2]

[1]*Faculty of Fisheries, Sher-e-Kashmir University of Agricultural Sciences &Technology of Kashmir, Rangil, Ganderbal, India*
[2]*Department of Fisheries, J&K, India*

Abstract

In the present study economically important and fast growing food fishes grass carp (*Ctenopharyngodon idella*) and silver carp (*Hypophthalmichthys molitrix*) were successfully spawned with Ovatide (combination of GnRH analogue with dopamine antagonist pimozide) in Kashmir. The preparations were administered by an intramuscular injection of single dose of 0.7 and 0.8-0.9 ml/kg body weight for female grass carp and silver carp and a single dose of 0.35 and 0.4-0.45 ml/kg body wt. body wt. for male fishes respectively. After dosing, the fishes were immediately carried to breeding pools in the Chinese Hatchery. After 14-16 hours of dosing spawning took place. After 10-12 hours of spawning, twitching movement started. Hatching occurred after 20-30 hours of fertilization at 24-26°C. Fecundity of grass carp and silver carp were recorded as 70000- 80000 and 1-1.10 lac eggs/kg body wt. of fish respectively. The fertilization percentage of grass carp and silver carp were recorded as 80.03% and 78.12% respectively. The hatching percentage of grass carp and silver carp were recorded as 70.10% and 69.71% respectively. And the fry survival percentage of grass carp and silver carp were recorded as 15. 21% and 14.56% respectively. The source of water was Manasbal Lake that favored most of the water quality parameters for spawning, hatching and survival of fry than ground water.

Keywords: Induced spawning; Brooder; Ovatide; GnRH; Pimozide; Fecundity; Grass carp; Silver car; Chinese hatchery

Introduction

Among the most significant advancements in the field of aquaculture during recent times is the development of techniques to induce reproduction in fish. These techniques have allowed farmers to profitably breed and raise species that do not naturally reproduce in captivity and to manipulate the timing of reproduction to suit production cycles. Some species will not readily breed in captivity due to environmental or culture conditions that are different from those found in nature, such as water temperature or substrate type. These conditions may cause stress or may not provide the signals needed to complete the reproductive process. Numerous hormones have been used to induce reproduction. Two methods have emerged over the past few years that seem to offer the best chance for success at the least expense. They are injection of a GnRH analog with dopamine antagonist and injection of gonadotropin. Chinese carps are quick growing economic food fishes extensively cultured in ponds in China, Japan and some of the South-East Asian countries. Grass carp is herbivorous and consume vegetation and increase natural food production in the pond by nutrient recycling and fecal production [1,2] On the other hand silver carp is predominantly a phytoplankton feeder and as such efficiently converts the food into fish flesh resulting in high yields. With a view to study the cultural possibilities of these carps in Indian waters and their compatibility with the cultivated species in India, namely the Indian major carps, experiment consignments of fingerlings of the Chinese grass carp (*Ctenopharyngodon idella)* and the silver carp (*Hypophthalmichthys molitrix*) were brought in 1959 [3]. Major breakthrough achieved in induced breeding of Indian major carps by administration of fish pituitary hormone by Chaudhuri and Alikunhi [4,5] Alikunhi et al., [6] and the technique has been further developed for attaining commercial production of fish seed of these economic species [3,4]. In the case of Chinese carps, successful spawning of the Grass carp and Silver carp by hormone injection has been reported by Aliev [7], Alikunhi et al. [8], Chaudury et al. [5] and Nayeem et al. [9]. In recent years, Human Chorionic Gonadotropin (HCG) has received some attention as a substitute for pituitary but has met with little success,except in the breeding of silver carp.

Lutinising releasing hormone (LH-RH), a mammalian hypothalamic peptide, has the capacity to release gonadotropin from pituitary gland. The Chinese report on successful use of mammalian based LH-RH analogue (D-Ala_6, Pro_9, Net) for induced breeding of carps created worldwide interest on the use of LH-RH for breeding various species of fish. A major breakthrough in fish breeding research was the finding that dopamine, acts as an inhibitory factor for synthesis of gonadotropin.

When LH-RH was used alone, without Pituitary Gland, spawning failure clearly indicates that dopamine blocks the action of LH-RH on the secretion of gonadotropin. Thus blocking of dopamine action with some antagonists like pimozide, potentiate the action of LH-RH resulting in successful spawning. There has been considerable research in India on spawning of carps with ovatide and ovaprim. Kaul and Rishi [10], reported the successful spawning of mrigal. Najar et al. [11], reported successful spawning and larval rearing of

***Corresponding author:** Mudasir Rashid, Professor, Faculty of Fisheries, Sher-e-Kashmir University of Agricultural Sciences & Technology of Kashmir, Rangil, Ganderbal, India, E-mail: mudasir.rather0@gmail.com

snow trout *(Schizothorax niger Heckel)* in Kashmir Himalaya with the application of ovatide. Khan, et al. [12], reported the successful spawning of rohu and mrigal with ovaprim (LH-RH analogue) at Fish Hatchery Islamabad, Pakistan. Ovatide is a new highly potent compound containing a synthetic peptide analog to the naturally occurring salmon GnRH –D.Arg[6] Pro Net. It also contains a dopamine antagonist pimozide, whereas GnRH analogs stimulates the pituitary to release gonadotropins and trigger the process of reproduction, the dopamine antagonist inhibits the release of dopamine and make sure that the secretion of gonadotropin is not inhibited. The use of ovatide, thus, constitutes the latest and the most advanced technology employed for induced breeding of fishes and production of high quality fish seed. A single dose of ovatide given intramuscularly to the brood fish leads to the production of an increased number of eggs through complete spawning with high fertilization and hatching percentage. The present study is the outcome of seed production trials carried out at National Fish Seed Farm, Manasbal (NFSFM), Kashmir on grass carp (*Ctenopharyngodon idella*) and silver carp(*Hypophthalmichthys molitrix).*

Materials and Methods

A total of 42 fishes consisting 21 grass carps (14 males and 7 females in 2:1 ratio) and 21 silver carps (14 males and 7 females in 2:1 ratio) were used for this study conducted from 20th July upto ist August and 15 July up to 10th of August 2014 respectively. The total wt. of female grass carps and silver carps were calculated as 15.504 and 16.949 kg respectively. For breeding, hatching and spawn collection, Chinese Circular Hatchery was used, which has four components for different operations: a) breeding pool or spawning pool, b) hatching pool or incubation pool, c) spawn collection pool and d) overhead storage tank. All the four components were meant for different operations like breeding, hatching, spawn collection and supply of water respectively. Hypodermic 2 ml syringe having 0.1 ml graduations with a needle no. 22 was used.

Selection of brooders

(1). Healthy, disease free, fully mature ripe fishes of the age group of 4-5 years (2-5 kg wt.) was selected. (2). Good characteristic traits of grass and silver carps of both sexes were selected. Secondary characteristics of both sexes of grass and silver carps that showed full maturity of these fishes during selection are (Table 1)

Care of brooders

Brooders of both grass and silver carps were reared in composite culture prior to inducing breeding. Grass carp liked clear and fresh water but their excreta proliferated large quantities of plankton. In order to control reproduction of plankton and water, planktivorus silver brood carps were reared with grass carp. But before a day to induce breeding, all the males and females were separated and were put in separate cemented raceways for conditioning. During this time feeding was stopped and special care was taken while drag netting and handling of brooders for induced breeding to avoid injuries to fishes. Brooders of grass carps were extensively fed with green fodder in the form of aquatic weeds like Vallisneria, Ceratophyllum and Hydrilla at the rate of double the wt. of their body so as to make them fully mature.

Method of injection

The fishes were held firmly and weighed cautiously, a calculated amount of single dose of ovatide injection to both sexes of grass carps and silver carps were given intramuscularly in the region of the caudal peduncle above the lateral line. The needle was inserted under the scale with hypodermic 2 ml syringe through to a depth of about 1.5 cm and injected the fluid slowly. Males of grass carp and silver carp were injected 0.35 and 0.4-0.45 ml/kg body wt., whereas females of grass carp and silver carp were given 0.7 and 0.8-0.9 ml/kg body of fish respectively.

Time of injection

The time of injection depends upon the water temperature. Due to low temperature in Kashmir valley as compared to other regions of India, a slightly high dose of ovatide injection was administered in the evening around 4-5 pm.

Handling and transfer of Brooders

After dosing, fishes in the 2:1 (male: female) ratios were immediately transferred to breeding pool of Chinese Hatchery in plastic bucket in the evening time (4-5 pm). Each plastic bucket carried 100 liters of water and about 10 kg parent fish. No anesthesia was given during transfer of brood fishes, as the place where dosing of ovatide was done was in close vicinity to Chinese hatchery. Acclimatization was maintained to avoid fish mortality.

Breeding and spawning

After single dose of injection to both males and females, they were put in the ratio of 2:1 (male: female) in the breeding pool of Chinese hatchery especially in the evening time. Then after about 4-5 hours, showering and water jets were started so as to create circular water motion, soon after about 7-8 hours of showering they got excited and showed sexual play. Males started to chase females and forced them to lay eggs. Spawning took place after 14-16 hours of injection at temperature 24-26°C which heavily depended on temperature. Eggs were washed with the solution of potassium per manganate, the colour of eggs were whitish muddy. The eggs of both grass and silver carps were demersal in nature, semi-pelagic were also seen.

Counting of eggs

The number of eggs was measured by volumetric method. A beaker was used in which number of eggs was counted in triplicate and the average of the three estimations was taken to know the numbers of eggs per beaker. Multiplying this with the number of beaker measured, total number of eggs was calculated.

Hatching and incubation

Hatching and incubation mainly depends on temperature, because of low temperature in Kashmir valley comparing to other regions of India , hatching period was recorded as 20-30 hours after fertilization at temperature 24-26°C. Newly hatched larvae remained in the circular incubation pool for four days until yolk was absorbed. 3 days later in incubation pool baby milk powder (lactogen) were started and on 5th day they were transferred to nursery ponds for further rearing. Water quality parameters during experiments are given in Table 2.

Calculation of fertilization rate, hatching rate and fry producing rate

When fish eggs have developed to the middle gastrula stage (after 16-30 hours of fertilization), about 100 eggs were collected with small net at random, they were put into a white dish and the eggs such as turbid eggs, white eggs, empty eggs and rotten eggs with naked eyes were given up. For more accuracy another method was operated to distinguish the fertilized and unfertilized eggs by simply dipping the

Species	Characteristics of the female	Characteristics of the male
Silver carp	Females with soft, distended belly and pink-red genital papilla, released ova when subjected to gentle pressure on the abdomen.	Released milt when subjected to gentle pressure on the abdomen Pectoral fin was rough.
Grass carp	Pectoral fin was soft. Abdomen round bulged with reddish fleshy vent. Ova oozed out on pressing abdomen.	Pectoral fin was rough, milt oozed out when pressed on abdomen.

Table 1: Secondary characteristics of both sexes

DO(mg/L)	6.5
pH	7.9
Water Temperature(°c)	24-26
Nitrate(mg/L)	0.3
Alkalinity(mg/L)	97.8
Co_2(mg/L)	9.8
Iron(mg/L)	0.05
Hardness(mg/L)	179
Phosphorous(mg/L)	0.09
Calcium(mg/L)	159
Carbonate(mg/L)	26.3
Bicarbonate(mg/L)	153
Total carbonate	181
TDS(mg/L)	588

Table 2: Phsico-chemical parameters of Chinese hatchery water.

eggs in a dilute $KMnO_4$ in a petri dish. The unfertilized eggs got colored whereas the fertilized eggs remained unchanged. Calculation of the fertilized eggs by percentage.

$$\text{Fertilization rate\%}=\frac{\text{no.of fertilized eggs}}{\text{total no. of eggs}}\times 100$$

$$\text{Hatching rate\%}=\frac{\text{no.of hatched fry}}{\text{no.of fertilized eggs}}\times 100$$

But in production, it was trouble-some to obtain the accurate figures of the actual hatching rate, therefore, it was important to calculate the fry producing rate in production.

$$\text{Fry stocking rate\%}=\frac{\text{no.of fry stocked in ponds}}{\text{no.of fertilized eggs}}\times 100$$

Results

Results of this study showed that successful induced spawning in grass carp and silver carp were achieved by using a single dose of ovatide. Certain drugs and different analogues of LH-RH are being tested for breeding fishes with varying degree of success. However, it was only when the dopamine inhibitory activity in the synthesis of gonadotropin was demonstrated that the reason behind the spawning failures became clear. Investigations have now clearly shown the potentiated actions of analogues when they are combined with dopamine antagonists like pimozide or domperidon. Based on the extensive research on Chinese carp, Peter, et al. [13] defined a new method of breeding called the Linpe in which LH-RH analogue is combined with a dopamine antagonist. It is major breakthrough in the history of aquaculture. Ovaprim and Ovatide are new drugs developed essentially on this new method combining releasing hormone with dopamine antagonist. Earlier studies conducted in India, Nandeesha et al. [14] and Khan et al. [12] have clearly demonstrated superiority of ovaprim in induced spawning of major carps.In this study both the grass carp and silver carp were successfully induced to spawn injected with single relatively high dose of ovatide (Tables 3 and 4). All the females and males of grass carp and silver carps (Total 42) showed positive response to ovatide and the ovulation of female fishes were 100%. The females of grass carps and silver caps were given 0.7 and 0.8-0.9 ml/kg body weight, whereas males were given 0.35 and 0.4-0.45 ml/kg body wt. respectively. In the present study grass carp showed full maturity and were ready to induce from 20th July upto 1st Aug 2014 whereas silver carps became fully mature and were ready to induce from 15th July upto 10th Aug.2014. Both males and females were put in 2:1 ratio in Chinese hatchery. The spawning and hatching periods for both grass carp and silver carp were prolonged and recorded as 14-16 hrs after injection and 20-30 hrs after fertilization at temperature 24-26°C respectively (Table 5). After 3 days in incubation pool, baby milk powder (lactogen) were started and on 5th day they were shifted to nursery ponds for further rearing.

The results obtained showed that both the grass carps and silver carps are not difficult to spawn with ovatide and the results obtained were satisfactory (Table 6). Though the number of eggs spawned by these carps in this study were relatively low than previously reported, but fecundity, fertilization and hatching to fry were comparatively high. The low fecundity may be due to low temperature and particularly not finding their natural productivity and the artificial feed given were not up to their standards, so these conditions might lower the nutritional intake of the brood fishes and finally might be cause of low fecundity.

Conclusion

The use of induced hormone "ovatide" was first time used in Kashmir valley at NFSF Manasbal to Chinese carps and the results obtained from the above experiment clearly demonstrated that the Ovatide was found effective in inducing spawning in Kashmir valley. It was also found that a slightly high dose of ovatide injection comparing to other regions to both grass carp and silver carp showed positive response. Although fecundity was low, fertilization and hatching period too was prolonged, yet fertilization, hatching and fry survival percentages were good. It

Females of the following fish species	Dosage of spawning agents
Catla	0.4-0.5 ml/kg body weight
Rohu	0.3-0.4 ml/kg body weight
Mrigal	0.25-0.3 ml/kg body weight
Fringe-lipped carp	0.3-0.4 ml/kg body weight
Catfish	0.6-0.8 ml/kg body weight
Silver carp	0.4-0.7 ml/kg body weight
Grass carp	0.4-0.8 ml.kg body weight
Bighead carp	0.4-0.5 ml.kg body weight
Mahseers	0.6-0.7 ml/kg body weight
Males of all species of carps	0.1-0.3 ml/kg body weight
Males of catfish	0.15-0.4 ml/kg body weight

Table 3: Dosage of ready-to-inject spawning agents (Ovaprim, Ovatide,WOVA-FH, etc.).

Fish species	Dose of ovaprim(ml)
Catla catla	0.4-0.5
Labeo rohita	0.3-0.4
Labeo rohita	0.4
Cirrhina mrigala	0.25-0.3
Cirrhina mrigala	0.4
Hypophthalmichthys molitrix	0.4-0.7
Ctenopharyngodon idella	0.4-0.8
Aristichthys nobilis	0.4-0.5
Hypophthalmichthys molitrix	0.6
Schizothorax niger	0.3-0.5
Ctenopharyngodon idella	0.7(ovatide)
Hypophthalmichthys molitrix	0.8 -0.9(ovatide)

Table 4: Dosage of Ovaprim-C and ovatide for carps at different locations.

	Grass carp	Silver carp
Parameters	Ovatide Treatment	Ovatide Treatment
Fecundity((body wt. of fish^{-1})	70000-80000	1-1.10 lac
No. of females treated	07	07
Totalwt. of females(kg)	15.504	16.949
Temperature(°c)	24-27	24-26
Spawning period(hours)	14-16	16-18
Total no. of eggs	1240320	1864390
Total no. of fertilized eggs	992628.096	1456461.468
Hatching period(hours)	20-30	20-30
Total no. of hatchling	695832.295	1015299.289
Fertilization percentage	80.03	78.12
Hatching percentage	70.10	69.71
Total no. of fry survival	150978.733	212060.79
Percentage of fry stocked	15.21	14.56

Table 5: Effect of Ovatide on grass carp and silver carp at NFSF Manasbal.

Shape	Circular
Diameter inner circle(m)	1.25
Diameter outer circle(m)	4
Depth of water(m)	0.75
Volume(L)	5107
Rate of water flow(m/sec)	0.2-0.4
Input of water (L/sec)	8
Duck mouth input pipe(nos.)	16
Nursing time(hour)	96
Egg loading capacity(lac)	30

Table 6: Incubation pool structure used for induced breeding at NFSF Manasbal.

was also seen use of ovatide reduced handlings of brood fish due to the single injection given to both the sexes simultaneously. These not only well decrease/avoid post spawning mortality of fish but also increase spawning response.

Further trials are now essential to standardize use of dosage and to gather additional information on the eggs and hatchlings produced through Ovatide treatment, such as their size, rate of growth, survival etc.

Acknowledgement

The authors like to thank Director Fisheries, Jammu and Kashmir for extending the technical support for the availability of fish and accessing their facilities for experiments.

References

1. Yang HZ, Fang YX, Liu ZY (1990) The biological effects of grass carp (Ctenopharyngodon idella) on filter-feeding and omnivorous fish in polyculture. In:R. Hirano and I. Hanyu (Eds.), The Second Asian Fisheries Forum. Asian Fisheries Society, Manila, Philippines: 197-200.
2. Li SF, Mathias J (1994) Fresh Water Fish Culture in China: Principles and Practices. Elsevier, Amsterdam.
3. Vijayalakshmanan MA, Pal RN, Sukumaran KK, Ramakrishna KV (1962) Experiments on inducing spawning of Indian carps by injection of fish pituitary hormones during 1959 at Cuttack, Proc. Indian Sci. Congr., Cuttack.
4. Chaudhuri H, Alikunhi KH (1957) Observations on the spawning in Indian carps by hormone injection. Current Science 26: 381-382.
5. Chaudhuri H (1960) Experiments on induced spawning of Indian carps with pituitary injections. Indian Journal of Fisheries 7: 20-49
6. Alikunhi KH, Vijayalakshmanan MA, Ibrahim KH (1960) Preliminary observations on the spawning of Indian carps, induced by injection of pituitary hormones. Indian J. Fish 7: 1–19.
7. Aliev DC (1961) Experiments of breeding from Far Eastern vegetarian fish in the conditions of Turkmenia (in Russian),Voprosi Ichtiologii.
8. Alikunhi KH, Sukumaran KK, Parameshwaran S (1963)Induced spawning of the Chinese carps Ctenopharyngodon idellus (C and V) and Hypophthalmichthys molitrix (Cand V) in ponds at Cuttack, India 32: 103-126.
9. Naeem M, Salam A, Diba F, Saghir A (2005) Fecundity and Induced Spawning of Silver Carp, Hypophthalmichthys molitrix by Using a Single Intramuscular Injection of Ovaprim-C at Fish Hatchery Islamabad, Pakistan. Pakistan Journal of Biological Sciences 8: 1126-1130.
10. Kaul M, Rishi KK (1986) Induced spawning of the Indian major carp Cirrhina mirgala (Ham.) with LH-RH analogue or pimozide. Aquaculture 54: 45-48.
11. Najar AM, Bhat F, Balkhi MH, Samoon MH, Dar Shabir A, et al. (2014) Induced breeding and larval rearing of snow trout (Schizothorax niger Heckel) in Kashmir Himalaya with the application of ovatide. Fishery technology 51: 8-12.
12. Khan MN, Janjua MY, Naeem M (1992) Breeding of carps with Ovaprim (LH-RH Analogue) at Fish Hatcheiy Islamabad, Proc. Pak. Cong. Zool 12: 545-552.
13. Peter RE, Lin HR, Van Der Kraak G (1988) Induced ovulation and spawning of cultured freshwater fish in China: Advances in application of GNRH analogues and dopamine antagonists. Aquaculture 74: 1-10.
14. Nandeesha MC, Rao KG, Javanna R, Parker R, Parker NS, et al. (1990) Induced Spawning of Indian Major Carps Through Single Application of Ovaprim-C, In: The Second Asian Fisheries Forum (Eds. Hirano.R. and I. Hanvu). Asian Fisheries Society, Manila, Philippines: 581-585.

6

A Study on the Determination of Heavy Metals in Sediment of Fish Farms in Bangladesh

Jahangir Sarker Md*, Indrani Kanungo, Mehedi Hasan Tanmay and Shamsul Alam Patwary Md

Department of Fisheries and Marine Science, Noakhali Science and Technology University, Noakhali-3814, Bangladesh

Abstract

Heavy metals in mud surface sediments have been determined to assess environmental pollution of the selected fish farms in Mymensingh, Bangladesh. Surface sediment samples (0-15 cm) from 20 ponds of a fish farm were collected in February 2014 with a single core sampler and were analysed to measure the concentrations (mg/Kg DW) of Zinc (Zn), Lead (Pb), Cadmium (Cd), Nickel (Ni) and Chromium (Cr) by atomic absorption spectrometry using a VARIAN model AA2407 in Bangladesh Agricultural Research Institute's laboratory. The measured concentration of heavy metals in the present study was in order of, Zn > Cr > Ni > Pb > Cd which are quite similar to the findings of other sediment of pond mud. The mean concentrations of Zn, Pb, Cd, Cr, Ni observed in the present study were 208, 14.845, 0.009, 63.054 and 58.665 mg/kg respectively. The ranges of the measured concentrations (mg/kg) in the total sediments were 11.5-18.5 for Pb, 0.007-0.011 for Cd, 53.5-77.3 for Cr, 50.8-66.4 for Ni and 100-250 for Zn. The concentrations of Zn and Ni in all sediment samples were above the US Environmental Protection Agency's (USEPA) guideline for severely polluted sediment and the concentration of Cr falls under the moderately polluted range. The metal Pb and Cd concentrations are well below the regulated level as per USEPA. Therefore, the study results revealed that the pond mud sediment quality in Mymensingh region might be considered as highly and moderately polluted for Zn, Ni and Cr respectively.

Keywords: Heavy metals; Sediment; Sediment quality guidelines; Mymensingh; Bangladesh

Introduction

Heavy metals are intrinsic, natural constituents of our environment and the term "heavy metals" refers to any metallic elements that have relatively high density and are toxic or poisonous even at low concentration [1]. Heavy metals include Lead (Pb), Cadmium (Cd), Zinc (Zn), Mercury (Hg), Arsenic (As), Silver (Ag), Chromium (Cr), Copper (Cu), Iron (Fe) and the platinum group elements. Moreover they are also known as trace elements because they occur in minute concentrations in biological systems. Sediments may become contaminated by the accumulation of heavy metals through various sources such as disposal of high metal wastes, land application of fertilizers, animal manures, sewage sludge, pesticides, wastewater irrigation [2,3]. The agricultural drainage water containing pesticides and fertilizers and effluents of industrial activities and runoff in addition to sewage effluents enter into the water bodies and sediment with huge quantities of inorganic anions and heavy metals [4]. Some animal wastes like livestock, poultry and pig manures created in agriculture and it also used as food in aquaculture pond and usually supplied to fish either in the form of solids or semi solids. The manures that are created from animals as a result of these diets possess greater amounts of Cu, As and Zn and if continually supplied as fish feed in pond, can result in reasonable accumulation of these metals in the longer period of time in these sediment [5].

Intensification of fresh water aquaculture is very high in the district of Mymensingh (Figure 1) and Jessore where fish farmers follow improved traditional techniques. Carp ploycultur and/mixed culture for 120 days (Thai koi, *Anabus testudineus*) to 300 days (Indian major carps, Pangus, *Pangasius pangasius*) with stocking density of 40-45/decimal fish fingerlings in Mymensingh region are common practices. Usually 150-250 kg/decimal fish feed are applied following very occasional water exchange during culture period. In Bangladesh fish and bone meal are added to the supplementary fish feed as a source of protein that are costly. To minimize cost or make the fish production cost effective farmers of this area uses tannery and poultry wastes that is believed as a cheaper protein source of fish feed. These sources caries high load of heavy metals. Therefore there might have a good chance to have high contamination of heavy metals on fish feed. As fish feed is directly consumed by the fish that ultimately accumulates on the sediment through deposition of fish excreta and excess feed waste as well. Theoretically these lead to the accumulation of toxic contaminants in cultured fish and present a food safety risk [6]. Also the heavy metal pollution of aquatic ecosystems is often most obvious in sediments despite there is scanty scientific works on river sediment and on fish in Bangladesh but unfortunately no work has yet been published particularly on heavy metal contamination of fresh water fish farm sediment. Therefore the present study is designed to conduct the present study in the district of Mymensingh, Bangladesh with a view to know the heavy metal concentration in sediments and to assess the sediment quality regarding the different status of pollution [7-12].

Materials and Methods

The research work was conducted in 20 different fresh water aquaculture ponds of Talukdar fish farm located at Shambhuganj Gouripur upazila of Mymensingh district in Bangladesh (Figure 1) during February, 2014. Different fish species such as Koi (*Anabas*

***Corresponding author:** Jahangir Sarker Md, Department of Fisheries and Marine Science, Noakhali Science and Technology University, Noakhali-3814 Bangladesh, E-mail: swaponj@yahoo.com

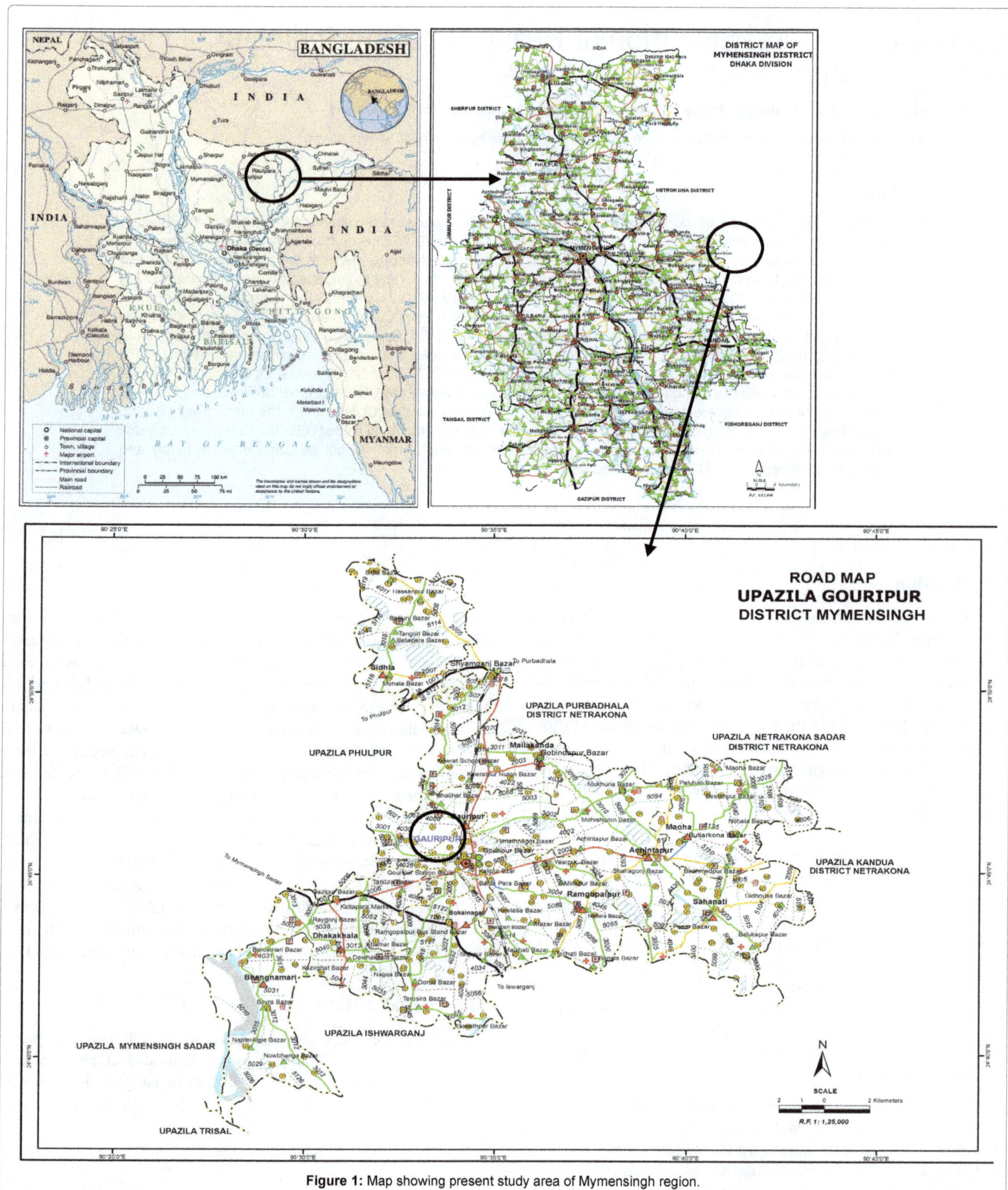

Figure 1: Map showing present study area of Mymensingh region.

testudineus), Shing (*Heteropneustes fossilis*), Magur (*Clarius batrachus*), Silver carp (*Hypothalmicthys molitrix*); Tilapia (*Tilapia mossumbica*), Rui (*Labeo rohita*) and Pangus (*Pangasius pangasius*) are cultured in the fish farm for 120 days to 300 days depending on the prices and demand of fish market. Total 60 sediment samples were taken from 20 ponds (3 replicates from each pond) by a single core sampler of (0-15) cm depth. The samples were then brought back to laboratory and were dried immediately at room temperature in a well aerated room and then grounded and screened to pass throw a 2.00 mm sieve and kept into plastic bags for further analysis.

In laboratory 1 g of dried sediment was weighed into 50 ml beakers, followed by the addition of 10 ml mixture of analytical grade acids HNO_3:$HClO_4$ in the ratio 5:1. The digestion was performed at a temperature of about 190°C for 1.5 h. After cooling, the solution was made up to a final volume (30 ml) with distilled water in a volumetric flask. The metal concentrations were determined by atomic absorption spectrometry (AAS) using a VARIAN model (AA2407) in the laboratory of Bangladesh Agricultural Research Institute. Analysis of each sample was carried out 4 times to obtain representative results and the data reported in mg/kg (on a dry matter basis). Baker and Amacher method for Cd, Cr, Ni, Zn and Burau method for Pb determination were followed [13,14].

Results and Discussion

The mean concentration of Zn, Pb , Cd, Cr, Ni of 20 different pond sediment samples are shown in Table 1 where the concentration levels of Cr, Ni and Zn are found higher than the concentration levels of Cd and Pb from same sampling pond. Metal contents were ranging over following intervals: Zn: (100-250) mg/kg, Pb: (11.5-18.5) mg/kg, Cd: (0.007-0.011) mg/kg, Cr: (53.5-77.3) mg/kg and Ni: (50.8-66.4) mg/kg. Mean concentrations of the metals were: Zn: (208 ± 31.388) mg/kg, Pb: (14.845 ± 1.995) mg/kg, Cd: (0.009 ± 0.001) mg/kg, Cr: (63.054 ± 6.922) mg/kg, Ni: (58.665 ± 4.617) mg/kg. On the basis of mean value, sediments were enriched with metals in following order:

Zn > Cr > Ni > Pb > Cd

The metal concentrations in the sediments were evaluated by comparison with the sediment quality guideline proposed by USEPA [15] (Table 2). According to these criteria our present study showed that all the ponds were heavily polluted for Zn and moderately polluted for Cr. The relatively high content of Cr and Zn may be due to use of $CuSO_4$, Hg $(NO_3)_2$, $K_2Cr_2O_7$ and $ZnSO_4$ especially that was not properly regulated. Fish culture in the present study area was completely dependent on the application of fish feed which are enriched with the concentration of Zn [16]. Therefore fish feed might lead the high

Table 1: Parametric presentation of heavy metal (mg/kg) in 20 sediment samples collected from a fish farm, Bangladesh.

Pond No:	Zn	Pb	Cd	Cr	Ni
P-1	210 ± 5	14.5 ± 0.2	0.008 ± 0.001	77.30 ± 0.22	50.80 ± 0.36
P-2	220 ± 4	18.2 ± 0.1	0.011 ± 0.003	67.11 ± 0.26	60.00 ± 1.64
P-3	100 ± 2	14.2 ± 0.1	0.007 ± 0.001	69.14 ± 0.34	55.30 ± 0.62
P-4	220 ± 4	13.5 ± 0.3	0.010 ± 0.002	57.21 ± 0.33	55.60 ± 0.51
P-5	200 ± 2	14.3 ± 0.3	0.008 ± 0.001	66.13 ± 0.36	54.60 ± 0.29
P-6	250 ± 4	13.6 ± 0.2	0.011 ± 0.002	64.04 ± 0.20	65.30 ± 0.37
P-7	220 ± 2	11.5 ± 0.2	0.011 ± 0.001	58.01 ± 0.16	56.30 ± 0.37
P-8	200 ± 4	12.3 ± 0.2	0.010 ± 0.002	55.02 ± 0.29	53.40 ± 0.41
P-9	210 ± 2	12.6 ± 0.2	0.008 ± 0.002	57.24 ± 0.28	57.80 ± 0.22
P-10	200 ± 5	12.8 ± 0.4	0.008 ± 0.002	54.09 ± 0.25	51.10 ± 0.22
P-11	200 ± 2	15.6 ± 0.2	0.010 ± 0.003	68.00 ± 0.40	62.50 ± 0.46
P-12	190 ± 4	14.3 ± 0.2	0.009 ± 0.003	54.60 ± 0.28	58.20 ± 0.29
P-13	180 ± 4	17.2 ± 0.3	0.008 ± 0.001	53.50 ± 0.28	60.50 ± 0.36
P-14	210 ± 5	18.5 ± 0.4	0.009 ± 0.002	75.60 ± 0.51	66.40 ± 0.37
P-15	220 ± 2	15.5 ± 0.2	0.010 ± 0.001	65.20 ± 0.37	59.80 ± 0.28
P-16	250 ± 5	16.2 ± 0.2	0.011 ± 0.002	60.50 ± 0.36	63.70 ± 0.43
P-17	230 ± 5	18.1 ± 0.2	0.008 ± 0.002	67.80 ± 0.36	56.50 ± 0.36
P-18	210 ± 3	14.5 ± 0.3	0.009 ± 0.002	59.40 ± 0.24	65.40 ± 0.43
P-19	200 ± 2	13.8 ± 0.5	0.010 ± 0.002	65.00 ± 0.43	57.80 ± 0.83
P-20	240 ± 5	15.5 ± 0.3	0.011 ± 0.002	66.20 ± 0.32	62.30 ± 0.37
Mean	208 ± 31.388	14.845 ± 1.995	0.009 ± 0.001	63.054 ± 6.922	58.665 ± 4.617
Range	100-250	11.5-18.5	.007-.011	53.5-77.3	50.8-66.4

Table 2: Different sediment quality guidelines of heavy metals (mg/kg).

Elements/Sediment Quality standard	Zn	Pb	Cd	Cr	Ni
(USEPA, 1991) Guidelines*					
Non-Polluted	<90	<40	-	<25	<20
Moderately Polluted	90 – 200	40 - 60	-	25 – 75	20 - 50
Heavily Polluted	>200	>60	>6	>75	>50
Present Study	208 ± 31.388	14.845 ± 1.995	0.009 ± 0.001	63.054 ± 6.922	58.665 ± 4.617
CBSOG SQG(2003)**					
Non-Polluted	<90	<40	<0.99	<43	<23
Moderately Polluted	90-200	40-70	0.993	43-76	23-36
Heavily Polluted	>200	>70	>3	>76	>36
Present Study	208 ± 31.388	14.845 ± 1.995	0.009 ± 0.001	63.054 ± 6.922	58.665 ± 4.617

- US Environmental Protection Agency [15].
- Consensus –Based Sediment Quality Guidelines [19].

concentration of Zn accumulated in the sediment of the present study area. High content of Cr in the present study area might be due to the fish feed as poultry and/tannery wastes are thought to add in fish feed where $K_2Cr_2O_7$ play key role in cleaning. Another source of heavy metal is the pellet feed that did not eaten by fish and settled in the sediment [17]. Where the mean levels of Pb (14.845 ± 1.995 mg/kg) of 20 samples (Table 1) was below the lowest effect level, threshold effect level, probable effect concentration and severe effect level (Figure 2) according to Sediment Quality Guideline [18] did not lie under the category of polluted area. In case of Cd (mg/Kg) which was below the lowest effect level, threshold effect level, probable effect concentration and severe effect level according to Sediment Quality Guideline (Figure 3) might be categorized the study area under non polluted state according to CBSQG SQG. The mean concentration of Cr recorded at the study area was 63.054 mg/kg (Table 1) which was below the severe effect level and probable effect concentration but the value exceeded the lowest effect level and threshold effect level according to Sediment Quality Guideline shown in (Figure 4). Moreover the mean concentration of Ni which was below the severe effect level but the value exceeded the lowest effect level, threshold effect level and probable effect concentration according to Sediment Quality Guideline (NOOA, 2009) shown in (Figure 5) [18,19].

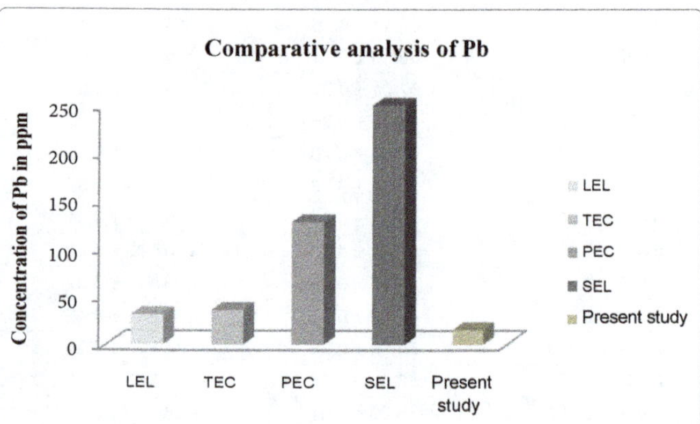

LEL=Lowest Effect level, TEC=Threshold Effect Level, PEC=Probable Effect Concentration and SEL=Severe Effect Level.

Figure 2: The mean concentration of Pb and comparison with Sediment Quality Guideline [18].

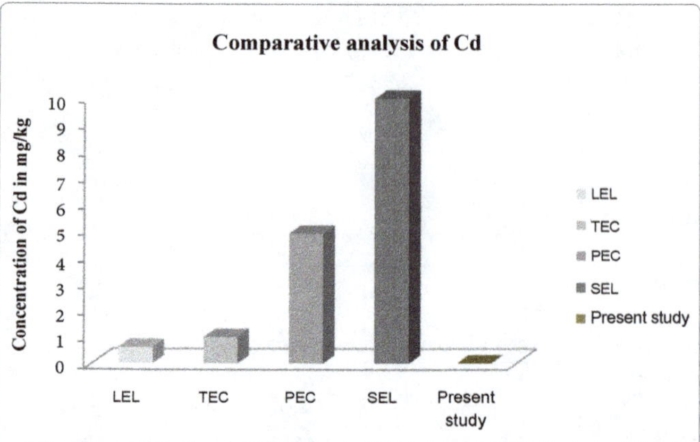

LEL=Lowest Effect level, TEC=Threshold Effect Level, PEC=Probable Effect Concentration and SEL=Severe Effect Level.

Figure 3: The mean concentration of Cd and comparison with Sediment Quality Guideline [18].

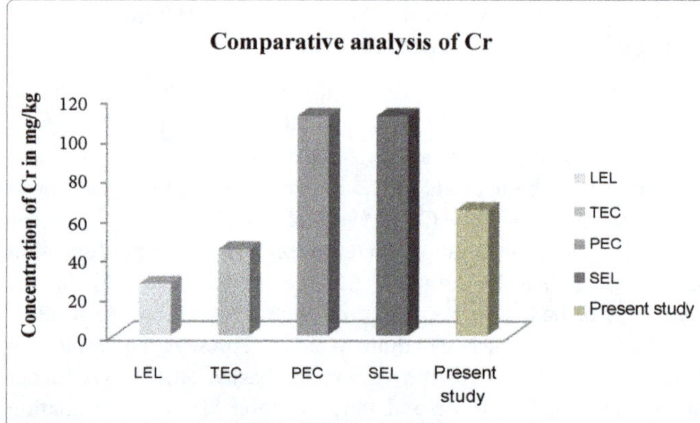

LEL=Lowest Effect level, TEC=Threshold Effect Level, PEC=Probable Effect Concentration and SEL=Severe Effect Level.

Figure 4: The mean concentration of Cr and comparison with Sediment Quality Guideline [18].

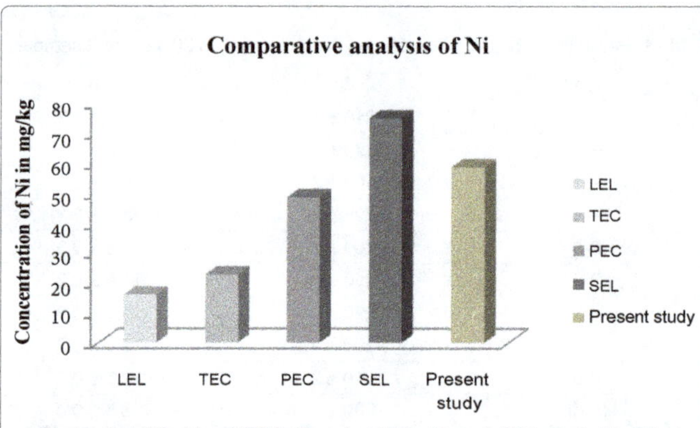

LEL=Lowest Effect level, TEC=Threshold Effect Level, PEC=Probable Effect Concentration and SEL=Severe Effect Level.

Figure 5: The mean concentration of Ni and comparison with Sediment Quality Guideline [18].

Conclusion

The main goal of this research work was to assess the concentration of some toxic heavy metals of sediment from several ponds of Mymensingh. A total of 20 fresh water aquaculture ponds sediment samples were collected and analyzed for five heavy metals namely Zn, Pd, Cd, Cr and Ni. The concentration of heavy metals was in order of, Zn > Cr > Ni > Pb > Cd. According to the present study, sediment can be categorized as heavily polluted for Zn and Ni whereas the areas can be categorized as moderately polluted for Cr according to USEPA. The concentrations of Pb and Cd in sediment were within the EPA sediment quality proposed and Consensus-Cased Sediment Quality Guidelines.

References

1. Lenntech (2004) Water Treatment and Air Purification. Water Treatment, Published by Lenntech, Rotterdamseweg, Netherlands.
2. Khan S, Cao Q, Zheng YM, Huang YZ, Zhu YG (2008) Health risks of heavy metals in contaminated soils and food crops irrigated with wastewater in Beijing, China, Environmental Pollution 152: 686-692.
3. Zhang MK, Liu ZY, Wang H (2010) Use of single extraction methods to predict bioavailability of heavy metals in polluted soils to rice. Communications in Soil Science and Plant Analysis 41: 820-883.

4. ECDG (2002) Heavy metals in waste. European Commission DG ENV. E3 Project ENV.

5. Basta NT, Ryan JA, Chaney RL (2005) Trace element chemistry in residual-treated soil: key concepts and metal bioavailability. Journal of Environmental Quality 34: 49-63.

6. Bartone CR, Benavides L (1997) Local Management of Hazardous Wastes from Small-Scale and Cottage Industries. Waste Management and Research 15: 3-21.

7. Linnik PM, Zubenko IB (2000) Role of bottom srdiments in the secondary pollution of aquatic environments by Heavy-metal compounds. Lakes and Reservoirs: Res and Man 5: 11-21.

8. Ahmad MK, Islam S, Rahman MS, Haque MR, Islam MM (2010) Heavy metals in water, sediment and some fishes of Buriganga River, Bangladesh. International Journal of Environmental Research 4: 321-332.

9. Islam MS, Ahmed MK, Raknuzzaman M, Habibullah-Al-Mamun M, Islam MK (2015) Heavy metal pollution in surface water and sediment: a preliminary assessment of an urban river in a developing country. Ecological Indicators 48: 282-291.

10. Rahman MS, Molla AH, Saha N, Rahman A (2012) Study on heavy metals levels and its risk assessment in some edible fishes from Bangshi River, Savar, Dhaka, Bangladesh. Food Chemistry 134: 1847-1854.

11. Islam MS, Ahmed MK, Habibullah-Al-Mamun M, Masunaga S (2015) Assessment of trace metals in fish species of urban rivers in Bangladesh and health implications. Environmental toxicology and pharmacology 39: 347-357.

12. Ahmed MK, Shaheen N, Islam MS, Habibullah-al-Mamun M, Islam S, et al. (2015) Dietary intake of trace elements from highly consumed cultured fish (*Labeo rohita*, *Pangasius pangasius* and *Oreochromis mossambicus*) and human health risk implications in Bangladesh. Chemosphere 128: 284-292.

13. Baker DE, Amacher MC (1982) Methods of Soil Analysis, Chemical and Microbiological Properties. ASA and SSSA, Madison, WI, USA pp: 323-336.

14. Burau RE (1982) Methods of Soil Analysis, Chemical and Microbiological Properties. ASA and SSSA, Madison, WI, USA pp: 347-366.

15. USEPA (1991) Sediment quality guidelines. Draft report. EPA Region V Chicago IL.

16. Kalantzi I, Shimmield TM, Pergantis SA, Papageorgiou N, Black KD, et al. (2013) Heavymetals, trace elements and sediment geochemistry at four Mediterranean fish farms, Science of the total Environment 1: 128-137.

17. Tao L, Zhang CX, Zhu JQ, Zhang SY, Li XL, et al. (2012) Improving Fishpond Sediment by Aquatic Vegetable Rotation. Adv Jour of Food Sci and Tech 4: 327-331.

18. National Oceanic and Atmospheric Administration (2009) Screening Quick Reference Tables for in Sediment, SQUIRT. Accessed on: March 23, 2009.

19. Concensus based sediment quality guideline (2003) Wisconsing Department of Natural Resources. Recommendation for use and application. Department of interior, Washington DC pp: 17.

In-season Forecast of Chum Salmon Return Using Smoothing Spline

Kyuji Watanabe*

Hokkaido National Fisheries Research Institute, Japan Fisheries Research and Education Agency, 2-2 Nakanoshima, Toyohira-ku, Japan

Abstract

We developed an in-season forecast model of return of chum salmon for the population off the Honshu region in the Sea of Japan using the smoothing spline based on catch data obtained in fishing season. The optimal in-season model was constructed using adult return in season 8 (middle October) as an explanatory variable. Residual sum of squares of the optimal in-season model was lower than that of the pre-season forecast (sibling) model, indicating the former was more accurate than the latter. The relationship between forecast error rate in the optimal model and the cumulative proportion of return until season 8 (middle October) was positive. Yearly variation in the forecast error rate may be affected by variability in the timing of return. We provide a new and accurate forecast model of chum salmon return.

Keywords: Chum salmon; Forecast; GAM; In-season; *Oncorhynchus*; Pacific salmon; Sibling; Sea of Japan; Smoothing spline

Introduction

Mature (adult return) chum salmon (*Oncorhynchus keta*) migrate from offshore into coastal waters and natal rivers to breed [1,2]. Spawning season spans approximately 7 months, from August to February [3]. In Japanese waters, returning adults are caught mainly by set-nets in coastal waters off natal rivers, and by gill- and purse-seine nets within rivers. Therefore, the number of returning adults is defined as the sum of numbers caught coastally (coastal catch) and numbers caught in rivers (river catch). As the fishing season proceeds the cumulative proportion of returning adults gradually increases (Figure 1). We can view in-season catch data as a kind of relative density index obtained by in-season survey.

At present, the pre-season forecast model of Japanese chum salmon, which is the sibling model, it has been often used to forecast age-specific return number. This pre-season forecast model requires many biological parameters and assumptions, which are age composition and constant rate of maturation through study period. Whereas, in-season forecast model using only in-season catch information could require no parameters and assumptions (Figure 1). In addition, a nonparametric approach such as smoothing spline is expressed with the added advantage of not having any a prior assumption of linearity [4]. Several studies have developed nonparametric models for in-season forecasts [5,6]. In-season forecast model using catch data and smoothing spline could ensure simplification of forecasting.

The in-season model in this study does not consider variability in salmon return timing, which is affected by both genetic and environmental factors [7,8]. The return timing affecting in-season catch, return, and the cumulative proportion of the adult return (Figure 1) could cause forecast error of the in-season model. We focused on the relationship between forecast error rate and the cumulative proportion of the adult return.

The purpose of this study was to develop in-season forecast models for chum salmon return by using smoothing spline. We compare forecast accuracy of the in-season forecast model with that of pre-season forecast model.

Materials and Methods

Study area and population

In this study, we subject data to the chum salmon population in the Sea of Japan off the Honshu region, Japan to forecast models (Figure 2). This regional population includes river stocks of chum salmon in the region (Figure 2). The region includes coastal areas in Aomori, Akita, Yamagata, Niigata, Toyama, and Ishikawa Prefecture.

Data source

The coastal catch in number, C, and river catch in number, D, in the regional population of the Sea of Japan off Honshu were used for 1993-2013 [9]. The catch data were collected by Aomori, Akita, Yamagata, Niigata, Toyama, and Ishikawa Prefecture. The catch data were summarized each season j (1st season is 10 days). The season j-specific return, $R_{t,j}$, and the total return in year t were defined as:

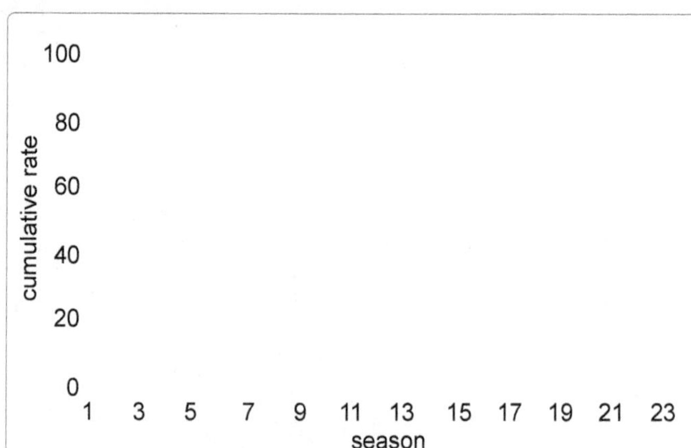

Figure 1: Season-specific cumulative proportions of return of chum salmon in the Honshu region, the Sea of Japan off, Japan, from 1993 to 2013. Season j is divided from 1 August-31 March in intervals of 10 days (j=1-24).

***Corresponding author:** Kyuji Watanabe, Hokkaido National Fisheries Research Institute, Japan Fisheries Research and Education Agency, 2-2 Nakanoshima, Toyohira-ku, Sapporo 062-0922, Japan, E-mail: watanabk@fra.affrc.go.jp

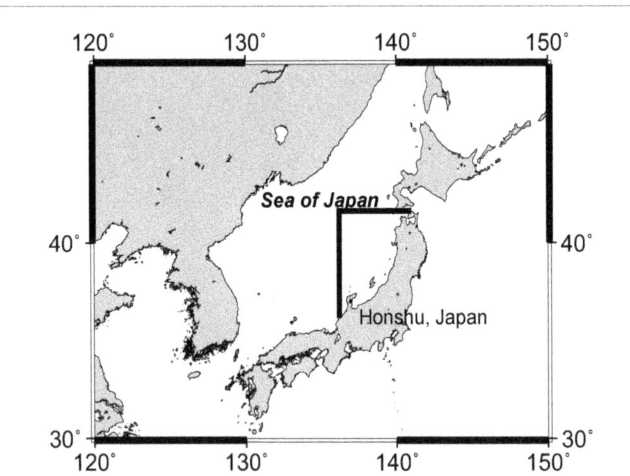

Figure 2: Summary of location. The solid line indicates the Honshu region in the Sea of Japan.

$$R_{t,j} = C_{t,j} + D_{t,j} \quad (1)$$

$$R_t = \sum_{j=1}^{24} R_{t,j} \quad (2)$$

where j is the season that is divided from 1 August-31 March in intervals of 10 days (j=1-24). For example, j=1, 2 and 3 represent periods from 1 to 10 August, 11 to 20 August, and 21 to 31 August, respectively.

Smoothing spline model

To evaluate relationships between log-transformed total return R_t and catch data in-season, models were constructed for the following four cases:

$$\ln(R_t) = s(\ln C_{t,j}) + \varepsilon \quad (3)$$

$$\ln(R_t) = s(\ln \sum_{j=1}^{\tau} C_{t,j}) + \varepsilon \quad (4)$$

$$\ln(R_t) = s(\ln R_{t,j}) + \varepsilon \quad (5)$$

$$\ln(R_t) = s(\ln \sum_{j=1}^{\tau} R_{t,j}) + \varepsilon \quad (6)$$

where s is the smoothing spline function and ε is assumed to be normal distribution. Function s is estimated by minimizing penalized residual sum of squares and equivalent 4 degrees of freedom in GAM by using S+ version 8.1 [10]. In Equations (3-6), explanatory variables are coastal catch, $C_{t,j}$, cumulative coastal catch-to-date, $\Sigma C_{t,j}$, return $R_{t,j}$, and cumulative return-to-date, $\Sigma R_{t,j}$, separately. In addition, to test explanatory variables in various fishing seasons, we used variables from fishing starting season 1 to 8 when the cumulative proportion of adult return met 0.5 (Figure 1). However, in seasons 1-5, return and catch were frequently 0. Therefore, to avoid values of 0 as explanatory variables, we used variables in seasons 6,7 and 8. Thus, season j was given as 6,7 and 8 in the case of Equations (3) and (5). In the case of Equations (4) and (6) τ is given as 6,7 and 8. Therefore, we generated 3 models for each case. The total number of models was 12 (Table 1).

In-season forecast procedure by cross-validation

Fitting of the above models is a part of forecast procedure. Log-transformed R was forecast by the following cross-validation:

Step 1: A smoothing spline was estimated by GAMs using data from 1993 to t, indicating fitting of the models (Equations: 3-6).

Step 2: An explanatory variable in t+1 is substituted into the estimated models and then $\ln(R_{t+1})$ is forecast.

This procedure was repeated from t=2000 to 2012. Therefore, year-specific forecasts, $\ln(R_{t+1})$, were provided for each year from 2001 to 2013.

To evaluate the accuracy of model forecasts, the residual sum of squares (RSS) was measured for each in accordance with:

$$\mathrm{RSS} = \sum_{t=2000}^{2012} (\ln R_{t+1} - \ln \hat{R}_{t+1})^2 \quad (7)$$

In addition, forecast error rate was calculated for each year. Note that the cumulative proportion of the adult return for any given season varied between years (Figure 1). Thus, the forecast error rate may be affected by return timing. We investigated relationships between forecast error rate and the cumulative proportion of returning adults.

Pre-season forecast

The sibling model, a traditional pre-season forecast model, was used to forecast the number of adult returns for specific age classes, as follows:

Step 1: Linear regression of log-transformed R at age a-1 in t-1 against log-transformed R at age a in t is estimated by using data from 1993 to t as:

$$\ln R_{t+1,a} = g_{t,a} \times \ln R_{t,a-1} + h_{t,a} \quad (8)$$

where $g_{t,a}$ is the regression slope and $h_{t,a}$ is the intercept from 1993 to t at ages a from 3 to 7. Calculations for 5 ages classes over 13 years produced 65 regressions. Note for ages 2 and 8 a regression (Equation 8) cannot be estimated. At ages 2 and 8, forecast values are given as the average of observed $\ln(R_{t,2})$ and $\ln(R_{t,8})$ from t to t-4, separately.

Step 2: When $g_{t,a}$ is significant (p<0.05), an explanatory variable, i.e., return at age a-1 in t, is substituted into the estimated regression (Equation 8) and then $\ln(R_{t+1,a})$ is forecast. If $g_{t,a}$ is not significant (p<0.05), $\ln(R_{t+1,a})$ is given as the average of observed $\ln(R_{t,a})$ from t to t-4. Finally, age-specific forecasts $\ln(R_{t+1,a})$ were combined by year. Age-combined forecast, $\ln(R_{t+1})$, were calculated for each year from 2001 to 2013.

This procedure was repeated from t=2000 to 2012. RSS between

Model	Explanatory variable	Season
1	C	6
2	C	7
3	C	8
4	Cumulative C	1-6
5	Cumulative C	1-7
6	Cumulative C	1-8
7	R	6
8	R	7
9	R	8
10	Cumulative R	1-6
11	Cumulative R	1-7
12	Cumulative R	1-8
13	Sibling	Pre-season

Table 1: In-season forecast models and sibling model. Coastal catch (C) and return (R) of chum salmon.

actual $\ln R_{t,+1}$ and age-combined $\ln\Sigma(R_{t+1,a})$ was calculated using Equation (7).

Results and Discussion

In catch models 1,2 and 3, RSS decreased as the season progressed from 6 to 8 (Figure 3). In cumulative catch models 4,5 and 6, and in return models 7,8 and 9, RSS also decreased in the same manner with catch models. To the contrary, in cumulative return models 10,11 and 12, RSS decreased as the season proceeded from 6 to 7 but increased in season 8. Of all 12 in-season models (Table 1), catch model 3 had the lowest RSS (Figure 3). RSS in model 3 was lower than that of the sibling model 13, indicating that, of all models examined, the optimal model was catch model 3.

The optimal model was particularly good at forecasting variation in the observed return in 2004-2008 compared with the sibling model (Figure 4). Ability of forecast the return of the optimal model (catch model 3) was better than that of the sibling model (Figure 3). Therefore, our result provides a new, simple, and accurate in-season forecast model compared with the sibling model.

The relationship between forecast error rate in catch model 3 and the cumulative proportion of return until season 8 (middle October) was positive (Figure 5). Yearly variation in the forecast error rate may be affected by variability in the timing of return [7,8]. Further study would need to incorporate variables associated with variability in the return timing into in-season forecast model. However, the forecast error rates of catch model 3 were relative low. In addition, smoothing splines of catch model 3, which were estimated by GAM framework, were soaring curves against coastal catch in season 8 (Figure 6). Although the end year for modeling, i.e., year t, changed from 2000 to 2012, the form of these curves did not change demonstrably. This result suggests that little the coastal catch in season 8 as explanatory variable in the optimal model is affected by changing of catch inducing by the return timing. Thus, the coastal catch in season 8 has robustness

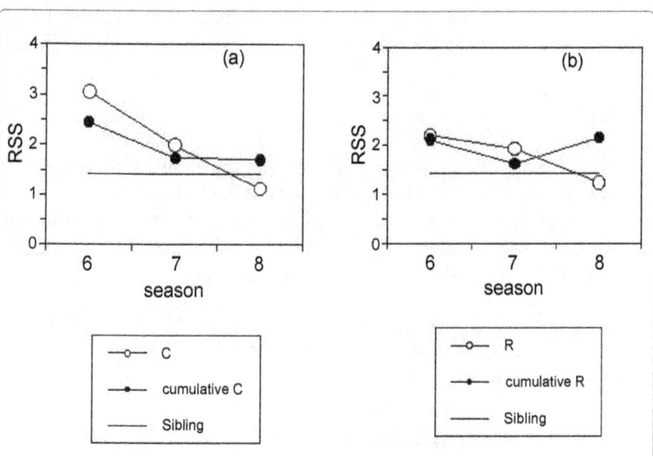

Figure 3: Residual sum of squares (RSS) of catch model and cumulative catch model (a) and return model and cumulative return model (b).

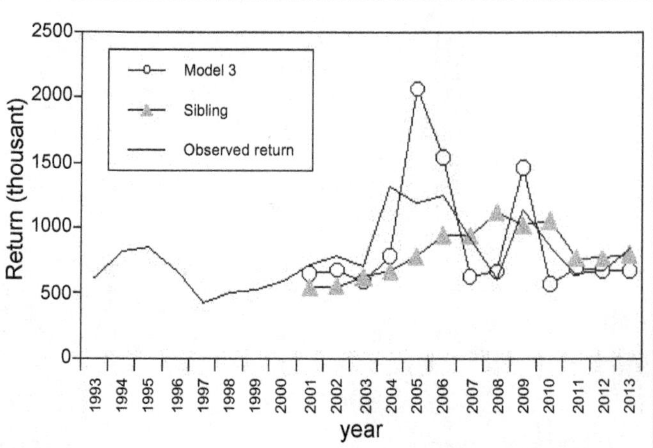

Figure 4: Forecast returns in model 3 and sibling model, and observed return.

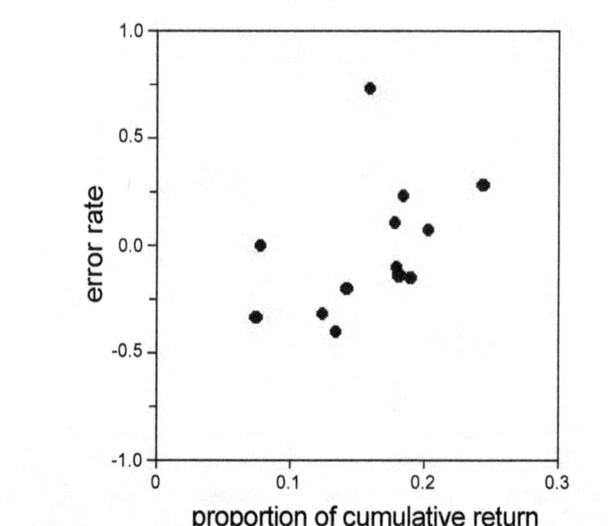

Figure 5: Relationship between proportion of cumulative return from season 1 to 8 and forecast error rate.

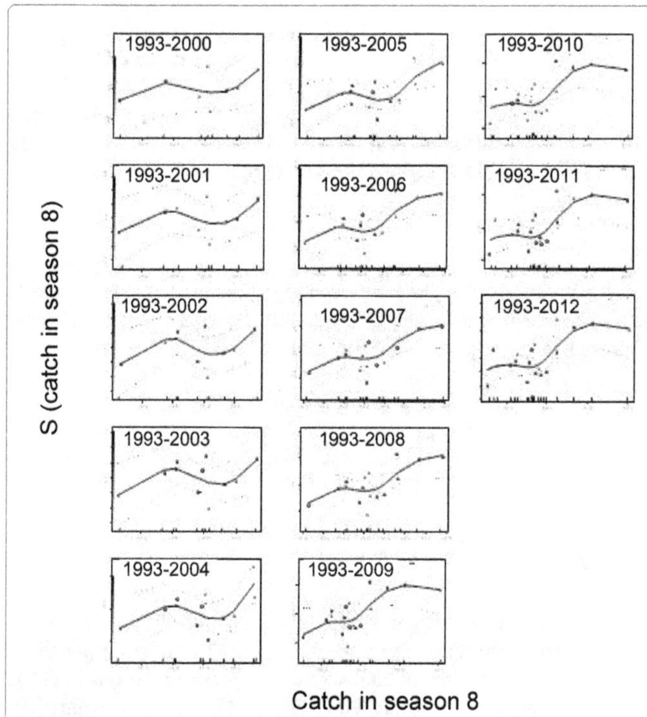

Figure 6: Estimated smoothing spline (s) of catch model 3 for each period for 1993 to *t*.

of variability of return timing. This model could explain variation in the observed return well.

Acknowledgement

I thank Fumihisa Takahashi and Yukihiro Hirabasyashi of the Hokkaido National Fisheries Research Institute, Kei Sasaki of Tohoku National Fisheries Research Institute, and anonymous reviewers, for their useful advice and support.

References

1. Neave F, Yonemori T, Bakkala RG (1976) Distribution and origin of chum salmon in offshore waters of the North Pacific Ocean. Int N Pac Fish Comm Bull 35: 1-72.
2. Salo EO (1991) Life history of chum salmon (Oncorhynchus keta) in Groot C, Margolis L, Pacific salmon life histories. UBC Press, Vancouver.
3. Okazaki T (1982) Geographical distribution of allelic variations of enzymes in chum salmon, Oncorhynchus keta, river populations of Japan and the effects of transplantation. Bull Japan Soc Sci Fish 48: 1525-1535.
4. Hastie TJ, Tibshirani RJ (1990) Generalized additive models. Chapman & hall/ CRC, New York.
5. Noakes DJ (1989) A nonparametric approach to generating in season forecasts of salmon returns. Can J Fish Aquat Sci 46: 2046-2055.
6. Chen XH, Shelton PA (1996) A nonparametric forecast model of inshore Atlantic cod (Gadus morhua) landing based on biomass, cumulative landings, and water temperature. Can J Fish Auat Sci 53: 558-562.
7. Quinn TP, Peterson JA, Gallucci VF, Hershberger WK, Brannon EL (2002) Artificial Selection and Environmental change: Countervailing factors affecting the timing of spawning by coho and chinook Salmon. T Am Fish Soc 131: 591-598.
8. Quinn TP, Unwin MJ, Kinnison MT (2000) Evolution of temporal isolation in the wild: genetic divergence in timing of migration and breeding by introduced chinook salmon populations. Evolution 54: 1372-1385.
9. Shinnkou-kai HSZ (1993-2013) Data. Honshu Sakemasu zousyoku shinnkou-kai, Tokyo.
10. Chambers J, Hastie T (1992) Statistical models in S. Chapman and Hall, New York.

8

Food and Feeding Habits of *Cyprinus carpio* Var. *communis*: A Reason that Decline Schizothoracine Fish Production from Dal Lake of Kashmir Valley

Gulzar Naik*, Mudasir Rashid, Balkhi MH and Bhat FA

Sher-e-Kashmir University of Agricultural Sciences and Technology of Kashmir, Rangil, Ganderbal, Jammu and Kashmir, India

Abstract

In this paper, we studied the food and feeding habits of exotic *Cyprinus carpio* Var. *communis*. The results obtained by analysing the gut contents of common carp showed that on an average basis, detritus formed 43.5% of total food, while the remaining food (56.5%) consisted of plant (31.21%) and animal matter (25.29%). The fish was designated as detri-omnivore with bottom feeding habit. Gastroosmatic index (Ga.S.I.) recorded its highest value during July (6.28), while lowest value was recorded in February (3.34).The index remained generally high during the warmer months, followed by a gradual decline with the approach of winter. On comparing, food and feeding habits and Ga.S.I. of exotic common carp with that of endemic schizothoracines was found almost similar and there might be existing a feeding competition between them, which might be one of the cause that declined endemic schizothoracine fish production from Dal Lake of Kashmir valley.

Keywords: *Cyprinus carpio*; Exotic; Endemic; Food and feeding habits; Schizothoracines.

Introduction

Valley Kashmir is bestowed with enormous and rich aquatic resources in the shape of rivers, lakes, streams, high altitude lakes, springs and low lying areas covering total water spread area of about 32765.3 hectares which is nearly 2% of total area of the Kashmir Valley. The major fish fauna of these water bodies comprises of exotic *Cyprinus carpio* and indigenous *Schizothorax* species. Other rarely found species are *Labeo, Glyptosternum, Puntius, Nemacheilus* e.t.c., *Schizothorax* is represented by many species v.i.z., *Schizothorax esocinus, Schizothorax curvifrons, Schizothorax niger, Schizothorax plageostomus, Schizothorax labiatus* e.t.c., which are commonly called as snow trouts. The Dal Lake is situated between 34°5' and 34°6'N latitude and 74°8' and 74°12' E longitude at an altitude of 1584 m above sea level. It is a shallow open drainage type water body spread over an area of 11.4 km^2. Till recent past, the Dal Lake was considered to be one of the finest lakes in the country and also as one of the most scenic spots in the world. However, due to over exploitation during the last fifty years this water body has turned into a highly polluted ecosystem. Addition of nutrients from anthropogenic perturbations in the catchment, creation of floating gardens and islands within its basins, anchoring of hundreds of house-boats within the lake and introduction of exotic common carp have changed the overall ecological setup of this water body. In the long run introduction of exotic species may turn out to be a deleterious problem as habitat loss and causes extinction of species [1]. The commercially important fish of the Dal lake are exotic *Cyprinus carpio-specularis* and *C. carpio- communis* and endemic ones are *Schizothorax niger, S. esocinus, S. micropogan* and *S. plagiostomus*.

The *Cyprinus carpio* (common carp) was brought to India in 1939 from Srilanka and introduced into the Nilgiris. Later in 1947 this species was introduced in Nainital and other lake of Kumaon and was carried to Bangalore. It is an ideal species for cold water of the hills and breeds in confined water. The common carp was introduced in Dal Lake of Kashmir in 1956 and since then this fish has shown remarkable adaptation in various water bodies of the state, and soon began to constitute a major fishery of flat land temperate waters of Kashmir [2]. The *Cyprinus carpio* formed almost 75% of the fish catch in Dal and Wular lakes of Kashmir [3]. The introduction of the exotic common carp caused a sharp decline in the population and almost exterminated the schizothoracine fishes in Kashmir valley [4,5]. The total fish production in Dal lake ranged from as low as 262 tonnes in 2007-08 to a maximum of 475 tonnes in 2003-2004 and fish production in 2010-2011 was 336 tonnes. The total fish production in Dal Lake was held up by increasing exotic carp fish production and the rate of decline in *Schizothorax* (local) fish production was steep as well as pronounced in variation when statistical models were applied [6] (Figure 1A-1C). So common carp caused a slow and steady decline in the initial phases and later an abrupt drop in the contribution of the *Schizothorax* species to the total fish production in Dal Lake of Kashmir.

The major reasons put forth by different zoologists for the predominance of common carp over the more prized endemic fish fauna in the lake were food competition due to more or less identical food spectra, higher fecundity, spawning facilities prevailing in the lake, shorter incubation period, better fertilization and better growth rate. The food and feeding habits of common carp and schizothoracines is almost identical, with many of the lacustrine species of schizothoracines feeding on detritus and benthos. So food competition is one of the important reason for declining the endemic fish production. It was with this background that a detailed study on food and feeding habits of exotic *Cyprinus carpio* in Dal Lake was undertaken and compared it with previous work based on food and feeding habits of endemic *schizothorax* species.

***Corresponding author:** Gulzar naik, Faculty of Fisheries, Sher-e-Kashmir University of Agricultural Sciences and Technology of Kashmir, Rangil, Ganderbal, Jammu and Kashmir, 190006, India, E-mail: mudasir.rathero@gmail.com

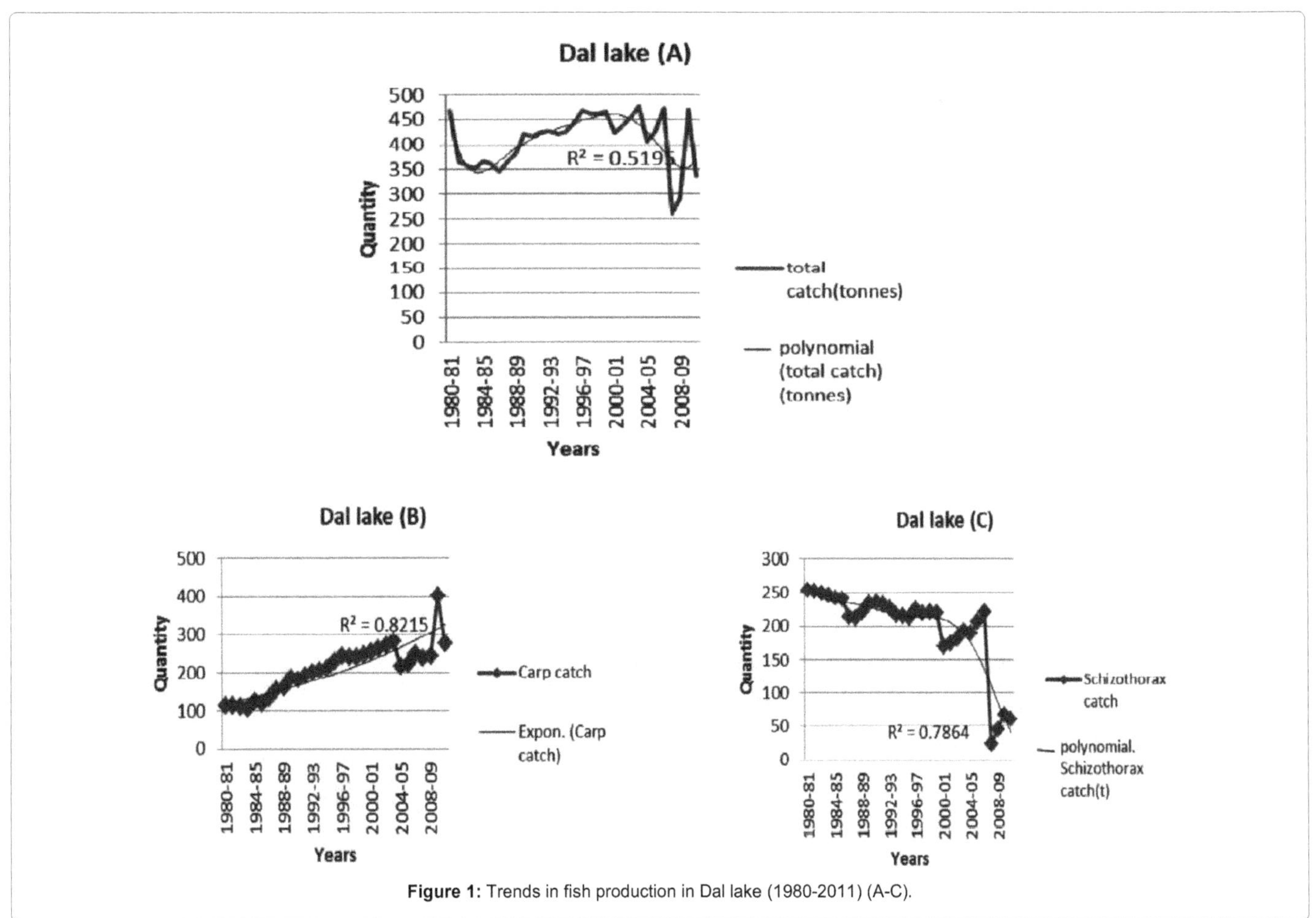

Figure 1: Trends in fish production in Dal lake (1980-2011) (A-C).

Material and Methods

In order to study the food and feeding habits of common carp, samples were collected from the commercial catcher during fishing in the year 2012 from January to December at Dal Lake Kashmir. All the fish specimens were weighed separately and then gutted for the collection of gut contents and preserved in 5% formalin. The collected guts were weighed and their content emptied in the watch glass. The same were analysed qualitatively as well as quantitatively by eye estimation, volumetrically [7] and occurrence method [8] for evaluating the relative importance of all food items. The various items were examined and sorted out using a binocular microscope and thus identified. Standard taxonomical keys were consulted for identification of plankton, oligochaetes, insects and other invertebrates [9,10]. The percentage occurrence of different items of food in different months was determined by summing the total number of occurrence of all items from which the percentage occurrence of each item was calculated. To find out the feeding rhythm of *Cyprinus carpio*, Gastrosomatic index was evaluated using the computational formula:

$$\textbf{Ga.S.I} = \frac{\text{total weight of full gut}}{\text{total weight of fish}} \text{X100}$$

Results

The analysis of gut contents of the fish revealed that on an average 43.50% of it was detritus and 56.5% included plant and animal matter. The animal food was contributed by crustaceans, oligochaetes, insect larvae, insect remains, fish remains, molluscan remains, rotifers and protozoans. On the average the total contribution of animal food was 25.29% of which crustaceans (copepods, cladocerans and ostracods) contributed 13.01%, oligochaetes 1.6%, insects 3.97%, fish remains 2.4%, molluscan remains 2.44%, protozoans 0.28% and rotifers 1.05% to the total food. The total contribution of vegetative matter was 31.21% consisted of macrophytic tissue and algae, former contributing 27.36% and the latter 3.85% on an average annual basis (Table 1).

The present study revealed that the fish is detri-omnivore in feeding habit, as on the whole 43.50% of gut contents were contributed by detritus and remaining by animal and plant matter. The detrital component revealed its peak contribution during December (53.9%), while minimum in March (31.7%) & May (34.9) (Table 1). Among the animal food, crustaceans were present throughout the year with maximum & minimum contribution in May (19.9%) and December (4.2%) respectively. Oligochaetes recorded peak contribution during February (6.9%), while in July and October it contributed only (0.1%) and was absent in November and December. The contribution of fish remains in the gut was maximum (7.9%) during December and minimum (0.1%) during June and August, however, during July these were absent from gut contents. Insects recorded their peak contribution in March (8.0%) but were absent from gut contents during November. Macrophytes recorded highest contribution in November (31.7%), and minimum (20.0) in January. The algae contributed the maximum (7.1%) in March, while minimum in August (1.7%).

Gastrosomatic index (Ga.S.I.) recorded its highest value during

Food ⇨ Month ⇩	CRT	INS	FR	MR	OLI	PROT	ROT	ALG	MT	MIS	DET
Jan.	9.9	3.50	4.3	2.1	1.9	0.2	1.7	6.5	20.0	2.8	47.1
Feb.	10.3	4.41	2.2	3.3	6.9	0.9	1.2	2.1	23.5	2.6	42.6
March	15.6	6.4	1.3	1.1	5.7	0.3	1.6	7.1	29.9	1.4	31.7
April	14.5	4.2	0.9	1.3	1.2	0.1	0.9	3.5	30.6	2.7	38.3
May	19.9	8.0	0.3	0.3	0.9	0.1	0.3	2.9	30.9	1.50	34.9
June	17.8	7.4	0.1	0.1	1.3	0.2	0.8	2.3	28.0	1.40	40.6
July	18.0	7.1	0.0	0.05	0.1	0.3	2.1	2.1	25.3	2.65	42.3
Augt.	16.3	2.2	0.01	0.0	0.9	0.01	1.1	1.7	28.9	2.9	46.0
Sept.	13.2	2.1	2.1	0.9	0.2	0.02	0.6	3.9	24.1	2.9	50.0
Oct.	8.4	2.1	2.3	0.2	0.1	0.05	0.8	4.2	31.1	1.45	49.3
Nov.	7.0	0.0	7.3	0.0	0.0	0.15	0.5	5.1	31.7	2.15	46.1
Dec.	4.2	0.5	7.9	1.0	0.0	0.19	1.0	4.8	24.3	2.21	53.9
Mean	**13.01**	**1.61**	**2.4**	**2.44**	**1.6**	**0.28**	**1.05**	**3.85**	**27.4**	**2.21**	**43.5**

CRT=Crustacea; INS=Insects; FR=Fish remains; MR=Molluscan remains; OLI=Oligochaetes; PROT=Protozoa; ROT=Rotifera; ALG=Algae; MT=Macrophyte tissue ; MIS=Miscellaneous; DET=Detritus.

Table 1: Monthly percentage of different food components in the gut of *Cyprinus carpio* at Dal Lake of Kashmir from January 2012 to December 2012.

Month	No. of fishes examined	Mean total wt. of fishes (g)	Mean wt. of gut (g)	Ga.S.I
Jan.	23	518	20	3.86
Feb.	16	910	31.4	3.34
March	14	1270	43.4	3.42
April	10	1800	88	4.89
May	12	1400	86	6.14
June	15	900	52	5.78
July	11	1450	91	6.26
Augt.	14	1190	70	5.89
Sept.	11	1350	83	6.14
Oct.	14	1430	87	6.10
Nov.	17	1080	61	5.65
Dec.	17	1020	50.5	4.95
Maximum	**23**	**1800**	**91**	**6.28**
Minimum	**10**	**518**	**20**	**3.34**

Table 2: Mean monthly variation of Gastrosomatic Index (Ga.S.I) of *Cyprinus carpio* at Dal Lake.

July (6.28), while lowest value was recorded in February (3.34). The fish showed a marked reduction in Ga.S.I values from January (3.86) to March (3.42) at a time when the abdominal cavity was filled with Gonadal mass. An improvement was recorded in April (4.89) and the index remained generally high during the warmer months, followed by a gradual decline with the approach of winter (Table 2).

Discussion

Endemic schizothoracines are fast losing their ground in Kashmir lakes due to various anthropogenic factors. One of the factor is introduction of exotic common carp that has higher fecundity and breeds in confined waters [11,12]. By contrast, schizothoracines undergo breeding migration for spawning in streams, and they also have a lower fecundity than common carp. Further the food and feeding habits of common carp and schizothoracines are almost identical. Knowledge of food and feeding habits of a fish is important for understanding its biology as well as for the successful management of its fishery. Nature offers a great diversity of food to fishes and accordingly various species are known to differ in their feeding habits, some being the predators like pikes, some others are omnivores like the gold fish, while many others are herbivores [13].

In this paper, we studied the food and feeding habits of exotic *Cyprinus carpio* from Dal Lake of Kashmir Valley and compared it with previous work based on the food and feeding habits of endemic *Schizothorax* species. The results obtained by analysing the gut contents of common carp showed that on an average basis, decayed organic matter (detritus) formed 43.5% of total food, while the remaining food (56.5%) consisted of plant (31.21%) and animal matter (25.29%). On the basis of gut content analysis, the fish was designated as detri-omnivore. Bottom feeding habit of common carp was supported by the present data as benthic organisms like oligochaetes, insect larvae and ostracods were recorded from their gut contents. Bottem feeding habit of common carp was also reported by other authors [11,14,15]. Earlier studies revealed herbivorous feeding habit of *C. carpio communis* and

C. carpio specularis from Dal Lake and indicated that 29% of the food of *C. c. specularis* was of animal nature, while in case of *C. c. communis* animal matter contributed 34% of the food [11,14]. Most of the authors reported omnivorous feeding habit of common carp [15-19] *Cyprinus carpio* was designated as detri-omnivore when 45% of detritus and remaining of both plant and animal matter was reported from the gut contents of fish [15].

The previous work based on food and feeding habits of endemic *Schizothorax* spp. of Kashmir revealed that the gut contents of *Schizothorax* spp is composed of detritus, vegetative and animal matter in varying quantities. Jan and Das [14] reported *Schizothorax niger* and *Schizothorax esocinus* as herbivorous fishes in which the contribution of average annual animal food was 33.5% and 30.5% and plant food was 61.0% and 63.5% respectively. The animal food contributed of protozoans, rotifers, zooplankters, insect adults, insect remains, fish scales, fish eggs etc., and plant food consisted of green algae, diatoms, macrophytes, besides some amount of detritus and sand. The gut contents of *Schizothorax esocinus* consisted of 63.5% of plant matter and 30.5% of animal matter [20]. On an average diet of *Schizothorax curvifrons* was composed of dissolved organic matter (40.33%), sand and mud (17.51%), phytoplankton (38.78%), zooplankton (2.00%) and miscellaneous matter (1.38%) [11,12]. By studying the food and feeding habits *Schizothorax curvifrons* and *Schizothorax esocinus,* the author reported former as a phytophagus fish with average contribution of animal matter (12.43%), vegetable matter (51.25%), unidentified animal matter (6.25%), unidentified vegetable matter (27.67%) and sand particles (2.595) and latter i.e, *Schizothorax esocinus* as an omnivorous fish as 57.09% of animal matter was recorded from the gut contents of the fish [21]. *Schizothorax* spp. are usually surface feeders but sometimes feed at the lower levels also due to the scarcity of the food or disturbance of the upper water strata [14]. So food of common carp and snow trouts was found to be almost similar composed of detritus, vegetable and animal matter in varying quantities. In almost all cases percentage of vegetable matter was found to be higher than that of animal matter.

During spawning season the size of ovaries increases and most of the abdominal cavity is occupied by it. So feeding (gastrosomatic index) of fishes is usually found to be related to their maturity stages (gonadosomatic index) to a great extent. The two indices has an inverse relationship to each other with the result the gastrosomatic index is low during spawning season. In common carp during present study Ga.S.I. recorded its highest value during July (6.28), while lowest value was recorded in February (3.34). The fish showed a marked reduction in Ga.S.I values from January (3.86) to March (3.42) due to spawning season. The index remained generally high during the warmer months, followed by a gradual decline with the approach of winter. Feeding (gastrosomatic index) of *Schizothorax curvifrons* and *Schizothorax esocinus* was also found to be related to their maturity stages and was low during April to June in former and March to May in latter [21]. So Ga.S.I values were low in all the three species in first few months i.e, January to March in common carp, April to June in *S. curvifrons* and March to May in cases *S. esocinus* and was higer during rest of the year in all cases. It is reported that during spawning season, feeding rate would be relatively lower and it increases immediately after spawning as the organisms feed voraciously to recover from fast [22-26].

Present study revealed that food and feeding habits of exotic common carp was almost identical to endemic snow trouts and there might existed a feeding competition between them. So besides various other reasons that declined endemic snow trout fish production in Dal Lake of Kashmir valley, overlapping of food and feeding habits with common carp might be one of the important reason.

Acknowledgement

The authors are thankful to the Faculty of Fisheries, Sher-e-Kashmir University of Agricultural Sciences and Technology of Kashmir for providing the necessary laboratory facilities to carry out this work.

References

1. Nyman L (1991) Conservation of fresh water fish, protection of biodiversity and genetic variability in aquatic ecosystems. Fisheries development series. P. 56.
2. Fotedar DN, Qadri MY (1974) Fish and fisheries and the impact of Cyprinus Carpio L. on endemic fish. J. Sci. univ. Kash. 2: 79-90.
3. Subla BA, Das SM (1970) Studies on the feeding habits, the food and the seasonal fluctuation in feeding in nine Kashmir fishes. Kash. Sci. 7: 25-44.
4. Yousf AR, Qadari MY (1992) "Current trends in Fish and Fishery Biology and Aquatic Ecology". University of Kashmir. pp: 43-52.
5. Zutshi DP, Gopal B (2000) State of bioderversity in lakes & wetlands of Kashmir Valley. Enviromental Biodiversity & Conservation. pp: 15-16.
6. Qureshi NW, Krishnan M, Sundaramoorthy C, Vasisht AK, Baba SH, et al. (2013). Truncated Growth and Compromised Sustainability: The Case of Lake Fisheries in Kashmir Agricultural Economics Research Review. 26: 57-66.
7. Pillay TVR (1952) Proc. Nat. Inst. Sci. India. 19: 777-827.
8. Hynes HBN (1950) The food of freshwater sticklebacks (*Gasterosteus aculeatus* and *Pygosteus pungitius*) with a review of methods used in studies of the food of fishes. J. Anim. Ecol. 19: 26-28.
9. Pennak RW (1978) Freshwater Invertebrates of United States. JohnWiley & Sons, N. Y.
10. Edmondson WT (1959) Freshwater Biology. JohnWiley, N.Y.
11. Sunder S, Kumar K, Raina HS (1984) Food and feeding habits and length weight relaitonship of *Cyprinus carpio specularis* of Dal Lake, Kashmir. Indian. J. Fish. 31: 90-99.
12. Sunder S, Subla BA (1985) Food of Juveniles of *Schizothorax curvifrons* (Heckel). Bull. Env. Sci. 2: 34-36.
13. Hoar WS, Randall DJ (1971) Fish Physiology. Environmental Relations and Behaviour. Academic Press, New York. pp: 55.
14. Jan NA, Das SM (1970) Qualitative and quantitative studies on the food of eight fishes of Kashmir valley. Ichthyologica. 10: 20-26.
15. Shafi S, Bhat FA, Yousuf AR, Parveen M (2012) Biology of Cyprinus carpio communis from Dal Lake, Kashmir with Reference to Food and Feeding Habits, Length-Weight Relationship, and Fecundity. Nature Environment and Pollution Technology. An International Quarterly Scientific Journal. 1: 79-87.
16. Spataru P, Hepher B, Halevy A (1980) The effect of the method of supplementary feed applications on the feeding habits of carp (*Cyprinus carpio* L.) with regard to the natural food in ponds. Hydrobiologia. 72: 171-178.
17. Soni DD, Shrivastava BK, Kalhal KA (1981) Environmental studies on the Sagar lake. Feeding spectrum of carps. Indian. J. Ecol. 8: 102-107.
18. Firdous G (1995) Fish and Fisheries of Anchar Lake, Kashmir. Ph.D. Thesis. Kashmir University.
19. Shukla SN, Patel (2012) Studies on Food and Feeding Behaviour of *Cyprinus Carpio* and their Gastrosomatic Index from Govindgarh Lake, Rewa (M.P.), India. International Interdisciplinary Research Journal.
20. Jhingran VG (1991) Fish and Fisheries of India, Hindustan Publishing Corporation, Delhi. P. 72.
21. Kausar N, Shah GM, Jan U (2010) Seasonal fluctuations in the gut contents of *Schizothorax esocinus* and *Schizothorax curvifrons.* I.J.S.N. 3: 928-930.
22. Rao LM (1998) Indian J. Fish. 45: 349-353.
23. Hatikakota G, Biswas SP (2004) Length-weight relationship and condition factor of *Oreochromis mossambicus* from a domestic pond, Nazira, upper Assam. Fishery Management, A.P.H. Publication Corporations, New Delhi. pp: 223-232.

24. Bhatnagar GK (1972) Maturity, fecundity, spawning season and certain related aspects of *Labeo fimbriatus*. Journal of Inland Fisheries Society of India. 4: 26-37.

25. Thakur NK (1978) On the food of air breathing cat fish, Clarius batricus (Linn.) occurring in the wild waters. Internationale Revue der Gesamten Hydrobiologie. 63: 421-431.

26. Malhotra YR (1967) On the relationship between feeding and ovarian cycle in *Schizothorax niger* Heckel and *Botia berdi*. Indian Journal of Fisheries. 14: 313-317.

9

Inventory of Ichthyofaunal Diversity, Fishing Gear and Craft in Turag River, Dhaka, Bangladesh

Naser Ahmed Bhouiyan, Mohammad Abdul Baki*, Anirban Sarker and Md. Muzammel Hossain

Department of Zoology, Jagannath University, Dhaka-1100, Bangladesh

Abstract

Biodiversity of many Bangladesh Rivers is seriously threatened by industrial and municipal pollution. The study was conducted in the Turag River starting from Amin Bazar bridge (23°47' N 90°20'E) to Kamarpara bridge (23°53' N 90°23'E). This inventory survey was sampled at a fortnightly interval usually between 7.00 am to 5.00 pm by a team using a boat from December 2012 to November 2013. Detailed information on catch by species, fish length and weight, different types of gear and craft were collected through direct observation. A total of 71 (65 indigenous and 6 exotic) fish species (under 25 families of 9 orders) have been identified. 17 different types of gears of two categories (active and passive gear) and 8 different types of crafts were observed to harvest fish in the study area. The survey revealed that rising floodwater stimulated an increase in fishing activities in the study area from July to October. Fish numbers were recorded lower from November to July (dry and pre-monsoon period) likely due to reduced water flow and adverse water quality of this river. A paired *t*-test indicate that fish species numbers were significantly difference between Dry and pre-monsoon (P=0.02), Dry and monsoon (P=0.02) and Dry and post-monsoon season (P=0.03) respectively. However, fisheries resources contribution is very limited for livelihood of the surrounding people.

Keywords: Fish species; Fishing activity; Flood water; Water quality; Extinct

Introduction

Population growth has resulted in increasing demand for the use of rivers to satisfy a diverse range of human needs, including solid waste disposal and the discharge of industrial, sewage and mining effluents. The modifications to rivers disrupt the aquatic ecosystem and diminish its integrity [1-3] affecting the capacity of fish and other organisms to survive. However, most of the wild populations have seriously declined in rivers and streams of Bangladesh due to over exploitation augmented by various ecological changes and degradation of the natural habitats [4]. Water quality has been affected by a combination of factors including sewage and industrial wastes and agricultural run-off [5]. The large input of organic matter to aquatic flood plain habitats may reduce dissolved oxygen and result in the emigration or death of a great number of fishes [6]. It has been established that pollution of the river impacts key physiochemical properties of water thereby causing reduced dissolved oxygen (DO) level [7]. Fishes are relatively sensitive to changes in their surrounding environment. The concept of using fish communities as biological indicator has been historically followed by several authors [8,9]. Their size, community composition and structure often reflect nutrient status of a water body. Fish health may therefore reflect and give a good indication of the status of specific aquatic ecosystem [10,11].

Turag River of Bangladesh is a tide-influenced River passing through west-north and north of Dhaka City [12]. In the recent past, the human population, different industries, agricultural land converted into industrial and housing development land, brick fields around the Turag river basin has increased tremendously caused serious environmental pollution through discharging their untreated effluents directly or indirectly into river water. Industrial area possesses about 29 heavy industries and this cluster of industries of the capital city generates 7,159 kg effluents daily discharge and pollutants enter freely into the river [13]. In September 2009, four rivers around the Dhaka city-the Buriganga, the Sitalakhaya, the Turag and the Balu, were declared as Ecologically Critical Areas (ECAs) by the Government of Bangladesh. Therefore, it is imperative to monitor the aquatic fauna of this river. However, the documented sources of pollution in this river are widely varied and range from Industrial Effluents; Solid Waste; Textile Dyeing Industries; Municipal and Sewerage Disposal; Heavy Metal in sediment and water; Oil discharge. These industries discharge untreated wastewaters into river containing various types of hazardous chemicals including enzymes, detergents, dyes, acids, alkalies, salts and toxic heavy metals [14-18].

Most of these wastes are non-biodegradable and continuously leaching pollutant into the water body. However, several studies indicated that the Turag river water and sediment are highly contaminated [5,19,20]. Therefore, the need for water body specific detailed biodiversity studies [21]. No quantitative data for assessing fish abundance is available for this river system. The objective of this study is to assess the ichyofaunal diversity of River Turag. We will classify fish species, how seasonal changes in water level impact the diversity of species.

Materials and Methods

Study area and period

The Turag is 75 km long of which only about 18.4 km are within the study area starting (Figure 1) from Amin Bazar bridge (23°47' N 90°20'E) to Kamar para bridge (23°53' N 90°23'E). Turag is the upper tributary of the Buriganga, a major river in Bangladesh. Turag River is supposed to derive massive pollutant loadings from industrial effluents directly as industries, textiles, dyeing and pharmaceuticals have

***Corresponding author:** Mohammad Abdul Baki, Assistant Professor, Department of Zoology, Jagannath University, Dhaka-1100, Bangladesh
E-mail: mabaki@gmail.com

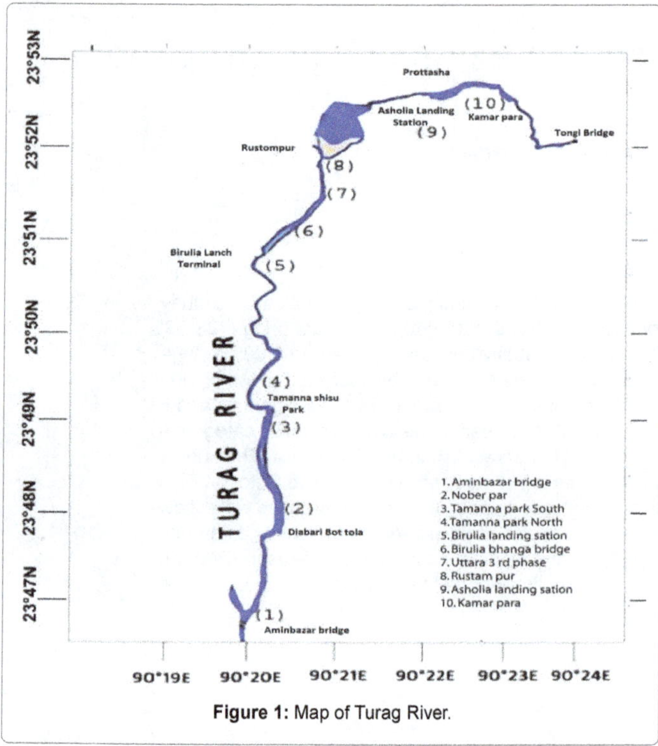

Figure 1: Map of Turag River.

clustered here. There are numerous canals, channels, and pipes directly discharging industrial, municipal and domestic sewage into the Turag, these observed by our study period (Figure 2). During the monsoon season, the water quality improves moderately, but on the advent of the dry season, pollution concentration increases abruptly because the water level of the rivers reduces a lot at this time, but the rate of pollutants released into the rivers remains identical. This inventory survey work of the Turag River was sampled inside at fortnightly interval for a total of 12 months from December 2012 to November 2013.

Sampling procedure

A team of two biologists carried out continuous survey using a boat. Detailed information on catch by species and different types of gear and craft were collected while fishermen were harvesting fish in the river. Survey procedure also included recording individual fish length and weight. Survey was usually made between 7.00 am to 5.00 pm. Materials were included digital camera, measuring tape, spring balance, polythene bags, data sheet, pencil, rubber band, map and other field logistics. The samples were photographed, immediately prior to preservation. The fish specimens caught by each fishing gears were also recorded separately.

Fish and gear identification

Fish identification, common and scientific names used throughout this study are in accordance with pictorial books and gear identified by Ahmed N [22-24].

Type of habitat preference categories

Fish species were divided into three categories according to [25] which are define below.

Riverine: Species usually found in rivers and estuaries throughout their life cycle with no dependence on the floodplain, although some of these species can be found more extensive floodplains.

Migratory: Species which move between river and floodplain during different stages in their life cycle. It remains unclear whether such movements are obligatory for their survival.

Floodplain resident (sedentary): Species which are generally sedentary and are capable of surviving in the perennial waters on the floodplain throughout the year. Many of these species also in habit a variety of other habitats including large rivers.

Hydrological year

Hydrological year can be divided into four seasons according to [25].

Rising flood (pre-monsoon): May-June.

Full flood (monsoon): July- September.

Flood drawdown (post monsoon): October-November.

Dry season (winter): December-April.

Bangladesh Water Development Board (BWDB) set up a water level monitoring station at Turag River for forecasting the flood situation of Dhaka city. This station was located at 23°78'33" and N 90°34'E for the daily monitoring of the water level of Turag River which included a staff gages. Therefore, this study collected the daily water depth data during study period from BWDB office, 72 Green Road, Farmgate, Dhaka, Bangladesh. Bangladesh metrological department showed that pre-monsoon, monsoon, post-monsoon and dry period in 2013 received average rainfall in Dhaka city was 339.9, 330.0, 103.35 and 54.3 mm respectively.

Statistical analysis

We used a paired t-test to test whether the fish species number in different seasons were significantly different between dry season and pre-monsoon, dry season and monsoon, dry season and post monsoon or not. Correlation analysis was also done among water depth, fish species and fishing activity.

Results

Hydrology

The measurement of water depth, increased and depletion of Turag River water in different months are shown in Figure 3. Depth of Turag

Figure 2: Different types of threats for fish in the Turag River.

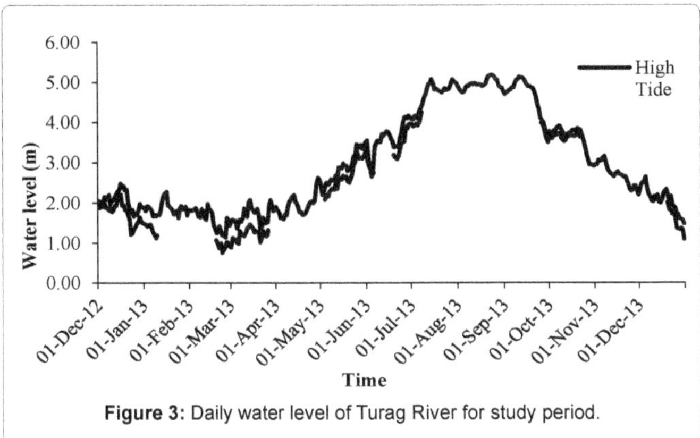

Figure 3: Daily water level of Turag River for study period.

River water starts to rise in May due to pre-monsoon water. This initial increase in discharge is followed by very sharp rise, usually occurring in July to reach flood peaks in August and September. This is result of monsoon. Depth of water normally decreases after peaks of September onwards, reaching a minimum level in March. Water depth data clearly show that water depth is lower in the winter and pre-monsoon (from December to June) periods compared to monsoon and post-monsoon period (July to November). There is no detectable change of water depth in Turag during winter period due to flow of water in this period.

Identification of fish species in Turag River

A total of 71 species of freshwater fishes (65 indigenous and 6 exotic species) belonging to 9 orders and included under 25 families were found in Turag River. Each of the individuals of all the species length and weight observations were recorded for the 71 fish species analyzed in this study also. Among fish species, 9 endangered, 5 critically endangered and 12 vulnerable species were classified respectively (Table 1).

Seasonal impact on fish distribution

Seasonal changes in the fisheries of rivers may be determined by fishing activities, cyclical changes in discharge, water velocity, water level and water pollution which in turns greatly influence the relative

Order	Family	Scientific name	English name	Local name	Length (cm)	Weight (gm)	Local Status
Osteoglossiformes	Notopteridae	*Chitala chitala*	Humped Featherback	Chital, Chetol	24	90	En
		Notopterus notopterus	Grey Featherback	Foli, Fholui	16	40	Vu
Cluperiformes	Clupeidae	*Tenualosa ilisha*	River Shad, Hilsa Shad	Ilish, Ilsha	10	10	
	Engraulidae	*Gudusia chapra*	Indian river shad	Chapila	10	10	
Channiformes	Channidae	*Channa punctata*	Spotted Snakehead	Taki, Lata, Lati	20	67	
		Channa striatus	Snakehead Murrel	Shol	13	48	
		Channa marulius	Great Snakehead	Gajar, Gajari	19	170	En
		Channa orientalis	Walking Snakehead	Gachua, Cheng	13	15	Vu
Cypriniformes	Cyprinidae	*Amblypharyngodon mola*	Mola carplet	Mola, Moa	5	5	
		Barbonymus gonionotus	Java Barb	Thai Sarpunti	27	300	
		Hypophthalmichthys molitrix	Silver Carp	Silver Carp	29	210	
		Aristichthys nobilis	Bighead Carp	Bighead	46	1250	
		Labeo calbasu	Black Rohu, Kalbasu	Kalibaus, Baus	23	200	En
		Catla catla	Catla	Catla, Katla	440	31	
		Cyprinus carpio	Common carp	Carpu	42	2450	
		Cirrhinus cirrhosus	Mrigal carp	Mrigal, Mirka	13	45	
		Labeo rohita	Rohu, Rohu Carp	Rui, Rohit	220	27	
		Labeo gonius	Kuria Labeo	Ghannya, Goni	22	520	En
		Labeo bata	Bata Labeo	Bata, Bhangan Bata	13	45	En
		Cirrhinus reba	Reba	Tatkini, Bata	10.5	15	Vu
		Labeo boggut	Boggut Labeo	Ghania , Gohria	14	50	
		Osteobrama cotio	Cotio	Keti, Dhela, Dhipali	4.5	2	En
		Puntius sarana	Olive Berb	Sar Punti	7	7	Cr
		Puntius sophore	Spotfin Swamp Barb	Punti, Jat Punti	6	5	
		Puntius chola	Swamp Barb, Chola Barb	Chalapunti, Punti	6	5	
		Puntius terio	One spot Barb	Teri Punti	6	6	Vu
		Puntius guganio	Grass barb	Mola punti	6	5	
		Puntius conchonius	Rosy Barb, Red Barb	Kanchan Punti	6	5	
		Rasbora daniconius	Common Rasbora	Darkina	6	1	
		Salmostoma phulo	Finescale Razorbelly Minnow	Fulchela	7	3	
		Salmostoma bacaila	Large Razorbelly Minnow	Narkalichela	6	4	
		Aspidoparia jaya	Jaya	Jaya, Peali	7	3	
	Cobitidae	*Botia dario*	Queen Loach, Bengal Loach	Rani	8	7	En
		Lepidocephalichthys guntea	Guntea Loach	Gutum	8	5	

Siluriformes	Bagridae	*Mystus bleekeri*	Stripped Dwarf catfish	Bajari Tengra, Bujri	11	9	
		Mystus tengara	Day's Mystus	Gulsha Tengra	6	4	
		Mystus cavasius	Gangetic Mystus	Kabashi Tengra,	8	7	Vu
		Mystus vittatus	Stripped Dwarf catfish	Tengra	7	8	
		Sperata aor	Long Whiskered	Ayre	21	120	Vu
	Siluridae	*Wallago attu*	Boal	Boal, Boali	14	15	
	Schilbeidae	*Ailia coila*	Gangetic Ailia	Kajuli, Bashpata	10	5	
		Ailia punctata	Jamuna Ailia	Kajuli, Bashpata	10	5	Vu
		Clupisoma garua	Garua Bacha, Gagra	Garua Bacha	18	50	Cr
		Eutropiichthys murius	Murius vacha	Muri bacha	15	30	
		Eutropiichthys vacha	Batchwa vacha, Bacha	Bacha, Garua Bacha	15	30	Cr
	Pangasiidae	*Pangaius pangaius*	Pungas	Pangas	10	15	Cr
	Sisoridae	*Bagarius bagarius*	Gangetic Goonch	Baghair	14.5	245	Cr
		Gagata cenia	Indian Gagata	Cenia, Jungla	7	8	
	Heteropneustidae	*Heteropneustes fossilis*	Stinging Catfish	Shing, Jiol	15	25	
	Loricariidae	*Hypostomus plecostomus*	Suckermouth catfish	Choshok machh	18	75	
Synbranchiformes	Synbranchidae	*Monopterus cuchia*	Cuchia	Kuchia, Kuicha	51	180	Vu
Perciformes	Ambassidae	*Pseudambassis lala*	Highfin Glassy Perchlet	Lal Chanda	3.5	1	
		Pseudambassis baculis	Himalayan Glassy Perchlet	Kata Chanda	3.5	1	
		Chanda nama	Elongate Glass-perchlet	Nama Chanda	5	2	Vu
		Pseudambassis ranga	Indian Glassy fish	Ranga Chanda	6.5	2	Vu
	Sciaenidae	*Otolithoides pama*	Pama Croaker, Pama	Poa, Poma	13	50	C
	Nandidae	*Nandus nandus*	Mottled Nandus	Bheda, Meni	13	50	Vu
	Cichlidae	*Oreochromis mossambicus*	Tilapia	Tilapia	21	200	
		Oreochromis niloticus	Nile Tilapia	Nilotica, Tilapia	26	325	
	Gobiidae	*Glossogobius giuris*	Tank Goby	Bele, Bailla	7	3	
	Anabantidae	*Anabas testudineus*	The Climbing Perch	Koi, Kai	17	60	
	Osphronemidae	*Colisa lalia*	Red Gourami	Lal khalisha	4.5	4	
		Colisa fasciata	Stripled Gourami	Khalisha, cheli	5.5	12	
		Ctenops nobilis	Indian paradisefish, Frail Gourami	Naftani, Napit khailsha	5	2	En
	Mastacembelidae	*Macrognathus pancalus*	Striped Spinyeel	Guchi Baim	10	10	
		Macrognathus aculeatus	Lesser Spiny Eel	Tara Baim	25	20	Vu
		Mastacembelus armatus	Tire-track Spiny Eel	Sal Baim, Bro Baim	28	70	En
	Mugilidae	*Rhinomugil corsula*	Corsula Mullet	Khalla	4	8	
Beloniformes	Belonidae	*Xenentodon cancila*	Needle Fish	Kankila, Kakila	18	10	
Tetraodontiformes	Tetraodon	*Tetraodon cutcutia*	Ocellated pufferfish	Tepa, Potka	9	6	
		Tetraodon fluviatilis	Green puffer fish	Potka	3.5	4	

*(C=Common, Cr=Critical endangered, En=Endangered and Vu=Vulnerable).

Table 1: Identification of Fish species in the Turag River.

abundance of different species of fish. Clear seasonal patterns in the variation of total number of species recorded in this study area were evident (Figure 4). Most of the species was observed from August to November (during monsoon and post monsoon period) for 4 months only. It can be seen that the higher species numbers were captured from July to November with two peaks in August and October (Figure 4) respectively. Correlation analysis between water depth and fish species number (r=0.74) and fishing activities (r=0.96) showed strong correlation. A paired *t*-test indicate that fish species numbers were significantly difference between dry and pre-monsoon (P=0.02), dry and monsoon (P=0.02) and dry and post-monsoon season (P=0.03) respectively. Fish species numbers rose fairly sharply from July when floodwaters also rose during monsoon (July-September) (Figure 4). So peak observed in August may be associated with monsoon because there is different kind of fishes which breeding cycle and migrations up and down river related with monsoon. Whilst second and highest peak in October was associated with flood drawdown (October-November) coincided with the entry of floodplain fishes into the river. The importance of the flood drawdown period to the catch of other species can clearly be seen as number of species increased (Figure 4) which had migrated from the rapidly drying floodplains. However, highest fish diversity was observed in October compared to August peak. These results support that the fish species composition was greatly influences by the flood water situation. Also is showed that the study proportion of the length of the rivers is fish less during this period. Despite this, water level and flow also sharply reduced in this period (Figure 3).

Gear and its distribution, number of species in gear

List of gears, trap and hooks are presented for this river in Figure 5. A total of 17 different types of fishing gears of two categories (active and passive gear) were observed to harvest fish in the study area. Dominant gear was cast net observed for 10 months followed by lift net (khora jal) observed for 7 months. Higher numbers (7-14) of the gears were used from July to November while extremely lower numbers (1-3) from December to June (Figure 6). The highest numbers of fish species were found in lift net (khora jal) and the lowest number of fish species was found in Box trap (Chai).

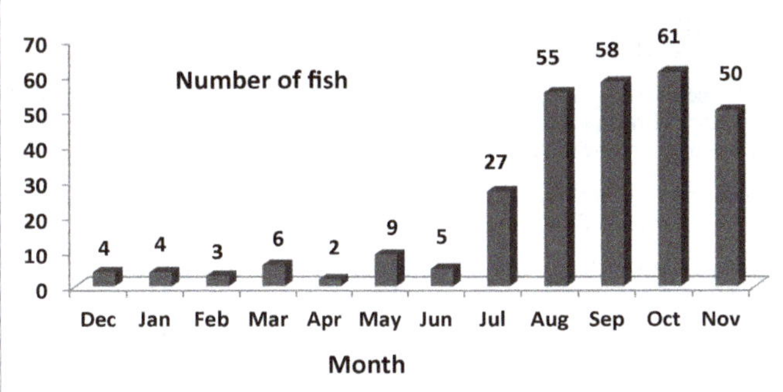

Figure 4: Monthly distribution of Fish species.

Figure 5: Different types of gears used for fishing in the Turag river: 1. Bel jal/Khora jal (Lift net, active gear) 2. Bash jal (Drag net, active gear) 3. Borshi (Hand line) 4. Borshi (Long line) 5. Carrent jal (Gill net, passive gear) 6. Uthar jal (Cast net, active gear) 7. Dharma jal/ toni jal (Lift net, active gear) 8. Moi jal (Drag net, active gear) 9. Jhaki jal Cast net) 10. Ber jal (Seine net, active gear) 11. Chai (Box trap) 12. Anta (Box trap). 13. Felun jal (Triangle trap, active gear).

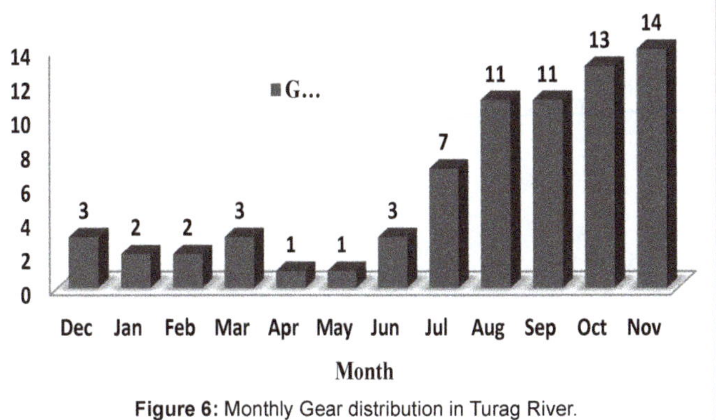

Figure 6: Monthly Gear distribution in Turag River.

Discussion

No previous statistics of fish fauna in this river was found and thus comparison of the present findings with previous one was not possible. This problem seemed not new in Bangladesh while working with fish diversity [21,26] and indicates the need for water-body specific fish diversity study in Bangladesh. The fish species of study area has been classified in terms of "endangered", "critically endangered", or "vulnerable" fish species by IUCN Bangladesh 2000 [27]. This same characteristic was noted in rivers Jamuna and Padma [25]. However, fish species numbers gradually decrease from October to November when gear number gradually increased in these months. This results indicated that reduce number of fish in these months may be associated with increased fishing activities. But fish species and gear numbers were sharply decreased starting from November. This continues till June with more or less constant number of fish and gear respectively. Our data indicated that there was almost zero catches during these periods.

Very low dissolved oxygen (DO) 1.9 mg/l to 0.7 mg/l) were recorded in this river from November to June (Dry and Pre-monsoon period) by Sharmin [28]. Furthermore, Rahman measured the DO concentration of Turag was lower from December to April and lowest value was 0.11 mg/l [5]. When DO goes below 4 to 5 mg/l, the survival of water organisms begin to go down, when anaerobic condition exists, higher life form like fish may be driven out. Furthermore, our data indicated that only *Channa puctata, Heteropneustes fossilis* and *Anabas testudinus* were observed during Dry and Pre-monsoon periods in the study area. *Heteropneustes fossilis* can respire aerially by gulping in air at various intervals when the oxygen content of water is low, [29]. The air-breathing apparatus of these species enables it to exist in almost any kind of water. Ahmed mentioned that Black fish have a broad environmental tolerance and can sustain the harsh conditions during the dry season [30]. Black fish include members of the Clariidae, Siluridae and Ophiocephalidae. However, only presence of these species during Dry and Pre-monsoon periods indicated that the health of river is highly polluted. Coates indicated that environmental degradation and habitat loss, not excessive fishing effort, is reported as the major cause of declining fisheries in most rivers under stress situation [31]. Furthermore, Naidu mentioned that the amount of catch depends upon its productivity of the fishing grounds [32]. Therefore, the extreme significantly lower number and diversity of fishes (almost zero) were recorded in Dry and Pre-monsoon period mainly due to adverse water quality of this river not for increased fishing activities. The lowest quality in fish assemblages occurred near cities that receive large amount of organic and industrial pollutants [33,34]. Considering the mentioned fact, it is noted that observed almost zero catch from December to June caused by reduced water flow and adverse water quality which may lead towards extinct of fishes from this river at least in this period if something is not done for their conservation.

In conclusion, this study provides the first basic and baseline information on ichyofaunal diversity, fishing Gear and Craft in the Turag river that would be beneficial for fishery biologists and conservationists to impose adequate regulations for sustainable fishery management and conservation of biodiversity for the river as well as for other rivers in Bangladesh.

Acknowledgment

Special thanks to professional staff associated with the Turag River Biological Survey, for field collection assistance and for professional training and other courtesies. Funding for this study was provided Jagannath University in support of the Baki's Lab as research grant 2012-2013.

References

1. Bretschko G, Moog O (1990) Downstream effects of intermittent power generation. Water Science and Technology 22: 122-135.
2. Moog O (1993) Quantification of daily peak hydropower effects on aquatic fauna and management to minimize environmental impacts. Regulated Rivers: Research and Management 8: 5-14.
3. Morgan RP, Jacobson RE, Weisley SB, McDowell LA, Willson HT (1991) Effects of low alteration on benthic macro-invertebrate communities below the Brighton hydroelectric dam. Journal of Freshwater Ecology 6: 419-429.

4. Hossain MY, Rahman MM, Fulanda B, Jewel MAS, Ahamed F, et al. (2012) Length-weight and length-length relationships of the five threatened fishes from the Jamuna (Brahmaputra River distributary) River, Northern Bangladesh. J. Appl. Ichthyol 28: 275-277.

5. Rahman AK, Lutfor M, Islam M, Hossain MZ, Ahsan MA (2012) Study of the seasonal variations in Turag river water quality parameter. Afric Journ of Pur and Appl Chem 6: 144-148.

6. Yousafzai AM, Khan AR, Shakoori AR (2010) Pollution of Large, Subtropical Rivers-River Kabul, Khyber-Pakhtunkhwa Province, Pakistan: Physico-chemical Indicators. Paki Jour Zoo 42: 795-808.

7. Winemiller KO (1989) Patterns of variation in life history among South American fishes in seasonal environments. Oecologia 81: 225-241.

8. Karr JR (1981) Assessment of Biotic Integrity Using Fish Communities. Fisheries 6: 21- 27.

9. Fausch KD, Lyons J, Karr JR, Angermeier PL (1990) Fish Communities as Indicators of Environmental Degradation. Ameri Fishe Soc Symp 8: 123-144.

10. Gupta A, Rai DK, Pandey RS, Sharma B (2009) Analysis of some heavy metals in the riverine water, sediments and fish from river Ganges at Allahabad. Environ Monito Assess 157: 449-458.

11. Mokhtar MB, Ahmad ZA, Vikneswaran M, Sarva MP (2009) Assessment level of heavy metals in Paenaeus mondon and Oreochromis mossambicus spp. In selected Aquaculture ponds of high densities development area. Intern Jour of Pharmac Invent, Euro Jr Sci Res 30: 348-360.

12. Alam K (2003) Cleaning up of the Buriganga River, Integrating the Environment into decision making. Ph. D. dissertation. Australia: Murdoch University.

13. Institute of Water Modelling (IWM) (2008) Mathematical Modelling for Planning and Design of Beel Kapalia Tidal River Management (TRM) and Sustainable Drainage Management, Dhaka, Bangladesh.

14. Islam MM, Mahmud K, Faruk O, Billah MS (2011) Textile Dyeing Industries in Bangladesh for Sustainable Development. Internat Journ of Environ Sci and Devel 2: 6.

15. Islam MS, Tusher TR, Mustafa M, Mahmud S (2012) Effects of Solid Waste and Industrial Effluents on Water Quality of Turag River at Konabari Industrial Area, Gazipur, Bangladesh. J. Environ. Sci. & Natural Resources 5: 213-218.

16. Naser HM, Sultana S, Haque MM, Akhter S, Begum RA (2014) Lead, Cadmium and Nickel Accumulation in Some Common Spices Grown in Industrial Areas of Bangladesh. The Agriculturists 12: 122-130.

17. Zoynab B, Chowdhary MSA, Hossain MD, Nakagami K (2013) Contamination and Ecological Risk Assessment of Heavy Metal in the Sediment of Turag River, Bangladesh: An Index Analysis Approach. Jour of Water Reso and Protec 5: 239-248.

18. Roy R, Fakhruddin ANM, Khatun R, Islam MS, Ahsan MA, et al. (2009) Characterization of textile industrial effluents and its effects on aquatic macrophytes and algae. Bangla Journ of Sci and Indus Rese 45: 79-84.

19. Zakir HM, Sharmin S, Shikazono N (2006) Heavy metal pollution assessment in water and sediments of Turag River at Tongi area in Bangladesh. Internat Journ Lake Riview 1: 85-96.

20. Mohiuddin KM, Ogawa ZHM, Otomo K, Shikazono N (2011) Heavy metals contamination in water and sediments of an urban river in a developing country. Interna Journ Environ Sci Tech 8: 723-736.

21. Imteazzaman AM, Galib SM (2013) Fish fauna of Halti Beel, Bangladesh. Internat Journ of Current Rese 5: 187-190.

22. Rahman AKA (1989) Freshwater Fishes of Bangladesh. Dhaka: Zoological Society of Bangladesh.

23. Siddiqui KU, Islam, Kabir MA, Ahmed SMH M, Ahmed ATA, et al. (2007) Encyclopedia of Flora and Fauna of Bangladesh. Freshwater Fishes. Dhaka: Asiatic Society of Bangladesh.

24. Ahmed N (1962) Fishing gear of East Pakistan, Directorate of Fisheries. Dacca: East Pakistan Govt. Press.

25. Flood Action Plan-17 (1994) Fisheries studies and pilot project. Prepared for the Government of Bangladesh. UK: Overseas Development administration 9: 68-70.

26. Mohsin ABM, Haque ME (2009) Diversity of fishes of Mahananda River at Chapai Nawabganj district. Resear Journ of Biolo Sci 4: 828-831.

27. IUCN Bangladesh (2000) Red Book of Threatened Fishes of Bangladesh. Dhaka: IUCN-The World Conservation Union.

28. Sharmin FR, Reza AHMM, Hossen MS, Zakir H (2013) Alterations in histopathological features and brain acetylcholinesterase activity in stinging catfish Heteropneustes fossilis exposed to polluted river water. Intern Aqua Rese 5: 7.

29. Munshi JSD (1993) Structure and function of the air breathing organs of Heteropneustes fossilis: In Advances in fish research, Singh BR Delhi: Narendra Publishing House. pp: 99-138.

30. Ahmed MS (2008) Assessment of Fishing Practices on the Exploitation of the Titas Floodplain in Brahmanbaria, Bangladesh. Turki Journ of Fisher and Aqua Sci 8: 329-334.

31. Coates D, Boivin T, Darwall WRT, Friend R, Hirsch P, et al. (2004) Information, Knowledge and Policy. Proceedings of the Second International Symposium on the Management of Large Rivers for Fisheries. RAP Publication. FAO Regional Office for Asia and the Pacific, Bangkok, Thailand.

32. Naidu MR (1939) Report on survey of the fisheries of Bengal. Calcutta Govt. Printer, India.

33. Araújo FG (1998) Uso da taxocenose de peixes como indicadora de degradaçãoo ambiental no Rio Paraíba do Sul, Rio de Janeiro, Brasil. Brazi Archi of Biol and Tech 41: 370-378.

34. Araújo FG, Fichberg I, Pinto BCT, Peixoto MG (2003) A preliminary index of biotic integrity for monitoring the condition of the Rio Paraiba do Sul, Southeast Brazil. Environ Manag 32: 516-526.

Effect of *Daucus carota* and *Beta vulgaris* on Color of *Anabus testudineus*

Chandasudha Goswami[1]* and Zade VS[2]

[1]*Department of Zoology, Govt. Institute of Science and Humanities, Amravati-444604, India*
[2]*Govt. Institute of Science and Humanities, VMV Road, Amravati-444604, India*

Abstract

Ornamental fish keeping is one of the most popular hobbies in the world today and rapidly gaining importance for their aesthetic value as well as trade value. The knowledge of nutritional requirement in ornamental fish species is essential to improve productive development and also for color improvement. The Climbing Perch, *Anabas testudineus* (Bloch) is a highly priced air breathing, freshwater food fish species which belongs to the family Anabantidae and order Perciformes. This paper deals with effect of feed; formulated from Natural plant products viz. carrot (*Daucus carota*) and beetroot (*Beta vulgaris*) on color improvement of *Anabus testudineus*. The feed and water environment changed the color of *Ananus testudineus* by 80% during the experiment. In practice, Fishery business has enormous potential to accelerate Indian Economy by earning foreign currency as well as it may also reopen a door for young entrepreneurs to do fishery business using natural plant products as feed. Also Ornamental fish feed from natural plant product will make its culture and rearing easy and less expensive and makes this business vibrant and native fishes will get the level of demand they deserve.

Keywords: Ornamental fishes; Aesthetic value; Nutrition; Color improvement; Feed formulation

Introduction

The production and trade of ornamental fish is a profitable alternative in the aquaculture activity. Feeding habit of the fish is very difference in the form of Carnivorous, Herbivorous, Omnivorous and also there is a large diversity in their feeding patterns. Like farmed fish, some aquarium fish are surface feeders, some mid-water or bottom feeders Diets for aquatic animals can only be effective if they are formulated to contain the full array of necessary nutrients at appropriate concentrations relative to each other along with appropriate factors inducing rapid consumption on a consistent basis. Some fishes depend mostly on natural feed. Obviously it is not possible to supply their native food and the varieties that they need to survive and grow but by analyzing the requirement the food factories try to prepare the best food for aquarium fish. Also, flavor and taste, sound (vibrations in water), smell, color and buoyancy of food are also important aspects. As ornamental fishes are characterized by a wide diversity of colors and color patterns; success in the ornamental fish trade is very much dependent on the vibrant color of the fish (World Journal of Fish & Marine Sciences) [1].

In fish, correct formulations of the diet improve the nutrient digestibility, supply the metabolic needs, reducing the maintenance cost, and at the same time the water pollution. Nutrients essential to fish are the same as those required by most other animals. Foods such as meal powder, flakes, milk powder, bovine heart and liver, tubifex worms, as well as live food including *Artemia* sp., rotifers and Moina have been used extensively in ornamental fish feeding. In their natural environment fish have developed a wide variety of feeding specializations (behavioral, morphological, and physiological) to acquire essential nutrients and utilize varied food sources. In past decade the nutritional requirements of various fish species have understood and technological advances in feed manufacturing have been obtained [2].

As ornamental fishes are characterized by a wide diversity of colors and color patterns; success in the ornamental fish trade is very much dependent on the vibrant color of the fish (World Journal of Fish & Marine Sciences). Pigments are responsible for the wide spectrum of color in fishes which is an essential prerequisite for the quality as they fetch higher price in the commercial market. As fishes cannot synthesize their own coloring pigments denovo, the coloring agents which are synthesized by some plants, algae and microorganisms need to incorporated in their diet [3,4]. Various coloring agents are used in aqua industry to impart color for the muscle & skin of fishes. Thus, pigmentation is an important criterion for fishes, since their color affect commercial acceptability. One of the greatest challenges in the ornamental fish industry is to replicate the accurate natural color of the fish in the captive environment. Numerous operations that have been propagated failed to successfully market fish due to faded color. Various products have been introduced to alleviate this problem, but none has performed so effectively. Hence, in the present work an effort is made in this direction.

Materials and Methods

Study area

Study area for collection of fishes was fixed as the Southern part of Kamrup District of Assam covering the water bodies Chandubi (Highest loop containing water body), Kulsi, Beeldora etc. These were prime wetlands connected to the river Brahmaputra; hence the sample covered most of the freshwater ornamental fish species found of Assam [5,6].

Sample collection

Samples were collected at random intervals by bag nets, scoop nets, cast nets or by hand picking. For feeding and experiment few fishes were

***Corresponding author:** Chandasudha Goswami, Research Scholar, Dept. of Zoology, Govt. Institute of Science and Humanities, Amravati-444604, India
E-mail: goswami.chandasudha@gmail.com

taken for the experiments and were cultured for a stipulated period of 6 months [7]. The experiments were carried out in aquariums, cement tanks and buckets.

Experimental setup

The experimental setup consisted of 3 aquariums, 2 cement tanks and 3 buckets. Aquariums and buckets were used for rearing of fishes whereas cement tanks were used as reserves and also for acclimatization [8].

Water collection and maintenance of quality in rearing tank

Water used here was not from eutrophicated pond, but it was normal clean well/tap water. Water was being changed in aquariums and buckets in every 2 days to maintain a hygienic condition. Tanks were just used as reserves and water was changed in 7 to 10 days [9].

Rearing

The tanks were covered with metallic net to prevent the escape of fish. The 1/4th surface area of the tanks were covered with *Azolla* to provide shelter and shadow to the fishes. Bacteria accumulated in the inner sides of the tanks were removed mechanically at regular interval. Faeces, waste particles of food and dead bodies of fish were siphoned or collected with small net at regular interval. Some *Anabus testudineus* were treated in clear water aquaria. They were fed one time/day with meal.

Diet

For the present investigation, the feed was prepared in the form of dry pellets [10]. The experimental diet contains the *Daucus carota* (carrot) and *Beta vulgaris* (beetroot) as basic ingredients (sometimes mixed with a portion of rice grain/meal powder). Carrot and beetroots were farmed and also purchased from the local market. The prepared foods were provided one time daily to the experimental fishes. Three aquariums: control, positive control (commercial feed added), and experimental tank were design. During the experiment the fishes were fed with prepared feed at a constant rate a time a day. And the experiment was continued for a period of 6 months though coloration we got on 3 months. After that till the end of 6 months there was a very slight variation in color with the same feed [11].

Coloration Judgment

Test panels of persons randomly recruited judged color. The treatments were not revealed to the individuals who will be asked to rank the fish according to intensity of color. Color ranking were scored by a score of 1-4 (one being the lowest) for each treatment groups [12].

Result and Discussion

Results

Effect of formulated feed on Coloration: During Coloration Judgment color ranking were scored by a score of 1-4 (one being the lowest) for each treatment groups. From Table 1 we got to know about the effect of formulated food on 4 different random species. The result shows highest coloration on *Anabus testudineus* (Koi). Koi fishes are black in color in eutrophilated or polluted ponds. When it is brought and cultured in clean water its color changes from dark to white tone gradually. Again, after *feeding Daucus carota* (carrot) and *Beta vulgaris* (beetroot), color changes gradually giving the fish a orange look in fins and heads and the color continues till it cultured in aquarium with clean water with formulated diet.

Fish species	Feed	Colour rank (out of 5)
Anabus testudineus	1. Daucus carota (carrot) 2.Beta vulgaris (Beet root) 3.Rice Meal Power	4
Puntius sp.		2.8
Channa sp.		2.5
Botia Sp .		1

Table 1: Color ranking score 1-4 (one being the lowest) for each treatment groups out of 5.

✓ In first month of culture fish turns to white shade from black shade.

✓ Upto 3rd month fish gradually develop in color in one of the dorsal and caudal fin.

✓ In 4th month coloration change widens to all fins (orange in dorsal & caudal fins) and above the head we got red color.

✓ After that the color improvement shows very slow pattern but retaining color continues.

✓ 5th and 6th month shows slight brightening in colors.

Discussion

Fishes cannot synthesis the carotenoid denovo). As per Sinha and Asimi [13] Pigmentation is one of the quality attributes of the fish for consumer acceptability. Carotinoids are responsible for pigmentation of muscle in food fish, and skin colour in ornamental fish. As fish is not capable of synthesizing carotenoids de novo there is a need to incorporate carotenoids in the diet of cultured species. Since synthetic carotenoids are known to have deteriorating effects on the environment, there is a great demand for inclusion of natural carotenoids in aqua feed to achieve bright coloration in fish. Carotenoids are absorbed in animal diets, sometimes transformed into other carotenoids, and incorporated into various tissues. In earlier days Ali and Salim [14] also established that fish do not possess the ability to synthesize carotenoids. Hence the carotenoid pigmentation of fish results depends upon the supplementary feed contains the carotenoid amount. The micro algae *Chlorella vulgaris* has become a potent pigment source which imparts yellow/blue hues [15]. Here in our study we tried *Daucus carota* (carrot) and *Beta vulgaris* (beetroot) instead of *Chlorella vulgaris* or other carotenoids to make a different wayout with low cost easily available natural products.

Conclusion

Pigmentation is an important criterion for fishes, since their color affect commercial acceptability. One of the greatest challenges in the ornamental fish industry is to replicate the accurate natural color of the fish in the captive environment. Numerous operations that have been propagated failed to successfully market fish due to faded color. Various products have been introduced to alleviate this problem, but none has performed so effectively. Hence, in the present work an effort was made in this direction using Carrot and beetroot as source of carotenoids to impart color on few randomly selected species. Here *Anabas testudineus* shows greatest improvement in color by 80%.

Acknowledgements

I want to extend deep sense of gratitude to Dr K. M. Kulkarni sir for his support and guidance. Also we want to offer thanks to Mr. Amol Nale, Mr. Jujusman, Dr. Dinesh Dabhadkar, Dr. Vaibhao Thakare, Ved Patki, Akanksha Mahajan and all supportive people we get on the way for providing us colloquial knowledge and information about the river, fish and feed.

References

1. Rainboth WJ (1996) Fishes of the Cambodian Mekong: FAO Species Identification Field Guide for Fishery Purposes. FAO, Rome.

2. Wang TY, Tzeng CS,Shen SC (1999) Conservation and phytogeography of Taiwan paradise fish: *Macropodusoper cularis* Linnaeus. Acta Zool Taiwan.

3. Kottelat M (2001) Fishes of Laos: WHT Publications Ltd, Colombo.

4. Tan HH, Lim KKP (2004) Inland fishes from the Anambas and Natuna Islands, South China Sea, with description of a new species of *Betta*(Teleostoi: Osphronemidae): Raffles Bull. Zool. Suppl, 11: 107-115.

5. Iwata *A,* Ohnishi N, Kiguchi Y (2003) Habitat use of fishes and fishing activity in plain area of southern Laos: Asian Afr. AreaStud. 3: 51-86.

6. Sarkar UK, Deepak PK, Kapoor D, Negl RS, Paul SK, et al. (2005) Captive breeding of Climbing Perch, *Anabus testudineus* (Bloch, 1792) with Wova-FH for conversation and aquaculture: Aquacult. Res 36: 941-945.

7. Mustafa Md G, Alam Md J, Islam Md M (2010) Effects of some artificial diets on feed utilisation and growth of the fry of Climbing Perch, *Anabas testudineus* (Bloch, 1972): Department of Fisheries, University of Dhaka, Bangladesh.

8. Graham JB (1997) Air-breathing fishes: Evolution, Diversity and Adaptation. Academic Press, London, UK.

9. Petiyagoda R (1991) Freshwater fishes of Sri Lanka: The Wildlife Heritage of Sri Lanka, 362.

10. Sterba G (1983) The Aquarium Fish Encyclopaedia: The MIT Process. Cambridge, Massachusetts, 605.

11. Sakurai A, Sakamato Y, Mori F (1993) Aquarium Fish of the World: The Comprehensive Guide to 650 Species. Chronicle Books. San Francisco: 288.

12. Liem KF (1987) Functional design of the air ventilation apparatus and overland excursions by Teleosts.Fieldiana: Zoology 37: 1-29.

13. Davenport J, Matin AKMA (2006) Terrestrial locomotion in the Climbing Perch, *A. testudineus* (Bloch) (Anabantidea, Pisces): Journal of Fish Biology 37: 175-184.

14. Joseph B, Sujath S, Shalin JJ, Palavesam A (2011) Influence of Four ornamental flowers on the growth and colouration of orange sword tail Chicilidae fish *(Xiphophorus hellerei*, Heckel, 1940): Int J Biol Med Res 2: 621-626

15. Meyers SP (1994) Developments in world aquaculture, feed formulation and role of carotenoids. Pure applied Chem 66: 1069-1076.

A CYP19 Based Sex Determination and Monosex Production in Aquaculture Species *Oreochromis niloticus* L. and a Cyprinid *Cyprinus carpio* L.

Atul K.Singh[1]* and Srivastava PP[2]

[1]*Exotic Fish Germplasm Section of Fish Health Management Division of National Bureau of Fish Genetic Resources, Canal Ring Road, P.O. Dilkusha, Lucknow-226002 (Uttar Pradesh), India*
[2]*Central Institute of Fishery Education, Mumbai-400061*

Abstract

The efficacy and effect of tamoxifen and letrozole on sex reversal in common carp and Nile tilapia was studied using free swimming fry of *Cyprinus carpio* and tilapia *Oreochromis niloticus*. Treatment of letrozole in the dose of 100 mg kg^{-1} feed brought about 79.39 ± 1.09% masculinisation in *C. carpio* and 87.91 ± 1.39% masculinisation in *O. niloticus*, while larger dose (200 mg kg^{-1} feed) produced 98.47 ± 1.34% male *C. carpio* and 99.65 ± 0.72% male *O. niloticus*. In the control group, there was 48.28% male and 50.78% female *C. carpio* while 46.38% male and 53.616% female *O. niloticus*. Letrozole significantly increased serum testosterone (T) level suppressing 17β-estradiol (E_2) production and the androgenizing action was more potent when compared with tamoxifen treatment. The histological examination of letrozole treated *C. carpio* and *O. niloticus* revealed increased number of irregularly spread spermatids and there was hardly any difference in control testes and letrozole treated gonads. The results showed potent and complete action of letrozole (*Cyp19*) on sex reversal for monosex male fish production delineating its commercial application in aquaculture.

Keywords: Masculinisation; Sex steroids; Tamoxifen; Letrozole; Nile tilapia; Common carp

Introduction

Early maturation and frequent spawning are management challenges when working with tilapia and common carp. Sex control for aquaculture and research purpose has been attempted in many fish species by administration of exogenous hormones such as several androgens and estrogens for production of all-male or all-female populations [1-3]. At the same time, use of aromatase inhibitors (AIs) for sex reversal have also been attempted in mammals [4], birds (chickens) [5,6], amphibians [7], reptiles [8] and fishes [9]. However, the efficacy of aromatase inhibitors (AIs) is yet to be evaluated in important aquaculture species such as tilapia and common carp for effective sex reversal and monosex production. It is important to mention that male tilapia is preferred for culture because of faster growth while reverse is true for common carp [10]. The common carp *C. carpio* L. has a XX/XY type chromosomal genetic sex determination [1,11] and inverted XX males, produced by endocrine sex reversal technique, has been used for mating with normal females (XX) so as to achieve all-female population [1,11] which have a faster growth rate than mixed populations, and prevent uncontrolled reproduction.

It has been hypothesized that male and female sex differentiation is determined by aromatase activity in fish as *Cyp19* gene is highly conserved in fish [12]. Estrogen biosynthesis is mediated by the steroidogenic enzyme *Cyp19*, which converts androgens to estrogens [12,13]. The key role of aromatase in ovarian differentiation is supported by the fact that the inhibition of estrogen synthesis in the differentiating ovaries can trigger partial or complete masculinisation in fish [12]. Tamoxifen and letrozole are known to influence the production of estradiol in mammals [14,15] as well as in fishes [9,16-18]. The mode of action of tamoxifen has been described to exert its main anti-estrogenic effect by competing with estradiol for its own receptor [9,14-15]. A likely mechanism of tamoxifen action has also been reported through highly reactive *Cyp19*-mediated tamoxifen metabolites [14-15]. However, letrozole has been reported as a highly potent inhibitor of aromatase (*Cyp19*) *in vitro* and *in vivo* in fishes [12], and in humans [14]. Tamoxifen is a known selective estrogen receptor modulator (SERM) and works through blockage of estrogen receptors [16] while letrozole is supposed to effectively down regulate *Cyp19a1a* or *Cyp19a1b* genes acting directly on aromatase production [12]. In this study, we have investigated the comparative performance of tamoxifen and letrozole on sex determination in a gonochorist cichlid *O. niloticus* and a cyprinid *Cyprinus carpio* hypothesizing that in these fish the sex determination is *Cyp19* gene dependent and they will respond similarly for the two drugs. The present investigation was designed to better understand the aromatase based sex determination in fish using two functional treatments of tamoxifen (11β-hydroxyandrostenedione) an estrogen receptor blocker and letrozole (*Cyp19*) which is a known aromatase inhibitor.

Materials and Methods

Six hundred free swimming fry of 30 days post fertilized (30 dpf) *C. carpio* and *O. niloticus* (15 dpf) (length, 0.68 ± 0.21 cm and weight, 0.37 ± 0.08 g) were procured and equally divided into six glass aquaria (150 L). The reason of procuring different age groups of *C. carpio* and *O. niloticus* was simply because both were sexually undifferentiated and were able to actively accept external feeding. The *C. carpio* as well as *O. niloticus* were acclimatized to laboratory condition for one week maintaining in separate aquaria. Water temperature in the aquaria was maintained with the help of thermo-controlled heaters (300 W Risheng, Made in China) with filters at 25 ± 1°C. Tamoxifen (SIGMA Chem. Co. USA) and letrozole (USP, Evalet™, India) were dissolved in

***Corresponding author:** Atul K.Singh, Exotic Fish Germplasm Section of Fish Health Management Division of National Bureau of Fish Genetic Resources, Canal Ring Road, P.O. Dilkusha, Lucknow-226002 (Uttar Pradesh), India
E-mail: aksingh56@rediffmail.com

minimal volume of 95% absolute alcohol and the required volume of these chemicals was sprayed over a commercial pellet feed (Tiayo Pvt. Ltd.) to make them in the doses of 100 and 200 mg kg^{-1} feed and then dried in an oven. The doses of tamoxifen and letrozole were decided in consultation with available literature. This tamoxifen and letrozole mixed feed [Tamoxifen{100 mg kg^{-1} (T_1) and 200 mg kg^{-1} (T_2)} as well as Letrozole {100 mg kg^{-1} (L_1) and 200 mg kg^{-1} (L_2)] were given to the fingerlings of *C. carpio* and *O. niloticus* daily twice a day up to satiation for sixty days. The commercial pellet diet was consisting of crude protein (32%), crude fat (4%), crude fiber (5%) and moisture (10%). The control group of fingerlings was given tamoxifen or letrozole free similar diet. Since there was no circulatory water system, water of all experimental aquaria was completely changed every third day to eliminate the metabolites, possibility of accumulation of any steroid and to maintain the water quality.

At the end of 60 days, fishes were anesthetized with tricaine methanesulfonate (MS-222) (Merck, Germany) dissolved in water (100 mg L^{-1}), buffered with sodium bicarbonate (100 mg L^{-1}). Blood of anesthetized fish was drawn from caudal vein and pooled for 5 live specimens of different treatment groups as well as control separately since blood drawn from individual fish was insufficient for hormonal examination. About 2 ml blood was taken and centrifuged at 3000 rpm using an Eppendorff Centrifuge for 5 min to obtain serum which was stored at -80°C until used for hormonal assay. Serum 17β- estradiol and testosterone were estimated using enzyme linked immune absorbent assay (ELISA) technique by using kit provided by Enzo-Life Sciences, India. The optical density (OD) was read on ELISA reader and the observation were plotted using logit- log paper. The limit of detection for testosterone assay was 5.67 pg mL^{-1} while it was 14.0 pg mL^{-1} for E2.

For all experimental groups, sex differentiation was studied using histological technique while sex frequency and sex change was assessed using acetocarmine squash method [19]. For histological examination, fishes were cut ventrally using a scalpel, from the genital papilla to the base of the pectoral fin. A window on the lateral side was opened and the viscera were removed, leaving gonads, swim bladder and kidneys in place. The pair of gonad was identified based on the morphology and location of the gonadal tissue. After macroscopic examination, the anterior and posterior ligaments were cut, and both gonads were removed using a forceps and fixed in 10% natural buffer formalin solution for histological examinations. Fixed tissue was dehydrated in different grades (30%, 70%, 90% and absolute) of ethanol and embedded in paraffin wax. Sections (5-7 μm) of gonadal tissues were cut, de-waxed and dehydrated with different grades of ethanol, stained with Haematoxylin and Eosin, dehydrated and mounted in Canada balsam (Merck, Darmstadt, Germany). Sections were examined under a compound microscope (Olympus Co., Japan).

Statistical Analysis

All data were calculated and presented as mean ± standard deviation (X ± SD). Student's t- test was calculated for observing the comparison between control and treatment groups. The coefficient of variance was also calculated so as to know the efficacy of the treatment groups. Further, 'Chi square' (χ^2) test was done for calculating the variation in sex ratio in the treatment groups from the expected Mendelian sex ratio of 1:1. Data were analyzed by one way analysis of variance (ANOVA) using a computer programmed software, statistical package for Social science (SPSS).

Results

Experimental fishes of control, tamoxifen and letrozole treated groups remained healthy and mortality ranged from 8.5% to 14.5% throughout the observation period of 60 days. Higher mortality was mainly in the tamoxifen treated fish particularly at the end of 60 days. Sex ratio of the experimental fishes was found to respond well with tamoxifen and letrozole treatments. There was significantly enhanced masculinisation both in *C. carpio* and *O. niloticus*. Low dose tamoxifen treated (T_1) *C. carpio* showed 63.21% male and 36.79% female while in *O. niloticus*, there were 78.36% male and 35.80% female (Figure 1). Larger dose of tamoxifen (T_2) brought about 84.56% male and 15.44% female in *C. carpio* and 90.01% male and 9.73% female populations of *O. niloticus* (Figures 1 and 2). Low dose of letrozole treated group showed 79.39% male and 20.61% female *C. carpio*; 87.91% male and 11.86 % female *O. niloticus*. Larger dose of letrozole (L_2) brought about 98.47% male and 1.53% female *C. carpio* and 99.65% male and 0.52% female *O. niloticus*. Percentage of masculinisation was comparatively more in letrozole treated fish as compared to tamoxifen treatment. In the control group, there was 48.28% male and 50.78% female *C. carpio* and 46.38% male and 53.616% female *O. niloticus* (Figures 1 and 2). Some intersex fish was also observed in low dose of tamoxifen (T_1) and letrozole treatments (L_1) to fish. In *C. carpio*, tamoxifen treatment (T_1) showed 0.30 % and letrozole treatment (L_1) showed 0.40 % intersex (Figures 1 and 2). In *O. niloticus*, low dose of tamoxifen (T_1) showed 1.0% intersex and low dose of letrozole (L_1) showed 0.85% intersex. The statistical coefficient of variance (CV) for male was 20.56 and 8.76 for female after tamoxifen treatment to tilapia and it was 7.33 for male and 18.73 for female in letrozole treated *O. niloticus*. The calculated χ^2 value for sex ratio was 17.64 which was close to degree of freedom (df) 15.51 showing that the deviation in sex ratio from the expected Mendelian sex ratio was highly significant (*P<0.001*). Similarly, the statistical CV for male was 7.97 and 3.83 for female common carp after tamoxifen treatment. In letrozole treated *C. carpio*, the CV was 4.07 for male and 18.27 for female. The calculated χ^2 value was 33.65 which was much less than the chi-square Table value and was 26.13. Hence, there was highly significant (*P<0.001*) variation in sex ratio from the expected Mendelian sex ratio of 1:1.

In control Nile tilapia and common carp, the level of 17β-estradiol (E_2) was 162.66 pg mL^{-1} and 196.5 pg mL^{-1} and testosterone (T) level was 610.23 pg mL^{-1} and 578.56 pg mL^{-1} respectively. The 17β-estradiol and testosterone (E_2/T) ratio was 0.266 for tilapia and 0.339 for common carp. However, E_2/T declined significantly after tamoxifen and letrozole treatments (Figures 3 and 4). Larger dose of letrozole treatment (L_2) to fish brought about more pronounced decline in E_2 level as compared to tamoxifen treated fish consequently declining the $E_{2:}$T ratio remarkably which was 0.076 for tilapia and 0.085 for common carp (Figures 3 and 4). However, it was observed that some E_2 level was present even after suppressing the aomatase activity by both tamoxifen and letrzole. The lowest value of E_2 was 85.83 pg mL^{-1} in tilapia while it was 56.0 pg mL^{-1} for common carp. The coefficient of variation (CV) for E_2 in tamoxifen treated *O. niloticus* was 16.0 while for testosterone, it was 10.44. The CV for E_2 in letrozole treated *O. niloticus* was 8.67 and it was 11.18 for testosterone (T) level. In case of common carp, the coefficient of variance for E_2 was 8.81 in tamoxifen treated *C. carpio* and 6.64 for letrozole treatment. The CV for T for tamoxifen treated *C. carpio* was 15.9 and 11.34 for letrozole treatment.

Histological examination revealed that there was no difference in gonadal tissues between control males and letrozole treated *C. carpio* except localized presence of spermatozoa (Figure 5a). The paired testis of *O. niloticus* was elongated milky white organ covered by a thin connective tissue capsule (tunica albuginea). The histological sections, testis of *O. niloticus* showed organized branching lobules of

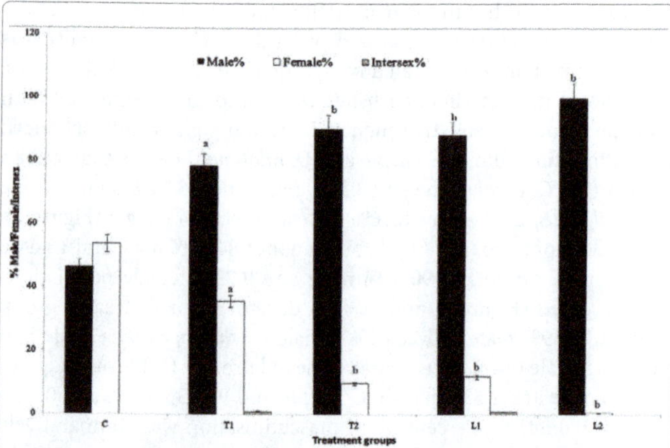

Figure 1: Effect of different concentration of tamoxifen (T_1 and T_2) and letrozole (L_1 and L_2) on sex ratio (% male, female and intersex) in *Cyprinus carpio* (Level of significance: a=*P*<0.005; b= *P*<0.001 as compared to control)

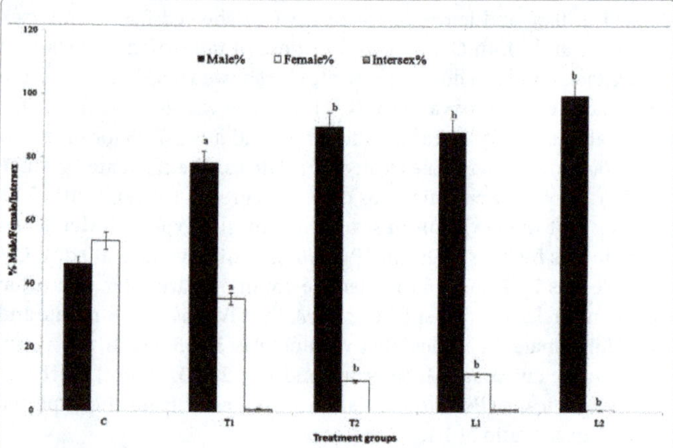

Figure 2: Effect of different concentration of tamoxifen (T_1 and T_2) and letrozole (L_1 and L_2) on sex ratio (% male, female and intersex) in *Oreochromis niloticus* (Level of significance: a=*P*<0.005; b= *P*<0.001 as compared to control)

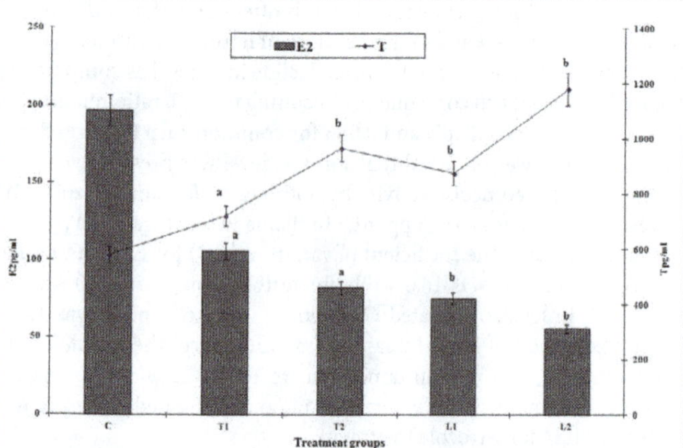

Figure 3: Tamoxifen and letrozole induced changes in serum 17β estradiol (E_2) and testosterone (T) concentrations in *Cyprinus carpio* (Level of significance: a=*P*<0.005) ; b= *P*<0.001 as compared to control).

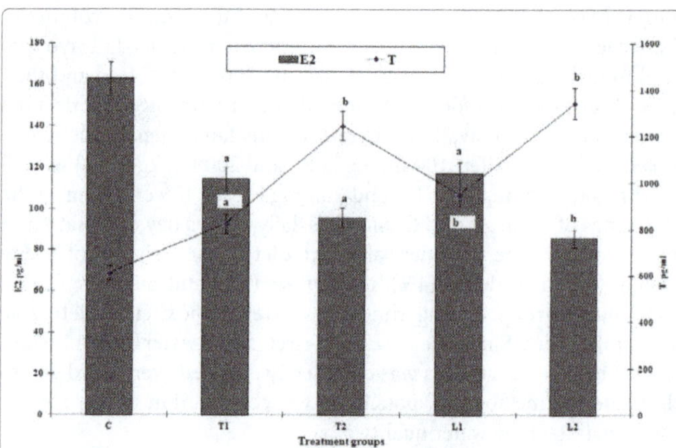

Figure 4: Tamoxifen and letrozole induced changes in serum 17β estradiol (E_2) and testosterone (T) concentrations in *Oreochromis niloticus* (Level of significance: a=*P*<0.005; b= *P*<0.001 as compared to control)

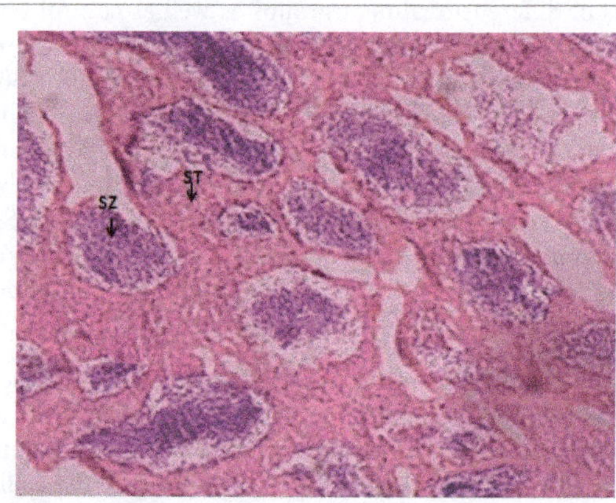

Figure 5a: Gonadal structure after high dose of letrozole treated (L_2) *Cyprinus carpio* showing clustered spermatids (ST) and spermatozoa (SZ), stain = H & E.. Bar scale = 60 μm

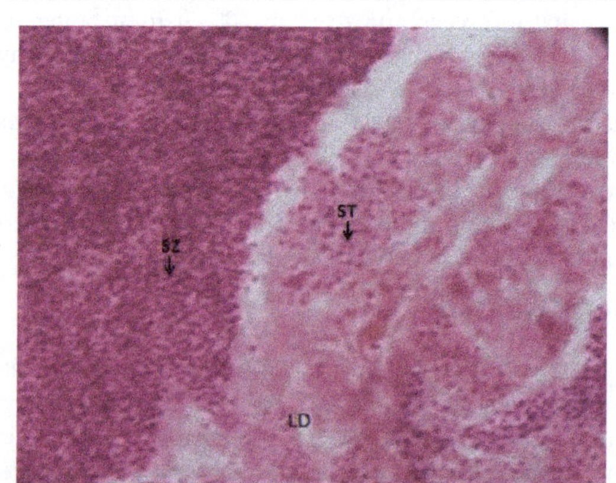

Figure 5b: Gonadal structure after high dose tamoxifen treatment (T_2) to *Oreochromis niloticus* showing leyding cells (LD), spermatids (ST) and spermatozoa (SZ), stain = H & E. Bar scale = 60 μm

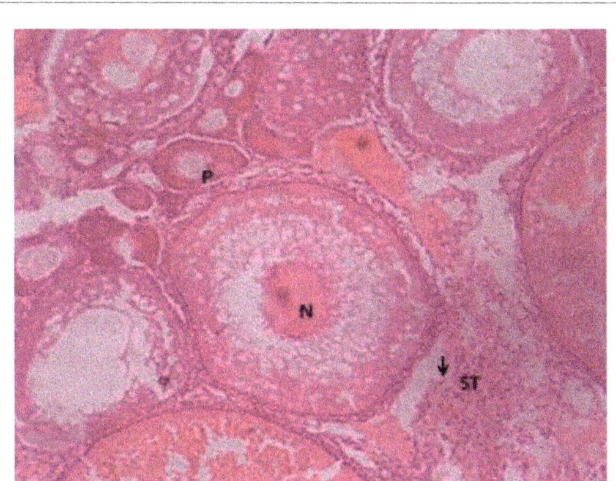

Figure 5c: Intersex *O. niloticus* after low dose of tamoxifen treatment (T_1) showing big nucleolus (N), Perinuclear cell (P), and spermatids (ST), stain = H & E. Bar scale = 60 µm

the unrestricted spermatogonial type (Figure 5b). The ovarian cavity, oocytes and perinuclear oocytes were clearly seen in the ovary of female *C. carpio* and *O. niloticus*. Transitional phase of intersex gonad had arrested oocytes and spermatogenic germ cells at the late stages of spermatogenesis. In masculinising fish after tamoxifen treatment to *C. carpio*, testis contained spermatozoa accumulated in the seminiferous tubules. Histological examination of gonads revealed that few specimens exhibited intersex (ovotestis) i.e. presence of ovarian as well as testicular structures together in fishes treated with 100 mg kg^{-1} of letrozole and tamoxifen (Figure 5c). In the intersex fish induced by tamoxifen treatment, some spermatids were similar to those observed in the control males along with numerous arrested perinuclear oocytes in the same gonad. The intersex fish was partly masculinized and was having male as well as female cells in the same gonad (Figure 5c).

Discussion

The results of this study revealed that tamoxifen and letrozole in the tested doses had potent action on sex reversal and gonadal differentiation bringing about significantly high production of male *C. carpio* as well as *O. niloticus*. However, level of masulinization was found higher in the larger dose of letrozole which was more pronounced in case of *O. niloticus* as compared to common carp where almost all male tilapia was obtained. In our earlier study, we obtained 90% male tilapia when treated with tamoxifen in the dose of 200 mg kg^{-1} without any significant mortality [9]. Reports on other fish species have reported 100% male production by the use of AI treatment in the dose of 100 and 1000 mg kg^{-1} in the zebra fish *Danio rerio* [20]; and 500 mg kg^{-1} in the golden rabbit fish *Siganus guttatus* [21]. Our findings corroborates with earlier report [22] where fadrozole in the doses of 200 mg kg^{-1} to 400 mg kg^{-1} produced approximately 97% male population of *C. carpio*. However, it has been reported [23,24] that tamoxifen or 17α-methyltestosterone (MT) masculinized 45-50% *C. carpio* at 100 mg kg^{-1} experimental feed which has much lower level of sex reversal % than the obtained value in this study. Effect of tamoxifen on sex differentiation has been reported in Zebra fish, *Danio rerio* [25], sea scallop, *Placopecten magellanicus* [26], and tilapia [9]. It has also been reported that oral administration of tamoxifen (200 ppm) effectively acts for sex reversal in the bagrid catfish during the labile period of sex differentiation [27] and in *Tinca tinca* [28]. Tamoxifen, as an anti-estrogen analogue has also been evaluated as a feed additive to regulate the sex ratio in rainbow trout *Oncorhynchus mykis* [29], *O. niloticus* (Hines and Watts 1995), and Japanese medaka *Paralichthys olivaceus* [30].

The analysis of estradiol (E_2) levels in the blood serum showed declined values after letrozole and tamoxifen treatments in comparison with the controls, whereas significantly high serum testosterone (T) level was detected in all treatment groups. However, there was some estradiol level left in the circulation even after aromatase inhibition which could be due to two reasons; firstly the blood sample was pooled where specimens of both the genetic sexes might have been present and alternately some estradiol level may be needed for maintenance of testicular activity since *Cyp19* gene is highly conserved in fish [12]. Our findings support the earlier findings of [31] where significant differences in concentrations of E_2 in the blood plasma and hypothalami between fadrozole (5, 50 or 500 µg kg^{-1}) treatment groups of goldfish and *C. carpio* has been reported. Fadrozole in the doses of 0, 100, 500 and 1000 µg/fish has been reported to decrease the plasma E_2 in honeycomb grouper *Epinephelus merra* and Nile tilapia [32-34] which further supports our findings. Decreased serum E_2 level has also been reported after the treatment of fadrozole (10.0 mg kg^{-1}) to red-spotted grouper *E. akaara* [35].

Complete masculinisation in *O. niloticus* and *C. carpio* significantly accounted for suppressed E_2 level which consequently elevated T level suggesting that aromatase based sex determination occurs in Nile tilapia and common carp. Our findings strongly supported that letrozole directly down regulate *Cyp19* gene thereby inhibiting the aromatase production [12]. However, the masculinisation level with tamoxifen was bit different because anti-estrogenic action [36] and selective estrogen receptor modulating (SERM) action through blockage of estrogen receptors [9,16]. Since the masculinisation level was potent with letrozole treatment, our findings suggested that masculinising with commercially available letrozole (USP, Evalet™, India) could be a cheap method of inducing sex reversal in tilapia and common carp. This drug is non-steroidal aromatase inhibitor; inhibitor of estrogen biosynthesis and anti-tumour agent available in market for breast cancer treatment.

Histological examinations of the tamoxifen and letrozole treated gonads showed well differentiated testicular cells showing spermatozoa in seminiferous tubules suggesting that the sex reversal was male hormone (testosterone) based since letrozole blocked the aromatization process [12,37] while tamoxifen inhibited receptor binding. In this study, there was a notable, dose-dependent enlargement of the seminiferous tubules accompanied by an abundant accumulation of spermatozoa in the lumina after letrozole treatment. The marked accumulation of spermatozoa in testes after latrozole treatment could be due to enhanced sperm production related to the increase in plasma testosterone [38]. However, presence of intersex gonads could be a result of low dose of tamoxifen or letrozole treatment. Further, results of this study suggested that 60 days exposure of AIs in the dose of 200 mg kg^{-1} produced complete sex reversal in tilapia and common carp.

We conclude that letrozole at a dose 200 mg kg^{-1} significantly inhibit estrogen production thereby increasing level of circulating testosterone via suppressed action of *Cyp19* which significantly influenced sex differentiation for masculinisation in *C. carpio* and *O. niloticus*. The masculinised fish had well-developed testes containing accumulated spermatozoa in the seminiferous tubules. The putative action of tamoxifen and letrozole was found more or less similar for their masculinizing effects; however, the effect of letrozole was more potent suggesting that the chemical could be used in reproductive management and sex reversal of such important aquaculture species. It is to mention here that the letrozole exposed fishes retained the sex reversed masculinity until six months of observation even after withdrawal of the treatment suggesting that gonochoristic fish particularly common carp and tilapia maintain their sexual plasticity until adulthood [39]. There is hardly any report to document that exogenous AI treated fishes are not safe for consumption [40,41]. Rather most potent estrogen blockers and aromatase inhibitors are reported to occur naturally as Chrysin (5'7-Dihydroxyisoflavone) a flavonoid from blue passion flower and White Button Mushrooms with great real life results having therapeutic values [42]. It is therefore, the results of this study to produce monosex tilapia and common carp using AI has strong potential having avenues in aquaculture industry.

Acknowledgements

Authors are grateful to Dr J.K.Jena, Director NBFGR for extending all support and encouragements. We thankfully acknowledge the help received by the Institute of Toxicological Research (IITR), Lucknow for reading out our samples over ELISA reader. We also thankfully acknowledge JVD Pharma Pvt.Ltd., New Delhi for providing us letrozole tablets (Evalet™) for our experimental use.

References

1. Devlin RH, Nagahama Y (2002) Sex determination and sex differentiation in fish: an overview of genetic, physiological, and environmental influences. Aquaculture 208: 191-364.
2. Nakamura Masaru (2013) Morphological and physiological studies on gonadal sex differentiation in teleost fish. Aqua-BioScience Monographs 6: 1-47.
3. Kobayashi Y, Nagahama Y, Nakamura M (2013) Diversity and plasticity of sex determination and differentiation in fishes. Sexual Development 7: 115-25.
4. Lawson C, Gieske M, Murdoch B, Ye P, Li Y, Hassold T, et al. (2011) Gene expression in the fetal mouse ovary is altered by exposure to low doses of bisphenol A. *Biol Reprod.* 84: 79-86.
5. Hudson QJ, Smith CA, Sinclair, AH (2005) Aromatase inhibition reduces expression of FOXL2 in the embryonic chicken ovary. Developmental Dynamics 233: 1052-1055.
6. Yang X, Zheng J, Na R, Li J, Xu G, Qu L, et al. (2008) Degree of sex differentiation of genetic female chicken treated with different doses of an aromatase inhibitor. Sexual Development 2: 309-315.
7. Olmstead Allen W, Patricia A Kosian, Joseph J Korte, et al. (2009) Sex reversal of the amphibian, *Xenopus tropicalis*, following larval exposure to an aromatase inhibitor. Aquatic Toxicology 91: 143-150.
8. Richard Shine, Daniel A, Warner, Rajkumar Radder (2007) Windows of embryonic sexual lability in two lizard species with environmental sex determination. *Ecology* 88: 1781-1788 Ecological Society of America.
9. Singh R, Singh AK, Madhu Tripathi (2012) Effect of an aromatase inhibitor analogue tamoxifen on the gonad and sex differentiation in Nile tilapia *Oreochromis niloticus*. Journal of Environmental Biology 33: 799-803
10. Taranger GL, Carrillo M, Schulz RW, Pascal F, Zanuy S, et al. (2010) Control of puberty in farmed fish. General and Comparative Endocrinology 165: 483-515.
11. Gomelsky B (2003) Chromosome set manipulation and sex control in common carp: a review. Aquatic Living Resources 16: 408-415.
12. Guiguen Y, Fostier A, Piferrer Francesc, Chang CF (2010) Ovarian aromatase and estrogen: pivotal role for gonadal sex differentiation and sex change in fish. General and Comparative Endocrinology 165: 352-366.
13. Simpson ER, Clyne C, Rubin G, Boon WC, Robertson K, et al. (2002) Aromatase-a brief overview. Annals of Reviews in Physiology 64: 93-127.
14. Bhatnagar AS, Brodie AMH, Long BJ, Evans DB, Miller WR (2001) Intracellular aromatase and its relevance to the pharmacological efficacy of aromatase inhibitors. Journal of Steroid Biochemistry & Molecular Biology 76: 199-202.
15. Leaños CO, Glen Van K, Rossanna Rodríguez C, Gold G (2007) Endocrine disruption mechanism of o,p'-DDT in mature male tilapia (*Oreochromis niloticus*). Toxicological Application & Pharmacology 221: 158–167
16. Kwon JY, Haghpanah V, Kogson-Hurtado LM, McAndrew BJ, Penman DJ (2000) Masculinization of genetic female Nile tilapia *(O. niloticus)* by dietary administration of an aromatase inhibitor during sexual differentiation. Journal of Experimental Zoology 287: 46-53.
17. Sun L, Zha J, Spear PA, Wang Z (2007a) Toxicity of the aromatase inhibitor letrozole to Japanese medaka (*Oryzias latipes*) eggs, larvae and breeding adults. Comparative Biochemistry and Physiology-C Toxicology and Pharmacology 145 (4): 533-41.
18. Sun L, Zha J, Spear PA, Wang Z (2007b) Tamoxifen effects on the early life stages and reproduction of Japanese medaka (*Oryzias latipes*). Environtal Toxicology and Pharmacology 24: 23-29
19. Guerrero RD, WL Shelton (1974) An acetocarmine method for sexing juvenile fishes. Progressive Fish Culturists 36: 56
20. Uchida D, Yamashita M, Kitano T, Iguchi T (2004) An aromatase inhibitor or high water temperature induces oocyte apoptosis and depletion of P450 aromatase activity in the gonads of genetic female zebrafish during sex-reversal. Comparative Biochemistry and Physiology- Part A: Molecular and Integrative Physiology 137: 11-20.
21. Komatsu T, Nakamura S, Nakamura M (2006) Masculinization of female golden rabbit fish *Siganus guttatus* using an aromatase inhibitor treatment during sex differentiation. Comparative Biochemistry and Physiology Part C 143: 402-409.
22. Tzchori I, Zak T, Sachs O (2004) Masculinization of genetic females of C. carpio (*Cyprinus carpio*) by dietary administration of an aromatase inhibitor-Israeli Journal of Aquaculture Bamidgeh 56: 239-246.
23. Hulak M, Rodina M, Gela D, Kocour M, Linhart O (2008b) Sex control strategies for the masculinization of *Cyprinus carpio* L. and tench (*Tinca tinca*). Cybium 32: 100-101.
24. Hulak M, Psenicka M, Gela, D, Rodina M, Linhart O (2010) Morphological sex change upon treatment by endocrine modulators in meiogyngenetic tench (*Tinca tinca* L). Aquaculture Research 41: 233-239.
25. Leo TM, Ven VN, Van EJ, Brandho Yang, X, Zheng J, et al. (2008) Degree of sex differentiation of genetic female chicken treated with different doses of an aromatase inhibitor. Sexual Development 2: 309-315
26. Wang Chunde, Croll RP (2004) Effects of sex steroids on gonadal development and gender determination in the sea scallop, *Placopecten magellanicus.* Aquaculture. 238: 483-498.
27. Park IS, Kimb JY, Choa SH, Kim DS (2004) Sex differentiation and hormonal sex reversal in the bagrid catfish *Pseudobagrus fulvidraco* (Richardson). Aquaculture 232: 183-193.
28. Hulak M, Psenicka M, Coward K, Linhart O (2008a) A quantitative study of testicular germ cell populations in masculinized neomale common carp (*Cyprinus carpio* L.). Cell Biology International 32: 515-524.
29. Elias EE, Kalombo E, Mercurio SD (2007) Tamoxifen protects against 17α-ethynylestradiol-induced liver damage and the development of urogenital papillae in the rainbow Darter (*Etheostoma caeruleum*). Environmental toxicology and Chemistry 26: 1879-1889.
30. Kitano T, Yoshinaga N, Shiraishi E, Koyanagi T, Abe S (2007) Tamoxifen induces masculinization of genetic females and regulates P450 aromatase and Mullerian inhibiting substance mRNA expression in Japanese founder (*Paralichthys olivaceus*)- Molecular Reproduction and Development 74: 1171-1177.
31. Mikolajczyk T, Sokolowska-Mikolajczyk M, Chyb J, Szczerbik P, Socha M, et al. (2007) LH secretion and 17β-estradiol concentration in the blood plasma and hypothalamus of goldfish (*Carassius auratus gibelio* B.) and common carp (*Cyprinus carpio* L.) treated with fadrozole (aromatase inhibitor) and GnRH analogues- Czech. Journal of Animal Science 52: 354-362.
32. Bhandari, RK, Komuro H, Higa M, Nakamura M (2004) Sex Inversion of Sexually Immature honeycomb grouper (*Epinephelus merra*) by Aromatase Inhibitor. Zoological Science 21: 305-310.
33. Bhandari RK, Alam MA, Higa M, Soyano K, Nakamura M (2005) Evidence that estrogen regulates the sex change of honeycomb grouper (*Epinephelus merra*), a protogynous hermaphrodite fish. Journal of Experimental Zoology- A Comparative Experimental Biology. 303: 497-503.
34. Bhandari RK, Nakamura M, Kobayashi T, Nagahama Y (2006) Suppression of steroidogenic enzyme expression during androgen-induced sex reversal in

Nile tilapia (*Oreochromis niloticus*). General and Comparative Endocrinology 145: 20-24.

35. Li GL, Liu XC, Zhang Y, Lin HR (2006) Gonadal development, aromatase activity and P450 aromatase gene expression during sex inversion of protogynous red-spotted grouper *Epinephelus akaara* (Temminck and Schlegel) after implantation of the aromatase inhibitor, fadrozole. Aquaculture Research 37: 484-491.

36. Wu F, Zhang X, Huang B, Liu Zhihao, Hu C (2009) Expression of the gonadotropin subunits in southern catfish gonad and their possible role during early gonadal development. Comparative Biochemistry and Physiology, Part A 153: 44-48.

37. Chikae M, Ikeda R, Hasan Q, Morita Y, Tamiya E (2004) Effects of tamoxifen, 17α-ethynylestradiol, flutamide, and methyltestosterone on plasma vitellogenin levels of male and female Japanese medaka (*Oryzias latipes*). Environmental Toxicology and Pharmacology 17: 29-33.

38. Afonso LOB, Iwama GK, Smith J, Donaldson EM (2000) Effects of the aromatase inhibitor fadrozole on reproductive steroids and spermiation in male coho salmon (Oncorhynchus kisutch) during sexual maturation. Aquaculture 188: 175-187.

39. Paul-Prasanth B, Bhandari RK, Kobayashi T, Horiguchi R, Kobayashi Y (2013) Estrogen oversees the maintenance of the female genetic program in terminally differentiated gonochorists. Scientific Reports 3: 2862.

40. Ghosh-Choudhury T, Mandal CC, Woodruff K, St Clair P, Fernandes G (2009) Fish oil targets PTEN to regulate NFkappaB for down regulation of anti-apoptotic genes in breast tumor growth. Breast Cancer Res Treat 118(1): 213-28.

41. Mandal CC, Ghosh-Choudhury T, Yoneda T, Choudhury GG, Ghosh-Choudhury N (2010) Fish oil prevents breast cancer cell metastasis to bone. Biochemical Biophysical Research Communication. 26: 402: 602-7.

42. Chen S, Oh SR, Phung S, Hur G, Ye JJ (2006) Anti-aromatase activity of phytochemicals in white button mushrooms (*Agaricus bisporus*). Cancer Research 66: 12026-12034.

Is There Any Species Specificity in Infections with Aquatic Animal Herpesviruses? – The Koi Herpesvirus (KHV): An *Alloherpesvirus* Model

Sven M Bergmann[1]*, Michael Cieslak[1], Dieter Fichtner[1], Juliane Dabels[2], Sean J Monaghan[3], Qing Wang[4], Weiwei Zeng[4] and Jolanta Kempter[5]

[1]*FLI Insel Riems, Südufer 10, 17493 Greifswald-Insel Riems, Germany*
[2]*University of Rostock, Aquaculture and Sea Ranching, Justus-von-Liebig-Weg 6, Rostock 18059, Germany*
[3]*Aquatic Vaccine Unit, Institute of Aquaculture, School of Natural Sciences, University of Stirling, Stirling, FK9 4LA, UK*
[4]*Pearl-River Fisheries Research Institute, Xo. 1 Xingyu Reoad, Liwan District, Guangzhou 510380, P. R. of China*
[5]*West Pomeranian Technical University, Aquaculture, K. Królewicza 4, 71-550, Szczecin, Poland*

Abstract

Most diseases induced by herpesviruses are host-specific; however, exceptions exist within the family Alloherpesviridae. Most members of the Alloherpesviridae are detected in at least two different species, with and without clinical signs of a disease. In the current study the Koi herpesvirus (KHV) was used as a model member of the Alloherpesviridae and rainbow trout as a model salmonid host, which were infected with KHV by immersion. KHV was detected using direct methods (qPCR and semi-nested PCR) and indirect (enzyme-linked immunosorbant assay; ELISA, serum neutralization test; SNT). The non-koi herpesvirus disease (KHVD)-susceptible salmonid fish were demonstrated to transfer KHV to naïve carp at two different temperatures including a temperature most suitable for the salmonid (15°C) and cyprinid (20°C). At 20°C KHVD was induced in carp cohabitated with infected trout. KHV was also detected virologically and serologically at the end of the experiment in both rainbow trout and carp.

Keywords: KHV; KHVD; Transmission; Rainbow trout

Introduction

Herpesviruses constitute a diverse family of widespread pathogens inducing severe diseases of veterinary importance in all animals including animals used for human consumption. It is likely that they evolve within their host species over long periods of time and most of them induce very host-specific disease, although sometime infections occur in the absence of clinical disease signs. As a rule herpesviruses have large genomes and all of them induce a latent or a persistent phase of infection, sometimes in different hosts. For the family Alloherpesviridae it has been demonstrated that most of the virus species have developed a high level of host specificity. In the majority of animal herpesviral infections only mild symptoms (internally and externally) appear under natural conditions. Virulence associated with herpesviral infection is often initially displayed in immunologically weakened hosts or during primary infection of a naïve host. The characteristics of KHV, including the morphology investigated by electron microscopy [1,2], and phylogeny meets the taxonomical criteria for viruses of the family Alloherpesviridae [3], which have been shown to be non-host-specific, but sometimes causing disease in one species. Within this family are four genera grouped as the *Batrachovirus, Salmonivirus, Ictalurivirus* and *Cyprinivirus*. In *Batrachovirus*, the *ranid herpesvirus* 1 *(RaHV-1)* can be found in different leopard frog (*Rana pipiens*) (sub) species and has been identified as the causative agent of renal adenocarcinoma [4], and the *ranid herpesvirus* 2 (*RaHV-2*), which can cause infections in a lot of species of the family Ranidae (pets and wild) leading to skin lesions and sometimes inducing tumors [5]. The members of the family ictaluriviridae, such as acipenserid herpesvirus 1 (*AcHV-1*), infect sturgeon spp. [6] and at least two sturgeon sp. can be infected with *AcHV-2* [7]. The most previously researched Alloherpesvirus, channel catfish herpesvirus (*Ictalurid herpesvirus* 1 (*IcHV-1*), CCV) infects channel catfish but also blue catfish [8] and possibly other catfish species or subspecies while *IcHV-2* was detected in black bullhead (*Ameiurus melas*) and in channel catfish (Ictalurus punctatus) [9]. Members of the *salmonivirus* group are known to often infect more than one species: *salmonid herpesvirus* 1 (*SaHV-1*) have been detected in rainbow trout (*Ocorhynchus mykiss*) [3] but also in Chum salmon (*O. keta*) and Chinook salmon (*O. kisutch*) [10]. *SaHV-2* was considered to infect all salmonid species [3] and *SaHV-3* was detected in different fish and hybrids of the genus *Salvelinus* [3,11,12]. In addition to *cyprinid herpesviruses* 1,2 and 3 the eel herpesvirus (HVA, *AngHV-1*) is also included in the genus *Cyprinivirus*. HVA was detected during disease outbreaks in Japanese eel (*Anguilla japonica*) and European eel (*A. anguilla*) in Japan [13] and in American eel (*A. rostrata*) in Poland [14]. A variant of the HVA in eels in Taiwan (eel herpesvirus Formosa, FEHV) has also been found to induce mortality in common carp (*Cyprinus carpio*) as well [15]. From investigations conducted in our Lab it was confirmed that HVA may at least persist in artificially infected common carp (immersion) for up to six weeks, which was detectable directly by PCR and indirectly by serum neutralization assay (Bergmann, unpublished data). The cyprinid herpesvirus 1 (*CyHV-1*, carp pox virus) induced typical clinical signs in carp or koi (*C. carpio*), in golden ide (*Leuciscus idus*) [16] or in other cyprinids (carps and minnows) [17]. *CyHV-2* or goldfish herpesvirus was first detected in goldfish (*Carassius auratus)*, inducing severe disease outbreaks in those affected populations [18,19]. Recently severe outbreaks of disease with high mortalities were observed in crucian carp (*Carassius carassius*) [20] and Prussian carp (*C. c. gibelio*) [21] where *CyHV-2* was identified as the disease causing agent. In Germany severe mortality rates were also induced by *CyHV-2* in a wild Prussian carp population

***Corresponding author:** Sven M Bergmann, FLI Insel Riems, Südufer 10, 17493 Greifswald-Insel Riems, Germany, E-mail: sven.bergmann@fli.bund.de

(Bergmann, pers. obs.). For *CyHV-3* or koi herpesvirus (KHV), although KHV disease (KHVD) is induced only in the species *C. carpio* (common carp and koi), it has been demonstrated that this virus can be present, and sometimes replicated, in a lot different fish species living in fresh and brackish water. KHV DNA has been identified by different PCR methods in a lot of members of the families Acipenseridae [22], Cyprinidae, including common carp or koi hybrids [23-25], and Percidae in many wild fish populations within close vicinity to, or in carp ponds [26,27]. Furthermore KHV has also been detected in fresh water mussels and crustaceans [28]. In this study we have shown that a common salmonid species, rainbow trout, is able to transfer KHV to naive carp inducing KHVD under certain circumstances.

Materials and Methods

Fish

Rainbow trout, steelhead strain (n=30, 80-100 g) were obtained from a commercial farm and kept under a re-circulating system with a volume of 450 l and fed once a day with commercial trout food. The daily water exchange was 100-120 l/h. The rainbow trout were divided into four groups including fish kept at 15°C (n=10) and 20°C (n=10) water temperature for the infection trial and two additional tanks containing fish (n=5, respectively) kept as negative controls at the same temperatures. The fish were confirmed free from notifiable pathogens like viral hemorrhagic septicemia virus (VHSV) and infectious hematopoietic necrosis virus (IHNV) by RT-PCR [29] and additionally with the conventional PCR and nested PCR [30] according to Engelsma et al.

Carp (n=15, 80-100 g) were obtained from a commercial farm in German federal state Thuringia and tested free for CyHV-1, -2 and -3 by PCR [24,30] for 10 years. They were divided into three groups (2×n=5, for the infection trial at 15°C and 20°C, respectively) and as a negative control (n=5). They were kept at 18°C in the same re-circulating system as the rainbow trout and fed with commercial carp food.

Scheme of animal experiment

After a 14 day adaptation period rainbow trout were kept at 15 and 20°C for one further week. Samples were then collected non-lethally (swabs, leukocyte separation, serum) and rainbow trout were infected with KHV by immersion at 15 and 20°C, respectively, for 1 h with 10^4 $TCID_{50}$/ml in water (v/v) in a separate aquarium. After the challenge period fish were caught and replaced in to the aquaria and maintained there for another 7 days. After 7 days post infection (dpi) samples were collected non-lethally but also lethally (gill and kidney tissue parts) from two rainbow trout from both aquaria (15°C and 20°C) as well as from negative controls from tanks set at both temperatures. All rainbow trout were caught and transportation was stimulated (30 seconds in the air, placed in a new aquarium for 10 minutes and returned to the original aquaria). After a water exchange the following day, the aquaria were divided using a water permeable wall with a water flow from the rainbow trout to the carp in circulation. Before carp were placed into the aquaria, samples were collected non-lethally. All fish were observed for another 21 days. At the end of the experiment samples were collected lethally and non-lethally.

Virus and cells

Koi herpesvirus isolate KHV-E (D 132), kindly provided by Dr. Keith Way (CEFAS, UK), was allowed to replicate in CCB cells [30,31]. The virus was characterized at the protein level by immunofluorescence assays [24], at the genomic level for quantification by qPCR [32] modified according [31] and at the infectious dose level by titrations onto CCB cells according to Spearman and Kärber [33].

Direct methods of KHV detection in fish samples

Non-lethally collected samples (gill and skin swabs, separated leukocytes) and lethally taken samples (gill and kidney tissue) were investigated regularly by qPCR at different time points of the experiment: rainbow trout on 1st day shortly before infection, on day 7 dpi and 37 dpi and carp on 1st day shortly before cohabitation with rainbow trout and on day 21 post cohabitation (dcoh).

Indirect methods of KHV detection in serum (plasma) samples

Fish (rainbow trout and carp) were bled from the caudal vein at the same time points of the experiment described for direct detection methods. Blood was allowed to clot at 4°C in a BD Microtainer® tube (Becton and Dickinson) over night. The tubes were then centrifuged for serum separation at 1000 × g for 1 min at 4°C. Sera was stored at -20°C until use.

For serum neutralization assays (SNT) all sera obtained from rainbow trout and from carp were pre-diluted 1:4 and 1:16 with CCB cell culture medium, respectively and then 2-fold diluted to 1:1048. Those dilutions were mixed with the same volume of KHV with a titre between $10^{0.5}$ and $10^{1.5}$ $TCID_{50}$/ml. These mixtures were incubated overnight at 4°C. The next day those preparations were adsorbed for 1 h at 20°C directly on the CCB cell monolayer. The medium was re-supplemented and the cell culture plates were incubated at 20°C for 10 days. After 5, 7 and 10 days post inoculation the plates were assessed for an occurring cytopathic effect (CPE) due to KHV replication in CCB cells (this study).

The KHV antibody enzyme-linked immunosorbant assay (ELISA) was prepared in combination of the assays according to [34,35] with some modifications. Briefly, the Medisorp ELISA plate (Nunc) was used for coating the plate with 3 µg purified KHV/ml [36], blocked with Roti Block® (Roth), and as secondary antibody a monoclonal anti-carp IgM (F 16, Aquatic Diagnostics) at a dilution 1:64 and the anti-trout IgM monoclonal antibody 4C10 at a dilution of 1:1000 were used. As ELISA substrate ready to use TMB (Pierce) was used. The plates were measured in an ELISA Reader (iMark, BioRad) at 450 nm. All reactions were carried out at room temperature in duplicates. All dilution and washing steps were proceeded with PBS-Tween 20 (0.05%) without any additional substances.

Results

Infection experiment at 15°C

At 15°C water temperature neither clinical signs of KHVD nor mortality were observed in rainbow trout or in cohabitated carp. The behaviour of both fish species was typical at this temperature. All animals stayed completely healthy throughout the period of the experiment.

Infection experiment at 20°C

At 20°C rainbow trout always stayed away from the heating source in the aquarium. The majority of the rainbow trout exhibited darkening of their skin and moved much slower compared to the animals in the 15°C experiment. While rainbow trout did not show any clinical signs characteristic of an infectious disease, all five carp became sick on 5th d pcoh. On the 10th d pcoh one carp died with severe symptoms of

KHVD. All surviving carp recovered till the 18th d pcoh. At the end of the experiment carp still displayed signs of KHVD like patches in the skin but their behaviour returned back to a normal including typical swimming activity and food uptake.

Direct detection of KHV in rainbow trout samples

On the first dpi prior to immersion no KHV DNA was detected in the samples from the 15°C and 20°C aquarium (gill swabs and separated leukocytes) from five rainbow trout, respectively (Table 1). On the 7th dpi before simulated transportation, KHV DNA was detected in leukocyte preparations but also in samples from organ tissues of two fish from both aquaria. In the aquarium with the KHV infection at 15°C ct values between 28 and 33 were detected, and at 20°C ct values between 24 and 28 were observed (Table 1). In both cases the gill swabs were negative by qPCR [32] but positive by the semi-nested PCR [31]. Those two fish from each aquarium were anesthetized, killed by decapitation and gill and kidney tissues were collected for qPCR. In samples from both aquaria a similar ct values between 27 to 33 were detected. At the end of the experiment on the 29th dpi, KHV DNA was detected in all remaining rainbow trout samples. There was no difference between the samples collected lethally or non-lethally and no differences in samples from the aquaria with the different temperatures. The values varied only between ct 27 and ct 33 corresponding to 100.000 to 50 genomic equivalents compared to the positive KHV plasmid controls [31].

Direct detection of KHV in carp samples

Samples were collected from carp before cohabitation to rainbow trout and on 21st dcoh. While samples were negative on 1st dcoh, on 21st dcoh in samples from carp of the aquarium with 15°C four of five (ct 27-ct 33) fish were positive and from aquarium wit 20°C three of four (ct 27-ct 33) were positive for KHV DNA (Table 1). There was no difference in the results from lethal and non-lethal sample collection. All samples obtained from carp that had died from KHVD were positive for KHV DNA (ct 23 in gill tissue-ct 26 in kidney tissue and gill swab) by qPCR.

Detection of KHV antibodies (indirect detection of KHV)

Aquarium with KHV infection at 15°C water temperature: While the sera obtained from rainbow trout were negative for anti-KHV antibodies on 1st and 7th dpi, on 29th dpi four of eight sera were positive for anti-KHV antibodies by ELISA (titres 1:100 to 1:400) and, interestingly, six of eight by SNT (titres 1:16 to 1:32) (Table 2). All sera obtained from carp contained KHV antibodies in large quantities as detected by ELISA but only in one serum sample were neutralizing antibodies detected by SNT (titre 1:64).

Aquarium with KHV infection at 20°C water temperature: All rainbow trout developed antibodies against KHV at this temperature as determined by ELISA (Table 2). However, at this temperature no neutralizing antibodies against KHV were detectable. Also all carp developed antibodies against KHV as detected by ELISA. Interestingly those titres were much lower compared to the aquarium with 15°C. Additionally, in three of the four sera from the remaining carp, neutralizing antibodies were found by SNT (titres between 1:64 to 1:128) (Table 2). In the serum from the carp that had died of KHVD, no antibody was found neither by ELISA nor by SNT.

All sera were tested with both secondary antibodies, anti-carp-IgM and anti-trout-IgM. No cross-reaction occurred by ELISA when carp sera were used for the rainbow trout ELISA with anti-trout IgM antibodies nor when trout sera were used for the carp ELISA with anti-carp IgM antibodies.

Discussion

The strong host specificity of Herpesviruses may be dependent on the infection and disease that they cause. This is indeed the case for most members of the family Alloherpesviridae, but based on the results of the current study, the Alloherpesviridae appear to deviate from this rule. While the diseases induced by members of the Alloherpesviridae occur in most cases in only a single species, infection with the virus in the absence of clinical signs has been be detected in many more species of the same family, e.g. Cyprinidae [23], but also in other fish families, e.g. acipenseridae [26] or percidae [26,27]. Earlier studies on KHV led [37] and [38] to conclude that KHVD cannot be transferred to fish other than *C. carpio*. The virus was therefore defined as host specific. The fact that the disease has never been found within other fish than *C. carpio*, namely in common carp and koi, did not detract from the

Species		dpi	15°C						20°C					
			qPCR (ct)			Semi-nested PCR			qPCR (ct)			Semi-nested PCR		
	Fish		s[3]	L[4]	o[5]	s	L	o	s	L	o	s	L	o
rt[1]	1-5[2]	1st	-	-	-	-	-	-	-	-	-	-	-	-
	1	7th	-	30	28	+	+	+	-	25	28	+	+	+
	2		-	31	29	+	+	+	-	24	28	+	+	+
	1	29th	33	-	32	+	-	+	27	32	31	+	+	+
	2		31	-	-	+	-	-	-	-	33	+	-	+
	3		29	nt[6]	33	+	nt	+	29	nt	30	+	nt	+
	4		29	nt	31	+	nt	+	28	nt	31	+	nt	+
	5		32	nt	-	+	nt	-	29	nt	29	+	nt	+
	6		29	nt	30	+	nt	+	29	nt	31	+	nt	-
	7		28	nt	30	+	nt	+	-	nt	-	-	nt	-
	8		32	nt	-	+	nt	-	30	nt	33	+	nt	+
Carp	1-5[2]	1st	-	-	-	-	-	-	-	-	-	-	-	-
	1	21st	29	-	32	+	-	+	-	-	-	-	-	+
	2		-	-	-	-	-	-	-	28	28	-	+	+
	3		29	-	30	+	-	+	31	29	29	+	+	+
	4		31	-	-	+	-	-	30	28	28	+	+	+
	5		28	-	-	+	-	-	-	-	-	-	-	-

[1]rt: Rainbow trout; [2]Tested before immersion with KHV; [3]s: Gill swab; [4]L: Leukocyte separations; [5]o: Organ tissues; [6]nt: Not tested.

Table 1: Direct detection of KHV DNA by qPCR and semi-nested PCR in samples from rainbow trout and carp [31,32].

Species		dpi	15°C		20°C	
	Fish		SNT[3/4]	ELISA[5/6]	SNT[3/4]	ELISA[5/6]
rt[1]	1-5[2]	1st	-	-	-	-
	1	7th	-	-	-	-
	2		-	-	-	-
	1	29th	1:16	1:400	-	1:800
	2		1:16	1:400	-	1:200
	3		1:16	1:200	-	1:200
	4		-	-	-	1:200
	5		1:16	-	-	1:100
	6		1:16	-	-	1:200
	7		-	-	-	-
	8		1:16	1:100	-	1:200
Carp	1-5[2]	1st	-	-	-	-
	1	21st	-	1:4.800	1:64	1:1.200
	2		-	1:4.800	1:128	1:300
	3		1:64	1:9.600	1:64	1:600
	4		-	1:4.800	-	1:300
	5		-	1:2.400		

[1]rt: Rainbow trout; [2]Tested before immersion with KHV; [3]SNT basic dilution rainbow trout sera 1:4; [4]SNT basic dilution carp sera 1:32; [5]Basic dilution of rainbow trout sera 1:100 and anti-trout-IgM mab 4C10 (FLI Insel Riems) used as secondary antibody for ELISA; [6]Basic dilution of carp sera 1:300 and anti-carp-IgM mab F 16 (Aquatic Diagnostics) used as secondary antibody for ELISA.

Table 2: Indirect detection of KHV by ELISA and SNT for antibody detection in serum samples from rainbow trout and carp.

possibility of virus replication in other fish species and investigations were not conducted to investigate the role of KHVD transmission through vector fish species, i.e., its transfer as an infectious agent to naive carp detected by a combination of virological and serological methods other than in carp or koi [38]. It has since been found that a lot of other fish species such as goldfish, crucian carp, different sturgeon species, grass carp (*Ctenopharyngodon idella*), silver carp (*Hypophthalmichthys molitrix*) or tench (*Tinca tinca*) may be act as a healthy appearing carrier hosts which can transfer the KHV to naive carp or koi [22,24,26,27,39]. KHVD has never occurred in any of those fish species when inoculated with KHV. In contrast, carp hybrids with goldfish and crucian carp were also found to be susceptible for the virus infection but expressed a severe KHVD [31]. In all those studies the results following virological examination has never been compared with the serological/antibody response against the virus. The first indication of vector-facilitated KHVD was apparent when fish such as tench or roach (*Rutilus rutilus*) from ponds with KHVD outbreaks, occurring in cohabitated common carp, were tested for the presence of antibodies against KHV by SNT. The antibody titres against this virus ranged between 1:64 to 1:128 (Fichtner, pers. comm.).

In this study rainbow trout was selected as a model member of the family salmonidae to determine it is susceptibility to KHV infection and to investigate for the first time the potential for trout to transfer infectious virus to common carp by virological and serological methods. After immersion KHV was detected after 7 dpi inside but not outside the rainbow trout (leukocytes and gill swabs). Furthermore, the faeces were tested every 7 days for the presence of KHV by qPCR, but were always negative. Whilst at 15°C, the most suitable temperature for rainbow trout which is also tolerated by carp, rainbow trout developed antibodies against KHV after 29 dpi. Amazingly, higher titres were observed by SNT compared to antibody ELISA. This phenomenon was previously found in carp sera in the field where sera samples always contained neutralizing antibodies against KHV, but a lack of detectable antibodies by ELISA, even where the sera was strongly SNT positive. This also was reported in Italy where the same immunological difference was observed, but in sera obtained from mirror carp and scaled carp where the scaled carp sera reacted positive in SNT but not by ELISA (Vendramin, pers. comm.). To determine the underlying factors behind this phenomenon basic immunological research is required. Alternatively, it is possible that the immunological differences are influenced by temperature and/or the genetic background of the fish. At 15°C water temperature carp do not normally exhibit a serologically strong response due to the fact that their optimum water temperatures, especially under European conditions, is between 18 and 22°C. This was confirmed by the fact that all carp sera reacted positive by ELISA but only one by SNT, as at this sub-optimal temperature, limited neutralizing antibodies were produced.

Nonetheless, it was concluded from the study that at 15°C, a water temperature suitable for salmonids, that rainbow trout are not only able to replicate KHV, provoking a serological and immunological reaction at this temperature, but they are also capable of transferring it without clinical signs to carp. This was confirmed by the transfer of infectious KHV from rainbow trout to carp at 20°C where clinical symptoms with mortality in carp were induced. Temperatures of 20°C may be tolerated by rainbow trout, where they are also able to react serologically against KHV. This was detected by ELISA where all rainbow trout sera contained antibodies against the virus. In the same sera no neutralizing antibodies were found by SNT. This confirmed the earlier hypotheses that rainbow trout may not produce an optimal antibody response above 17°C, which was in fact around 2-fold lower. In the challenge experiment at 20°C rainbow trout developed antibodies detectable only by ELISA. In contrast, to the observed responses of rainbow trout, the majority of carp developed neutralizing antibodies and other antibodies detected by ELISA and SNT following the disappearance of clinical signs. Carp do not develop antibodies when there is a productive infection with KHV ensuing, which was evident by the negative reaction found with carp serum from the animal that had died of KHVD. Moreover, those reactions were confirmed with 20 sera from carp that displayed severe clinical signs after infection with KHV. No antibodies against KHV were found on the 7th and 14th dpi, neither by ELISA nor by SNT. After an additional 14 days, all survivors had produced antibodies against KHV, as indicated by SNT and ELISA at 22°C water temperature (Fichtner and Bergmann, pers. comm).

In conclusion, from the experiments presented in the current study, the previous observations reported, and the assessment of the literature, it has to be stated that the majority of the members of the family Alloherpesviridae are not species-specific in terms of infection, but are specific with regards to disease. Additionally, it has been demonstrated that rainbow trout is able to replicate KHV and transmit the virus to naive carp at different temperatures. This information will be vital for controlling outbreaks of KHVD. Further investigations on immunogenesis after infection with KHV are urgently needed.

Acknowledgement

The authors thank the invaluable technical assistance of Irina Werner (FLI). This study was funded and conducted as part of the EU project Epizone "WP 4.5. KHV-Sero" and the EU project "Moltraq" (EMIDA-ERA-net).

References

1. Miwa S, Ito T, Sano M (2007) Morphogenesis of koi herpesvirus observed by electron microscopy. J Fish Dis 30: 715-722.
2. Miyazaki T, Yoshitaka K, Shinya Y, Masahiro Y, Tatsuya K (2008) Histopathological and ultrastructural features of koi herpesvirus (KHV)-infected carp *Cyprinus carpio*, and the morphology and morphogenesis of KHV. Dis Aqua Org 80: 1-11.
3. Waltzek TB, Kelley GO, Alfaro ME, Kurobe T, Davison AJ, et al. (2009) Phylogenetic relationships in the family Alloherpesviridae. Dis Aqua Org 84: 179-194.
4. Lunger PD, Darlington RW, Granoff A (1965) Cell-virus relationships in the lucké renal adenocarcinoma: An ultrastructure study. Annual New York Acad Sci 126: 289-314.
5. Bennati R, Bonetti M, Lavazza A, Gelmetti D (1994) Skin lesions associated with herpesvirus-like particles in frogs (Rana dalmatina). Vet Rec 135: 625-626.
6. Kurobe T, Kelley GO, Waltzek TB, Hedrick RP (2008) Revised Phylogenetic Relationships among Herpesviruses Isolated from Sturgeons. J Aquat An Health 20: 96-102.
7. Hua YP, Wang D (2005) A review on sturgeon virosis. J For Res 16: 79-82.
8. Silverstein PS, Bosworth BG, Gaunt PS (2008) Differential susceptibility of blue catfish, Ictalurus furcatus (Valenciennes), channel catfish, I. punctatus (Rafinesque), and blue channel catfish hybrids to channel catfish virus. J Fish Dis 31: 77-79.
9. Goodwin AE, Marecaux E (2010) Validation of a qPCR assay for the detection of Ictalurid herpesvirus-2 (IcHV-2) in fish tissues and cell culture supernatants. J Fish Dis 33: 341-346.
10. Davison A (1998) The Genome of Salmonid Herpesvirus. J Virol 72: 1974-1982.
11. McAllister PE, Herman RL (1989) Epizootic mortality in hatchery-reared lake trout salvelinus namaycush caused by a putative virus possibly of the herpesvirus group. Dis Aquat Org 6: 113-119.
12. Bradley TM, Medina DJ, Chang PW, McClain J (1989) Epizootic epitheliotropic disease of lake trout (salvelinus namaycush): History and viral etiology. Dis Aquat Org 7: 195-201.
13. Sano M, Fukuda H, Sano T (1990) Isolation and characterization of a new herpesvirus from eel. In: Perkins TO, Cheng TC, Pathology in marine science, Academic Press, New York pp: 15-31.
14. Kempter J, Hofsoe P, Panicz R, Bergmann SM (2013) First detection of anguillid herpesvirus (AngHV-1) in European eel (Anguilla anguilla) and imported American eel (Anguilla rostrata) in Poland. Bull Eur Ass Fish Pathol 34: 87-94.
15. Ueno Y, Kitao T, Chen SN, Aoki T, Kou FH (1992) Characterization of a Herpes-like virus isolated from cultured Japanese eels in Taiwan. Fish Pathol 27: 7-17.
16. McAllister PE, Lidgerding BC, Herman RL, Hoyer LC, Hankins J (1985) Viral diseases of fish: first report of carp pox in golden ide (Leuciscus idus) in North America. J Wildl Dis 21: 199-204.
17. Carp pox, Wikivet.
18. Jung SJ, Miyazaki T (1995) Herpesviral haematopoietic necrosis of goldfish, Carassius auratus (L) J Fish Dis 18: 211-220.
19. Goodwin AE, Merry GE, Sadler J (2006) Detection of the herpesviral hematopoietic necrosis disease agent (Cyprinid herpesvirus 2) in moribund and healthy goldfish: Validation of a quantitative PCR diagnostic method. Dis Aquat Org 69: 137-143.
20. Fichi G, Cardeti G, Cocumelli C, Toffan A, Eleni C, et al. (2013) Detection of Cyprinid herpesvirus 2 in association with an Aeromonas sobria infection of Carassius carassius (L.), in Italy. J Fish Dis 10: 823-830.
21. Danek T, Kalous L, Veselý T, Krásová E, Reschová S, et al. (2012) Massive mortality of Prussian carp Carassius gibelio in the upper Elbe basin associated with herpesviral hematopoietic necrosis (CyHV-2). Dis Aquat Org 102: 87-95.
22. Kempter J, Sadowski J, Schütze H, Fischer U, Dauber M, et al. (2009) Koi herpes virus: do acipenserid restitution programs pose a threat to carp farms in the disease-free zones? A Icht et Pisca 39: 119-126.
23. Bergmann SM, Schütze H, Fischer U, Fichtner D, Riechardt M, et al. (2009) Detection KHV genome in apparently health fish. Bull Eur Ass Fish Pathol 29: 145-150.
24. Bergmann SM, Lutze P, Schütze H, Fischer U, Dauber M, et al. (2010a) Goldfish (Carassius auratus auratus) is a susceptible species for koi herpesvirus (KHV) but not for KHV disease (KHVD). Bull Eur Ass Fish Pathol 30: 74-84.
25. Bergmann SM, Sadowski J, Kielpinski M, Bartlomiejczyk M, Fichtner D, et al. (2010b) Susceptibility of koi x crucian carp and koi x goldfish hybrids to koi herpesvirus (KHV) and the development of KHV disease (KHVD). J Fish Dis 33: 267-270.
26. Kempter J, Kielpinski M, Panicz R, Sadowski J, Mysłowski B, et al. (2012) Horizontal transmission of koi herpes virus (KHV) from potential vector species to common carp. Bull Eur Ass Fish Pathol 32: 212-219.
27. Fabian M, Baumer A, Steinhagen D (2012) Do wild fish species contribute to the transmission of koi herpesvirus to carp in hatchery ponds? J Fish Dis 36: 505-514.
28. Kielpinski M, Kempter J, Panicz R, Sadowski J, Schütze H, et al. (2010) Detection of KHV in freshwater mussels and Crustaceans from ponds with KHV history in Common Carp (Cyprinus carpio). Is J Aqua Bamidgeh 62: 28-37.
29. Miller TA, Rapp J, Wastlhuber U, Hoffmann RW, Enzmann PJ (1998) Rapid and sensitive reverse transcriptasepolymerase chain reaction based detection and differential diagnosis of fish pathogenic rhabdoviruses in organ samples and cultured cells. Dis Aquat Org 34: 13-20.
30. Engelsma MY, Way K, Dodge MJ, Voorbergen-Laarman M, Panzarin V, et al. (2013) Detection of novel strains of cyprinid herpesvirus closely related to koi herpesvirus. Dis Aquat Org 107: 113-120.
31. Bergmann SM, Riechardt M, Fichtner D, Lee P, Kempter J (2010) Investigation on the diagnostic sensitivity of molecular tools used for detection of koi herpesvirus. J Virol Meth 163: 229-233.
32. Gilad O, Yun S, Zagmutt-Vergara FJ, Leutenegger CM, Bercovier H, et al. (2004) Concentrations of a koi herpesvirus (KHV) in tissues of experimentally infected Cyprinus carpio koi as assessed by real-time TaqMan PCR. Diseases of Aquatic Organisms 60: 179-187.
33. Mayr A, Bachmann PA, Bibrack B, Wittmann G (1974) Quantitative Bestimmung der Virusinfektiosität (Virustitration). In: Virologische Arbeitsmethoden, Band I (Zellkulturen - Bebrütete Hühnereier-Versuchstiere).
34. Adkison MA, Gilad O, Hedrick RP (2005) An Enzyme Linked Immunosorbent Assay (ELISA) for Detection of Antibodies to the Koi Herpesvirus (KHV) in the serum of Koi Cyprinus carpio. Fish Pathol 40: 53-62.
35. St-Hilaire S, Beevers N, Joiner C, Hedrick RP, Way K (2009) Antibody response of two populations of common carp, Cyprinus carpio L, exposed to koi herpesvirus. J Fish Dis 67: 15-23.
36. Gray WL, Mullis L, LaPatra SE, Groff JM, Goodwin A (2002): Detection of koi herpesvirus DNA in tissues of infected fish. J Fish Dis 25: 171-178.
37. Perelberg A, Smirnov M, Hutoran M, Diamant A, Bejerano Y, et al. (2003) Epidemiological description of a new viral disease afflicting cultured Cyprinus carpio in Israel. Isr J Aqua-Bamidgeh 55: 5-12.
38. Ronen A, Perelberg A, Abramowitz J, Hutoran M, Tinman S, et al. (2003) Efficient vaccine against the virus causing a lethal disease in cultured Cyprinus carpio. Vaccine 21: 4677-4684.
39. Meyer K (2007) Asymptomatic carriers as asymptomatic carriers as spreader of koi herpesvirus.

13

Effects of Different Prey and Rearing Densities on Growth and Survival of *Octopus* Maya Hatchlings

Carlos Rosas[1]*, Maite Mascaró[1], Richard Mena[1], Claudia Caamal-Monsreal[1] and Pedro Domingues[2]

[1]*Unit Multidisciplinary Teaching and Research, Faculty of Sciences UNAM, Puerto de Abrigo s/n Sisal Yucatán, México*
[2]*Spanish Institute of Oceanography, Oceanographic Centre of Vigo, Stay out, Canido, 36390 VIGO, Spain*

Abstract

The present study aims to determining the isolated and combined effects of stocking and prey densities on growth and survival of *Octopus maya* hatchlings both at experimental level and in a pilot scale system (8 m^2; 2700 L). Octopus survival was not related to prey density. Gained wet weight resulted in a significant interaction between initial stocking density and prey density indicating that octopus growth under low and high density was affected in a different manner depending on the density in which prey were offered. Prey density did not have a significant effect on growth and octopus fed with all three prey densities gained wet weight in a similar way. Results indicate the use of culture densities of 140 octopus m^{-2}, and at least 0.27 g prey $octopus^{-1}$ d^{-1} can be used to cultivate octopuses in small tanks. In tanks of $8m^2$ a higher growth rate was obtained with both 25 and 50 octopus m^{-2} densities were used. Survival was not affected by stocking density between 25 to 75 octopus m^{-2}.

Keywords: Cephalopods; Cannibalism; Culture densities; Growth rate; *Octopus maya;* Prey densities

Introduction

The Yucatan octopus, *Octopus maya* [1], is an endemic species from the Yucatan peninsula. Distribution ranges from Ciudad del Carmen, in the North of the Yucatan peninsula, to Isla Mujeres [2]. It has direct embryonic development with high hatchling survival [3], and easily adapts to laboratory conditions, accepting dead prey or prepared diets immediately after hatching [4,5]. *O. maya* has been cultured in the laboratory [6-9] up to four [7] or five [10] generations. The Yucatan octopus is characterized by fast growth rates (up to 8% body weight, [BW] d^{-1}), high feeding rates and food conversions that vary between 30% and 60% [4,10]. This species grows to 1 kg in 4 months at 25ºC, attaining maximum weight (>3 Kg) at 9 months old [10].

The determination of optimum culture densities is a major aspect for the optimization of large scale culture of cephalopods [11,12]. There is information reported on density studies for several cephalopod species, but none for *O. maya.*

It is known that both juvenile and adult *Sepia officinalis* tolerate high stocking densities well [13,14], and juvenile *O. vulgaris* also tolerates high densities, with similar growth rates at lower and higher stocking densities, although a higher mortality was observed in groups maintained at high density [11]. Although density studies have been conducted for cephalopods, [12-15], the combined effects of culture and prey density have never been analyzed

The effects of prey density on cephalopod growth and behaviour has been poorly studied and understood, especially when using hatchlings. To our knowledge, a pioneering investigation by Borer [16] was partly devoted to study the functional response of adult octopus (*Octopus briareus*), while Koueta & Boucaud-Camou [17] reported data on the relationship between the amounts of food offered and ingested in *S. officinalis* juveniles. Nevertheless, prey density is a key factor in optimizing growth and survival. In fact, an interval of prey densities for which density-dependent prey mortality is probable, was reported by Marquez et al. [18] when studying the functional response curve of *S. officinalis* hatchlings preying on live mysids. Rearing density studies for the cuttlefish were also reported by Correia, Domingues, Sykes & Andrade [19]. The effects of prey density were reported for *O. vulgaris* paralarvae [20] while growth and prey consumption by *S. officinalis* fed different densities of live mysid shrimp was reported by Sykes [14].

During the first 15 days after hatch, O. maya hatchlings pass through a transition process characterized by physiological and morphological changes associated with the maturity of the digestive gland, usc of remaining yolk and changes in proportion of arms related with total length [21]. This period was also identified as a "no net growth" on squid [22]. For that reason the pilot scale culture of O. maya has been divided in two parts: pre-fattening and fattening, the first in hatchery/ nursery facilities and the second in outside ponds [23]. During the pre-fattening period octopuses are maintained in 8 m^2 ponds for 60 days, where they are fed with crab paste and *Artemia* adults.

The present study was divided in three experiments in an attempt to

i) Determine the isolated and combined effects of stocking and prey densities on growth and survival of O. maya hatchlings (Experiments I and II), and

ii) To evaluate the effect of stocking densities on growth and survival of O. maya hatchlings maintained in a pilot scale system (Experiment III). All the animals used in the experiments were hatchlings of 2 days old.

Material and Methods

The study was carried out in the Experimental Cephalopod Production Unit (EPHAPU) at the UMDI-UNAM, Sisal, Yucatan, Mexico, following the procedures of Rosas and Moguel [5,21] for

***Corresponding author:** Carlos Rosas, Unit Multidisciplinary Teaching and Research, Faculty of Sciences UNAM, Puerto de Abrigo s/n Sisal Yucatán, México E-mail: crv@ciencias.unam.mx

collecting and maintenance of egg-laying females, as well as rearing, and maintenance, of *Octopus maya* embryos. The octopuses were obtained from spawning of females copulated in controlled conditions [23].

Experiments I and II were carried out in tanks with seawater maintained at 26 ± 2°C; pH between 8.0 and 8.2; dissolved O_2>5.5 mg/L; and a natural photoperiod of 10-14 h light-darkness (300 ± 50 lux cm^{-2}). Live shrimp (*Palaemonetes sp.)* was used as prey, which were collected in a daily bases in ponds around the research facilities (Sisal, Yucatan, Mexico). Individual shrimp wet weight varied between 0.1 and 1.3 g with a mean individual value of 0.60 ± 0.01 g (n = 500). Food was provided daily at 10:00 h and 14:00 h, and uneaten live shrimp and shrimp leftovers were removed and weighed every day prior to the first morning feed, in order to determine feeding rates. Thorough tank cleaning was performed on a daily basis.

Experiment I: Effect of initial stocking density and prey density on survival and growth

A two-way factorial design was used to test the effect of two octopus stocking densities (140 and 280 octopuses m^{-2}, corresponding to 50 and 100 octopus $aquaria^{-1}$, respectively) in combination with three prey densities (2, 4 and 6 prey $octopus^{-1}$ d^{-1}, corresponding to 0.09, 0.18 and 0.27 g prey $octopus^{-1}$ d^{-1}, and 71, 141 and 226% of initial body weight, respectively) on octopus survival, growth and feeding rate. Three replicate tanks (70.4 L; 76 cm x 47.5 cm x 19.5 cm) with re-circulating sea water were randomly allocated in each of the 6 treatment combinations. A total of 1350 hatchlings of O. maya (2 days old) were used in this experiment (Table 1). Each tank had empty shells of adults *Melongena corona bispinosa* (3 shells per octopus) to reduce stress by providing shelter for experimental animals to hide. A total of 7100 g of shrimp were used as live prey throughout the duration of the experiment (30 days).

All animals were individually weighed using an Ohaus® semi micro scale (± 0.001 g) at the beginning and at the end of the experiment. Individual mean wet weight gained was calculated by subtracting the mean final weight of the mean initial weight in each tank.

Specific growth rate (SGR, % day^{-1}) was calculated as:

$$SGR, \% \ day^{-1} = [(Ln\ Wf - Ln\ Wi)/T] \times 100$$

Where Wf is the final weight (g), Wi is the initial weight (g), and T is time in days.

Survival in each tank was obtained as the difference between the number of octopus at the start and end of the experiment. Ingestion rate was calculated as a percentage of food provided (IR_{fpt}) and as a percentage of final biomass production (IR_{tbp}) obtained in each tank:

$$IR_{fpt} = (\text{Ingested food, g } tank^{-1}\ d^{-1}/\text{Food provided, g } tank^{-1}\ d^{-1}) \times 100$$

Using ingestion rate and biomass production per tank the food conversion index (FCI_t) was calculated as:

$$FCI_t, \% = \text{Total ingested food } (g\ tank^{-1}\ 30\ d^{-1}) / \text{Total octopus production } (g\ tank^{-1}\ 30\ d^{-1}) \times 100$$

The tank gross growth efficiency (GGE_t,%) was defined as the fraction or the percentage of food intake by animals in the tank for 30 d [24] that was converted into body mass in the tank (ΔBW_t) during experimental time:

$$GGE_t = \Delta BW_t/FI_t$$

Survival, %IR_{fpt}, FCI, %GGE and SGR (% day^{-1}) were analysed by means of two-way ANOVAs with octopus stocking density and prey density as the two main (fixed) factors (n = 3 replicates). An arc-sin transformation of data was performed prior to statistical analysis [25]. Comparisons amongst means were carried out using the Student - Newman - Keuls procedure for balanced designs. Statistical differences were assumed when a P<0.05 was obtained [26]. Growth rates were finally expressed as specific growth rate (SGR % bw d^{-1}) for 30 experimental days for comparative purposes. Statistical analysis was done using R software.

Experiment II: Effect of prey density on growth

A total of 30 hatchlings of O. maya (2 days old) were used in this experiment (Table 1). Animals were individually placed in 30 plastic 500 ml chambers (10 cm diameter) connected to re-circulating seawater. Each chamber had one empty shell of the mollusk *M. corona bispinosa* in order to provide shelter and reduce stress. Similar prey densities as for experiment I were tested (0.09, 0.18 and 0.27 g prey $octopus^{-1}$ d^{-1}), using n = 10 individual replicates in each treatment.

All animals were weighed at the beginning and at the end of the experiment, and individual wet weight gained was calculated. Specific growth rate (SGR, % day^{-1}) was calculated as:

$$SGR, \% \ day^{-1} = [(Ln\ Wf - Ln\ Wi)/T] \times 100$$

where Wf is the final weight (g), Wi is the initial weight (g), and T is time in days.

Ingestion rates were calculated as percentage of food provided per animal (IR_{fpa}):

$$IR_{fpa} = (\text{Ingested food, g } animal^{-1}\ d^{-1}/\text{Food provided, g } animal^{-1}\ d^{-1}) \times 100$$

Using ingestion rate and biomass production per tank the food conversion index (FCI_i) was calculated as:

$$FCI_i, \% = \text{Total ingested food } (g\ animal^{-1}\ 30\ d^{-1}) / \text{Total octopus growth } (g\ animal^{-1}\ 30\ d^{-1}) \times 100$$

The individual gross growth efficiency (GGE_i, %) is defined as the fraction or the percentage of individual food intake (FI_i: g $animal^{-1}$ d^{-1}) that is converted into body mass and can be expressed as:

$$GGE_i = \Delta BW_i/FI_i$$

where ΔBW_i is individual growth (g $animal^{-1}$ d^{-1}).

One-way ANOVAs [25] were performed to determine differences in gained weight and food ingestion rates. Mean values were then compared using Tukey's HSD test for unequal samples. Individual growth rates were, here again, expressed as SGR (%bw d^{-1}) for 30 experimental days. Statistical differences were assumed when a P<0.05 was obtained.

Experiment I		Prey density (g $octopus^{-1}$ d^{-1})		
		0.09	0.18	0.27
Stocking density(number of hatchlings m^{-2})	140	0.136 ± 0.012; n = 150	0.130 ± 0.010; n = 150	0.124 ± 0.018; n = 150
	280	0.134 ± 0.051; n = 300	0.124 ± 0.021; n = 300	0.115 ± 0.015; n = 300
Experiment II		0.124 ± 0.004; n = 10	0.125 ± 0.010; n = 10	0.125 ± 0.006; n = 9

Table 1: Initial wet weight of Octopus maya hatchlings (2 days old) that were randomly allocated to different stocking and prey density treatments in experiments I and II. Values are mean ± standard deviation; n is the number of weighed individuals.

Experiment III: Effect of stocking densities on growth and survival of O. maya hatchlings maintained in a pilot scale system

A total of 4800 hatchlings of O. maya (2 days old) with similar initial weight as for experiment I (0.13 ± 0.01 g, n=500) were used in this experiment. Animals were randomly placed in ten 8 m^2 black tanks (2,400 L; 2 × 4 × 0.3 m) under the following stocking densities: 3 tanks with 25, 50, and 75 octopuses m^{-2} each, and one tank with a density of 150 octopuses m^{-2}. Each tank was connected to a re-circulatory sea-water system coupled to anthracite vertical filter and protein skimmer [23]. Each tank was provided with 3 *M. corona bispinosa* shell per animal. During the 30 days of the experiment, animals were fed *ad libitum* two times a day (09:00 and 18:00 h) with crab paste (95% liophylized crab meat plus 5% gelatin without flavor; [5], at ration of 150% of octopus wet weight. Also, *Artemia* adults (25 g wet weight per tank d^{-1}) were provided during the first 15 days [21]. *Artemia* were collected in salty ponds from the Celestun coastal lagoon, Yucatán, México. Before feeding (08:00 h), tanks were cleaned and remaining food removed using a siphon. Seawater in the tanks was maintained at 26 ± 2°C, 8 ± 0.5 pH, dissolved O_2>5.5 mg/L, nitrite<0.05 mg/L and ammonia<0.5 mg/L; A natural photoperiod of 10-14 h light-darkness, respectively was maintained during the experiment. At the end of the experiment, all animals were weighed and counted to calculate the final wet weight biomass and survival obtained in each experimental condition.

The relationship between biomass (g) of O. maya juveniles after 30 days of culture and initial density (number of octopuses m^{-2}) was explored using additive modelling (GAM). This was done after data exploratory analysis revealed a non-linear relationship between response and explanatory variables. Biomass data was previously square root transformed to comply with the Gaussian family of distribution curves. Exploration analysis used point graphs to identify extreme data; histograms and percentile graphs to assess normality; and X-Y graphs to verify linear relationships. Since there were only 4 values for the covariate (25, 50, 75 and 150 octopuses m^{-2}), the model with the maximum amount of smoothing used 3 degrees of freedom (d. f.), whereas, the model with the minimum amount of smoothing used d. f. = 2. The difference amongst these was assessed by means of AIC (Akaike Information Criteria = -2 log (Likelihood) +2df) and hypothesis testing using the *F* statistic. The model was validated by visual inspection of residuals [27].

Results

Experiment I:

The effect of octopus and prey density on survival and growth

Although during experiment I casual cannibalism was observed (but not quantified), hatchling survival was not related to prey density (pP>0.05) or initial stocking density (P>0.05) either through main terms or the interaction term (P>0.05), and hatchlings from all treatment combinations had similar survival (Figure 1). The two way ANOVA on SGR resulted in a significant interaction between initial stocking density and prey density (p<0.05), indicating that octopus growth under low and high density was affected in a different manner depending on the density in which prey were offered (Figure 2).

Mean growth rates expressed as SGR (%bw d^{-1}) for octopus in experiment I are shown in Table 2.

The two way ANOVA on ingestion rates showed that prey (p<0.05) and initial stocking density (p<0.05) independently affected the IR_{fpt} %, and no significant effect was found in the interaction (P>0.05) (Figure 3). Mean ingestion rate as percentage of food provided per tank (IR_{fpt}) showed a negative relationship to prey density, with higher values as prey density decreased for both low and high initial stocking densities (Figure 3A). Ingestion rate expressed in this manner was also higher in treatments amongst all treatments with high than with low octopus density (Figure 3B).

Food conversion index (FCI_t) varied with prey density, but did so in a different way depending on octopus initial density (p<0.01). Whilst FCI_t in all treatments with low stocking density were statistically similar amongst each other, values increased with prey density amongst those treatments with high initial octopus density (Figure 4A). In contrast a reduction on GGE_t, % was observed with octopus density, with high values on animals maintained at lower than higher densities (Figure 4B).

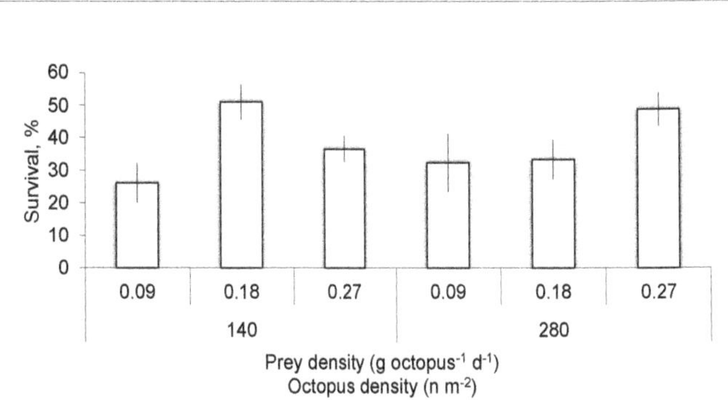

Figure 1: Survival (%) of two days old O. maya hatchlings cultured at two stocking densities (140 and 280 octopuses m^{-2}) and three prey densities (0.09, 0.18 and 0.27 g shrimp octopus^{-1} d^{-1}). Bars indicate standard deviations and letters indicate significant differences.

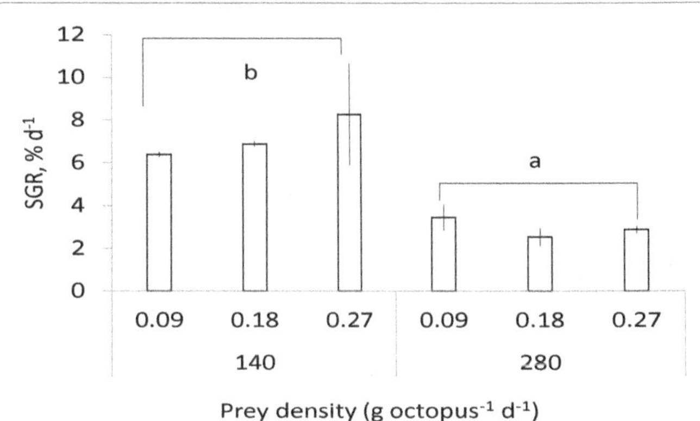

Figure 2: Specific growth rates (SGR, %bwd^{-1}) of two days old O. maya hatchlings cultured at two stocking densities (140 and 280 octopuses m^{-2}) and three prey densities (0.09, 0.18 and 0.27 g shrimp octopus^{-1} d^{-1}). Bars indicate standard deviations and letters indicate significant differences.

	Prey density (g octopus^{-1} day^{-1})		
Stocking density	**0.09**	**0.18**	**0.27**
140	6.39 ± 0.1	6.89 ± 0.1	8.26 ± 2.35
280	3.46 ± 0.58	2.54 ± 0.40	2.87 ± 0.17

Table 2: SGR (% bw d^{-1}) of Octopus maya hatchlings (2 days old) after 30 days in experiment I using different stocking (number of hatchlings m^{-2}) and prey density (number of prey octopus^{-1} d^{-1}) treatments. Values are mean ± SD; n = 3 in all cases.

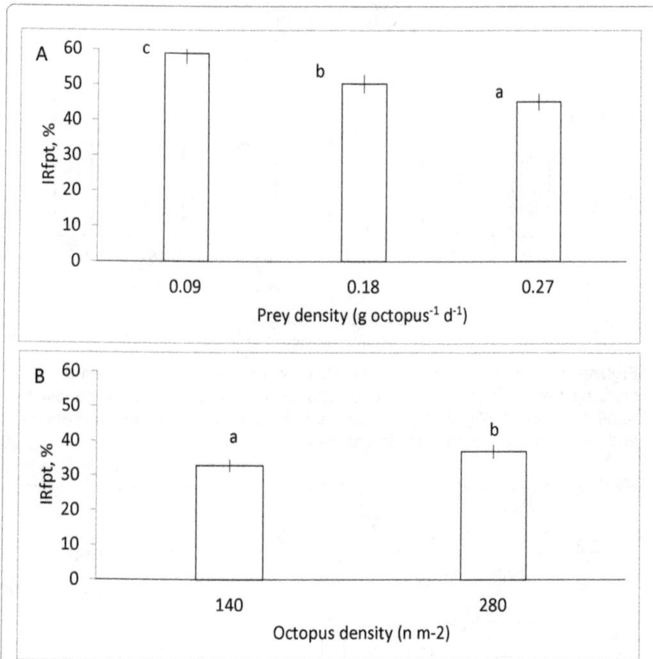

Figure 3: Ingestion rate as percentage of food ingested (IRfpt, %Fpt) of two days old *O. maya* hatchlings cultured for 30 days at two stocking densities (140 and 280 octopuses m^{-2}) and three prey densities (0.09, 0.18 and 0.27 g shrimp octopus^{-1} d^{-1}). Mean ± standard deviations; letters indicate significant differences P < 0.05.

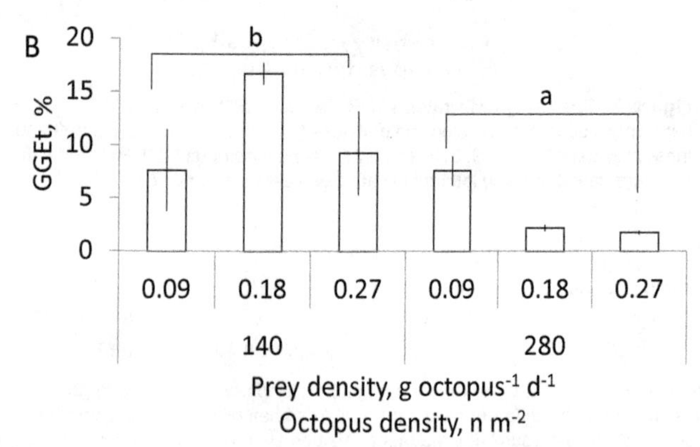

Figure 4: Food conversion efficiency (FCI$_t$) (A), and Gross growth efficiency (GGE$_t$%) (B), obtained for tank of two days old *O. maya* hatchlings cultured for 30 days at two stocking densities (140 and 280 octopuses m^{-2}) and three prey densities (0.09, 0.18 and 0.27 g shrimp octopus^{-1} d^{-1}). Mean ± standard deviations; letters indicate significant differences P<0.05.

Experiment II

The effect of prey density on growth of individualized *O. maya* hatchlings.

Prey density did not have a significant effect on growth (P>0.05), and octopus fed with all three prey densities gained wet weight in a similar way (Figure 5). Survival of individualized animals was 60, 80 and 67% for treatments with 0.09, 0.18 and 0.27 g prey octopus^{-1} d^{-1}.

Ingestion rate as percentage of food provided per animal (IR$_{fpa}$, %) showed a negative relation with prey density (p<0.001), with the highest ingestion rate at the lowest prey density (0.09 g prey octopus^{-1} d^{-1}), but statistically similar values for 0.18 and 0.27 g prey octopus^{-1} d^{-1} (Figure 6). The Food Conversion Index (FCI$_t$) and GGE$_t$, % did not show changes according to prey density (p>0.05) with a mean value of 11.6 ± 8.25 and 11.9 ± 6.06 for animals fed three prey densities, respectively (Figure 7).

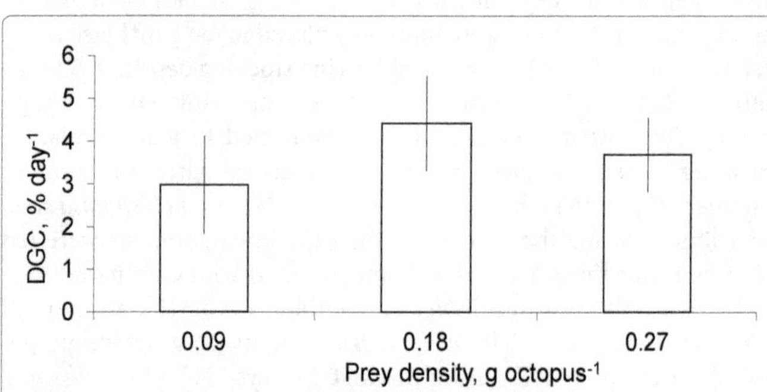

Figure 5: Growth rates (%bwd^{-1}) of two days old *O. maya* hatchlings individually cultured and fed three prey densities (0.09, 0.18 and 0.27 g shrimp octopus^{-1} d^{-1}). Mean ± standard deviations.

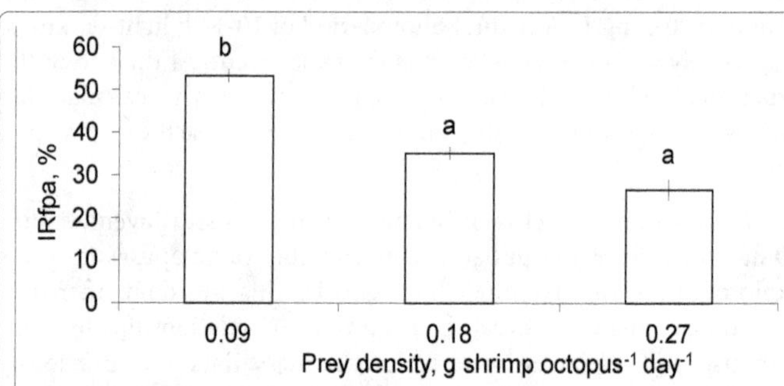

Figure 6: Ingestion rates as percentage of food provided per animal (IRfpa, %Fpa) of two days old *O. maya* hatchlings individually cultured for 30 days, and fed three prey densities (0.09, 0.18 and 0.27 g shrimp octopus-1 d-1). Mean ± standard deviations; letters indicate significant differences P < 0.05.

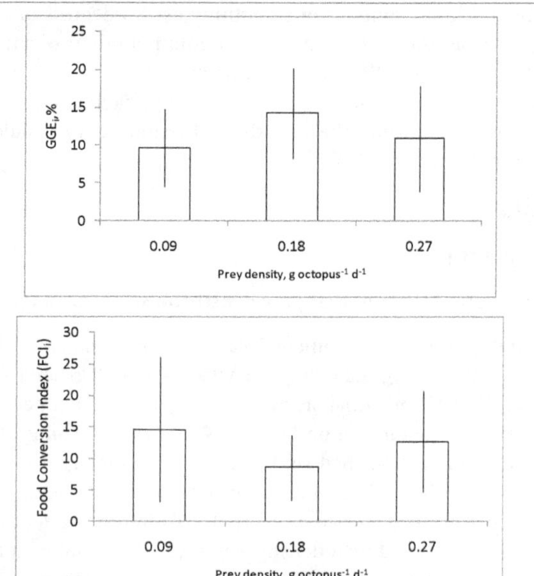

Figure 7: Ingestion rate as a percentage of final octopus body weight (IRbw, %bw d-1) of two days old O. maya hatchlings individually cultured for 30 days, and fed three prey densities (0.09, 0.18 and 0.27 g shrimp octopus-1 d-1). Mean ± standard deviations; letters indicate significant differences P<0.05.

Experiment III

Moderate cannibalism was observed during this experiment. Nevertheless, hatchling survival was similar amongst densities from 25 to 75 octopus's m^{-2} (mean survival: 58 ± 2.93%). Contrary, survival decreased in octopus stocked at 125 hatchlings m^{-2} (31%; Table 3). Final octopus weight ranged between 0.41 and 0.58 g, with low values for animals stocked at 125 octopuses m^{-2} and high values for animals maintained at 25 octopuses m^{-2}, however, the negative relationship between growth and stocking density was not linear. Density affected the SGR (% day^{-1}) values with high values in animals maintained in low density in comparison to that obtained in the highest density (Table 3).

The high variability on wet weight at the end of the experiment did not allow a conventional statistical analysis. For that reason, GAM was applied to analyze the effect of animal density on final wet weight obtained during the experiment (Figure 8). The model with d.f.=3 had a lower AIC value than that with d. f.=2 (-1549.0, -1538.8, respectively) and proved to explain significantly more variability in final biomass ($p<0.001$). The estimated variance in this model was $\sigma^2 = 0.01$ (n = 854), and only a 25.5% of the total deviance was explained, nevertheless indicating large amount of unexplained variation in the data. The smoothing term was significant ($p<0.001$) indicating an important effect of density on final body octopus weight (Figure 8). The estimated intercept in the model was 0.69 ± 0.003 (± standard error.) and it was significantly different from 0 ($p<0.001$). Thus, octopuses at an initial density of 50 octopuses m^{-2} are predicted to weigh 0.42 g after 30 days of culture, whereas those at an initial density of 25 octopuses m^{-2} are predicted to weigh of 0.59 g in the same period of time (Table 4) [27].

Animals m^{-2}					
	25	50		75	150
Initial wet weight, g	0.13 ± 0.02	0.13 ± 0.02		0.13 ± 0.02	0.13 ± 0.02
N[a]			500		
Final weight, g	0.58 ± 0.02	0.53 ± 0.19		0.4 ± 0.11	0.41 ± 0.18
N[b]	3	3		3	1
N[c]	150	150		150	150
Days	30	30		30	30
SGR, % day^{-1}	±	±		±	±
Survival, %	61 ± 5	55 ± 7		58 ± 9	31

a. Number of hatchlings weighed at the beginning of the experiment
b. Number of tanks of each experimental density
c. Number of octopuses weighed at the end of the experiment per treatment

Table 3: Effects of animal density on growth and survival of O. maya hatchings cultured during the first 30 days after hatch on $8m^2$ dark tanks. Values as mean ± SE ±.

Density Animal m^{-2}	Smoother value	Predicted biomass (g)
25	0.08	0.59
	0.06	0.56
	0.04	0.53
	0.02	0.50
	0	0.48
	-0.02	0.45
50	-0.04	0.42
	-0.06	0.40
	-0-08	0.37

Table 4: Predicted biomass (g) of juvenile O. maya after 30 days of culture at different initial density (ind. $/m^2$). Biomass values were estimated as: (smoother value + estimated intercept)^2.

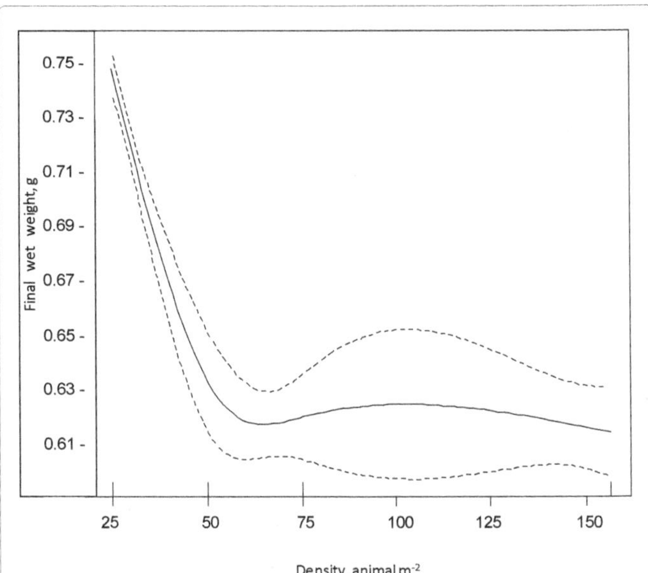

Figure 8: Additive modelling (GAM) curve of the final wet weight (g) in relation with stocking density of Octopus maya cultured in tanks of 8 m^2. Dashed lines indicate the interval of confidence of the model.

Discussion

Experiments I and II:

Results obtained in the present study showed that the relationship between mortality and growth rates had a different behavior depending on combinations between octopus and prey density; survival was not affected by prey or octopus density while a low growth was registered on animals maintained on high stocking density, independently of prey density.

Marquez *et al.* [18] suggests a type III functional response for cuttlefish hatchlings preying on live mysids, since a peak is reached and afterwards the increase in prey density did not lead to higher consumptions. This occurred in the present study when octopus density changed in animals cultivated in groups; animals maintained in low densities showed an increment on prey consumption, while animals cultured at high density showed an inverse reduction on prey consumption according with prey density increment. In individualized animals, ingestion rate increased constantly with prey density suggesting that, when octopuses had no school pressures, prey consumption increases in direct relation to prey density provoking higher growth rates.

During experiment I, casual cannibalism was observed although not quantified. Cannibalism has been reported for several cephalopod species such as *O. vulgaris* [28] and *Loligo vulgaris* [29]. The higher stocking densities increase the rates of encounters between hatchlings, which most likely increases cannibalism. In general, the behavior of the victim (as that of the aggressor) can be the reason for cannibalism and as behavior often is density-related, the rates of cannibalism are often related to the number of encounters [30]. Results from experiment I suggest that *O. maya* hatchlings do not support culture densities higher than 140 octopuses m^{-2}, even when ideal live prey is provided. Not only cannibalism is likely enhanced, but growth was markedly reduced when animals were cultivated at 280 octopuses m^{-2}, probably due to stress and competition for territory or food [31]. In fact, it was frequent to observe that in this high density, hatchlings chasing each other, or trying to steal a shrimp captured by other octopuses, which probably generated great stress and the loss of considerable amounts of energy. Boal et al. [32] studied the effects of crowding in *S. officinalis* with two different culture densities, and suggested that cuttlefish cultured at lower densities were less stressed, and because of this ate and grew more, while the ones cultured at higher densities showed signs of stress.

S. officinalis presents a much less cannibalistic behaviour, are very tolerant to high stocking densities [14,33], and most likely can be cultured at higher densities than O. maya during the first part of their life cycle. Although maximum densities of about 1000 cuttlefish m^{-2} for small juveniles (<3 cm mantle length, ML) were suggested by Forsythe et al. [34] it is necessary to consider differences between benthic and nektonic species; cuttlefish densities should consider the volume while octopus species should be considered in relation with the area of the benthic environment.

Results from the IR_{fpt}% clearly indicate that hatchlings cultured at 140 octopuses m^{-2} were less stressed and fulfilled their nutritional requirements before and more efficiently than those cultured at 280 octopuses m^{-2}. This is explained by the gradual decrease between all treatments in IR_{fpt}% with increasing prey density. Contrary to results obtained in the present study, Forsythe et al. [34] reported that the effects of culture density in juvenile (<1.5 g) *S. officinalis* (100 and 400 cuttlefish m^{-2}) had no statistical significance in food consumption or growth rates, but suggests that stocking densities of 400 cuttlefish m^{-2} may be approaching levels that affect feeding and growth. Warnke [35] reported that group reared cuttlefish ate more and grew larger (up to 9%) compared to others placed in isolation, while Dickel [36,37] indicated that *S. officinalis* reared in isolation, or in poor environments (e.g. simple tank with no substrate or hiding places) grew less than others placed in groups, or in enriched environments (e.g. tanks with shells, sand and shelters). Similarly, [38] indicated that *S. officinalis* can be cultured in isolation only for short-term experiments. The comparison between IR from experiments I and II in the present study does not support these findings, since octopuses placed in isolation ate more than those cultured in groups. Again, this could be due to the fact that *O. maya* is more territorial than *S. officinalis*, and therefore isolation provides less stress and the absence of competition for food or territory. Domingues et al. [39] also observed that isolated cuttlefish inked more and appeared to be more stressed than group cultured ones, although this was not reflected in growth or survival.

Experiment III:

Results obtained in small tanks suggest that for *O. maya* culture densities lower than 140 octopuses m^{-2} must be used during first 30 days. For that reason experiment III tested densities between 25 to 75 octopus m^{-2}. According to the GAM model used, a density between 25 to 50 octopuses m^{-2} is adequate to culture *O. maya* hatchlings on 8 m^2 tanks, since with such densities the higher final weight and survival were obtained. Between 25 and 75 octopus m^{-2}, survival was not affected, suggesting that in this stock density range cannibalism should not be an important factor. As During this part of the study hatchlings were fed crab paste and *Artemia* adults during first 15 days followed by other 15 days with only crab paste and using a ratio of 150% of octopus wet weight [5]. Apparently, those rations and diets covered the nutritional requirements of hatchlings cultured between 25 to 75 octopus m^{-2}, and can be recommended as culture conditions for *O. maya* juveniles in experimental pilot scale. Lower growth at higher densities was reported for the cuttlefish, *Sepia officinalis* [14], while Otero reported differences in food conversion and growth for juveniles of *Octopus vulgaris* stocked at 10 and 20 Kgm^{-3}, but reaching final culture densities up to 45.5 Kg m^{-3}, in a small experiment, at temperatures varying between 13 and 16ºC. Domingues et al. [12] also reports different food conversions for juvenile octopus, but reported no differences in growth rates with initial densities between 4 and 16 animals m^{-2}. These authors suggest that initial density should not be higher than 10 Kg m^{-3} for *O. vulgaris* juveniles.

The statistical model used here also showed the necessity to evaluate octopus growth rate in a different form, because in it was included variability. Variability of growth rates on cephalopods has been widely recognized and, as in the present study, must be considered in all studies where octopus culture needs to be applied [40-45]. Variability is the key of aquaculture, because through variability it is possible design process to enhance survival, reducing at the same time, heterogeneity of cultured population. Considering this, *O. maya* juveniles production has incorporated the variability evaluation every 30 days of culture to separate animals with different sizes, delivering higher survival and growth rates at pilot scale culture conditions.

Acknowledgement

The present study was partially financed by CONACYT through the project CB-2010-01 150810. Also thanks are given to DGAPA-UNAM Papitt program for financial support to CR on the project IN 212012. Thanks to SCPP Moluscos del Mayab for maintenance the octopuses on experimental pilot scale system.

References

1. Voss GL, Solís-Ramírez M (1966) Octopus maya, a new species from the Bay of Campeche, México. Bulletin of Marine Science 16: 615-625.
2. Solis M (1998) Aspectos biológicos del pulpo Octopus maya Voss y Solis. Contribuciones de Investigaci¢n Pesquera, Instituto Nacional de la Pesca 7: 1-38.
3. Rosas C, Gallardo P, Mascaró M, Caamal-Monsreal C, Pascual C (2014) Octopus maya. In: Cephalopod Culture. Edited by Iglesias J, et al. Dordrecht: Springer Science+Business Media: DOI 10.1007/1978-1094-1017-8648-1005_1020.
4. Domingues P, López N, Muooz JA, Maldonado T, Gaxiola G, Rosas C (2007) Effects of an artificial diet on growth and survival of the Yucatan octopus, Octopus maya. Aquaculture Nutrition 13: 1-9.
5. Rosas C, Tut J, Baeza J, Sánchez A, Sosa V, Pascual C, Arena L, Domingues P, Cuzon G (2008) Effect of type of binder on growth, digestibility and energetic balance of Octopus maya. Aquaculture 275: 291-297.
6. DeRusha R, Forsythe JW, DiMarco FP, Hanlon RT (1989) Alternative diets for maintaining and rearing cephalopods in captivity. Lab. Anim. Sci 39: 312.
7. Hanlon RT, Forsythe JW (1985) Advances in the laboratory culture of Octopuses for biomedical research. Laboratory Animal Science 35: 40.
8. Van Heukelem WF (1976) Growth, bioenergetics and life span of Octopus cyanea and Octopus maya. PhD Thesis, Univ. Hawaii.
9. Van Heukelem WF (1977) Laboratory maintenance, breeding , rearing and biomedical research potential of the Yucatan octopus (Octopus maya). Lab. Anim. Sci 27: 852-859.
10. Van Heukelem WF (1983) Octopus maya In: Cephalopod Life Cycles. Academic Press, London 1: 311-323.
11. Forsythe JW, Van Heukelem WF (1987) Growth. In: Cephalopod life cycles. Edited by Boyle PR. Comparative Reviews. London: Academic Press 2: 135-155.
12. Domingues P, Garcia S, Garrido D (2010) Effects of three culture densities on growth and survival of Octopus vulgaris. Aquaculture International 18: 165-174.
13. Rosas C, Cuzon G, Pascual C, Gaxiola G, López N, Maldonado T, Domingues P (2007) Energy balance of Octopus maya fed crab and artificial diet. Marine Biology 152: 371-378.
14. Sykes A, Domingues P, Andrade JP (2014) Sepia officinalis In: Cephalopod Culture. Edited by Iglesias J, Fuentes L, Villanueva R. Dordrecht: Springer Science + Busines Media 175-205.
15. Villanueva R, Sykes A, Vidal AT, Rosas C, Nabhitabhata J, Fuentes L, Iglesias J (2014) Current Status and Future Challenges in Cephalopod Culture. In:

Cephalopod culture. Edited by Iglesias J, Fuentes L, Villanueva R. New York: Springer Science+Business Media: 479-489.

16. Borer KT, Lane CE (1971) Oxygen requirements of Octopus briareus Robson at different temperatures and oxygen concentrations. J Exp Mar Biol Ecol 7: 263-269.
17. Koueta N, Boucaud-Camou E (1999) Food intake and growth in reared early juvenile cuttlefish *Sepia* officinalis *L.Mollusca* Cephalopoda. J Exp Mar Biol Ecol 240: 93-109.
18. Marquez L, Caballos M, Domingues P (2007) Functional response of early stages of the cuttlefish Sepia officinalis preying on the mysid Mesopodopsis slabberi. Mar Biol Res 3: 462-467.
19. Correia M, Domingues P, Sykes A, Andrade JP (2005) Effects of culture density on growth and broodstock management of the cuttlefish, Sepia officinalis (Linnaeus, 1758). Aquaculture 245: 163-173.
20. Marquez L, Quintana D, Almansa E, Navas JI (2007) Effects of visual conditions and prey density on feeding kinetics of paralarvae of Octopus vulgaris from a laboratory spawning. J. Mollus. Stud 73: 117-121.
21. Domingues P, Márquez L (2010) Effects of culture density and bottom area on growth and survival of the cuttlefish Sepia officinalis (Linnaeus, 1758). Aquacult Int 18: 361-369.
22. Moguel C, Mascaró M, Avila-Poveda O, Caamal C, Sánchez A, Pascual C, Rosas C (2010) Morphological, physiological, and behavioural changes during post-hatching development of Octopus maya (Mollusca: Cephalopoda) with special focus on digestive system. Aquatic Biology 9: 35-48.
23. Vidal E, DiMarco FP, Wormuth JH, Lee PG (2002) Influence of temperature and food availability on survival, growth and yolk utilization in hatchling squid. Bull mar Sci 71: 915-931.
24. Wigglesworth JM, Griffith DRW (1994) Carbohydrate digestion in Penaeus monodon. Mar. Biol 120: 571-578.
25. Zar JH (1999) Biostatistical Analysis. New Jersey: Prentice-Hall.
26. Underwood AJ (1997) Experiments in ecology: Their logical design and interpretation using analysis of variance, Cambridge University Press, Cambridge.
27. Montgomery DC, Peck EA (1992) Introduction to Linear Regression Analysis (2nd edtn). New York: John Wiley.
28. Vaz-Pires P, Seixas P, Barbosa A (2004) Aquaculture potential of the common octopus Octopus vulgaris (Cuvier, 1797) a review. Aquaculture 238: 221-238.
29. Summer WC, McMahon JJ (1970) Survival of unfed squid, Loligo pealei in an aquarium. Biol Bull 138: 389-396.
30. Fox LR (1975) Cannibalism in natural populations. Ann Rev Ecol Syst 6: 87–106.
31. Ibañez C, Chong JV (2008) Feeding ecology of Enteroctopus megalocyathus (Cephalopoda: Octopodidae) in southern Chile. J Mar biol Ass UK 88:793-798.
32. Boal JG, Hylton RA, Gonzales SA, Hanlon RT (1999) Effects of crowding on the social behaviour of cuttlefish (Sepia officinalis). Contemporary Topics in Laboratory Animal Science 38: 49-55.
33. Forsythe JW, Lee G, Walsh L, Tara C (2002) Effect of crowding on growth of the European cuttlefish, Sepia officinalis Linnaeus, 1758 reared at two temperatures. J Exp Mar Biol Ecol 269: 175-183.
34. Forsythe JW, DeRusha RH, Hanlon RT (1994) Growth, reproduction and life span of Sepia officinalis (Cephalopoda: Mollusca) cultured through seven consecutive generations. J Zool 233: 175-192.
35. Warnke K (1994) Some aspects of social interaction during feeding in Sepia officinalis (Mollusca: Cephalopoda) hatched and reared in the laboratory. Vie et Millieu 44: 125-131.
36. Dickel L (1999) Effects de l'expérience précoce sur la maturation des capacités mnésiques au cours de l'ontogenèse post-emryonnaire ches la seiche (Sepia officinalis). Ann Fond Fyssen 14: 86-94.
37. Dickel L, Boal JG, Budelmann BU (2000) The effect of early experience on learning and memory in cuttlefish. Develop Psychobiol 36: 101-110.
38. DeRusha RH, Forsythe JW, DiMarco FP, Hanlon RT (1989) Alternative diets for maintaining and rearing cephalopods in captivity. Laboratory Animal Science 39: 306-312.
39. Domingues P, Poirer R, Dickel L, Almansa E, Sykes A, Andrade JP (2003) Effects of culture density and live prey on the growth and survival of juvenile cuttlefish, Sepia officinalis. Aquacult Int 11: 225-242.
40. Pecl G, Steer MA, Hodgson K (2004) The role of hatchling size in generating the intrinsic size-at-age variability of cephalopods: extending the Forsythe hypothesis. Mar Freshw Res 55.
41. Doubleday Z, Semmens JM, Pecl G, Jackson G (2006) Assessing the validity of stylets as ageing tools in Octopus pallidus. J Exp Mar Biol Ecol 338: 35-42.
42. Leporati S, Pecl GT, Semmens JM (2007) Cephalopod hatchling growth: the effects of initial size and seasonal temperatures. Mar Biol 151: 1375-1383.
43. André J, Grist EPM, Semmens JM, Pecl G, Segawa S (2009) Effects of temperature on energetics and the growth pattern of benthic octopuses. J Exp Mar Biol Ecol 374: 167-179.
44. Briceño-Jacques F, Mascaró M, Rosas C (2010) GLMM-based modelling of growth in juvenile Octopus maya siblings: does growth depend on initial size? ICES J Mar Sci 67: 1509-1516.
45. Briceño-Jacques F, Mascaró M, Rosas C (2010) Energy demand during exponential growth of Octopus maya: exploring the effect of age and weight. ICES J Mar Sci 67: 1501-1508.

Determination of Protein, Lipid and Carbohydrate Contents of Conventional and Non-Conventional Feed Items Used in Carp Polyculture Pond

Asadujjaman M[1], Shahangir Biswas[2], Manirujjaman M [3], Matiar Rahman[2], Hossain MA[1] and Islam MA[2]*

[1]*Department of Fisheries, University of Rajshahi, Rajshahi, Bangladesh*
[2]*Department of Biochemistry and Molecular Biology, University of Rajshahi, Rajshahi, Bangladesh*
[3]*Department of Biochemistry, Gonoshasthaya Samajvittik Medical College and Hospital, Savar, Dhaka-1344, Bangladesh*

Abstract

A study was conducted during April'2010-September'2010 with a view to compare the protein, lipid and carbohydrate contents in conventional and non-conventional feed items and to recommend suitable strategy in selecting feed item for the development of weed based fish farming in carp polyculture pond. The experiment was carried out at the Protein and Enzyme Research Laboratory, Department of Biochemistry and Molecular Biology, Rajshahi University, Rajshahi. Six different conventional and non-conventional fish feed items like rice bran, wheat bran, mustard oilcake, *Azolla*, grass and banana leaves were tested to determine the nutrient contents under 6 treatments as T_1, T_2, T_3, T_4, T_5 and T_6, respectively. In this study, nutrient contents (protein, lipid and carbohydrate) were monitored monthly. Significant variations ($P<0.05$) were found in the mean values of nutrient contents with different treatments of feed items but in case of same feed item no significant difference was found in the nutrient content at different months. Among the non-conventional feed items treatment T_4 (*Azolla*) varied more significantly ($P<0.05$) for the mean values of protein content. Findings indicated that *Azolla* was more nutritive and low cost effective diets for fish farming in Bangladesh.

Keywords: *Azolla,* Conventional and non-conventional feed; Carp polyculture; Bangladesh

Introduction

The technique of polyculture of fish is based on the concept of utilization of different trophic and spatial niches of a pond in order to obtain maximum fish production per unit area. Different compatible species of fish of different trophic and spatial niches are raised together in the same pond to utilize all sorts of natural food available in the pond [1]. Supplementary feed plays an important role in achieving higher fish production. Unfortunately lack of low cost supplementary feed is found as one of the major problems in aquaculture in Bangladesh [2]. Commercial fish feeds are not easily available and unaffordable to poor fish farmers in Bangladesh. Consequently, there is no regular organized supplementary feeding practice and the fish production is found as low as 0.5-1.5 t/ha/year [3]. It was thus considered necessary to look for cheaper and locally available materials as substitutes.

The optimal protein requirements of carp are affected by the nutritional value of the dietary protein and level of non-protein energy in the carp diet. When sufficient energy sources such as lipids and carbohydrates are available in the diet, most of the ingested protein goes to protein synthesis. Adult Indian major carps require 30% dietary protein for proper growth and survival. Lipids or fats are required as sources of energy and essential fatty acids, and serve as carriers for fat-soluble vitamins. The gross lipid requirement of Indian major carp is 7-8% of the diet, and young fish require relatively more fat and protein than adults. Carbohydrate is the least-expensive nutrient and also a less expensive energy source for carp. Indian major carp, being herbivorous/ omnivorous feeders, easily digest appreciable quantities of carbohydrates in their diets. A dietary level up to 30% carbohydrate does not affect the growth of carp and growth retardation and reduced feed efficiency are observed, however, when carbohydrate levels exceeded 35% of diet [4]. Fish culture is induced primarily by the need for increased protein supply. One of the most essential prerequisites for the successful management of fish culture programme is a comprehensive understanding of feeding [5]. The increase in cost and demand of feed protein from conventional sources necessitates fish culturists of the developing countries to incorporate cheap and locally available ingredients in fish feeds. Recently the utilization of aquatic plants having high food value are used to supplement fish food has taken a new dimension for producing the much required animal protein at low cost [6].

Aquatic macrophytes have been known to have potential food value. A perusal of the available literature shows that some of the aquatic weeds are highly nutritive and, therefore, one alternative solution to check the massive population of these weeds might be their utilization through incorporation as components of feedstuff for fish. In fact, significant effort has been directed towards evaluating the nutritive value of different non-conventional feed resources, including terrestrial and aquatic macrophytes, to formulate nutritionally balanced and cost-effective diets for fish and poultry [7-10]. Most of these nutritional studies are carried out abroad and no comprehensive studies are found in comparing the nutritional quality of both conventional and non-conventional feeds for fish farming in Bangladesh. However, before advocating the utilization of these aquatic weeds for supplementation of fish feeds, there is an urgent need to explore their nutritional quality, throughout the major culture season in ponds under carp polyculture system. Therefore, the present study aimed at evaluating the protein, lipid and carbohydrate content in conventional and non-conventional

***Corresponding author:** Mohammad Amirul Islam, Professor, Department of Biochemistry and Molecular Biology, University of Rajshahi, Rajshahi, Bangladesh E-mail: maislam06@gmail.com

feed items used for carp polyculture system in Bangladesh.

Materials and Methods

Duration and location of the study

The study was conducted for a period of six months from April 2010 to September 2010. Feed items were collected from the fish farming study site located at Alampur village under Kushtia district of Bangladesh. Whereas nutrient analysis was done at the Protein and Enzyme Research Laboratory under the Department of Bio-Chemistry and Molecular Biology, Rajshahi University, Rajshahi, Bangladesh.

Experimental design

The current experiment was carried out under six treatments of feed items each with three replications. The treatment assignments were designated as T_1, T_2, T_3, T_4, T_5 and T_6 for rice bran, wheat bran, mustard oilcake, grass and banana leaves, respectively. Conventional feed items (rice bran, wheat bran, mustard oilcake) were collected from local market during the experimental period. Non-conventional feed item like was collected from ponds adjacent to the research area whereas grass and banana leaf were collected from adjacent grass field and banana garden. Both conventional and non-conventional feed items were collected once a month for nutritional analysis during the experimental period.

Nutrient analysis of the collected samples

Total protein, total lipid and total carbohydrate of the collected samples were determined by the micro-kjeldahl method [11,12] method and Anthrone method [13] respectively.

Statistical analysis

All the data were subjected to ANOVA (analysis of Variance) using computer software SPSS (Statistical Package of Social Science). The mean values were also compared to see the significant difference from the DMRT (Duncan Multiple range Test) [14].

Results

Monthly variations

Protein content significantly varied from 6.05 ± 0.45% with T_6 (banana leaf) at 6th month (September, 2010) to 31.20 ± 0.32% with treatment T_3 (mustard oilcake) at 2nd month (May, 2010). Lipid content significantly varied from 2.95 ± 0.21% with treatment T_6 (banana leaf) at 5th month (August, 2010) to 13.72 ± 0.36% with treatment T_3 (mustard oilcake) at 4th month (July, 2010). Carbohydrate significantly varied from 32.85 ± 0.14% with treatment T_3 (mustard oilcake) at 4th month (July, 2010) to 66.35 ± 0.32% with T_2 (wheat bran) at 3rd month (June, 2010). In the same feed item no significant difference in the nutrient content was found during the study period (Tables 1-6).

Mean variations

The variations in the mean values of nutrient contents (protein, lipid and carbohydrate) with different treatments of feed items are presented in Table 7 and Figure 1. Protein content significantly varied from 6.18 ± 0.13% with treatment T_6 (banana leaf) to 30.53 ± 0.40% with treatment T_3 (mustard oilcake). Lipid content significantly varied from 3.06 ± 0.09% with treatment T_6 (banana leaf) to 13.33 ± 0.10% with treatment T_3 (mustard oilcake). Carbohydrate significantly varied from 32.95 ± 0.29% with treatment T_3 (mustard oilcake) to 66.12 ± 0.47% with treatment T_2 (wheat bran).

Discussion

Monthly variations of the nutrient contents

Protein content varied from 6.05 ± 0.45% with (T_6 at 6th month) to 31.20 ± 0.32% (T_3 at 2nd month). Lipid content ranged from 2.95 ± 0.21% (T_6 at 5th month) to 13.72 ± 0.36% (T_3 at 4th month). Carbohydrate content ranged from 32.85 ± 0.14% (T_3 at 4th month) to 66.35 ± 0.32% (T_2 at 3rd month). Suresh and Mandal [3] worked on the determination of nutritive value of rice bran, mustard oil cake and *Azolla* for a period of 4 months from July to October. In rice bran they found crude protein and crude fibre as 12.6% and 21.9%, respectively. In mustard oilcake, crude protein and crude fibre was 38.6% and 6.8%, respectively and in *Azolla*, crude protein and crude fibred was 26.5% and 20.4%, respectively. Sithara and Kamalaveni [15] worked on the formulation of low cost fish feed using *Azolla* as a protein supplement during September to March and reported 20-25.5% protein in *Azolla*. Ebrahim, et al. [16] used *Azolla* as tilapia diet for a period of 90 days in summer season and reported 20% protein in *Azolla*. Fasakin and Balogan [17] worked on the nutritional aspects of *Azolla* in August, 1997 and reported 20.9% protein in *Azolla*. Present findings also indicated that in case of same feed item, no significant difference was found in the nutrient content at different months (Table 1 to 6). This might be due to no major change in the temperature was found to affect the growth and composition of *Azolla* during the study period. This statement was almost agreed with Lumpkin and Plucknett [18] who reported that change in *Azolla* composition was subjected to change in environment. Statement also agreed with Van-Hove et al. and Ebrahim et al. [7,19] who reported that change in *Azolla* composition was subjected to change in species.

Mean variation of the nutrient contents

In the present study the protein content varied from 6.18 ± 0.13% (T_6, banana leaf) to 30.53 ± 0.40% (T_3, mustard oilcake), lipid content varied from 3.06 ± 0.09% (T_6, banana leaf) to 13.33 ± 0.10% (T_3, mustard oilcake) and carbohydrate content varied from 32.95 ± 0.29% (T_3, mustard oilcake) to 66 .12 ± 0.47% (T_2, wheat bran). The highest protein and lipid content was found in treatment T_3 (mustard oilcake) whereas the highest carbohydrate content was found in treatment T_2, wheat bran (66.12 ± 0.47%) followed by T_4, *Azolla* (50.21 ± 0.54%), T_6, banana leaf (48.50 ± 0.51%), T_5, grass (46.36 ± 0.16%), T_1, rice bran (44.09 ± 0.67%), T_3, mustard oilcake (32.95 ± 0.29%). Hepher [20] reported the protein content of ricebran, wheat bran, oil cake and *Azolla* as 11.88%, 14.57%, 30-33% and 19.27%, respectively. Banerjee and Matai [21] determined the nutritive status of and reported protein as 21.9% and Lipid as 3.8%. Gavina [22] reported crude protein of 20.98%, crude fat of 5.17% and crude fiber of 19.30% in *Azolla*. Tavares [23] observed 38.8% crude protein, 3.8% crude fat and 13.2% crude fiber in dried duck weed. They also reported that the protein content of duckweeds growing on nutrient poor and nutrient rich water varied between 15-25% and 35-45% (Dry matter basis), respectively. In case of conventional feed items the major nutrient like protein varied from 14.40 ± 0.32% (rice bran) to 30.53 ± 0.40% (mustard oilcake). Whereas in case of non-conventional feed items the protein varied from 6.18 ± 0.13% (banana leaf) to 18.58 ± 0.09% (*Azolla*). Being an omnivore, the fish can also feed on vegetation [24] and may be able to assimilate *Azolla* in the diets.

The chemical composition of *Azolla* species varies with ecotypes

Nutrients	Months					
	April	May	June	July	August	September
Protein (%)	14.60 ± 0.22ᵃ	13.92 ± 0.19ᵃ	14.65 ± 0.19ᵃ	14.50 ± 0.36ᵃ	14.22 ± 0.28ᵃ	14.50 ± 0.24ᵃ
Lipid (%)	10.42 ± 0.31ᵃ	10.50 ± 0.25ᵃ	10.64 ± 0.25ᵃ	10.20 ± 0.21ᵃ	10.24 ± 0.15ᵃ	10.45 ± 0.26ᵃ
Carbohydrate (%)	44.25 ± 0.41ᵃ	43.72 ± 0.19ᵃ	43.85 ± 0.19ᵃ	44.20 ± 0.24ᵃ	44.32 ± 0.20ᵃ	44.20 ± 0.16ᵃ

Figures bearing common letter(s) in a row as superscript do not differ significantly (P<0.05)

Table 1: Monthly variations in nutrient (protein, lipid and carbohydrate) contents with treatment T_1 (Rice, *Oryza sativa* bran).

Nutrients	Months					
	April	May	June	July	August	September
Protein (%)	17.20 ± 0.05ᵃ	17.05 ± 0.12ᵃ	17.25 ± 0.12ᵃ	16.95 ± 0.24ᵃ	17.10 ± 0.34ᵃ	17.22 ± 0.18ᵃ
Lipid (%)	6.75 ± 0.41ᵃ	6.66 ± 0.69ᵃ	6.80 ± 0.69ᵃ	7.12 ± 0.46ᵃ	6.47 ± 0.32ᵃ	6.32 ± 0.38ᵃ
Carbohydrate (%)	66.20 ± 0.36ᵃ	65.75 ± 0.32ᵃ	66.35 ± 0.32ᵃ	66.32 ± 0.26ᵃ	66.12 ± 0.15ᵃ	65.99 ± 0.23ᵃ

Figures bearing common letter(s) in a row as superscript do not differ significantly (P <0.05)

Table 2: Monthly variations in nutrient (protein, lipid and carbohydrate) contents with treatment T_2 (Wheat, *Trticum aestivum* bran).

Nutrients	Months					
	April	May	June	July	August	September
Protein (%)	30.65 ± 0.18ᵃ	31.20 ± 0.32ᵃ	30.50 ± 0.32ᵃ	30.25 ± 0.15ᵃ	30.15 ± 0.11ᵃ	30.45 ± 0.17ᵃ
Lipid (%)	13.34 ± 0.31ᵃ	13.24 ± 0.47ᵃ	13.25 ± 0.47ᵃ	13.72 ± 0.36ᵃ	13.22 ± 0.18ᵃ	13.20 ± 0.19ᵃ
Carbohydrate (%)	32.86 ± 0.18ᵃ	32.90 ± 0.25ᵃ	33.10 ± 0.25ᵃ	32.85 ± 0.14ᵃ	32.98 ± 0.31ᵃ	33.02 ± 0.46ᵃ

Figures bearing common letter(s) in a row as superscript do not differ significantly (P<0.05)

Table 3: Monthly variations in nutrient (protein, lipid and carbohydrate) contents with treatment T_3 (Mustard, *Brassica napus* Oilcake).

Nutrients	Months					
	April	May	June	July	August	September
Protein (%)	18.65 ± 0.08ᵃ	18.45 ± 0.41ᵃ	18.35 ± 0.41ᵃ	18.45 ± 0.32ᵃ	18.75 ± 0.24ᵃ	18.80 ± 0.26ᵃ
Lipid (%)	3.25 ± 0.09ᵃ	3.15 ± 0.12ᵃ	3.12 ± 0.12ᵃ	3.35 ± 0.18ᵃ	3.14 ± 0.34ᵃ	3.10 ± 0.41ᵃ
Carbohydrate (%)	50.36 ± 0.75ᵃ	50.45 ± 0.61ᵃ	50.20 ± 0.61ᵃ	50.15 ± 0.54ᵃ	50.20 ± 0.17ᵃ	49.88 ± 0.27ᵃ

Figures bearing common letter(s) in a row as superscript do not differ significantly (P<0.05)

Table 4: Monthly variations in nutrient (protein, lipid and carbohydrate) contents with treatment T_4 (*Azolla pinnata*).

Nutrients	Months					
	April	May	June	July	August	September
Protein (%)	7.28 ± 0.35ᵃ	7.32 ± 0.25ᵃ	7.45 ± 0.25ᵃ	7.15 ± 0.14ᵃ	7.25 ± 0.19ᵃ	7.12 ± 0.23ᵃ
Lipid (%)	6.35 ± 0.05ᵃ	6.28 ± 0.06ᵃ	6.45 ± 0.06ᵃ	6.23 ± 0.12ᵃ	6.21 ± 0.18ᵃ	6.32 ± 0.28ᵃ
Carbohydrate (%)	46.58 ± 0.12ᵃ	46.30 ± 0.41ᵃ	45.95 ± 0.41ᵃ	46.85 ± 0.38ᵃ	46.70 ± 0.19ᵃ	45.76 ± 0.14ᵃ

Figures bearing common letter(s) in a row as superscript do not differ significantly (P<0.05)

Table 5: Monthly variations in nutrient (protein, lipid and carbohydrate) contents with treatment T_5 (Grass, *Cynodon dactylon*).

Nutrients	Months					
	April	May	June	July	August	September
Protein (%)	6.25 ± 0.11ᵃ	6.20 ± 0.21ᵃ	6.32 ± 0.21ᵃ	6.12 ± 0.31ᵃ	6.14 ± 0.36ᵃ	6.05 ± 0.45ᵃ
Lipid (%)	3.05 ± 0.04ᵃ	3.12 ± 0.11ᵃ	3.10 ± 0.11ᵃ	3.20 ± 0.17ᵃ	2.95 ± 0.21ᵃ	2.96 ± 0.41ᵃ
Carbohydrate (%)	48.85 ± 0.36ᵃ	47.98 ± 0.26ᵃ	48.10 ± 0.26ᵃ	48.30 ± 0.31ᵃ	48.90 ± 0.35ᵃ	48.85 ± 0.24ᵃ

Figures bearing common letter(s) in a row as superscript do not differ significantly (P<0.05)

Table 6: Monthly variations in nutrient (protein, lipid and carbohydrate) contents with treatment T_6 (Leaf of banana, *Musa acuminata*).

Treatments	Nutrient content		
	Protein (%)	Lipid (%)	Carbohydrate (%)
T_1 (Rice bran)	14.40 ± 0.32ᵈ	10.41 ± 0.31ᵇ	44.09 ± 0.67ᵉ
T_2 (Wheat bran)	17.13 ± 0.07ᶜ	6.69 ± 0.30ᶜ	66.12 ± 0.47ᵃ
T_3 (Oilcake)	30.53 ± 0.40ᵃ	13.33 ± 0.10ᵃ	32.95 ± 0.29ᶠ
T_4 (*Azolla pinnata*)	18.58 ± 0.09ᵇ	3.19 ± 0.10ᵈ	50.21 ± 0.54ᵇ
T_5 (Grass- *Cynodon dactylon*)	7.26 ± 0.18ᵉ	6.31 ± 0.13ᶜ	46.36 ± 0.16ᵈ
T_6 (Leaf of *Musa acuminata*- Banana leaf)	6.18 ± 0.13ᶠ	3.06 ± 0.09ᵈ	48.50 ± 0.51ᶜ
F value	16.42	13.88	114.85
P value	0.002	0.004	0.0000008

Figures bearing common letter(s) in a column as superscript do not differ significantly (P<0.05)

Table 7: Variations in the mean values of protein, lipid and carbohydrate contents in different fish feed items.

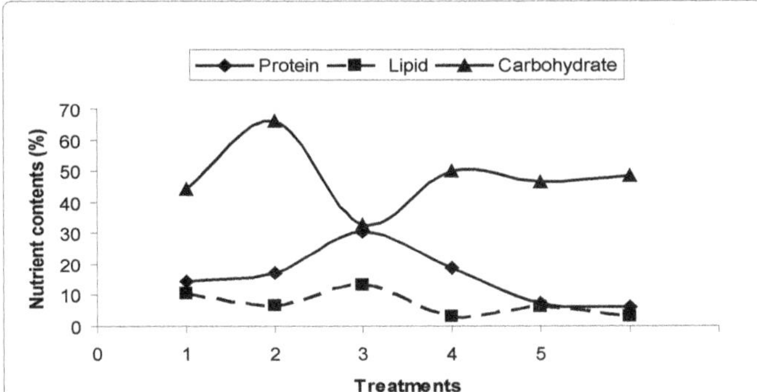

Figure 1: Variations in the mean values of nutrient contents under different fish feed items.

and with the ecological conditions and the phase of growth. The crude protein content is about 19-30 percent dry matter basis during the optimum conditions for growth [25,26]. The protein contents of *Azolla* species are comparable to or higher than that of most other aquatic macrophytes. Aquatic weeds' are highly nutritious with protein content of 20-30%, when cultivated in nutrient rich waters [27] Importantly, they are preferred food of a wide range of herbivorous fish such as grass carp (*Ctenopharyngodon idella*), silver barb (*Barbonymus gonionotus, Puntius jerdoni*), tilapias (*Oreochromis niloticus, Tilapia rendalli, Tilapia zillii*) and rohu (*Labeo rohita*) [28,29].

Overall findings indicated that inspite of having variations in nutrient contents, monthly supply of nutrients was almost same respective feed item under non-conventional feeds as with conventional feeds. Mean values of the nutrient contents under non-conventional feed items are found potentials for the development of low cost aquaculture.

Fish feed generally constitutes 60-70% of the operational cost in intensive and semi- intensive aquaculture system [30]. The fish feed used in aquaculture is quite expensive, irregular and short in supply in many third world countries. These feeds are sometimes adulterated, contaminated with pathogen as well as containing harmful chemicals for human health. Naturally there is a need for the development of healthy, hygienic fish feed which influences the production as well as determines the quality of cultured fish. Considering the importance of nutritionally balanced and cost-effective alternative diets for fish, almost similar expression to evaluate the nutritive value of different non-conventional feed resources, including terrestrial and aquatic macrophytes was found with Wee and Wang [10,31]. However potentials roles of aquatic and terrestrial macrophytes as supplementary feeds in fish farming were also found to be expressed with Bardach [32] and Edwards [33].

Conclusion

In case of conventional feed items, protein, lipid and carbohydrate varied from 14.40 ± 0.32% to 30.53 ± 0.40%, 6.69 ± 0.30% to 13.33 ± 0.10% and 32.95 ± 0.29% to 66.12 ± 0.47%. In case of non-conventional feed items, protein, lipid and carbohydrate varied from 6.18 ± 0.13% to 18.58 ± 0.09%, 3.06 ± 0.09% to 6.31 ± 0.13% and 46.36 ± 0.16% to 50.21 ± 0.54%. Inspite of variations weeds are moderately nutritive and low cost effective diets for fish. However, the present study did not evaluate the fish production and economics of feed and weed based systems.

Recommendation

Present findings explored the nutritive aspects of both conventional and non-conventional feed items and question raised about the response of utilizing the feed specially of aquatic weeds to fish growth and economics. Therefore, it is recommended to conduct further study on the evaluation of fish production and economics under different feed and weed based systems in polyculture ponds.

Acknowledgement

The research work was conducted under a financial support by the Ph. D. Fellowship Programme of Ministry of Science and Technology, Govt. of the People's Republic of Bangladesh which is gratefully acknowledged.

References

1. Rahman MM, Varga I and Chowdhury SN (2011) Manual on polyculture and integrated fish farming in Bangladesh. Project report of BGD/87/045/91/11, Food and Agriculture Organization (FAO), Rome, Italy
2. DoF (2011) National Fish Week 2011 Compendium (in Bengali), Department of Fisheries, Ministry of Fisheries and Livestock, Bangladesh 136.
3. Suresh VR Mandal BK (2000) Growth response and nutritive value of Azolla and Alternanthera incorporated pelleted feeds on fingerlings of Cyprinus carpio var. communis: a preliminary study. Indian J. Fish. 47: 225-229.
4. Halver JE (1972) Fish nutrition. Academic Press, NewYork: 713.
5. Lakshmanan MAV, Murthy DS, Pillai KK, Banerjee (1967) On a new artificial feed for carp. FAO Fisheries Report : 373-387.
6. Edwards P (1980) Food potential of aquatic macrophytes. ICLARM Studies and Reviews ICLARM Manila 5: 51
7. Edwards P, Kamal M, Wee KL (1985) Incorporation of composted and dried water hyacinth in pelleted feed for the tilapia Oreochromis niloticus (Peters). Aquaculture and Fisheries Management 16: 233-248.
8. Patra BC, Ray AK (1988) A preliminary study on the utilization of the aquatic weed Hydrilla verticillata Rayle as feed by the carp, Labeo rohita (Hamilton): growth and certain biochemical composition of flesh. Indian Biology : 44–50.
9. Ray AK, Das I (1995) Evaluation of dried aquatic weed, Pistia stratiotes meal as a feedstuff in pelleted feed for rohu, Labeo rohita fingerlings. Journal of Applied Aquaculture 5: 35-44.
10. Wee KL, Wang SS (1987) Nutritive value of Leucaena leaf meal in pelleted feed for Nile Tilapia. Aquaculture 62: 97-108.
11. Rangama S (1979) Manual of analysis of Fruits and vegetable products, Tata Mc Graw- Hill Publishing Company Ltd. New Delhi.
12. Bligh EG, Dyer W (1989) Total Lipid Extraction and Purification, Can. Jour. Biochem. Physiol, 37: 911.
13. Boel E, Huge-Jensen B, Christensen M, Thim L and Fill NP (1988). Lipids: 701.
14. Gomez KA, Gomez AA (1984) Statistical Procedure for Agricultural Research. 2nd Ed. John Wiley & Sons: 697 .
15. Sithara K, Kamalaveni K (2008) Formulation of low-cost feed using Azolla as a protein supplement and its influence on feed utilization in fishes. Current Biotica 2: 212-219.
16. Ebrahim MSM, Zeinhom MM Abou-Seif RA (2007) Response of Nile tilapia (Oreochromis niloticus) fingerlings to diets containing Azolla meal as a source of protein. Journal of Arabian Aquaculture Society 2: 54-68.
17. Fasakin EA, Balogun AM (2001) Nutritional and anti-nutritional analyses of Azolla africana Desv. and Spirodela polyrrhiza L. Schleiden as feedstuffs for fish production. In: 14th Annual Conference of the Fisheries Society of Nigeria (FISON): 31-39.
18. Lumpkin TA, Plucknett L (1982) Azolla as a green manure: use and management in crop production. Westview press Boulder, Colorado. Westview Tropical Agriculture: 230
19. Van-Hove C, Waha Baillonville T, Diara HF, Godard P, Mai Kodomi Y, et al. (1987) Azolla collection and selection. Azolla Utilization. In: Proceedings of the Workshop on Azolla Use, Fuzhou, Fujian, China,Int. Rice Res. Inst,Los Banos, Philippines: 77-87.
20. Hepher B (1988) Nutrition of Pond Fishes. Cambridge University Press, UK: 180
21. Banerjee A, Matai S (1990) Composition of Indian aquatic plants in relation to

utilization as animal forage. J. Aquat. Plant Manage 28: 69-73.

22. Gavina LD (1994) Pig-Duck-Fish-Azolla integration LA Union Philippines. Naga The ICLARM Quarterly:18-20.

23. Tavares FA, Roudrigues JSR, Fracalossi DM, Esquivel J and Roubach R (2008) Dried duckweed and commercial feed promote adequate growth performance of tilapia fingerlings. Biotemas 21: 91-97.

24. Santhanam R, Sukumaran N, Natarajan P (1990) A Manual of Fresh Water Aquaculture. Oxford and IBH Publishing Go. Pvt. Ltd., New Delhi: 193

25. Becking JH (1979) Environmental requirements of Azolla for use in tropical rice production. In Nitrogen and Rice, Los Banos, Laguna, International Rice Research Institute: 345-373.

26. Peters GA, Mayne BC, Ray TB, Toia RE (1979) Physiology and biochemistry of the Azolla-Anabaena symbiosis. In Nitrogen and Rice. Los Baños, Laguna, Phillipines, International Rice Research Institute: 325-344.

27. Culley DD, Rejmankova E, Koet J, Prye JB (1981) Production, chemical quality and use of duckweeds (Lemnaceae) in aquaculture, waste management and animal feeds. J. World Maricult. Soc 12: 27-49.

28. Singh SB, Pillai KK, Chakraborty PC (1967) Observation on the efficacy of grass carp in controlling and utilizing aquatic weeds in ponds in India. Proc. Indo-Pacific Fish Counc 12: 220-235

29. Gaiger IG, Porath D, Granoth G (1984) Evaluation of duckweed (Lemna gibba) as feed for tilapia (Oreochromis nilotieus cross Oreochromis aureus) in a recirculating unit. Aquacultre 41: 235-244.

30. Singh PK, Gaur SR, Chari MSP (2006) Growth Performance of Labeo rohita (Ham.) Fed on Diet Containing Different Levels of Slaughter House Waste, J. Fish. Aquat. Sci 1: 10-16.

31. Mondal, TK Ray AK (1999) The nutritive value of Acacia auriculiformis leaf meal in compounded diets for Labeo rohita fingerlings. The Fourth Indian Fisheries Forum Proceedings 1996, Kochi: 295–298.

32. Bardach JE, Ryther JH, MeLarney WO (1972) Aquaculture: The Farming and Husbandry of Freshwater and Marine Organisms, Wiley-Interscience, New York.

33. Edwards P (1990) Use of terrestrial vegetation and aquatic macrophytes in aquaculture. In: Detritus and microbial ecology in aquaculture. ICLARM. Conf Proc 14, International Aquat Living Resour Manag Cent: 311-335.

Influence of Dietary *Sorghum* Starch on Growth Performance, Digestibility Coefficient and Some Hepatic Enzyme Activities in Hybrid Red Tilapia (*Oreochromis mossambicus* × *Oreochromis niloticus*) Fingerlings

Abdel Moneim Yones M* and Atallah Metwalli A

National Institute of Oceanography and Fisheries (NIOF), Shakshouk Aquatic Research Station, El-Fayoum, Egypt

Abstract

A 120-day feeding trial was conducted to investigate the effects of dietary sorghum starch on growth performance, feed utilization, apparent digestibility coefficient (ADC) and some hepatic enzyme activities regulating glycolytic and gluconeogenic metabolic pathways of fingerlings hybrid red tilapia (*Oreochromis mossambicus* × *O. niloticus*) with mean initial body weight of 10.9 ± 0.2 g. Five diets containing graded levels of sorghum starch (15%, 20%, 25%, 30% and 35%) were formulated. The results demonstrated that weight gain (WG), specific growth rate (SGR), protein efficiency ratio (PER) and net protein utilization (NPU) values increased with increasing dietary sorghum starch up to 30%. Hepatosomatic index, plasma glucose, triglycerides, liver glycogen and liver lipid concentration of fish significantly increased with increasing dietary sorghum starch level ($P<0.05$). ADC of starch decreased significantly with increasing sorghum starch level over 30%. However, whole body compositions and ADC of protein and lipid showed no significant differences. Dietary sorghum starch supplements tended to enhance gluconokinase and pyruvate kinase activities of the liver but insignificant differences were observed in activities of hexokinase, phosphofructokinase-1, fructose-1, 6-bisphosphatase and glucose-6-phosphatase in the liver for all dietary treatments. Based on WG and FCR results, the appropriate dietary sorghum starch supplementations of fingerlings hybrid red tilapia (*O. mossambicus* × *O. niloticus*) can be incorporated up to 30% of diet.

Keywords: Hybrid red tilapia; Growth performance; Whole body composition; Digestibility coefficient; Sorghum starch; Hepatic enzyme activities

Introduction

Carbohydrates are the most economical source of energy available in abundant quantities at low prices and have a protein-sparing effect in some low-protein diets and for binding other ingredients [1,2]. Feed supply and feed costs are amongst the greatest challenges for the development of sustainable fish farming. Therefore, the aquaculture industry is searching for feed ingredients that can be used to formulate cheap fish feed [2]. It was noticed that fish meal and fish oil contribute 75% of the protein and 35% of the energy in aquaculture feed [3]. It has been estimated that the cost of feed constitutes 74% of total costs for farm-made feeds and 92% for manufactured pellet feeds [4]. The cost of aquaculture production can be reduced by efficient feed formulation [5].

Dietary carbohydrate inclusion in several fish species appears to produce positive effects on growth and digestibility [6-8]. However, using the appropriate level of carbohydrates in aqua feed is of great importance, because if the appropriate amount of carbohydrates is not provided, this may have negative effects on nutrient utilization, growth, metabolism and health [9,10]. Several studies have reported that an increase in dietary carbohydrate content improves metabolism and growth in tilapia [11-13]. Similarly, improved growth was observed in tilapia fed diets with 10%-40% inclusion of starch [14,15].

In most aquaculture feeds, complex carbohydrates such as starch have been introduced as the principal source of carbohydrate due to the fact that, dietary monosaccharide's are rapidly absorbed but poorly utilized [16,17]. There are still many inter-specific differences in the utilization and metabolism of dietary CHO by fish. For example, there is a strong relationship between the natural trophic level of fish species and their ability to utilize CHO, with herbivorous and omnivorous species usually better at digesting and utilizing CHO than carnivorous species [18-21]. Omnivorous fish species as Nile tilapia and common carp, which feed at low trophic levels, can efficiently utilize high dietary levels of carbohydrates (30%-50%) in comparison to the high trophic level carnivorous fish species [18-23].

In addition, digestibility of CHO by different species appears to be dependent on the complexity of the CHO as well as its dietary content with simpler CHO being more easily digested while digestibility of CHO generally decreases as dietary content increases [20,24-26]. Knowledge of the optimum inclusion of dietary carbohydrate is indispensable for improving growth performance of fish with reducing the amount of nitrogen waste and diet costs [27]. However, feeding excessive dietary carbohydrates of fish may have harmful effects on growth, feed efficiency, physiological dysfunction, and fat deposition by stimulating lipogenic enzymes [28,29].

Processing CHO by cooking-extrusion has been found to increase digestibility of CHO in most fish species, largely by breaking down the molecular structure of starch and increasing the degree of gelatinization [24]. Digestion is thought to be the primary limiting step in the utilization of starch for growth [20]. The inability of fish to utilize digested CHO is reflected in reduced growth, inferior feed conversion

***Corresponding author:** Abdel Moneim Yones M, Professor of Fish Nutrition, National Institute of Oceanography and Fisheries (NIOF), Shakshouk Aquatic Research Station, El-Fayoum, Egypt,
E-mail: yones_55200010@yahoo.com

ratio (FCR) and lower protein retention efficiency. Excessive dietary CHO may also decrease the palatability of feeds leading to reduced feed intake. However, fish which can tolerate and utilize high levels of CHO in their aqua-feed allow feed manufacturers much greater flexibility to explore reductions in dietary protein and lipid content of complete feeds and therefore feed cost.

The digestion and absorption of nutrients are mostly dependent on enzyme activities involved in the breakdown and assimilation of food [30]. Therefore, analysis of enzyme activities is a convenient and reliable technique that can provide comprehensive information relating to digestive physiology and nutritional conditions in the fish [31]. Digestible efficiency of digestible and non-digestible carbohydrates varies in herbivorous and carnivorous fish species [32-34]. The herbivorous fish species can utilize part of the non-starch carbohydrates in their diet due to symbiosis with the gut microbiota. However, most fish species are unable to utilize non-starch carbohydrates properly because they lack the adequate gut microbiota in their gut [26].

The optimal dietary carbohydrate level of hybrid red tilapia (*O. mossambicus* × *O. niloticus*) has not been reported yet. Therefore, this study was designed to evaluate the effect of dietary sorghum starch on the growth, feed utilization, body composition, apparent digestibility, and hepatic enzyme activities of carbohydrate metabolism on fingerlings hybrid red tilapia (*O. mossambicus* × *O. niloticus*).

Materials and Methods

Experimental diets

Five semi-purified experimental diets were formulated varying only in their dietary sorghum starch level (15%, 20%, 25%, 30% and 35%) (Table 1). The dietary 30% protein and 10% lipid level was selected because it is recommended to cover the requirements of this specie [20]. Fish meal and casein were used as protein sources where, fish oil and sunflower oil were used as lipid sources. Chromic oxide (Cr_2O_3) was used as an inert marker to determine the apparent digestibility of nutrients. Ingredients were ground into fine powder through a 150-µm mesh before pelleting and an appropriate amount of water was added to produce stiff dough. The dough was pelleted using California pelleting machine with 2 mm diameter.

Fish and experimental design

Hybrid red tilapia (*O. mossambicus* × *O. niloticus*) fingerlings were obtained from the Kilo 21 hatchery belonging to General Authority for fish resource development, Alexandria road, Egypt. Fish were acclimated to the system and fed with the experimental diet twice daily for 2 weeks before the trial. After 24 hour of starvation, 1500 fish (initial body weight=10.9 ± 0.2 g) were randomly selected from the acclimatized fish and allocated into 15 circular cement ponds (size of each pond was 2 m^3) in equal number (n=100 with stocking rate of 50 fish/m^3). During the experiment, fish were hand-fed the experimental diets to apparent satiation twice daily (10:00 pm and 4:00 am) and weighting every two weeks to adjust the amount of feed consumption. The system contained two water pumps and upstream sandy filter units at a point between the water source (Lake Qaroun) and tanks. Each pump was drowning the water to the storage tanks and forced it through polyvinyl chloride (PVC) tubes into the rearing tanks in open system. The experimental period lasted 120 days after start. Physicochemical properties of water tanks were examined every week according to [35].

Sample collection and chemical analysis

Before the experiment, 20 fish from the same population were randomly selected for determination of initial whole-body proximate composition. At the end of the feeding trial, fish were starved for 24 hours prior to sampling. Fish in each tank were weighed and counted for information on growth, feed efficiency and survival. Twenty fish from each tank were randomly selected and anesthetized with tricaine methanesulfonate (MS-222, 50 mg/L) for individual weight measurements. Blood was collected from the caudal vein of individuals using 2.5 mL sterile syringes. Plasma samples were collected after centrifugation at 3000 g for 20 min at 4°C and stored at 80°C prior to biochemical analysis. Then, the fish were quickly dissected for organ and tissue sampling. Liver and dorsal muscles were stored at 80°C immediately before further analysis. Finally, twenty fish per tank were randomly collected for determination of final whole-body proximate composition. After the sample collection described above, the remaining fish were fed with the same diets after adding 0.5% chromic oxide (Cr_2O_3) to determine the apparent digestibility coefficients (ADCs) for dry matter, crude protein, crude lipid, and starch. Faecal collection was conducted 5-6 h after the first meal at 10:00 pm. Fish from each replicate were anesthetized with MS-222 (50

Ingredients	Sorghum starch levels				
	15%	20%	25%	30%	35%
Fish meal	10	10	10	10	10
Poultry-by product meal[1]	15	15	15	15	15
Casein[2] 26	15	15	15	15	15
Sorghum starch[3]	15	20	25	30	35
Microcrystalline cellulose	34	29	24	19	14
Oil mix[4]	6	6	6	6	6
Choline chloride	0.5	0.5	0.5	0.5	0.5
Ascorbyl-2-monophosphate	0.4	0.4	0.4	0.4	0.4
Vitamin. mineral mix[5]	2.0	2.0	2.0	2.0	2.0
Chromic oxide	0.1	0.1	0.1	0.1	0.1
Sodium alginate	2.0	2.0	2.0	2.0	2.0
Proximate analysis					
Dry matter	91.2	91.4	91.5	91.6	91.4
Crude protein	30.0	30.4	30.8	31.2	31.6
Crude lipid	10.2	10.1	10.2	10.1	10.1
Starch	15.0	19.9	25.1	30.2	34.9
Crude fiber	4.5	4.6	4.8	5.0	5.2
Ash	12.6	12.2	12.4	12.5	12.4
Tannin[6]	0.25	0.3	0.32	0.36	0.42
Gross energy(MJ kg^{-1} diet)[7]	13.86	14.77	15.82	16.77	17.7
ME (MJ kg^{-1}diet)[8]	11.51	12.26	13.14	13.92	14.7

Table 1: Ingredients and proximate composition of the experimental diets (%DM basis).

1. Poultry-by product meal: crude protein-55%; crude lipid-14.4% (poultry production comp.cairo, Egypt).
2. Casein: crude protein-93.1%; crude lipid-1.5% (Gannanzhou Kerui Dairy Products Development Co., Ltd., Gansu, China).
3. International starch institute, Denmark.
4. Oil mixture: fish oil and sunflower oil were mixed as a ratio of 1:1.
5. Vitamin, mineral premix (g/kg of mixture): L-ascorbic acid monophosphate-120.0; L-α-tocopherylacetate-20.0; thiamin hydrochloride-4.0; riboflavin-9.0; pyridoxine hydrochloride-4.0; niacin-36.0; D-pantothenic acid hemicalcium salt-14.5; myoinositol-40.0; D-biotin-0.3; folic acid-0.8; menadione-0.2; retinyl acetate-1.0; cholecalciferol-0.05; cyanocobalamin-0.01; $MgSO_4{\cdot}7H_2O$-80.0; Na $H_2PO_4{\cdot}2H_2O$-370.0; KCl-130.0; $FeSO_4{\cdot}7H_2O$-40.0; $ZnSO_4{\cdot}7H_2O$-20.0; Ca-lactate-356.5; $CuSO_4$0.2; $AlCl_3{\cdot}$6H2O-0.15; $Na_2Se_2O_3$-0.01; $MnSO_4{\cdot}H_2O$-2.0; $CoCl_2{\cdot}6H_2O$-1.0
6. Tannin=percent tannin on a catechin equivalent basis.
7. Gross energy (MJ Kg^{-1} diet) was calculated by using the following calorific values: 23.9, 39.8 and 17.6 KJ g^{-1} diet for protein, ether extract and nitrogen free extract, respectively [33].
8. The metabolizable energy (MJ Kg^{-1} diet) of the experimental diets were calculated as 18.9, 35.7 and 14.7 KJ g^{-1} diet for protein, lipid and nitrogen free extract, respectively [34].

mg/L) and manually stripped of faeces by applying gentle pressure in the anal area according to the method described by Ren [36]. Faeces were collected once a fortnight until sufficient dried faeces had been collected for analysis. Pooled faeces from each replicate were freeze-dried and stored at 20°C until analysis of nutrient contents. Analyses of ingredients, diets, faecal samples, whole body and muscle composition were made following the usual procedures [37]. Dry matter was determined by drying samples in an oven at 105°C until constant weight, crude protein was determined by measuring nitrogen (N × 6.25) after acid digestion using the Kjeldahl method, crude lipid was determined by petroleum ether extraction using the Soxhlet method, ash was determined by incineration in a muffle furnace at 550°C for 16 h and starch was determined using an enzymatic method as described by Hemre [25]. Tannin content of sorghum starch was determined using a modified version of Price's vanillin-Hcl assay [38]. One gram of sorghum starch was placed in a 50 ml conical flask and 50 ml of analytical grade methanol was added. The flask was covered with a cork stopper, shaken thoroughly every few minutes for 2 hours and then left to stand at room temperature for an additional 20 h. Two ml of 2% vanillin, 4% Hcl were added to one of the test tube and 5 ml of 4% Hcl to other. The differences in two optical densities (the 4%Hcl acting as the blank) was read on a Beckman spectrophotometer at 500 nm, then compared to catchin standard curve. Diets and faeces chromic oxide were determined using an inductively coupled plasma-atomic emission spectrophotometer (IRIS Advantage [HR], Thermo Jarrell Ash, Woburn, MA, USA) after perchloric acid digestion, triplicate analyses were conducted for each sample.

Determination of enzyme activities

Liver samples were homogenized in four volumes of ice-cold 100 mM Tris–HCl buffer containing 0.1 mM EDTA and 0.1% Triton X-100 (v/v), p^H 7.8. Homogenates were centrifuged (Kubota model 6900, Kubota Corporation, Tokyo, Japan) at 30,000 g at 4°C for 30 min and the resultant supernatants divided in aliquots and stored at 80°C for further enzyme assays. All enzyme activities were performed at 25°C and absorbance read at 340 nm in a micro plate reader (ELx808TM, Bio-Tek Instruments, USA). Hexokinase (HK; EC 2.7.1.1) and glucokinase (GK; EC 2.7.1.2) activities were measured as described by Enes using a reaction mixture containing 50 mM imidazole-HCl buffer (p^H 7.4), 2.5 mM ATP, 5 mM $MgCl^2$, 0.4 mM NADP, 2 units mL^1 G6PDH and 1 mM (HK) or 100 mM (GK) glucose [23]. Pyruvate kinase (PK; EC 2.7.1.40) activity was measured according to Panserat [39] with a reaction mixture consisting of 50 mM imidazole-HCl buffer (p^H 7.4), 5 mM $MgCl^2$, 100 mM KCl, 0.15 mM NADH, 1 mM ADP, 2 units mL^1 LDH and 2 mM PEP. Fructose 1,6-bisphosphatase (FBPase; EC 3.1.3.11) activity was measured as described by Foster using a reaction mixture consisting of 50 mM imidazole-HCl buffer (p^H 7.4), 5 mM Mg Cl^2, 12 mM 2-mercaptoethanol, 0.5 mM NADP, 2 units mL^1 G6PDH, 2 units mL^1 PGI and 0.5 mM fructose 1,6-bisphosphate [40]. Glucose 6-phosphate dehydrogenase (G6PDH; EC 1.1.1.49) activity was measured as described by Metón and Panserat, using a reaction mixture containing 50 mM imidazole–HCl buffer (p^H 7.4), 5 mM Mg Cl^2, 2 mM NADP and 1 mM glucose-6-phosphate [41,42]. The reaction mixture containined 50 mM imidazole–HCl buffer (p^H 7.4), 5 mM Mg Cl^2, 0.4 mM NADP and 2 mM L-malate. Protein concentration in liver crude extracts was determined at 600 nm according to the Bradford method using bovine serum albumin as a standard [43]. All enzyme activities were expressed as per mg of hepatic soluble activity was defined as the amount of enzyme that catalyzed the hydrolysis of 1 μmol of substrate per minute at assay temperature. Plasma glucose and triacylglycerol concentration were determined using commercial kits from Enzyline, Biomerieux, Linda-A-Velha, Portugal (ALAT/GPT, ref. 63313; ASAT/GOT, ref. 63213). Liver and muscle glycogen concentration were determined at 620 nm using the anthrone reagent method [44].

Growth performance

The following growth performance parameters were calculated as follows:

a) Specific growth rate (SGR)=100 × (Ln final weight-Ln initial weight)/120

b) Condition factor (CF g/cm^{-3})=(wet weight) / (total length-3) × 100

c) Feed conversion (FCR)=(feed given per fish) / (weight gain per fish)

d) Protein efficiency ratio (PER)=(weight gain per fish) / (protein intake per fish)

e) Net protein Utilization (NPU%)=100 (Final body protein-initial body protein / protein intake)

f) Hepatosomatic index (HSI%)=(liver weight) / (fish weight) × 100

g) ADC of dry matter (%)=100–(100% Cr_2O_3 in diet/%Cr_2O_3 in faeces)

h) ADC of nutrients (%)=100–(100 × %nutrient in faeces / %nutrient in diet × %Cr_2O_3 in diet/%Cr_2O_3 in faeces)

Statistical analyses

The results are presented as means ± SE of three replications. All data were subjected to one-way analysis of variance and tested. One way Analysis of Variance (ANOVA) was applied to test the effect of different sorghum starch levels on various growth parameters, nutrient utilization, chemical composition and hepatic enzyme activity of experimental fish according to Snedecore [45]. Duncan Multiple Range test was used to detect the significant differences between the means of treatments [46]. All analysis were performed using SAS (version 6, 2004 SAS Institute, Cary, NC, USA) [47].

Results

Water physico-chemical properties (Table 2) revealed that water temperature, salinity, p^H, dissolved oxygen and unionized ammonia are within the optimum ranges for rearing red tilapia. Similar physico-chemical condition was observed in all tanks of the present study as presented in Table 2.

After 120-days growth trial, survival rate of red hybrid tilapia (*Oreochromis mossambicus* × *O. niloticus*) was not affected by dietary starch levels (Table 3). As presented in the same table, averages of initial weights ranged between 10.85 to 10.96 g/fish with insignificant differences among the dietary groups indicating the random distribution of the experimental fish among treatment groups (Table 4). Fish in all dietary treatments survived well during the trial (97%), indicating that the tested diets had no effects on red hybrid tilapia survival rates, thus all mortalities were due to accidental factors during the samples collection every two weeks to adjust the feed amounts. Significant differences in weight gain (WG), specific growth rate (SGR), protein efficiency ratio (PER), feed conversion ratio (FCR), net protein utilization (NPU) and

Parameters	Sorghum starch levels				
	15%	20%	25%	30%	35%
Temperature (°C)	29.4 ± 0.3	29.5 ± 0.4	29.3 ± 0.2	29.4 ± 0.4	29.5 ± 0.3
Salinity (g/l)	32.2 ± 1.2	32.1 ± 1.1	32.2 ± 1.2	32.1 ± 1.1	32.1 ± 1.2
p^H	7.8 ± 0.3	7.6 ± 0.4	7.7 ± 0.2	7.6 ± 0.4	7.6 ± 0.2
Dis.Oxy[1] (mg/l)	7.2 ± 0.1	7.1 ± 0.2	7.2 ± 0.4	7.1 ± 0.2	7.0 ± 0.4
Uni.Am[2] (mg/l)	0.03 ± 0.01	0.03 ± 0.01	0.04 ± 0.01	0.03 ± 0.01	0.04 ± 0.01

Table 2: An average of water physicochemical characteristic parameters during experimental period.
1. Dissolved oxygen.
2. Unionized ammonia.
3. Mean value ± SE.

Parameters	Sorghum starch levels				
	15%	20%	25%	30%	35%
Initial weight (g/fish)	10.85^a ± 0.32	10.91^a± 0.32	10.96^a ± 0.32	10.94^a ± 0.32	10.88 ± 032
Final weight (g/fish)	61.6^b ± 2.47	64.4^b ± 1.8	78.5^a ± 1.6	81.0^a ± 2.4	50.4^c ± 2.2
Total gain (g/fish)	50.75^b ± 1.4	53.49^b ± 1.8	67.54^a ± 2.2	70.06^a ± 1.6	39.52^c ± 1.8
Average gain (g/fish/d)	0.42^b ± 0.1	0.44^b ± 0.1	0.53^a ± 0.1	0.58^a ± 0.1	0.32^c ± 0.1
Specific growth rate	1.74^b ± 0.1	1.78^b ± 0.2	1.97^a ± 012	2.0^a ± 0.2	1.27^c ± 0.1
Condition factor (g/cm^{-3})	1.8^b ± 0.2	1.9^b ± 0.3	2.32^a ± 0.3	2.4^a ± 0.1	1.5^c ± 02
Survival rate%	97	97	97	97	97
Feed consumed (g/fish)*	90.0	92.0	98.0	100.0	110
Feed conversion ratio	1.8^b ± 0.3	1.71^b ± 0.1	1.42^a ± 0.3	1.42^a ± 0.2	2.78^c ± 0.1
PER	1.87^b ± 0.2	1.85^b ± 0.1	2.23^a ± 0.2	2.24^a ± 0.1	1.13^c ± 0.2
NPU (%)	30.66^b ± 2.4	29.77^b ± 2.8	35.77^a ± 3.2	36.16^a ± 2.8	17.94^c ± 1.8
HSI (%)	1.84^b ± 0.1	1.76^b ± 0.4	2.20^a ± 0.2	2.32^a ± 0.4	2.75^c ± 0.2

Table 3: Growth performance mean values (Mean ± S.E *n=3*) of red tilapia fed on different experimental diets for 120 days.
Means in the same row with different superscript letters are significantly different (P<0.05).
*Non-consumed portion of food was collected, dried and deducted from total given ration.

Items	Sorghum starch levels					
	Initial	15%	20%	25%	30%	35%
Dry matter	27.3 ± 1.4	27.5^a ± 1.6	27.5^a ± 1.8	27.2^a ± 1.5	27.1^a ± 1.4	27.4^a ± 1.6
Protein	16.8 ± 1.2	16.4^a ± 1.4	16.2^a ± 1.2	16.1^a ± 1.5	16.2^a ± 1.2	16.0^a ± 1.2
Lipid	5.0 ± 1.2	5.2^a ± 1.1	5.4^a ± 1.2	5.7^a ± 1.1	5.9^a ± 1.2	6.4^a ± 1.1
Ash	5.5 ± 1.1	5.7^a ± 1.2	5.7^a ± 1.2	5.5^a ± 1.1	5.6^a ± 1.2	5.8^a ± 1.2

Table 4: Carcass analysis of red tilapia fed on the experimental diets, %w/w basis (Mean ± S.E *n=3*).
Means in the same row with different superscript letters are significantly different (P<0.05).

ADCs%	Sorghum starch levels				
	15%	20%	25%	30%	35%
Dry mattser	70^c ± 2.4	71^c ± 1.2	74^b ± 1.6	76^b ± 1.5	80^a ± 1.2
Crude protein	91.5^a ± 2.2	91.6^a ± 1.8	91.4^a ± 2.5	91.6^a ± 2.4	91.2^a ± 1.8
Crude lipid	89.2^a ± 1.8	89.0^a ± 1.5	89.4^a ± 1.6	89.5^a ± 1.5	89.1^a ± 1.4
Starch	88.6^a ± 2.1	86.4^a ± 2.4	84.5^b ± 2.4	85.2^a ± 2.2	80.4^c ± 2.4

Table 5: Apparent digestibility coefficient (ADCs) of the experimental diets (Mean ± S.E *n=3*).
Means in the same row with different superscript letters are significantly different (P<0.05).

hepatosomatic index (HSI) were observed among dietary treatments (P<0.05). Fish fed with 30% sorghum starch diet had significantly higher WG, SGR, and PER, and lower FCR than those fish fed with diets containing 15%, 20%, 25% and 35% sorghum starch (P<0.05). NPU significantly increased with dietary sorghum starch level from 15 to 30% and then decreased with 35% level (P<0.05). Fish fed with diets containing 25%-30% sorghum starch had significantly higher HSI values than fish fed with 15 and 35% sorghum starch diet (P<0.05).

The ADCs of dry matter increased from 70%-80% with dietary sorghum starch levels increasing. Fish fed diets containing 30% and 35% sorghum starch were significantly higher than fish fed with 15% sorghum starch diet P<0.05 (Table 5). ADCs of starch were significantly lower when dietary sorghum starch level is more than 30% compared with fish fed with the other sorghum starch diet (P<0.05). On the other hand, the ADCs of crude protein and crude lipid were not significantly different among dietary treatments (P>0.05).

As it is demonstrated in Table 6, significantly higher plasma glucose and triglyceride concentrations were obtained in fish fed diets containing 30% and 35% sorghum starch than those fish fed with 15%, 20% and 25% sorghum starch diets (P<0.05). However, lower liver glycogen concentrations were observed in fish fed with 15% sorghum starch diet compared with fish fed the other diets (P<0.05). Fish fed with 35% sorghum starch diet showed higher liver lipid concentrations than fish fed with the other diets (P<0.05).

As presented in Table 7, activities of GK and PK in liver were significantly affected by dietary sorghum starch levels. GK and PK activities were significantly higher and positively correlated with

Parameters	Sorghum starch levels				
	15%	20%	25%	30%	35%
Plasma glucose	$3.12^c \pm 0.14$	$3.18^c \pm 0.18$	$3.45^b \pm 0.15$	$3.52^b \pm 0.12$	$3.86^a \pm 0.14$
Plasma triglycerides	$4.25^c \pm 0.16$	$4.32^c \pm 0.14$	$4.66^b \pm 0.18$	$4.72^b \pm 0.16$	$4.98^a \pm 0.11$
Liver glycogen	$26.5^c \pm 1.12$	$27.12^c \pm 1.15$	$32.52^b \pm 1.22$	$32.62^b \pm 1.44$	$34.44^a \pm 1.22$
Liver lipid	$322.12^c \pm 1.11$	$322.84^c \pm 1.16$	$333.65^b \pm 1.34$	$334.12^b \pm 1.42$	$342.15^a \pm 1.56$

Table 6: Plasma glucose and triglycerides, liver glycogen and lipid concentrations of red tilapia fed different dietary sorghum starch levels (Mean ± S.E *n*=3).
Means in the same row with different superscript letters are significantly different (P<0.05).

Parameters	Sorghum starch levels				
	15%	20%	25%	30%	3%5
HK	$2.98^a \pm 0.18$	$3.12^a \pm 0.14$	$3.22^a \pm 0.12$	$3.26^a \pm 0.15$	$3.52^a \pm 0.16$
GK	$24.16^c \pm 1.1$	$24.24^c \pm 1.4$	$26.15^b \pm 1.2$	$26.12^b \pm 1.5$	$29.46^a \pm 1.8$
PK	$26.12^c \pm 1.4$	$26.22^c \pm 1.2$	$29.14^b \pm 1.6$	$29.45^b \pm 1.4$	$34.16^a \pm 1.2$
PFK-1	$1.14^a \pm 0.16$	$1.8^a \pm 0.14$	$1.21^a \pm 0.11$	$1.25^a \pm 0.12$	$1.22^a \pm 0.15$
FBPase	$15.45^a \pm 1.6$	$16.12^a \pm 1.1$	$16.22^a \pm 1.2$	$16.66^a \pm 1.4$	$16.44^a \pm 1.1$
G6Pase	$11.14^a \pm 1.2$	$11.22^a \pm 1.4$	$11.16^a \pm 1.2$	$12.18^a \pm 1.1$	$12.46^a \pm 1.5$

Table 7: Activities of glucolytic and gluconeogenic enzymes in liver of hybrid red tilapia fed different dietary sorghum starch levels (Mean ± S.E *n*=3).
Means in the same row with different superscript letters are significantly different (P<0.05).
HK=Hexokinase, GK=Gluconokinase, PK=Pyruvate Kinase, PFK-1=6-Phosphofructo-1-Kinase, FBPase=Fructose-1, 6-Bisphosphatase, G6Pase=Glucose-6-Phosphatase.

dietary sorghum starch levels from 25-35 compared with fish fed with 15% sorghum starch diet (P<0.05). Insignificant differences were detected in activities of HK, PFK-1, FBPase, and G6Pase in liver among dietary treatments (P>0.05).

Discussion

In this study, the hybrid red tilapias were growing well under physico-chemical properties of water tanks. This important finding will improve the extension of red tilapia culture under scarce and restricted of fresh water supply to fish farms. These results are in agreement with that reported by Watanabe, where, this species can tolerate high salinity conduction [48,49].

The carbohydrate utilization varies greatly among fish species, where the appropriate dietary carbohydrate can improve growth and feed efficiency of fish [21,50]. The present study showed that WG, SGR and CF of hybrid red tilapia significantly increased with increasing dietary sorghum starch level from 15% to 30%, while FCR had a contrary tendency. In the same trend, PER and NPU showed a significance increase up to 30% sorghum starch level. Similar observations were also mentioned by Yones in red tilapia, Nile tilapia, in sea bream. However, in this study, fish fed diets with high sorghum starch contents (>30%) showed a decreased growth and feed utilization compared with those fed diets containing 30% or lower sorghum starch levels [51-53]. Same findings were also reported in sea bream, *Sparus aurata* [53], *Oncorhynchus mykiss* [54] and, *Carassius auratus* [28]. In experimental diets, it was clear that the maximum level of tannin was 0.42% in 35% sorghum starch diet and this value less than 0.59% in sea bream diet, which had no negative effects on growth performance, yones postulated same results [53]. In the present trial, noticed also that the depressed in growth performance in 35% sorghum starch diet, maybe due to the decreased in digestibility coefficient of nutrients in this diet.

Enlargement of liver size and glycogen concentration was increased with elevated levels of dietary carbohydrate in several fish [29]. Absorbed carbohydrate that is not used for energy usually accumulated in the liver of fish both as lipid and as glycogen after being converted [54]. This study showed that the value of HSI was increased with dietary sorghum starch levels. In the same manner, Tian reported that HSI, liver lipid, and glycogen concentrations increased with increasing in dietary wheat starch level, and demonstrated that grass carp, *Ctenopharyngodon idella*, had a very high capacity of transforming absorbed starch into tissue lipids [29]. For instance, plasma glucose, triglycerides, liver glycogen and lipid concentration also significantly increased with increasing dietary sorghum starch levels. These results found that excess dietary carbohydrate was deposited as lipid and glycogen in hybrid red tilapia, similar to those observed in European sea bass, *Dicentrarchus labrax* [55]. In the present trial, whole-body lipid content of hybrid red tilapia was positively related to dietary sorghum starch levels and reflected with different response of this specie to glucose metabolic of starch. These results were in agreement with the previous results in several fish species [50,52,56]. However, Mohanta reported that the body lipid content of gibel carp was stable as the dietary starch level increased from 24% to 28% and decreased as the starch level increased from 28% to 40% [27]. This study showed that the dry matter, protein and ash contents of whole body and muscle were not affected by dietary sorghum starch levels, which agreed with the findings in European sea bass [23], gilthead sea bream [23,53] and silver carp [57].

Gelatinization of starch can enhance its digestibility compared to the low digestibility of native starch [55]. The present results indicated that sorghum starch was very well digested by hybrid red tilapia (ADC of starch ≥ 85.2%) when their levels were not more than 30%. However, ADC of starch reduced significantly when dietary sorghum starch level was up to 30%, which is similar to the results of sea bream [53], cobia, *Rachycentron canadum* [36], and large yellow croaker, *Pseudosciaena crocea* [56]. The progressive enhancement in apparent dry matter digestibility concomitant with increasing dietary sorghum starch level is in line with findings in grass carp [29] and it may be explained that higher dietary cellulose level caused the lowered apparent dry matter digestibility. Dietary starch level is found to highly influence digestibility of other nutrients, especially lipid [56,58]. The decrease in protein digestibility with increasing dietary carbohydrate level was reported in white sea bream, *Diplodus sargus* [59] and large yellow croaker [56]. However, the apparent protein and lipid digestibility in this trial were not affected by dietary sorghum starch levels, in agreement with the reported results in hybrid tilapia [60], sea bream [53], cod, *Gadus morhua* [25,61] and Atlantic salmon, *Salmo salar* [5].

Carbohydrates are metabolized by glycolysis or the pentose phosphate pathway, leading to generation of energy transfer molecules in fish [62,63]. Also, dietary carbohydrates could depress the increase rate of amino acid metabolism and utilization by gluconeogenic pathways in salmon fish [64]. Hepatic gluconeogenesis is an important metabolic pathway in fish, science available scientific data on its regulation and effects by dietary carbohydrate is relatively scarce and somewhat discordant. There are a few studies on the key hepatic glycolytic (HK, GK, PFK-1, and PK) and gluconeogenic (G6Pase and FBPase) enzymes involved in glucose metabolic pathway in red tilapia. GK catalyzes the phosphorylation of glucose to glucose-6-phosphate and G6Pase hydrolyzes the glucose-6-phosphate to glucose, the two key enzymes also catalyzes the hepatic glucose/glucose-6-phosphate cycle and both play a major role in glucose homeostasis [64]. Our results showed that GK activities in liver were positively correlated with dietary sorghum starch levels, confirming that this enzyme could be regulated by dietary carbohydrates in hybrid red tilapia, as previously observed in rainbow trout [37], European sea bass [23] and gilthead sea bream [65]. On the other hand, HK activity was not affected by dietary sorghum starch levels, similar to the findings in rainbow trout, European sea bass and gilthead sea bream [23,53,66]. In the same manner, our data showed that the G6Pase activity was not affected by dietary sorghum starch levels in hybrid red tilapia. Similar findings were reported in rainbow trout, gilthead sea bream and European sea bass suggesting that G6Pase gene expression and activity were also unaffected by dietary carbohydrate levels and sources [67-69]. PK is a key glycolytic enzyme that catalyzes the last step in glycolysis, the conversion of phosphoenol pyruvate to pyruvate [70]. Present data showed also that PK activities in liver were positively correlated with dietary sorghum starch levels, comparable to those observed in rainbow trout [39] and European sea bass [23]. In this study, the dietary sorghum starch levels did not affect PFK-1 and FBPase activities in liver, it was similar to those mentioned in other fish species [39,70-72], where these enzymes activities are not regulated by dietary carbohydrates. This finding was reported in rainbow trout, where FBPase and G6Pase activities were regulated by dietary protein levels and their activities were significantly higher with fish fed on 68% protein diet rather than 48% protein diet [73].

Conclusion

Our data suggest that the dietary sorghum starch level can incorporated up to 30% of diet, without negative effects on growth performance, nutrients utilization, digestibility coefficients, body composition and hepatic enzyme activities of carbohydrate metabolism in hybrid red tilapia fingerlings. The present study encourages the use of saline water as alternative to the limited sources of fresh water in fish culture.

References

1. Keshavanath P, Manjappa K, Gangadhara B (2002) Evaluation of carbohydrate rich diets through common carp culture in manured tanks. Aquaculture Nutrition 8: 169-174.
2. Stone DAJ, Allan GL, Anderson AJ (2003) Carbohydrate utilization by juvenile silver perch, Bidyanus bidyanus (Mitchell). III. The protein-sparing effect of wheat starch-based carbohydrates. Aquaculture Research 34: 123-134.
3. Tacon (1999) Overview of world aquaculture and aquafeed production. Data presented at World Aquaculture 99, Sydney, April 27-May 2.
4. Da CT, Hung LT, Berg H, Lindberg JE, Lundth T (2011) Evaluation of potential feed sources and technical and economic consideration of small-scale commercial striped catfish (Pangasius hypothalamus) pond farming systems in the Mekong delta of Vietnam. Aquaculture Research 44: 427-438.
5. Ganguly S, Dora KC, Sarkar S, Chowdhury S (2013) Supplementation of prebiotics in fish feed: a review. Reviews in Fish Biology and Fisheries 23: 195-199.
6. Li XF, Wang Y, Liu WB, Jiang GZ, Zhu J (2013b) Effects of dietary carbohydrate/lipid ratios on growth performance, body composition and 51 glucose metabolism of fingerling blunt snout bream (*Megalobrama amblycephala*). Aquaculture Nutrition 19: 701-708.
7. Hung LT, Lazard J, Mariojouls C, Moreau Y (2003) Comparison of starch utilization in fingerlings of two Asian catfishes from the Mekong River (Pangasius bocourti) Sauvage, 1880, Pangasius hypophthalmus Sauvage, 1878). Aquaculture Nutrition 9: 215-222.
8. Watanabe T (2002) Strategies for further development of aquatic feeds. Fisheries Science 68: 242-252.
9. Xiang-Fei L, Wen-Bin L, Kang-Le L, Wei-Na X, Ying Wang (2012) Dietary carbohydrate/lipid ratios affect stress, oxidative status and non-specific immune responses of fingerling blunt snout bream (Megalobrama amblycephala). Fish & Shellfish Immunology 33: 316-323.
10. Erfanullah, Jafri AK (1998) Effect of dietary carbohydrate-to-lipid ratio on growth and body composition of walking catfish (Clarias batrachus). Aquaculture 161: 159-168.
11. Azaza MS, Khiari N, Dhraief MN, Aloui N, Kraïem MM, et al. (2013) Growth performance, oxidative stress indices and hepatic carbohydrate metabolic enzymes activities of juvenile Nile tilapia (Oreochromis niloticus), in response to dietary starch to protein ratios. Aquaculture Research 46: 14-27.
12. Shiau SY (1997) Utilization of carbohydrates in warm water fish-with particular reference to tilapia, Oreochromis niloticus × Oreochromis aureus. Aquaculture 151: 79-96.
13. Tung PH, Shiau SY (1993) Carbohydrate utilization versus body size in tilapia, Oreochromis niloticus × Oreochromis aureus. Comparative Biochemistry and Physiology part A: Physiology 104: 585-588.
14. Amirkolaie AK, Verreth JAJ, Schrama JW (2006) Effect of gelatinization degree and inclusion level of dietary starch on the characteristics of digesta and feces in Nile tilapia (Oreochromis niloticus). Aquaculture 260: 194-205.
15. Anderson J, Jackson AJ, Matty AJ, Capper BS (1984) Effects of dietary carbohydrate and fibre on the tilapia (Oreochromis niloticus). Aquaculture 37: 303-314.
16. Deng DF, Refstie S, Hung SSO (2001) Glycemic and glycosuric responses in white sturgeon Acipenser transmontanus after oral administration of simple and complex carbohydrates. Aquaculture 199: 107-117.
17. Tan Q, Xie S, Zhu X, Lei W, Yang Y (2006) Effect of dietary carbohydrate sources on growth performance and utilization for gibel carp (Carassius auratus gibelio) and Chinese longsnout catfish (Leiocassis longirostris Günther). Aquaculture Nutrition 12: 61-70.
18. Hemre GI, Mommsen TP, Krogdahl A (2002) Carbohydrates in fish nutrition: effects on growth, glucose metabolism and hepatic enzymes. Aquaculture Nutrition 8: 175-194.
19. Moon TW (2001) Glucose intolerance in teleost fish: face or fiction?. Comp Biochem Physiol B Biochem Mol Biol 129: 243-249.
20. NRC (2011) National Research Council, Nutrient requirement of fish and shrimps. National Academy Press, Washington.
21. Wilson R (1994) Utilization of dietary carbohydrate by fish. Aquaculture 124: 67-80.
22. Enes P, Panserat S, Kaushik S, Oliva-Teles A (2011) Dietary carbohydrate utilization by European sea bass (Dicentrarchus labrax) and gilthead sea bream (Sparus aurata) Juveniles. Reviews in Fisheries Science 19: 201-215.
23. Enes P, Panserat S, Kaushik S, Oliva-Teles A (2006) Effect of normal and waxy maize starch on growth, food utilization and hepatic glucose metabolism in European sea bass (Dicentrarchus labrax) juveniles. Comp Biochem Physiol A Mol Integr Physiol 143: 89-96.
24. Booth MA, Anderson AJ, Allan GL (2006) Investigation of the nutritional requirements of Australian snapper Pagrus auratus (Bloch & Schneider 1801): digestibility of gelatinized wheat starch and clearance of an intra-peritoneal injection of D-glucose. Aquaculture Research 37: 975-985.
25. Hemre GI, Lie Ø, Lied E, Lambertsen G (1989) Starch as an energy source in feed for cod (Gadus morhua): digestibility and retention. Aquaculture 80: 261-270.
26. Krogdahl A, Hemre GI, Mommsen TP (2005) Carbohydrates in fish nutrition: digestion and absorption in post larval stages. Aquaculture Nutrition 11: 103-122.

27. Mohanta KN, Mohanty SN, Jena JK (2007) Protein-sparing effect of carbohydrate in silver barb, Puntius gonionotus fry. Aquaculture Nutrition 13: 311-317.

28. Tan Q, Wang F, Xie S, Zhu X, Wu L, et al. (2009) Effect of high dietary starch levels on the growth performance, blood chemistry and body composition of gibel carp (Carassius auratus var. gibelio). Aquaculture Research 40: 1011-1018.

29. Tian LX, Liu YJ, Yang HJ, Liang GY, Niu J (2012) Effects of different dietary wheat starch levels on growth, feed efficiency and digestibility in grass carp (Ctenopharyngodon idella). Aquaculture International 20: 283-293.

30. Klein S, Cohn SM, Alpers DH (1998) The alimentary tract in nutrition. In: Shils ME, Olson AJ, Shike M, Ross AC (Eds.), Modern Nutrition in Health and Disease pp: 605-633.

31. Bolasina S, Perez A, Yamashita Y (2006) Digestive enzymes activity during ontogenetic development and effect of starvation in Japanese flounder (Paralichthys olivaceus). Aquaculture 252: 503-515.

32. Panserat S, Skiba-Cassy S, Seiliez I, Lansard M, Plagnes-Juan E, et al. (2009) Metformin improves postprandial glucose homeostasis in rainbow trout fed dietary carbohydrates: a link with the induction of hepatic lipogenic capacities? Am J Physiol Regul Integr Comp Physiol 297: 707-715.

33. Lozano NBS, Vidal AT, Martinez-Llorens S, Merida SN, Blanco JE, et al. (2007) Growth and economic profit of gilthead sea bream (Sparus aurata L.) fed on sunflower meal. Aquaculture 272: 528-534.

34. Jobling M (1994) Fish bioenergetics, Series-13 published by Chapman & Hall, Boundary R, London SBI 8HN p: 300.

35. APHA (1992) Standard methods for the examination of water and waste water. American Public Health Association, Washington, DC p: 1134.

36. Ren MC, Ai QH, Mai KS, Ma HM, Wang XJ (2011) Effect of dietary carbohydrate level on growth performance, body composition, apparent digestibility coefficient and digestive enzyme activities of juvenile cobia, Rachycentron canadum L. Aquaculture Research 42: 1467-1475.

37. AOAC (2006) Association of Official Analytical Chemists 14th edu. Assoc Office, Anal Chem Washington, Dc.

38. Price ML, Vanscoyoc S, Butler LG (1978) A critical evaluation of the vanillin reaction as an assay for sorghum grain. J Agric Food Chem 26: 1214-1218.

39. Panserat S, Médale F, Blin C, Breque J, Vachot C, et al. (2000a) Hepatic glucokinase is induced by dietary carbohydrates in rainbow trout, gilthead sea bream, and common carp. Am J Physiol Regul Integr Comp Physiol 278: 1164-1170.

40. Foster G, Moon T (1990) Control of key carbohydrate-metabolizing enzymes by insulin and glucagon in freshly isolated hepatocytes of the marine teleost Hemitripterus americanus. Journal of Experimental Zoology 254: 55-62.

41. Metón I, Fernández F, Baanante IV (2003) Short and long-term effects of re-feeding on key enzyme activities in glycolysis–gluconeogenesis in the liver of gilthead sea bream (Sparus aurata). Aquaculture 225: 99-107.

42. Panserat S, Capilla E, Gutierrez J, Frappart P, Vachot C, et al. (2001) Glucokinase is highly induced and glucose-6-phosphatase poorly repressed in liver of rainbow trout (Oncorhynchus mykiss) by a single meal with glucose. Comp Biochem Physiol B Biochem Mol Biol 128: 275-283.

43. Bradford MM (1976) A rapid and sensitive method for the quantization of microgram quantities of protein utilizing the principle of protein-dye binding. Analytical Biochemistry 72: 248-254.

44. Carroll NV, Longley RW, Roe JH (1956) The determination of glycogen in liver and muscle by use of anthrone reagent. Journal of Biological Chemistry 220: 583-593.

45. Snedecore WG, Cochran WC (1967) Statistical Methods. Iowa state University, USA.

46. Duncan DB (1955) Multiple ranges and multiple F test. Biometrics 11: 1-42.

47. SAS (2004) SAS User's Guide Version 6th Edition. SAS Institute, Cary, NC. USA.

48. Watanabe W, French KE, Emst DH, Olla B, Wicklund R (1989) Salinity during early development influence growth and survival of Florida red tilapia. Journal World Aquaculture Society 20: 134-142.

49. Garcia-ulloa M, Villa RL, Martinez TM (2001) Growth and feed utilization of the tilapia hybrid (Oreochromis mossambicus × O. niloticus) culture at different salinities under controlled laboratory conditions. Journal World Aquaculture Society 32: 117-121.

50. Wang Y, Liu YJ, Tian LX, Du ZY, Wang JT, et al. (2005) Effects of dietary carbohydrate level on growth and body composition of juvenile tilapia, Oreochromis niloticus × O. aureus. Aquaculture Research 36: 1408-1413.

51. Yones AM (2010) Effect of lupin kernel meal as plant protein sources in diets of red hybrid tilapia (Oreochromis niloticus × O. mossambicus) on growth performance and nutrients utilization. African J Biol Sci 6: 1-16.

52. Yones AM, Abdel-Hakim NF (2010) Studies on growth performance and apparent digestibility coefficient on some common plant protein ingredients used in formulated diets of Nile tilapia (Oreochromis niloticus). Egyptian J Nutrition and Feeds 13: 589-606.

53. Yones AM (2005) Effect of dietary sorghum as carbohydrate source and two lipid levels in feeds of gilthead sea bream (Sparus aurata) on its growth performance. Egypt J Aquat & Fish 9: 85-99.

54. Brauge C, Medale F, Corraze G (1994) Effect of dietary carbohydrate levels on growth, body composition and glycaemia in rainbow trout, Oncorhynchus mykiss, reared in seawater. Aquaculture 123: 109-120.

55. Peres H, Oliva-Teles A (2002) Utilization of raw and gelatinized starch by European sea bass (Dicentrarchus labrax) juveniles. Aquaculture 205: 287-299.

56. Cheng ZY, Mai KS, Ai QH, Li Y, He ZG (2013) Effects of dietary pregelatinized corn starch on growth performance, apparent digestibility coefficient and digestive enzyme activities of large yellow croaker fingerlings, Pseudosciaena crocea. Israeli Journal of Aquaculture-Bamidgeh 65: 843-850.

57. Mohanta KN, Mohanty SN, Jena JK, Sahu NP, Patro B (2009) Carbohydrate level in the diet of silver barb, Puntius gonionotus (Bleeker) fingerlings: effect on growth, nutrient utilization and whole body composition. Aquaculture Research 40: 927-937.

58. Hemre GI, Sandnes K, Lie Ø, Torrissen O, Waagbø R (1995) Carbohydrate nutrition in Atlantic salmon, Salmo salar L.: growth and feed utilization. Aquaculture Research 26: 149-154.

59. Sá R, Pousão-Ferreira P, Oliva-Teles A (2007) Growth performance and metabolic utilization of diets with different protein: carbohydrate ratios by white sea bream (Diplodus sargus, L.) juveniles. Aquaculture Research 38: 100-105.

60. Dong XH, Guo YX, Ye DY, Song WD, Huang XH, et al. (2010) Apparent digestibility of selected feed ingredients in diets for juvenile hybrid tilapia, Oreochromis niloticus × Oreochromis aureus. Aquaculture Research 41: 1356-1364.

61. Hemre GI, Karlsen Ø, Mangor-Jensen A, Rosenlund G (2003) Digestibility of dry matter, protein, starch and lipid by cod, Gadus morhua: comparison of sampling methods. Aquaculture 225: 225-232.

62. Polakof S, Panserat S, Soengas JL, Moon TW (2012) Glucose metabolism in fish: a review. J Comp Physiol B 182: 1015-1045.

63. Richard N, Kaushik S, Larroquet L, Panserat S, Corraze G (2006) Replacing dietary fish oil by vegetable oils has little effect on lipogenesis, lipid transport and tissue lipid uptake in rainbow trout (Oncorhynchus mykiss). Br J Nutr 96: 299-309.

64. Sanchez-Muros MJ, Garcia-Rejon L, Lupianez JA, De La Higuera M (1996) Long-term nutritional effects on the primary liver and kidney metabolism in rainbow trout (Oncorhynchus mykiss). Adaptive response of glucose 6-phosphate dehydrogenase activity to high-carbohydrate/low-protein and high-fat/non-carbohydrate diets. Aquaculture Nutrition 2: 193-200.

65. Enes P, Panserat S, Kaushik S, Oliva-Teles A (2008a) Growth performance and metabolic utilization of diets with native and waxy maize starch by gilthead sea bream (Sparus aurata) juveniles. Aquaculture 274: 101-108.

66. Kirchner S, Seixas P, Kaushik S, Panserat S (2005) Effects of low protein intake on extra-hepatic gluconeogenic enzyme expression and peripheral glucose phosphorylation in rainbow trout (Oncorhynchus mykiss). Comp Biochem Physiol B Biochem Mol Biol 140: 333-340.

67. Panserat S, Médale F, Breque J, Plagnes-Juan E, Kaushik S (2000) Lack of significant long-term effect of dietary carbohydrates on hepatic glucose-6-phosphatase expression in rainbow trout (Oncorhynchus mykiss). J Nutr Biochem 11: 22-29.

68. Caseras A, Metón I, Vives C, Egea M, Fernández F, et al. (2002) Nutritional

regulation of glucose-6-phosphatase gene expression in liver of the gilthead sea bream (Sparus aurata). Br J Nutr 88: 607-614.

69. Enes P, Panserat S, Kaushik S, Oliva-Teles A (2008) Hepatic glucokinase and glucose-6-phosphatase responses to dietary glucose and starch in gilthead sea bream (Sparus aurata) juveniles reared at two temperatures. Comp Biochem Physiol A Mol Integr Physiol 149: 80-86.

70. Enes P, Panserat S, Kaushik S, Oliva-Teles A (2009) Nutritional regulation of hepatic glucose metabolism in fish. Fish Physiol Biochem 35: 519-539.

71. Zhou P, Wang M, Xie F, Deng F, Zhou O (2016) Effect of dietary carbohydrate to lipid ratios on growth performance, digestive enzyme and hepatic carbohydrate metabolic enzyme activities of large yellow croaker (Lamichthys crocea). Aquaculture 452: 45-51.

72. Gao W, LiuYJ, Tian LX, Maj KS, Liang GY, et al. (2010) Effect of dietary carbohydrate to lipid ratios on growth performance, body composition, nutrient utilization and hepatic enzyme activities of herbivorous grass carp (Ctenopharyngodon idella). Aquaculture 16: 327-333.

73. Kirchner S, Kaushik S, Panserat S (2003) Low protein intake is associated with reduced hepatic gluconeogenic enzyme expression in rainbow trout (Oncorhynchus mykiss). J Nutr 133: 2561-2564.

16

Distribution and Abundance of Finfish Eggs from Muthupettai, South East Coast of India

J Selvam*, D Varadharajan, A Babu and T Balasubramanian

Faculty of Marine Sciences, Centre of Advanced Study in Marine Biology, Annamalai University, Parangipettai, Tamil Nadu, India

Abstract

The occurrence and distribution of finfish eggs and larvae is an integral part of a fishery research programme. Finfish eggs abundance data is an important for patterns of distribution, an areas providing information for nursery ground and a range of adult and spawning trends. The density of fin fish eggs at all stations showed a seasonal variation. The maximum number of eggs were recorded during post-monsoon followed by pre-monsoon, summer and monsoon seasons. The seasonal occurrence of finfish eggs did not follow a similar pattern during the two-year period of study. This might be due to the fluctuation in the environmental parameters. Environmental parameters such as rainfall, atmospheric temperature, water temperature, salinity, pH and dissolved oxygen were recorded and correlated with the distribution of fish eggs. It is evident from the present study that the water temperature and salinity appear to play a significant role in determining the distribution of fin fish eggs in the study area.

Keywords: Fin fish eggs; Distribution; Abundance; Physico-chemical

Introduction

Estimation of abundance of fish eggs and larvae helps to evaluate marine fishery resources. Most of the marine fishes spawn in the open sea and produce pelagic eggs and larvae [1,2]. By regular collection of plankton, it is possible to map the marine area with respect to the breeding of fish and the relative abundance of ichthyoplankton of commercial fish stock, and use the information as an index of fish abundance or for the prediction of year class strength. Generally fishes spawn during a definite time of the day and this has been found to be true in marine fishes of Porto-Novo region [3,4]. Investigation on the occurrence and distribution of finfish eggs and larvae is an integral part of a fishery research programme. Most of the eggs and almost the larvae are pelagic and it is easy to sample several species over a wide area with simple plankton nets. Regular sampling of ichthyoplankton is essential for locating shoals of adult fishes and their spawning grounds. Ichthyoplankton studies are extensively useful in fishery investigation. Information on fish eggs and larvae of a particular region is useful in understanding the spawning season of fishes of commercial importance. Studies on the early developmental stages of fish allow us to comprehend the biology of the species besides determining their spawning seasons and to estimate spawning stock abundance. Such a study is also an essential prerequisite in undertaking the spawning biomass of target species monitoring, changes in exploitable stocks and yields, forecasting trends of production etc. [3-5]. Distribution of the early developmental stages, in space and time is known with considerable precision so that sampling effort can be efficiently concentrated in areas and time periods when they will be most effective [6]. Generally, fishes spawn during a definite time, hence, studies on the seasonal occurrence of fish eggs and larvae are useful in locating shoals of fish and their breeding grounds. Till recently studies on the quantitative aspects of fish eggs and larvae in Indian seas were limited to studies on their taxonomy, seasonal abundance based on material from the inshore plankton and post larval fish collections from restricted localities.

Materials and Methods

The study was conducted at Muthupettai coast during between from January 2011 to December 2012. Finfish eggs were collected every month in the early hours of the day during high tide, with the help of plankton net of diameter 0.5 m made of bolting silk (No: 10 mesh size, 158 μm). Volume of water filtered was quantified with the help of a calibrated flow meter (General Oceanics, INC model) attached to it. The net was towed horizontally along the surface water at a constant speed of 1.0 km/hr for about 15-20 minutes in each station by adopting the method of Venkataramanujam, Ramamoorthi and Bensam [7,8]. Samples from all the stations were preserved onboard in 5% buffered formalin-seawater and sorted in the laboratory [9]. Fin fish eggs were sorted out from this sample and their abundance was expressed as number of eggs/100 m^3.

Description of the Study Area

Station I (Sethukuda)

This station is situated at 10°20'49.68" N Lat. and 79°32'13.23" E Long. The average depth of the station is about 1 metre. *Avicinnia* sp. is dominant in this station (Plate 1).

Station II (Lagoon)

This station is situated between 10°20'19.71" N Lat. and 79°31'52.64" E Long. The Lagoon is shallow with an average of 1 m depth. *Avicennia* sp. borders the Lagoon (Plate 1).

Station III (Chellimunai)

This station is situated at 10°19'42.56" N Lat. and 79°33'35.10" E Long. This station is shallow with an average depth of 1.5 m. The station is dominated by *Avicennia* sp (Plate 1).

***Corresponding author:** J Selvam, Faculty of Marine Sciences, Centre of Advanced Study in Marine Biology, Annamalai University, Parangipettai, Tamil Nadu, India
E-mail: jagaselvam@yahoo.com

Station IV (River Mouth)

This station is situated at 10°18'2076" N Lat. and 79°31'08.02" E Long (Plate 1).

Results and Discussion

Seasonal abundance and species/genus wise density of finfish eggs in the Muthupettai waters

Family: Ophichthidae

Ophichthys **sp.**: *Ophichthys* sp. single egg was observed during April-2011 at station II and maximum (29 eggs/100 m^3) observed during January-2011 at station IV. During second year also a single egg was noticed in the station II during July-2012 and Maximum 36 eggs/100 m^3 observed during January -2012 at station IV (Figure 1).

Considerable numbers of *Ophichthys* sp. eggs were observed during pre-monsoon, early month of monsoon and late month of post monsoon seasons. Present observation gives the picture about lengthened spawning period of these fishes. However, Ganapathi and Raju [10] observed only in post-monsoon season along Waltair coast. Manickasundaram [11] observed in early month of monsoon and pre-monsoon season at Coleroon estuary complex along the south east coast of India.

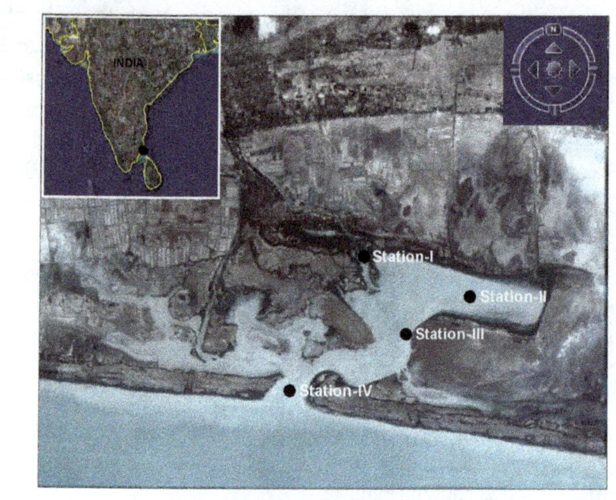

Plate 1: The study area (Muthupettai).

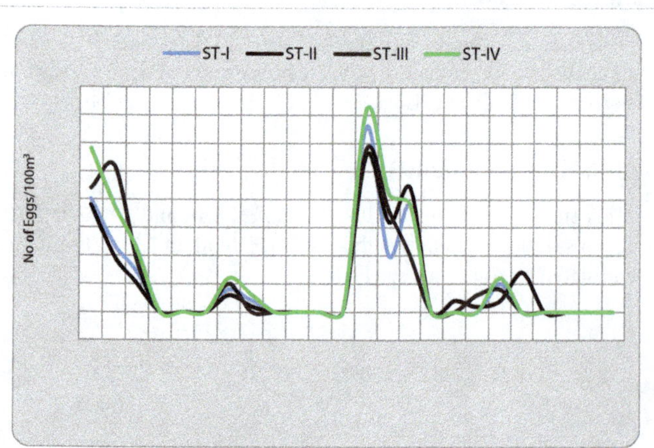

Figure 1: Distribution of *Ophichthys* sp eggs along Muthupettai waters during January-2011 to December-2012.

Family: Engraulidae

Setipinna taty: Minimum density (3 eggs/100 m^3) was observed in May 2011 at station II and maximum (30 eggs/100 m^3) in February 2011 at station III. In 2012 only single egg observed in the station III during December-2012 and maximum (32 eggs/100 m^3) at station III during the month of January (Figure 2).

Thangaraja [12] observed the occurrence of this species eggs during post-monsoon season in the Parangipettai coastal waters. The present observation in the Muthupettai waters confirms that the spawning activity of these fishes is during the post-monsoon season.

Stolephorus tri: *Stolephorus tri* 3 eggs/100 m^3 of were collected during May 2011 at station II and (54 eggs/100 m^3) in November at station IV. In 2012, the minimum density of 2 eggs/100 m^3 were observed in January-2012 at station I and maximum of 60 eggs/100 m^3 in October -2012 at station III (Figure 3).

Occurrences of the eggs of *Stolephorus tri* was observed almost throughout the year in Muthupettai waters. Bensam [8] recorded the eggs of this species during summer months from Parangipettai waters, Nair [13] observed December to January in the Madras coastal waters. Siraimeetan and Marichamy [14] observed biannual spawning season of this species along the Tuticorin coast, Ramaiyan et al. [15]

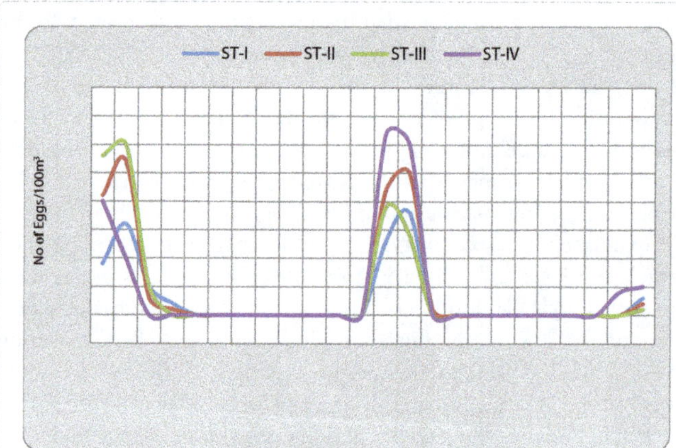

Figure 2: Distribution of *Setipinna taty* eggs along Muthupettai coastal waters during the study period (January-2011 to December-2012).

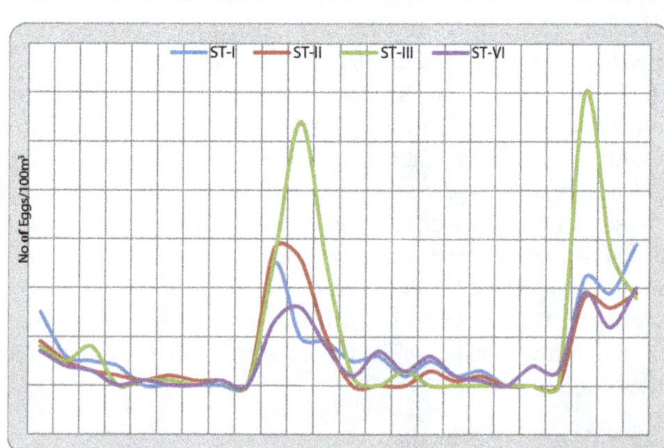

Figure 3: The monthly variation of *Stolephorus tri* eggs Muthupettai coastalwaters.

collected these eggs throughout the year from Parangipettai coast, these observations are in support of the present findings.

Stolephorus punctifer: Minimum number (1 egg/100 m^3) was observed in July 2011 at station I and maximum (35 eggs/100 m^3) in August 2011 at station III. During the second year 2012, the minimum density (1 egg/100m^3) was observed during May -2012 at station III and maximum density (31 eggs/100 m^3) in July at station III (Figure 4).

Eggs of *Stolephorus punctifer* were collected during the early month of pre-monsoon (July-September), post-monsoon (January-March) and summer (October-December) during the present investigation. Observation made during post-monsoon and summer from Madras and along Coleroon estuary and observed this eggs in Porto Novo waters during both pre and post-monsoon and also in summer are in support of the present investigation [11,13,15].

Stolephorus heterolobus: Minimum number (1 egg/100 m^3) was observed in April at station IV and maximum (60 eggs/100 m^3) in January at station III. In the second year these eggs were observed minimum density in the station IV (1 egg/100 m^3) during May 2012 and maximum (55 eggs/100 m^3) in station III during March 2012 (Figure 5).

In the present investigation, *Stolephorus heterolobus* eggs was observed only during post monsoon and summer seasons and is suggesting that the breeding of this species may be takes place only in these seasons. This observation coraboram works are made in the Coleroon estuarine complex and along the Parangipettai coastal waters [11,15].

Stolephorus macrops: Figure 6 shows the monthly variation of eggs of *Stolephorus macrops* along Muthupettai coastal waters. Minimum density (2 eggs/100 m^3) was observed in September at station II and maximum (24 eggs/100 m^3) in February at station II during the year 2011. In 2012, a single egg was observed in October at station IV and maximum density (20 eggs/100 m^3) was observed during February at station I.

Stolephorus macrops eggs were found to occur during postmonsoon and premonsoon, in the present investigation. In Parangipettai coastal waters, Thangaraja [12] found these eggs during February to March and Manickasundaram [11] observed during January to March with peak abundance in March along Coleroon estuary. Ramaiyan et al. [15] observed these eggs during post monsoon and summer season in Parangipettai waters.

Thryssa dussumieri: Figure 7 shows the distribution of the eggs of *Thryssa dussumieri* along Muthupettai waters. During the study period, the minimum density (1 egg/100 m^3) was observed during September at station III and maximum (38 eggs/100 m^3) in March at station IV.

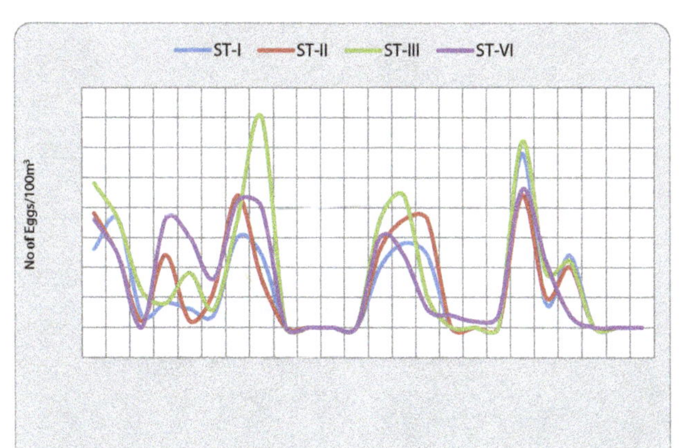

Figure 4: Distribution of *Stolephorus punctifer* eggs along Muthupettai coastal waters during the study period (January-2011 to December-2012).

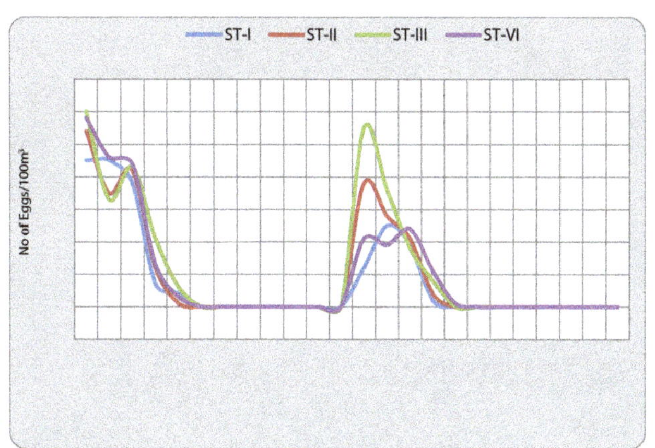

Figure 5: Distribution of *Stolephorus heterolobus* eggs along Muthupettai coastal waters (January-2011 to December-2012).

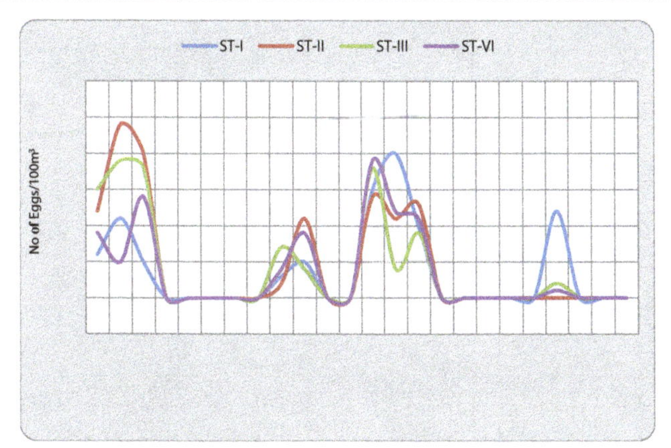

Figure 6: Distribution of *Stolephorus macrops* eggs along Muthupettai coastal waters (January-2011 to December-2012).

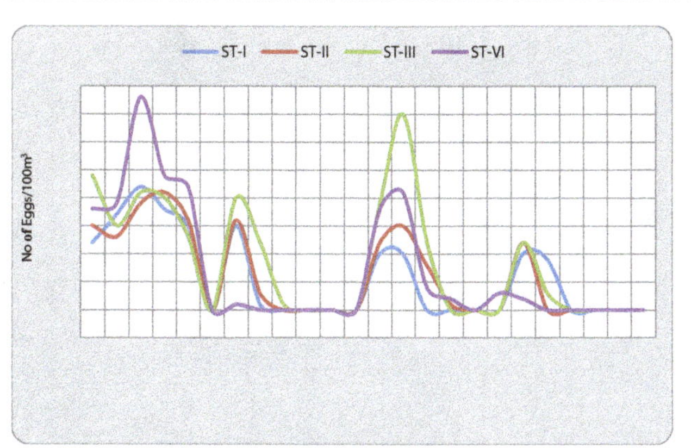

Figure 7: Distribution of *Thryssa dussumieri* eggs along Muthupettai coastal waters (January-2011 to December-2012).

In 2012, the minimum density (1 egg/100 m^3) was observed in April at station II and maximum (35 eggs/100 m^3) in August at station III.

Thryssa dussumieri eggs were found in pre monsoon, post monsoon and summer seasons during the present investigation. The present observations are in agreement with the earlier ones made by [7,8,12,15] along Parangipettai coastal waters.

Thryssa mystax: Figure 8 shows the monthly variation of eggs of *Thryssa mystax* along Muthupettai waters. In the year 2011, minimum density was observed during August (1 egg/100 m^3) at station I and the maximum (21 eggs/100 m^3) during January at station IV. In year 2012, the minimum density of these eggs (2 eggs/100 m^3) was noticed in August at station I and maximum (23 eggs/100 m^3) in February at station III.

Thryssa mystax eggs were found in this investigation during post monsoon, early months of summer and premonsoon seasons in Muthupettai waters. Similar observations were made from Coleroon estuarine, Parangipettai coastal waters and in Madras the coast [7,11,15,16].

Thryssa hamiltoni: Monthly variation in the distribution of *Thryssa hamiltoni* eggs along Muthupettai coastal waters is shown in Figure 9. The minimum density of these eggs (1 egg/100 m^3) were observed in April at station IV and the maximum density of (22 eggs/100 m^3) were observed in January at station II. In the year 2012, 3 eggs/100 m^3 were observed in July at station II and (22 eggs/100 m^3) in February 2012 at station I.

Eggs of *Thryssa hamiltonii* were found to be abundant during the post monsoon and summer months. The eggs of *Thryssa hamiltonii* were also observed [7,8,15].

Family: Pristigastridae

Opisthopterus tardoore: During the first year, the minimum density (1 egg/100 m^3) was observed in December at station I and maximum (42 eggs/100 m^3) in October at station IV. During the second year, the minimum density (2 eggs/100 m^3) was observed in October at station II and maximum (49 eggs/100 m^3) in March at station III (Figure 10).

Opisthopterus tardoore eggs were found to be abundant in the Muthupettai waters. These eggs were reported during post monsoon and summer in the Vellar estuary in the Coleroon estuarine [7,11,15,17].

Family: Chirocentridae

Chirocentrus dorab: The distribution of the eggs of *Chirocentrus dorab* is depicted in the Figure 11. During the first year of the study (2011), one was observed in September at station IV and 27 eggs/100 m^3 in September at station III. In the year 2012, the minimum density of 1 egg/100 m^3 was observed in December at station III and a maximum 23 eggs/100 m^3 in November at station III.

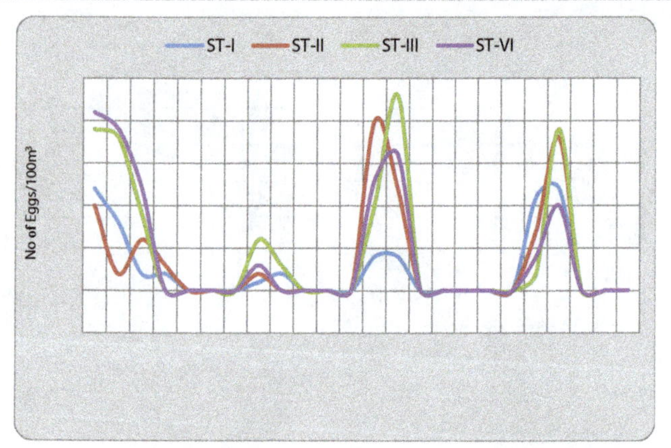

Figure 8: Distribution of *Thryssa mystax* eggs along Muthupettai waters (January-2011 to December-2012).

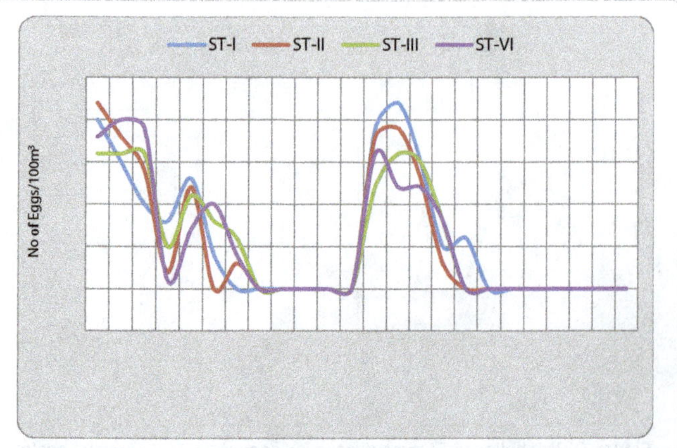

Figure 9: Distribution of *Thryssahamiltoni* eggs along Muthupettai coastal waters (January-2011 to December-2012).

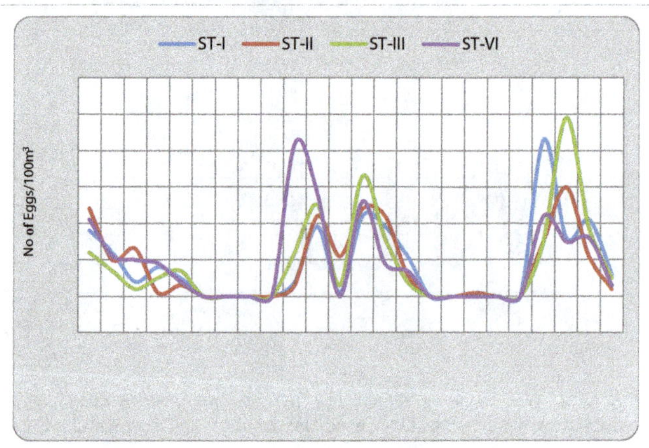

Figure 10: Distribution of *Opisthopterus tardoore* eggs along Muthupettai coastal waters (January-2011 to December-2012).

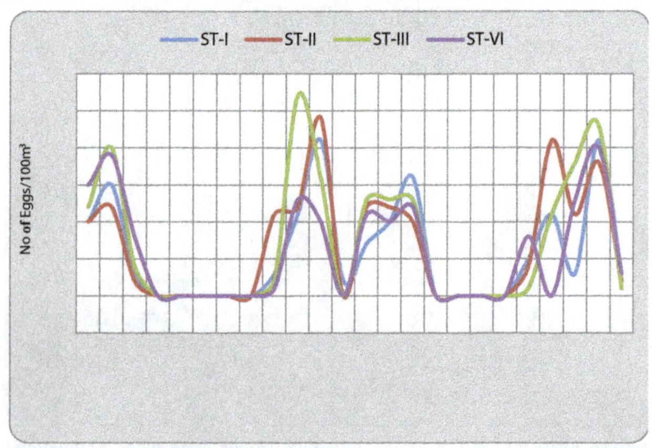

Figure 11: Distribution of *Chirocentrus dorab* eggs along Muthupettai waters (January-2011 to December-2012).

Chirocentrus dorab eggs were collected during premonsoon, monsoon and postmonsoon seasons from Muthupettai waters. Similar observations were made in the Coleroon estuarine system as well [11].

Family: Clupeidae

***Esculosa thoracata*:** A single egg observed in December at station IV and maximum

(25 eggs/100 m^3) in February at station IV. In the second year, the minimum density was in April (1 egg/100 m^3) at station II and maximum (55 eggs/100 m^3) in October at station II (Figure 12).

The abundance of *Esculosa thoracata* eggs was found during March, May to August from Mandapam and during March to October from Chilka Lake [18,19]. Their distribution during March to October from Parangipettai was reported [17]. These fishes appear to spawn over an extended period in Muthupettai waters in the Vellar estuary and in the Coleroon estuary [7,11].

***Sardinella fimbriata*:** During the first year of the study, the minimum density was in September (3 eggs/100 m^3) at station I and maximum (54 eggs/100 m^3) in February at station III. During the second year of the study, the minimum density (1 egg/100 m^3) was collected during April at station I and maximum in January (45 eggs/100 m^3) at station IV (Figure 13). *Sardinella fimbriata* eggs were collected during premonsoon, post monsoon and summer seasons. Similar observations were made along Coleroon estuary, Parangipettai coastal waters and along Madras coast [7,8,11,17,20].

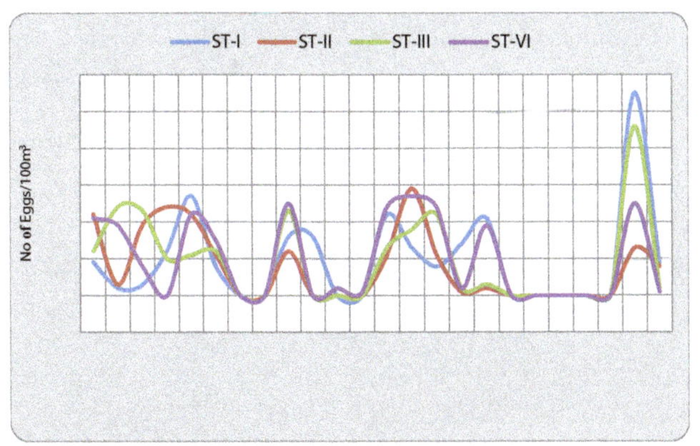

Figure 12: Distribution of *Esculosa thoracata* eggs along Muthupettai waters (January-2011 to December-2012).

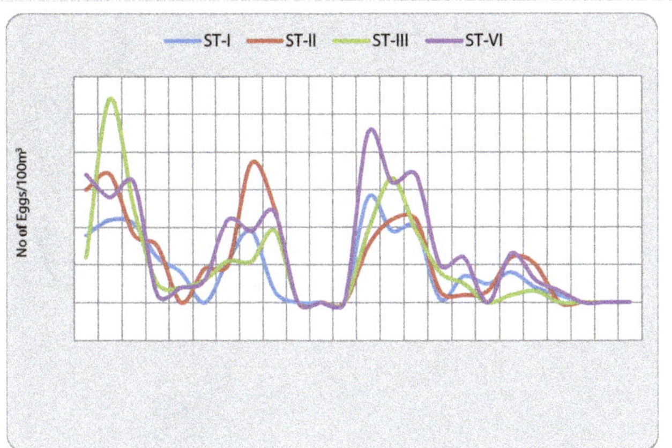

Figure 13: Distribution of *Sardinella fimbriata* eggs along Muthupettai waters during the study period (January-2011 to December-2012).

***Sardinella gibbosa*:** Figure 14 shows the monthly variation of *Sardinella gibbosa* eggs along Muthupettai coastal waters. During the first year of the study period the minimum density (8 eggs/100 m^3) were observed in October at station III and maximum (49 eggs/100 m^3) in January at station III. But during the second year, the minimum density of these eggs (3 eggs/100 m^3) was encountered during September at station I and maximum (45 eggs/100 m^3) in January in the same station.

The present findings revealed that the eggs of Sardinella gibbossa were noticed during monsoon, post monsoon and summer seasons with a peak spawning activity in January to May along Muthupettai waters. Similar observations were reported from Parangipettai coastal waters and along Coleroon estuary that agreed with the present observation [7,11,17].

***Sardinella longiceps*:** Figure 15 shows the monthly variation of *Sardinella longiceps* eggs along Muthupettai coastal waters. During the first year of the study, a single egg observed in May at station II and maximum (22 eggs/100 m^3) in March and April at station I. During the second year, single egg observed at station I during September and maximum (23 eggs/100 m^3) in station III during February.

Sardinella longiceps eggs were present during post monsoon, summer and early month of premonsoon seasons along Muthupettai waters. The present result confirms that the discontinuous spawning habit of these fishes. Findings of John and Devanesan [20,21] have

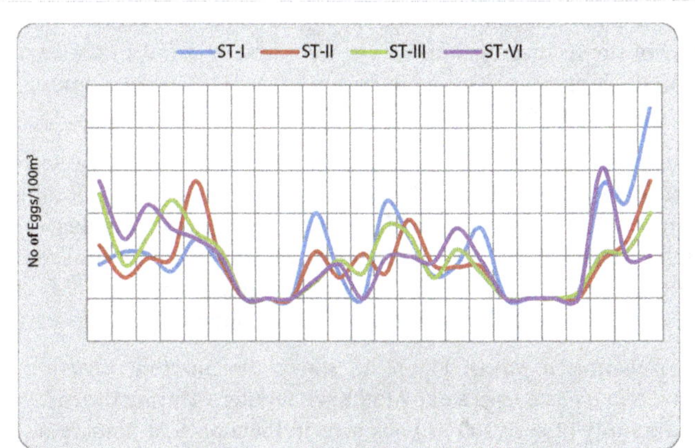

Figure 14: Distribution of *Sardinella gibbosa* eggs along Muthupettai coastal waters during the study period (January-2011 to December-2012).

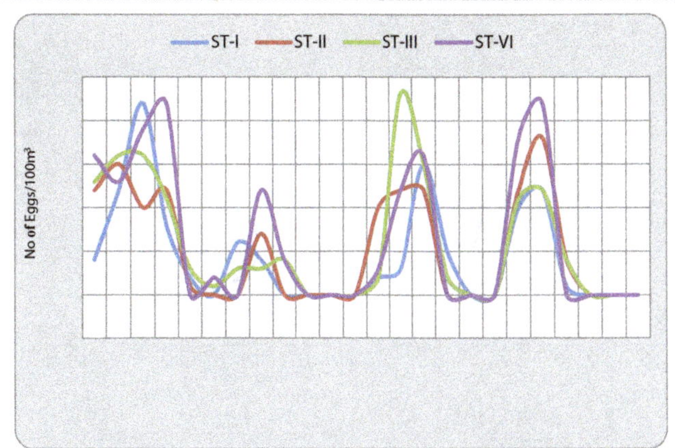

Figure 15: Distribution of *Sardinella longiceps* eggs along Muthupettai coastal waters during the study period (January-2011 to December-2012).

noticed these eggs in the plankton sample during post monsoon seasons; Lazarus [22] collected these eggs during premonsoon and monsoon seasons in Madras coast. Venkataramanujam [17] observed these eggs during monsoon, post monsoon and summer seasons and Venkataramanujam and Ramamoorthi [7] noticed during post monsoon and summer seasons along Parangipettai coastal waters. Ganapathi and Raju [10] observed during the premonsoon, post monsoon and summer seasons with peak abundance during February along Coleroon estuary were agreed with the present findings.

***Sardinella clupeoides*:** Figure 16 shows the monthly variation of *Sardinella clupeoides* eggs along Muthupettai coastal waters. During the first year of the study, the minimum numbers of eggs (2 eggs/100 m^3) were observed in May at station I and maximum (33 eggs/100 m^3) in February at station IV. In the second year, single egg was in May at station II and III and maximum (40 eggs/100 m^3) in the same month at station VI.

Eggs of *Sardinella clupeoides* were collected during post monsoon, summer, early month of premonsoon and monsoon seasons with a peak in February along Muthupettai waters. Similar findings were made and these eggs were observed during monsoon, post monsoon and summer seasons, these eggs appeared in post monsoon and summer seasons from Vellar estuary and were observed during monsoon, post monsoon and summer seasons with a peak during February which supports the present investigation [8,11,17].

Anadontostoma chacunda: Figure 17 shows the monthly variation of *Anadontostoma chacunda* eggs along Muthupettai waters. In the first year of the in investigation no egg were observed, but in the second year, single egg was observed in January at station II and maximum (11 eggs/100 m^3) in April at station III.

Eggs of *Anadontoma chacunda* were observed only in the second year of the study period in the post monsoon and the early months of summer. Annual spawning of these fish agrees with the previous works on the east coast where similar observation from Parangipettai coastal waters were reported [7,12,17,18,20]. Manickasundaram M [11] observed these eggs during February to May along Coleroon estuary.

***Nematolosa nasus*:** Figure 18 shows the monthly variation of *Nematolosa nasus* eggs along Muthupettai waters. During the first year of the study (2 eggs/100 m^3) observed in the month of January in the station I and maximum (8 eggs/100 m^3) were observed in February at Station I. During the second year of the study a single egg of this species observed during March at station I and Maximum (16 eggs/100 m^3) in the station IV in the month of December.

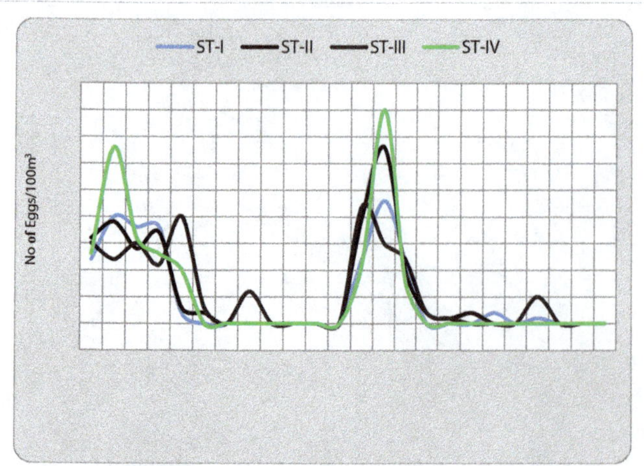

Figure 16: Distribution of *Sardinella clupeoides* eggs along Muthupettai coastal waters during the study period (January-2011 to December-2012).

Eggs of *Nematolosa nasus* were collected during late months of monsoon and post monsoon seasons during the present study. This present observation supported by the previous work noticed these eggs during the post monsoon season from Vellar estuary and this eggs were observed in the monsoon and post monsoon season in parangipettai waters [8,15].

Family: Synodontidae

*Saurida **gracilis***: Figure 19 shows the seasonal distribution of *Saurida gracilis* eggs along Muthupettai coastal waters. During the first year of the study, a single egg observed in the month of June in the station I and maximum (15 eggs/100 m^3) in February at station IV. In second year, the minimum number of egg (1 egg/100 m^3) was observed in January and February at station I and maximum (18 eggs/100 m^3) in January at station IV.

Saurida gracilis eggs were collected during post monsoon, summer, late months of premonsoon and late month of monsoon seasons gives the picture about the prolonged spawning of these fishes. It was noticed during post monsoon and summer along Parangipettai coastal waters, during post monsoon along Coleroon estuary and these eggs

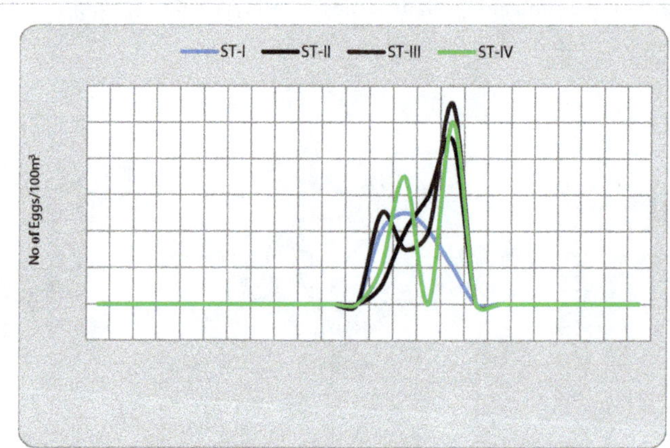

Figure 17: Distribution of *Anadontostoma chacunda* eggs along Muthupettai coastal waters during the study period (January-2011 to December-2012).

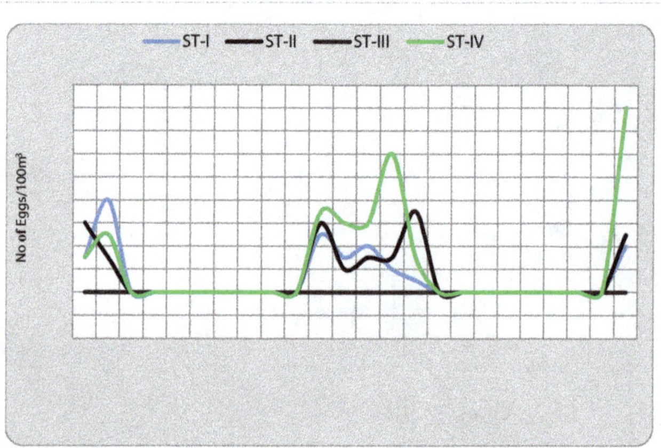

Figure 18: Distribution of *Nematolosa nasus* eggs along Muthupettai coastal waters during the study period (January-2011 to December-2012).

were observed in post monsoon and summer, seasons in Parangipettai waters in the support of the present investigation [7,11,15,17,23].

Saurida ***tumbil***: Figure 20 shows the monthly variation of *Saurida tumbil* eggs along Muthupettai coastal waters. During the first year of the study, the minimum density of these eggs (1 egg/100 m^3) was observed in April at station II and maximum (9 eggs/100 m^3) in May at station the same station. During the second year of study, a single egg was obtained in June at station IV and the maximum (6 eggs/100 m^3) in May at the same station.

Saurida tumbil eggs were collected during post monsoon, summer, premonsoon and late month of monsoon seasons along Muthupettai coastal waters. However, Venkataramanujam and Ramamoorthi [7] noticed during post monsoon and summer along Parangipettai coastal waters, Venkataramanujam [17] observed during post monsoon, summer and premonsoon seasons, Vijayaraghavan [24] observed during monsoon season along Madras coast.

***Saurida* sp.**: Eggs of *Saurida* sp. (5 eggs/100 m^3) observed collected in April at station IV and a single egg was in the same month at station II, in the second year of study period (9 eggs/100 m^3) eggs were obtained in the month of April at the station II and (3 eggs/100 m^3) collected in the same month at the III and IVth stations along Muthupettai coastal waters (Figure 21). Ramaiyan et al. [15] observed this egg during early months of summer along the Parangipettai coast is the similar to the present findings.

***Saurus* sp.**

A single egg of *Saurus* sp. was collected during April from station II and 5 eggs /100 m^3 eggs were obtained in the same month in the station IV. In the second year of the investigation minimum (3 eggs/100 m^3) were observed in the month of April in the station IV and maximum (6 eggs/100 m^3) were observed in the same month at the station II (Figure 22). Bapat [18] observed during summer season along Mandapam coast and Manickasundaram [11] noticed only during April along Coleroon estuarine are similar to the present findings.

Synodontid **Eggs:** Only few of *Synodontid* eggs observed in the first year of the investigation. Minimum (3 eggs/100 m^3) observed in April at Station II and maximum (5 eggs/100 m^3) in the same month of the station IV (Figure 23). Ramaiyan et al. [15] observed the eggs in the month monsoon and pre monsoon in the Parangipettai waters is the support of the present findings.

Family: Mugilidae

Liza dussumieri: The seasonal distribution of eggs of *Liza dussumieri* is depicted in Figure 24. The minimum density (2 eggs/100 m^3) was observed in April at station II and maximum (17 eggs/100 m^3) in January at station I. During the second year, the minimum number eggs (4

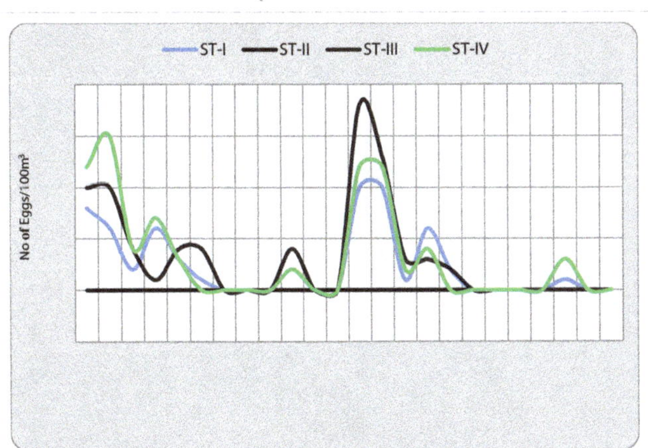

Figure 19: Distribution of *Saurida gracilis* eggs along Muthupettai coastal waters during the study period (January-2011 to December-2012).

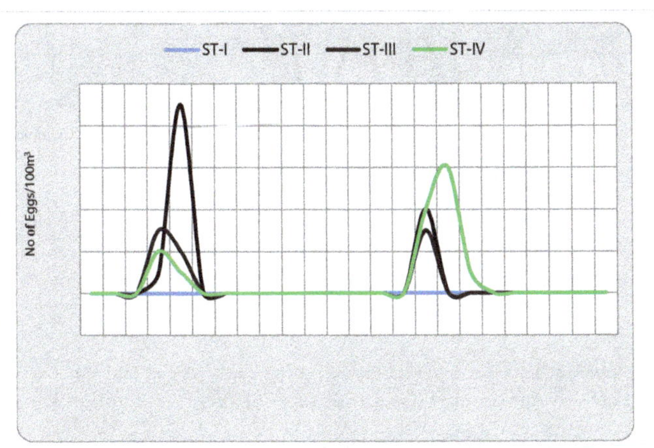

Figure 21: Distribution of *Saurida sp eggs* along Muthupettai coastal waters during the study period (January-2011 to December-2012).

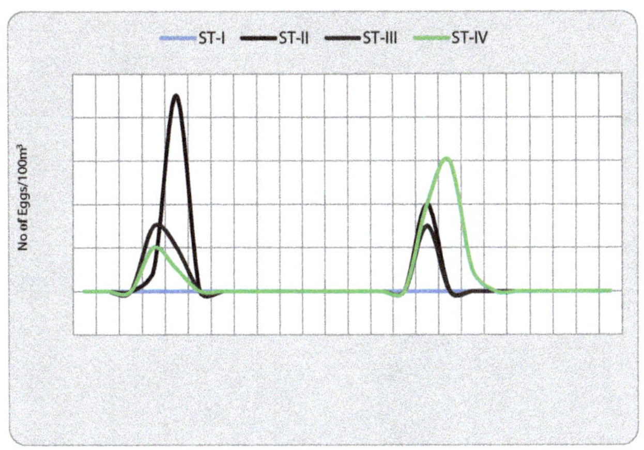

Figure 20: Distribution of *Saurida tumbil* eggs along Muthupettai waters (January-2011 to December-2012).

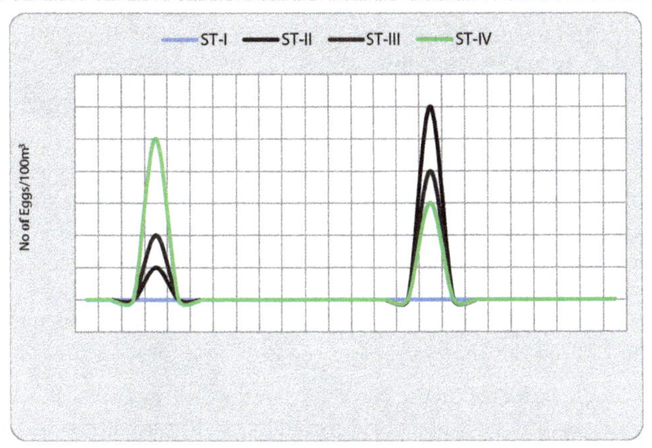

Figure 22: Distribution of *Saurus* sp eggs along Muthupettai coastal waters during the study period (January-2011 to December-2012).

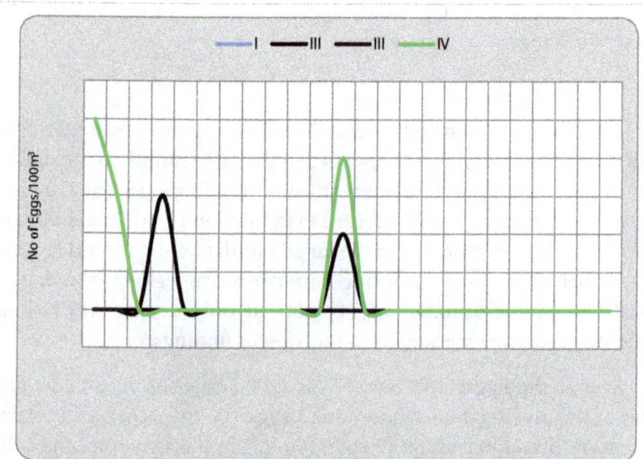

Figure 23: Distribution of *Sunodontid eggs* along Muthupettai coastal waters during the study period (January-2011 to December-2012).

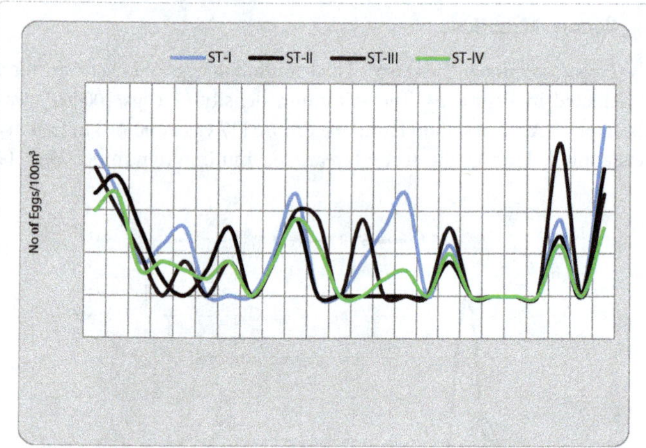

Figure 24: Distribution of eggs along Muthupettai coastal waters during the study period (January-2011 to December-2012).

eggs/100 m^3) were observed in December and May at station I and II respectively, maximum (20 eggs/100 m^3) in December at station I.

However it was observed during premonsoon season and also during post monsoon, summer, and early months of monsoon in the support of the present findings [8,15].

***Liza tade*:** Figure 25 shows the seasonal variation of *Liza tade* eggs along Muthupettai coastal waters. During the first year, the minimum density was observed in April

(2 eggs/100 m^3) at station II and maximum (20 eggs/100 m^3) in February and December at station I and IV respectively. During the second year, the minimum density (2 eggs/100 m^3) was in September at station II and maximum in January (22 eggs/100 m^3) at station IV. Eggs of *Liza tade* were noticed and these eggs were reported during premonsoon season along Parangipettai coastal waters, in the post monsoon and summer seasons along the Cuddalore coast, and its occurrence is reported throughout the year in Pitchavaram mangrove area [8,15,25].

***Mugil cephalus*:** Figure 26 shows the seasonal variation of *Mugil cephalus* eggs along Muthupettai coastal waters. The minimum density of these eggs was collected during April (5 eggs/100 m^3) at station II and maximum density was observed during February (21 eggs/100 m^3) at station IV during the study period of first year. In the second year single egg was obtained in March at the station II and maximum (20 eggs/100 m^3) in February at station III.

The present observation agrees with the previous findings during January to April and July to October along Coleroon estuary, during August from Vellar estuary, the eggs were recorded during premonsoon and post monsoon seasons and these eggs were noticed during July to August from Chilka lake [11,12,17,19].

Family: Hemiramphidae

***Hemiramphus* sp.:** A single egg of *Hemiramphus* sp. was observed in February at station II during the second year of study. Single egg observed in station III during May, and (2 eggs/100 m^3) observed in the same month at the station III. During the second year of the investigation single egg was observed in the month of June at the station III (Figure 27).

Another type of Hemiramphid eggs was reported during the summer season and this egg was observed during February along Parangipettai coastal waters [11,15].

Family: Atherinidae

Pranesus pinguis: Figure 28 gives the details of the distribution of

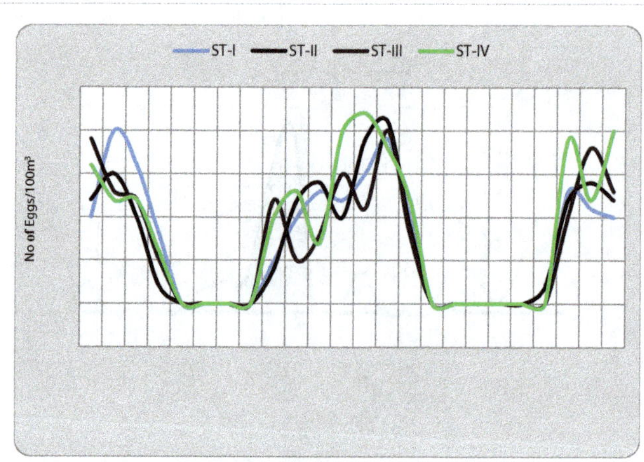

Figure 25: Distribution of *Liza tade* eggs along Muthupettai coastal waters during the study period (January-2011 to December-2012).

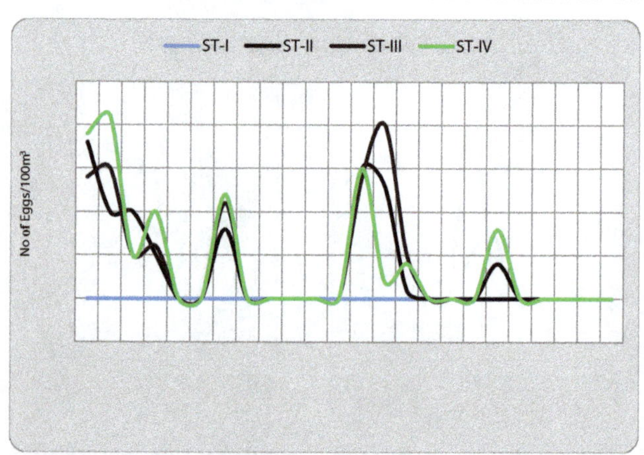

Figure 26: Distribution of *Mugilcephalus* eggs along Muthupettai coastal waters during the study period (January-2011 to December-2012).

Pranesus pinguis eggs along Muthupettai coastal waters. During the first year, single egg was observed in the month of February at station I and maximum (12 eggs/100 m^3) in June at station III. However, during the second year, the minimum density of single egg observed in September at station I and maximum (20 eggs/100 m^3) in February at station II.

Present observation supported by the previous works made by Thangaraja [12]. But it was recorded during March, May to September and October from Vellar estuary and reported during March to June and September to November along Coleroon estuary [11].

Family: *Carangidae*

***Carangoides malabaricus*:** The distribution of *Carangoides malabaricus* eggs is given in Figure 29. During the first year, the minimum density of single egg observed in May at the station III and IV and maximum (18 eggs/100 m^3) in December at station I. The minimum density (2 eggs/100 m^3) of these eggs was collected during April at station I and maximum (10 eggs/100 m^3) during September at the same station I in the second year.

Eggs of Perciformes were numerically abundant and collected almost throughout the year in the present investigation. Same observation was made along the Vellar estuary[17]. During the post monsoon, summer and premonsoon seasons the carangid eggs were more in number and they were collected in all the stations, as already reported during May to June from Coleroon estuary [11]. But only during January it was reported along Madras coast [16]. Venkataramanujam [17] from Vellar, Bapat [18] from Mandapam and George [26] along Cochin backwaters also. Eggs of *Carangoides malabaricus* were observed during premonsoon, monsoon and summer in the present study. Krishnamurthy and Prince Jeyaseelan [27] recorded only in summer from Pichavaram mangrove systems.

***Caranx* sp. 1:** Figure 30 shows the seasonal distribution of *Caranx* sp. 1 egg along Muthupettai coastal waters. During the first year, the minimum density (2 eggs/100 m^3) was recorded in July at station III and maximum (100 eggs/100 m^3) in January at station IV. During second year of the investigation single egg was observed in December at station I and maximum (99 eggs/100 m^3) in February at the same station I.

Caranx sp. 1 eggs were available throughout the year in the present study revealed some clear trends. This result obtained in the present study confirms that these fishes are continuous spawners. Similar observation made along Vellar estuary, Coleroon estuary, Mandapam region and from Chilka Lake [7,11,19].

***Caranx* sp. 2:** Figure 31 shows the seasonal distribution of the eggs of *Caranx* sp. 2 along Muthupettai coastal waters. During the first year

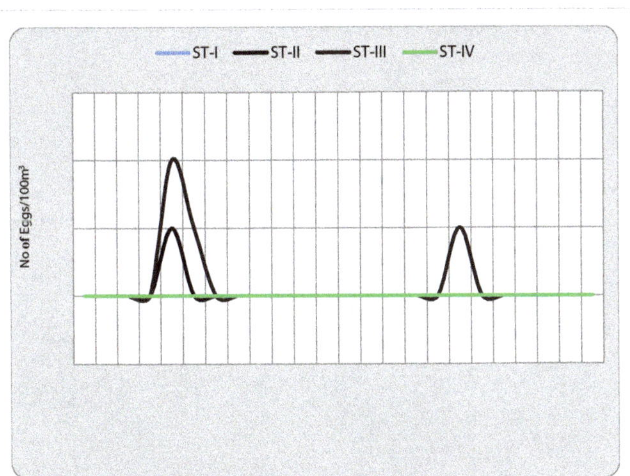

Figure 27: Distribution of *Hemiramphus* along Muthupettai waters during the study period (January-2011 to December-2012).

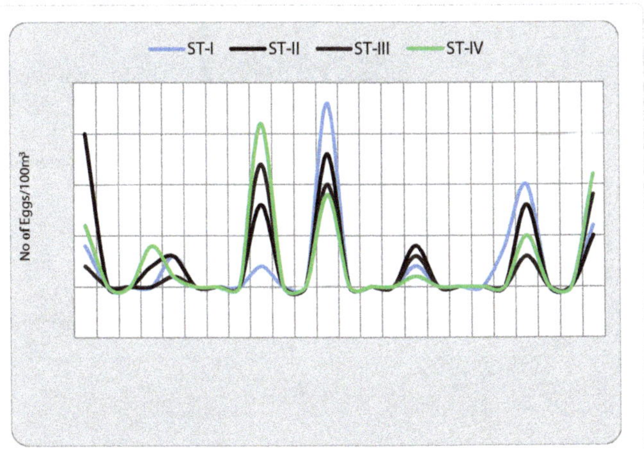

Figure 29: Distribution of *Carangoides malabaricus* eggs along Muthupettai coastal waters during the study period (January-2011 to December-2012).

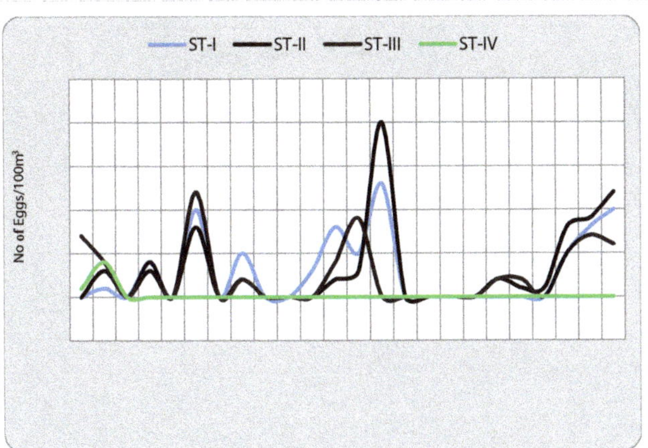

Figure 28: Distribution of *Pranesus pinguis* eggs along Muthupettai coastal waters during the study period (January-2011 to December-2012).

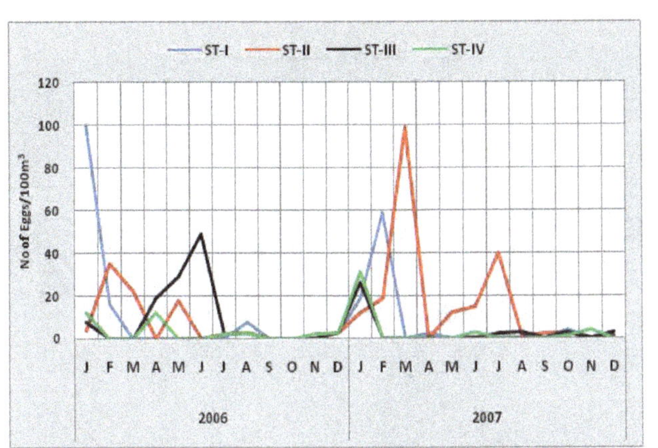

Figure 30: Distribution of *Caranx* sp 1 eggs along Muthupettai coastal waters during the study period (January-2011 to December-2012).

of the study period, the minimum density(2 eggs/100 m^3) were in July at station II and maximum (55 eggs/100 m^3) in August at station I. In the second year of the study a single egg observed in the month of February and March in the station I and maximum (21 eggs/100 m^3) in April at station IV.

Manickasundaram [11] observed throughout the year along Coleroon estuary was agreed and Venkataramanujam [17] observed during January to July and September to November along Parangipettai coastal waters with the present investigation.

Decapterus russelli: Figure 32 shows the seasonal distribution of the eggs of *Decapterus russelli* along Muthupettai coastal waters. During the first year, the minimum density (2 eggs/100 m^3) was observed in December at station II and maximum (23 eggs/100 m^3) in May station IV. During the second year, a single egg was observed in March at station I and maximum density (23 eggs/100 m^3) in January at station I.

Decapterus russelli eggs were observed during post monsoon, and monsoon seasons in the present study. However, Manickasundaram [11] noticed during post monsoon season along Coleroon estuary and Ramaiyan et al. [15] observed these eggs in the monsoon and post monsoon seasons was in agreement of the present findings.

***Scomberoides tol*:** Figure 33 shows the monthly variation of *Scomberoides tol* eggs along Muthupettai coastal waters. During the first year, a single egg was collected in September at station IV and maximum (23 eggs/100 m^3) in April at station III. During the second year, the minimum density (4 eggs/100 m^3) was in January at station II and maximum (22 eggs/100 m^3) in August at station III.

Scomberoides tol eggs were collected in this investigation during premonsoon, and post monsoon season. Similar observation made by Manickasundaram [11] along Coleroon estuary and Ramaiyan et al. [15] along Parangipettai waters.

Family: *Leiognathidae*

***Secutor ruconius*:** Figure 34 gives the details of the distribution of *Secutor ruconius* eggs along Muthupettai coastal waters. During the first year, the minimum density (2 eggs/100 m^3) was observed in February at station I and maximum (34 eggs/100 m^3) in February at station II. The minimum density (4 eggs/100 m^3) were in December at station I and maximum (25 eggs/100 m^3) in January and February at station I and II respectively in second year of investigation.

Venkataramanujam and Ramamoorthi [7] noticed during post monsoon and summer, Thangaraja [12] observed only during monsoon along Vellar estuary. Krishnamurthy and Prince Jeyaseelan [25] observed in monsoon, post monsoon and summer seasons along Pitchavaram mangrove systems support the present study.

Family: *Gerreidae*

***Gerrus oblongus*:** The distribution of the eggs of *Gerrus oblongus*

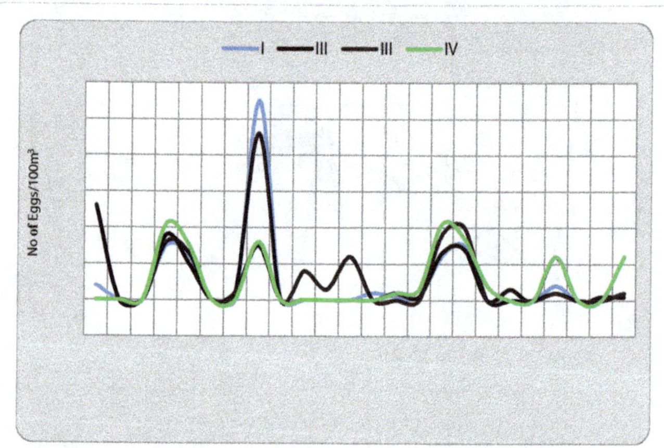

Figure 31: Distribution of *Caranx* sp 2 eggs along Muthupettai coastal waters during the study period (January-2011 to December-2012).

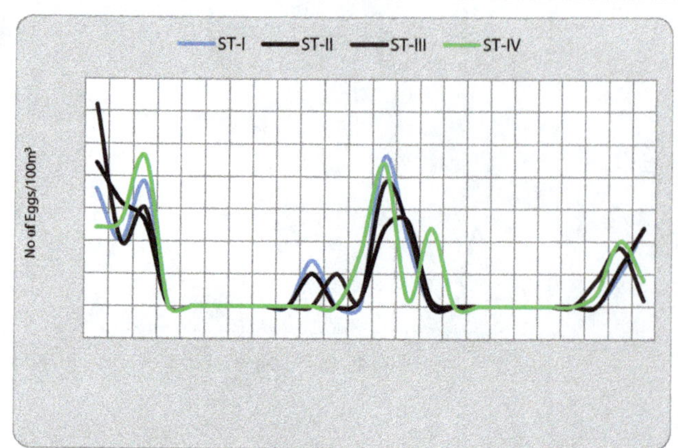

Figure 32: Distribution of *Decapterus russelli* eggs along Muthupettai coastal waters during the study period (January-2011 to December-2012).

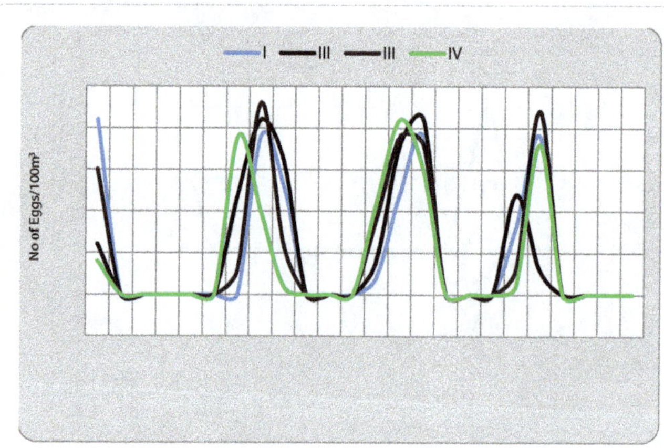

Figure 33: Distribution of *Scomberoides tol* eggs along Muthupettai coastal waters during the study period (January-2011 to December-2012).

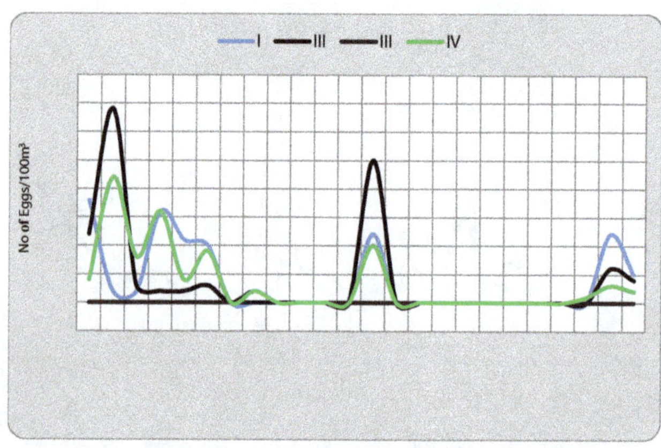

Figure 34: Distribution of *Secutor ruconius* eggs along Muthupettai coastal waters during the study period (January-2011 to December-2012).

is detailed below (Figure 35). During the first year, the minimum number of eggs (2 eggs/100 m^3) was collected in December at station IV and maximum (21 eggs/100 m^3) in October at station IV. During the second year, the minimum density (2 eggs/100 m^3) was in December at station I and maximum (12 eggs/100 m^3) in November and December at station II.

Gerrus oblongus eggs were observed monsoon season only in the study period. It is with the agreement of Bensam [8] along the Parangipettai waters. But Ramaiyan et al. [15] noticed this egg during monsoon and post monsoon season.

Family: *Teraponidae*

***Terapon jarbua*:** Figure 36 gives the details of the distribution of eggs of *Terapon jarbua* along Muthupettai coastal waters. The minimum number of eggs (2 eggs/100 m^3) was observed in March at station I and maximum (100 eggs/100 m^3) in October at the same station. In the second year minimum density (2 eggs/100 m^3) was observed in May at station I and maximum (42 eggs/100 m^3) in December at station II.

Venkataramanujam [17] observed during post monsoon and summer season and annual spawning of these fish agrees with the previous works on the east coast by Bensam [8] observed during post monsoon, Manickasundaram [11] noticed almost throughout the year along Coleroon estuary, Thangaraja [12] noticed almost throughout the year along Vellar estuary and this present study agreed with the observation already made by Krishnamurthy and Prince Jeyaseelan[25] along Pitchavaram mangrove systems and Thangaraja and Ramamoorthi [28] observed during April to June, August to October [18,19,20].

Family: *Scombridae*

***Scomberomorus* sp.:** The distribution of the eggs of *Scomberomorus* sp. is given in Figure 37. During the first year, the minimum density (3 eggs/100 m^3) was observed during November at station II and maximum (22 eggs/100 m^3) were in August at station II. During the second year, the minimum density (3 eggs/100 m^3) were observed in November at station III and maximum density (64 eggs/100 m^3) in November at station I.

Scomberomorus sp. eggs were collected during premonsoon, and monsoon months from Muthupettai coastal waters. But, Venkataramanujam [17] noticed only during November along Vellar estuary, Ramaiyan et al. [15] noticed these eggs during premonsoon and summer along Parangipettai waters.

Family: *Bothidae*: Figure 38 gives the summary of *Pseudorhambus javanicus* eggs observed during the study period along Muthupettai coastal waters. During the first year, the minimum density (3 eggs/100 m^3) was observed during February at station IV and maximum (32 eggs/100 m^3) in March at station IV. In the second year of the

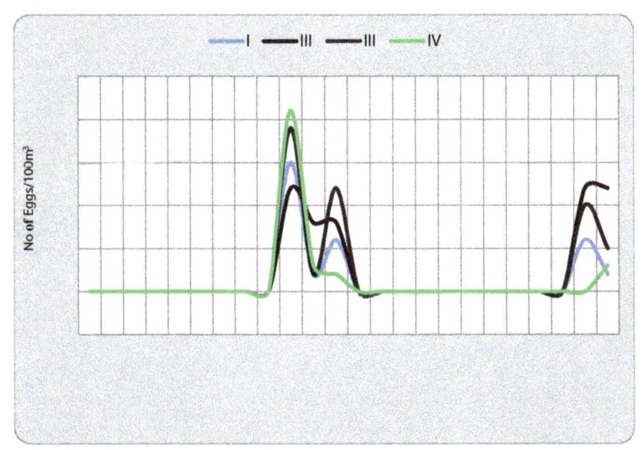

Figure 35: Distribution of *Gerrus oblongus* eggs along Muthupettai coastal waters during the study period (January-2011 to December-2012).

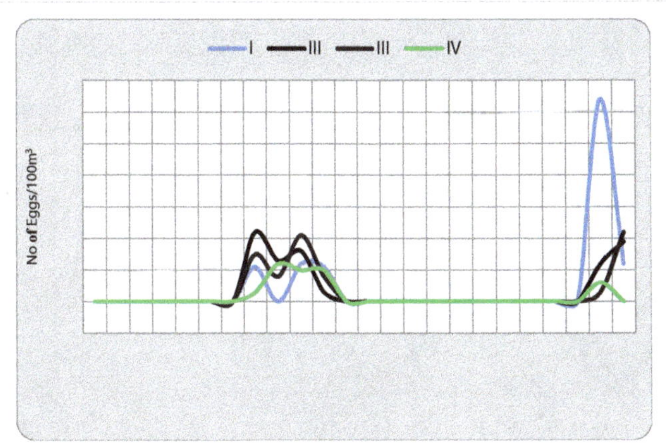

Figure 37: Distribution of *Scomberomorus* sp eggs along Muthupettai coastal waters during the study period (January-2011 to December-2012).

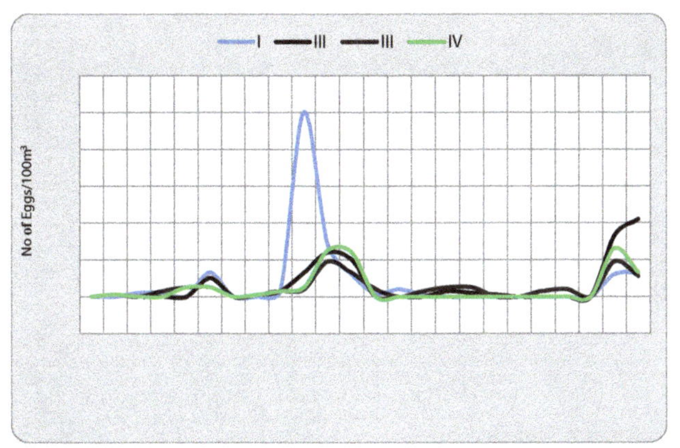

Figure 36: Distribution of *Teraponjarbua* eggs along Muthupettai coastal waters during the study period (January-2011 to December-2012).

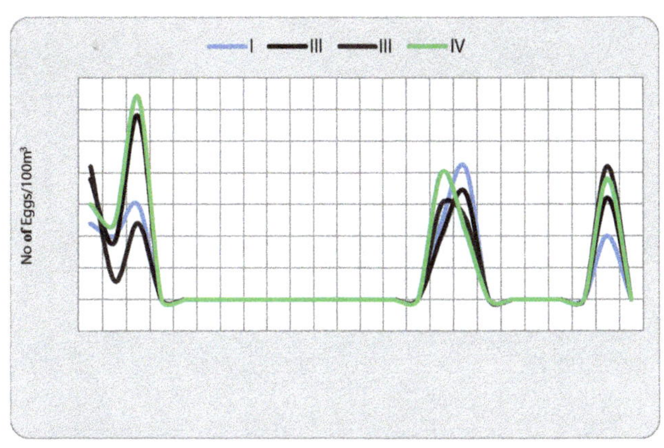

Figure 38: Distribution of *Pseudorhambusjavanicus* eggs along Muthupettai coastal waters during the study period (January-2011 to December-2012).

investigation minimum (10 eggs/100 m^3) observed in the month of April at the station II and maximum (21 eggs/100 m^3) observed May and November at the station IV and I, respectively.

Pseudorhambus javanicus eggs were collected only in the post monsoon and summer seasons from Muthupettai waters. Similar observations were made along Coleroon estuary and along Vellar estuary [7,11,17].

Family: *Pleuronectidae*: Distribution of *Pleuronectid* eggs in the study period (January-2011 to December-2012) along Muthupettai coastal waters is given below (Figure 39). *Pleuronectid* eggs were observed during April, minimum (1 egg/100 m^3) at station I and maximum (19 eggs/100 m^3) at station II, During the second year minimum (6 eggs/100 m^3) were observed in the month of October at the station I and maximum (12 eggs/100 m^3) collected in the month of September at station II.

Ramaiyan et al. [15] observed this egg during early month of summer is the support of the present investigation.

Family: *Soleidae*

Solea ovata: The details of the distribution of eggs of *Solea ovata* during the study period of two years along Muthupettai coastal waters are given in Figure 40. During the first year, single egg observed during October at station IV and maximum (10 eggs/100 m^3) was in December at station III. During the second year, the minimum density (2 eggs/100 m^3) was observed in November at station III and maximum (11 eggs/100 m^3) in October station IV.

Few *Solea ovata* eggs were noticed during the in monsoon and last months of premonsoon season. Thangaraja [12] observed these eggs in summer season along Parangipettai waters. Ramaiyan et al. [15] observed these eggs in monsoon and premonsoon seasons along the Parangipettai waters.

Family: *Cynoglossidae*

***Cynoglossus arel*:** The details of the distribution of eggs of *Cynoglossus arel* during the period of investigation (January-2011 to December-2012) along Muthupettai coastal waters are depicted in Figure 41. Single egg observed during the month of November at station IV and maximum (31 eggs/100 m^3) in February at station III in the first year and the minimum eggs (2 eggs/100 m^3) were observed in December at station I and maximum (29 eggs/100 m^3) in November at station IV during the second year.

Manickasundaram [11] noticed these eggs abundant in summer months along Coleroon estuary. Nair [13] observed these eggs only during monsoon and post monsoon season. Bapat [18] observed irregular appearance of these eggs along Mandapam region. These previous works agreed with the present investigation.

***Cynoglossus puncticeps*:** Figure 42 shows the monthly variation

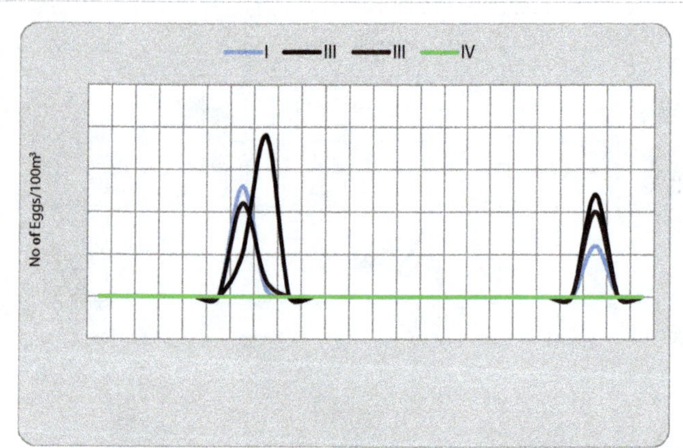

Figure 39: Distribution of *Pleuronectid* eggs along Muthupettai coastal waters during the study period (January-2011 to December-2012).

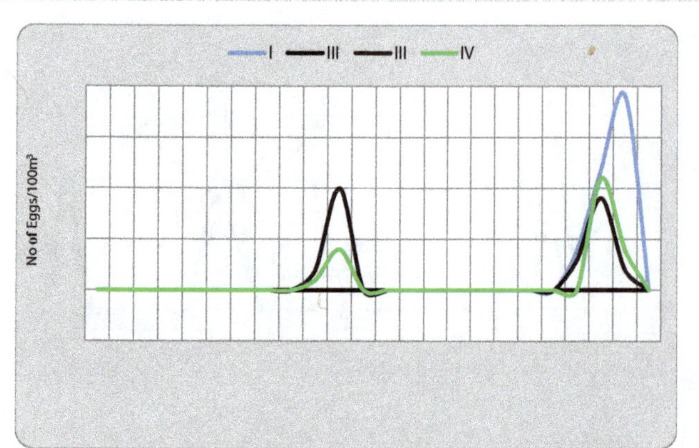

Figure 40: Distribution of *Soleaovata* eggs along Muthupettai coastal waters during the study period (January-2011 to December-2012).

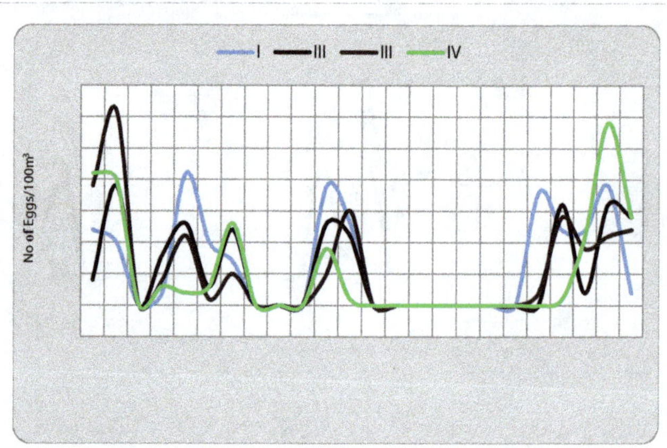

Figure 41: Distribution of *Cynoglossusarel* eggs along Muthupettai coastal waters during the study period (January-2011 to December-2012).

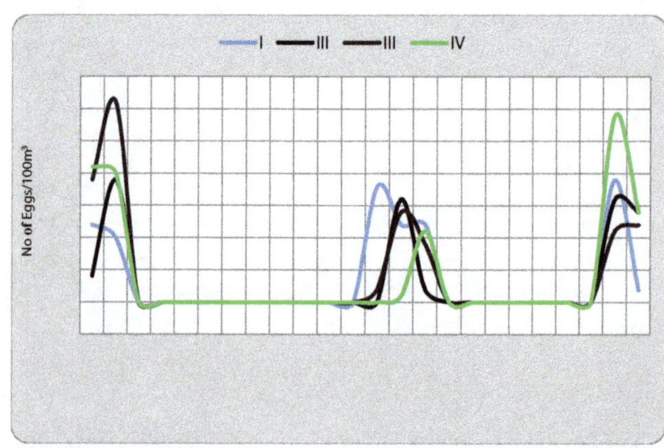

Figure 42: Distribution of *Cynoglossuspuncticeps* eggs along Muthupettai coastal waters during the study period (January-2011 to December-2012).

of the eggs of *Cynoglossus puncticeps* along Muthupettai coastal waters. During the first year of the study, these eggs were collected in January (4 eggs/100 m^3) at station II and maximum (30 eggs/100 m^3) eggs observed in month of February at station IV. During the second year, single egg observed in the month of February at station IV and maximum (29 eggs/100 m^3) in November at the same station IV.

Cynoglossus puncticeps eggs were collected post monsoon and late months of the monsoon season along Parangipettai coastal waters [12,15,17].

Family: *Tetraodontidae*

Arothron ***hispidus***: The distribution of eggs of *Arothron hispidus* along Muthupettai coastal waters is displayed in Figure 43. Minimum density (2 eggs/100 m^3) was available in November at station III and maximum density during February (31 eggs/100 m^3) at station III. During the second year minimum density (2 eggs/100 m^3) were observed in January at station III and maximum (18 eggs/100 m^3) were in January at station I.

Eggs of *Arothron hispidus* were collected during Monsoon, post monsoon and summer season in the present investigation but Venkataramanujam and Ramamoorthi [7] noticed during post monsoon and summer seasons, Thangaraja [12] observed during monsoon season and Thangaraja [29] collected during monsoon and post monsoon seasons along Parangipettai coastal waters. The present result exposed that the protracted spawning habit of these fishes.

***Arothron* sp.**: Figure 44 shows the seasonal distribution of *Arothron* sp. eggs along Muthupettai coastal waters. During the first year single egg observed at station I in July and maximum (31 eggs/100 m^3) at station III in the month of May, in the second year minimum (2 eggs/100 m^3) observed at station I in the month of April, and Maximum (34 eggs/100 m^3) collected at station III in the month of July.

Eggs of *Arothron* sp. were collected summer and premonsoon along Muthupettai coastal waters. The results of this present investigation were supported by the previous works made by Manickasundaram [11] from Coleroon estuary and Venkataramanujam [17] along Muthupettai coastal waters south east coast of India.

Population density of finfish eggs

The population density of fin fish eggs observed during the study period (January-2011 to December-2012) is given in Figure 45. During first year of observation, the minimum density of fish eggs observed during July (15 No/100 m^3 m^3) and maximum during January and October (100 No/100 m^3) at station I. However at station II, the minimum density was observed in June (99 No/100 m^3) and maximum during January (499 No/100 m^3). At station III, the minimum density was observed during December (152 No/100 m^3) and maximum during January (558/100 m^3).

At station IV, the minimum density was observed during August (96 No /100 m^3) and maximum during January (553 No/100 m^3). During the second year 2012, the population density of fish eggs ranged from 16 to 89 No/100 m^3 at station I. Minimum density was recorded during April and maximum in December. Station II, the minimum density of 45 and maximum of 472 No/100 m^3 were recorded during June and February, respectively. In station III, it varied from 23 to 472 No/100 m^3, the minimum during June and maximum during January. Fish eggs density recorded at station IV ranged from 14 to 464 No/100 m^3. Minimum density was observed during June and maximum in January.

Species composition

In total 43 species of fin fish eggs were collected and identified over a period of two years (January-2011 to December-2012) from the Muthupettai mangroves in the present investigation (Table 1). They

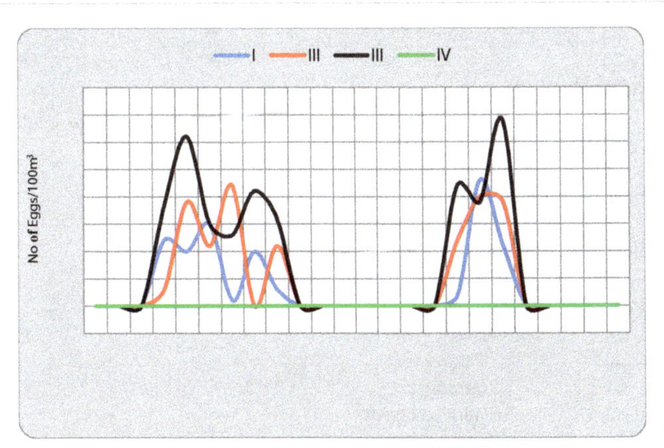

Figure 44: Distribution of *Arothron* sp eggs along Muthupettai coastal waters during the study period (January-2011 to December-2012).

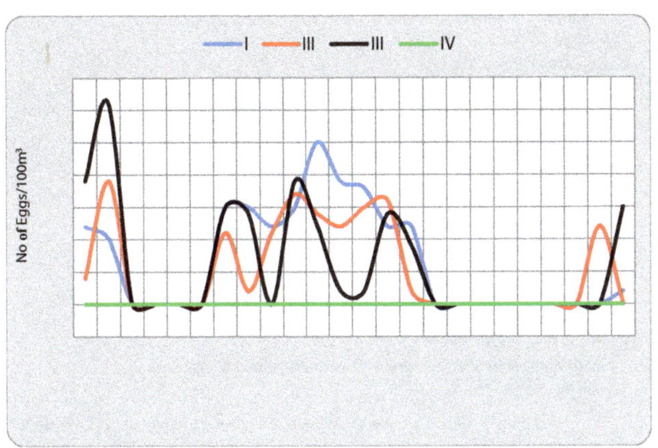

Figure 43: Distribution of *Arothronhispidus* eggs along Muthupettai coastal waters during the study period (January-2011 to December-2012).

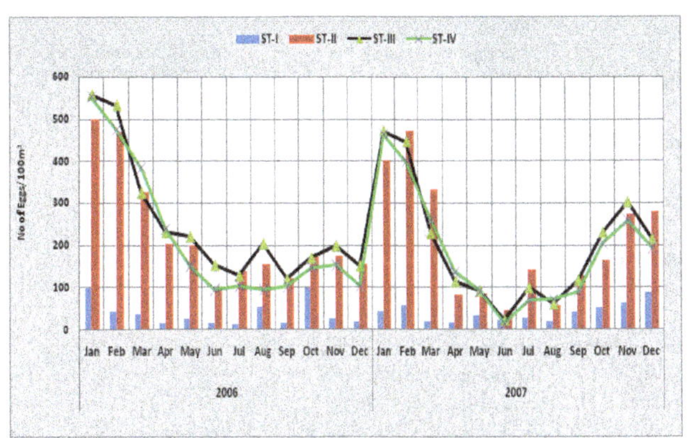

Figure 45: Monthly variation in density of finfish eggs recorded at stations I - IV along Muthupettai waters (January-2011 to Jun-2012).

S. No	Species	Station			
		I	II	III	IV
Family : Ophichthidae					
1	*Ophichthys sp.*	+	+	+	+
Family : Engarulidae					
2	*Setipinna taty*	-	+	+	+
3	*Stolephorus tri*	+	+	+	+
4	*S. punctifer*	+	+	+	+
5	*S. heterolobus*	+	-	+	+
6	*S. macrops*	+	+	+	+
7	*Thryssa dussumieri*	+	+	+	+
8	*T. mystax*	+	+	+	+
9	*T. hamiltoni*	+	+	+	+
Family : Pirstigastridae					
10	*Opisthopterus tardoore*	+	+	+	+
Family : Chirocentridae					
11	*Chirocentrus dorab*	+	+	+	+
Family Clupeidae					
12	*Esculosa thoracata*	+	+	+	+
13	*Sardinella fimbriata*	+	+	+	+
14	*S. gibbosa*	+	+	+	+
15	*S. longiceps*	+	+	+	+
16	*S. clupeoides*	+	+	+	+
17	*Anadontostoma chacunda*	+	+	+	+
18	*Nematolosa nasus*	+	+	-	+
Family : Synodontidae					
19	*Saurida gracilis*	+	-	+	+
20	*S. tumbiol*	-	+	+	+
21	*Saurus sp*	-	+	+	+
22	*Saurida sp*	+	+	+	+
23	*Synodontid egg*	+	+	+	+
Family : Mugilidae					
24	*Liza dussumieri*	+	+	+	+
25	*L. tade*	+	+	+	+
26	*Mugil cephalus*	-	+	+	+
Family : Hemiramphidae					
27	*Hemiramphus sp*	-	+	+	-
Family : Atherinidae					
28	*Pranesus pinguis*	+	+	+	+
Family : Carangidae					
29	*Carangoides malabaricus*	+	+	+	+
30	*Caranx sp 1*	+	+	+	+
31	*Caranx sp 2*	+	+	+	+
32	*Decapterus russelli*	+	+	+	+
33	*Scomberoides tol*	+	+	+	+
Family : Gerriedae					
34	*Gerrus oblongus*	+	+	+	+
Family : Teraponidae					
35	*Terapon jarbua*	+	+	+	+
Family : Scombridae					
36	*Scomberomorus sp*	+	+	+	+
Family : Pleuronectidae					
37	*Pleuronectid egg*	+	+	+	-
Family : Bothidae					
38	*Pseudorhambus javanicus*	-	-	+	+
Family : Soleidae					
39	*Solea ovata*	+	+	+	+
Family : Cynoglossidae					
40	*Cynoglossus arel*	+	+	+	+
41	*C. puncticeps*	+	+	+	+
Family : Tetraodontidae					
42	*Arothron hispidus*	+	-	+	+
43	*Arothron sp*	+	+	+	-
	Total	**37**	**39**	**42**	**40**

(+denotes presence, -denotes absence)

Table 1: Checklist of finfish eggs species recorded from four stations.

were found to be ubiquitously present in all the station, almost, station I and IV few species are absent.

Percentage composition

Eggs from Station-I (2011): At station I, *Clupeids* formed the dominant group contributing 24.60% and *Engralids* ranked next (17.84%) followed by *Caranjids* (14.22%), *Tetrodontides* (9.18%), *Teraponids* (7.14%), *Mugilids* (5.60%), *Cynoglassids* (4.54%), *Pritigristids* (2.42%), *Chirocentrides* (2.29%), *Scombrids* (1.83%), *Athernides* (1.59%), *Ambasids* (1.32%) and *Ophicthids* (1.29%) (Figure 46).

Eggs from Station-II (2011): At station II also *Clupeids* were the dominant group (29.36%) followed by *Engraulids* (19.64%), *Caranjids* (11.50%), *Tetrodontids* (7.73%), *Mugilids* (6.38%), *Cynoglossids* (4.02%), *Scombrids* (3.07 %), *Teraponids* (3.63%), *Chirocentrids* (2.92%), Pristigasterids (2.89%), *Ambassides* (1.67%), *Plueronectinids* (1.45%) and *Gerrides* (0.66%) (Figure 47).

Eggs from Station-III (2011): Station III, *Enguralids* ranked first (24.22%) followed by *Cluepeids* (24.07%), *Caranjids* (14.19%),

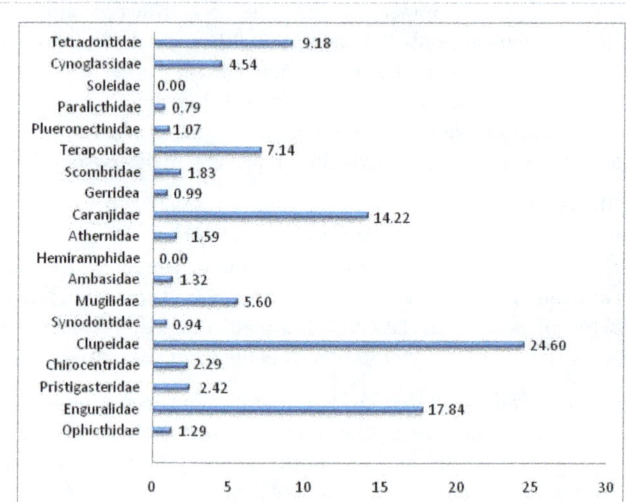

Figure 46: The percentage composition of fin fish eggs observed during 2011.

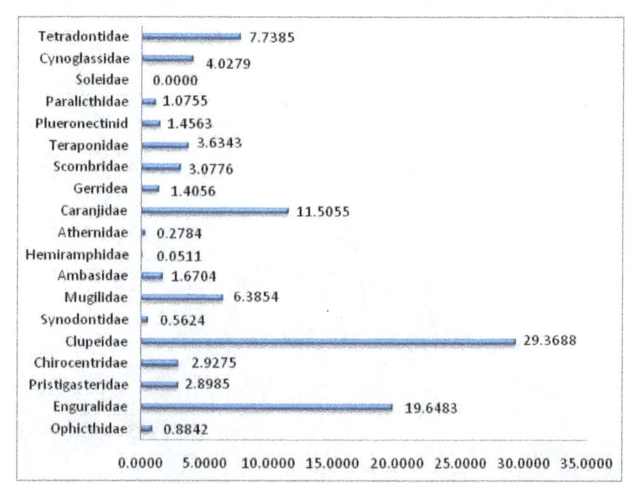

Figure 47: The percentage composition of fin fish eggs observed during 2012.

Tetraodontids (9.59%), *Mugilids* (6.77%), *Cynoglossids* (3.73%), *Pristigastrids* (2.94%), *Scombrids* (2.78%), *Teraponids* (2.59%), *Gerreids* (1.71%), *Synodontids* and *Atherinids* contributing 1.28% and 1.16 % respectively. *Ambassides* (0.88%), *chrocentrides* (0.74%) and others are contributing very lesser (Figure 48).

Eggs from Station-IV (2011): In station IV, more number of *Clupeids* eggs (31.04%) were collected followed *Engraulids* (21.4%), *Carangids* (01.81%), *Mugilids* (9.08%), *Pristigastrids* (5.73%), *Teraponids* (4.20%), *Cynoglossids* (3.53%), *Scombrids* (2.47%), *Chirocentrids* (2.24%), *Ophichthids* (1.77%), *Gerreids* (1.61%), *Synodontids* (1.59%) and *Ambassieds* (1.10 %,) (Figure 49).

Eggs from Station-I (2012): The details of the percentage composition of fin fish eggs observed during the second year of study period is shown in Figure 50. At station I, *Clupeids* ranked first contributing 31.97 %, followed by *Engraulids* (18.75%), *Caranjids* (11.59%), *Pristigastrids* (5.94%), *Cynoglossids* (5.00%), *Tetrodontids* (4.63%), *Paralicthids* (3.89%), *Chirocentrids* (2.86%), *Scombrids* (2.28%), *Ophicthids* (1.83%), *Synodontids* (1.38%), *Ambassides* (1.32%), *Soleids* (1.25%), *Atherinids* (1.10%) and *Teraponids* (0.91%).

Eggs from Station-II - (2012): Similarly at station II, the *Clupeids* (10.48%) were the dominant group (25.07%) followed by *Cynoglossids Carangids* (18.28%), *Engraulids* (16.24%), *Tetraodontids* (6.36%), *Mugilids* (4.53%), Plueuronectinids (3.94%), Pritigastrides and *Paralicthids* contributing 3.77% and 3.74% respectively. The eggs of *Chirocentrides* and Cynoglasides contributing 3.65% and 3.64% respectively followed by *Ambassides* (3.10%) (Figure 51).

Eggs from Station-III - (2012): At station also III, *Clupeids* ranked first contributing 26.39%, followed by *Engraulids* (21.0%), *Carangids* (10.85%), *Tetrodontids* (8.51%), *Mugilids* (5.14%), *Cynoglossids* (3.80%), *Paralicthids* (3.73%), *Chirocentrids* (3.32%), *Synodontids* (2.04%), *Ophicthids* (1.93%), *Ambassieds* (1.88%), *Atherinids* (1.55%), and Hemiramphids (0.108%) (Figure 52).

Eggs from Station-IV - (2012): At station IV also, *Clupeids* (38.06%), eggs were relatively more followed by *Engraulids* (20.45%), Carangid (13.57%), *Pristigastrids* (4.58%), *Tetraodontids* (4.06%), *Chirocentrids* (2.96%), *Ophichthids* (2.33%), *Ambassides* (1.98%), Plueoronectinids (1.48%) and *Soleids* (0.72%) (Figure 53).

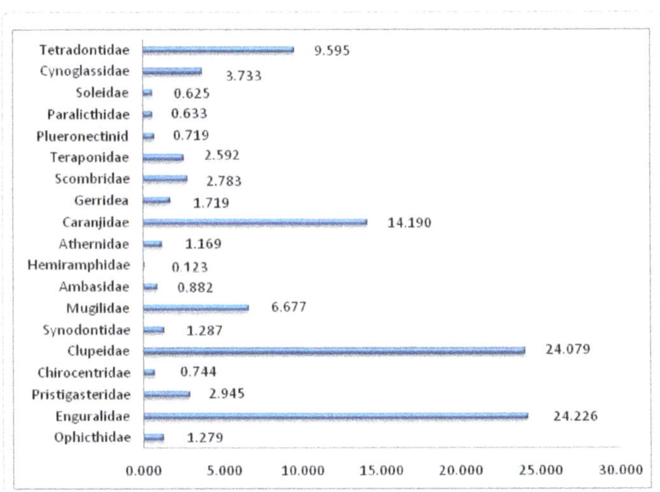

Figure 48: The percentage composition of fin fish eggs observed during 2011.

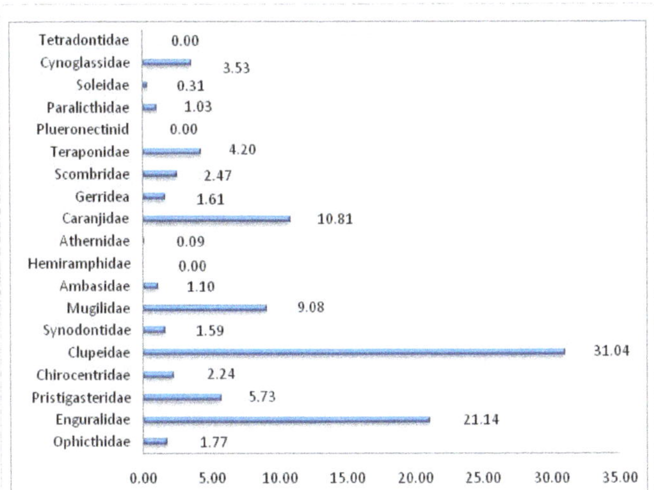

Figure 49: The percentage composition of fin fish eggs observed during the Second year.

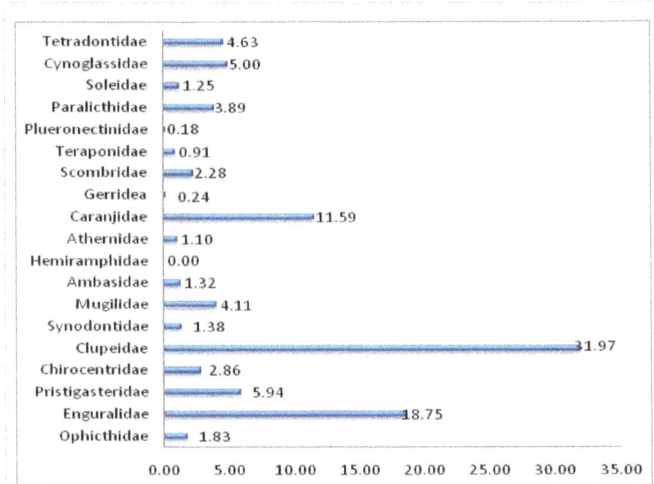

Figure 50: The percentage composition of fin fish eggs observed during the second year Station-I.

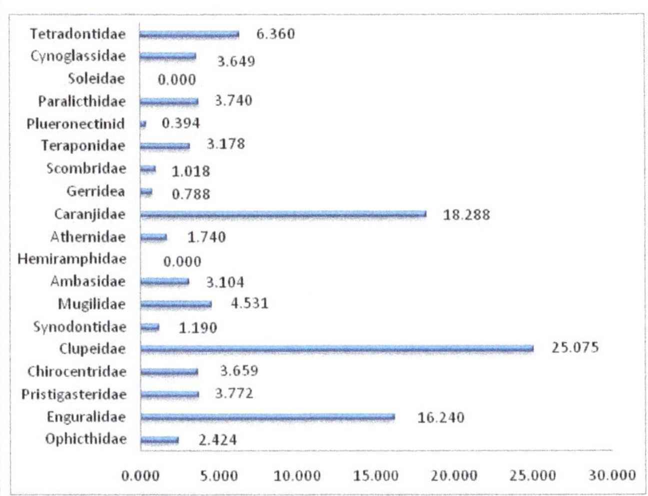

Figure 51: The percentage composition of fin fish eggs observed during the second year Station-II.

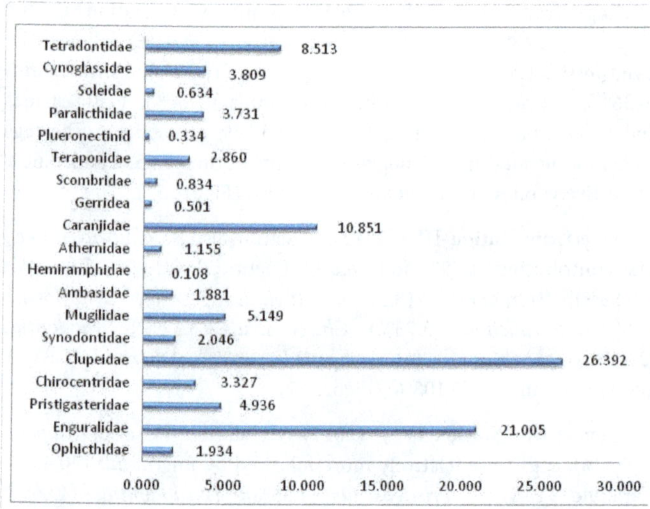

Figure 52: The percentage composition of fin fish eggs observed during the second year Station-III.

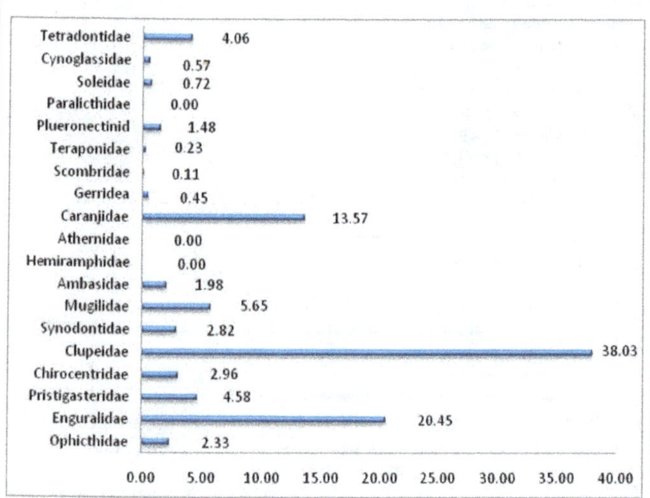

Figure 53: The percentage composition of fin fish eggs observed during the second year Station-IV.

General observations

As a result of the present observation over a period of two years, 43 forms of fin fish eggs were identified, of which, finfish eggs belonging to 38 species of fish were identified up to species level, 3 were identified up to the genus level and 2 were identified up to the family level.

Manickasundaram [11] documented the fish eggs belonging to 30 species from Coleroon estuary. Fish eggs belonging to 44 species from Parangipettai waters was also described [12,15]. Venkataramanujam [17] recorded 18 species fin fish eggs along Parangipettai waters. Similarly, Koteswarma [30] recorded fin fish eggs belonging to 12 species from Bapatla coast.

The present investigation confirms the relatively rich ichthyoplankton occurrence along Muthupettai waters, south east coast of India.

Clupeidae > Engraulidae > Carangidae > Cynoglossidae > Tetraodontidae Teraponidae > Mugilidae > Atherinidae > Synodontidae > Ophichthidae > Pristigastridae > Chirocentridae > Scombridae > Gerreidae > Soleidae > Bothidae > Pleuronectidae > Hemiramphidae

Minimum eggs were observed during Monsoon in all the stations along Muthupettai waters. Similar observations were made from Parangipettai coastal waters, Coleroon estuary, Indian Ocean, Baptla coast and from Cochin back waters [7,12,15,17,30,31,32].

Seasonal distribution of eggs was significant during post monsoon followed by premonsoon, summer and monsoon. Similar observations were made by earlier workers from Parangipettai coastal waters, Coleroon estuary, Vellar estuary and from Tuticorin coastal waters along south east coast of India, more number of fish eggs were also observed in premonsoon season along Cape Comarin area [7,11,12,14,17,33]. Common occurrence of fin fish eggs along Bombay coast during premonsoon has also been reported [34,35]. Abundant occurrence of eggs during post monsoon may be due to the peak spawning activities of these fishes along Parangipettai, Bapatla coast and Coleroon estuary along the south east coast of India [17,30].

Two-way analysis of variance showed significant differences in fin fish egg population density between seasons and stations in both the years of study period (Tables 2 and 3).

Pattern of distribution and abundance of fish eggs and larvae is associated with environmental factors and the environment may act either as a favorable factor for successful spawning by fish and survival during eggs and larval stages. The suitability of developmental stages in a spatial feature is a main characteristic of the life cycle of fish. For pelagic species the stability is related to hydrographic features. Thus the pattern of distribution and the area of availability of eggs may vary with the fluctuation in environmental conditions. Pattern of distribution and seasonal occurrence of eggs is not similar in the present study period of two years. This may be due to the changes in environmental conditions due to variation in the environmental factors. Two environmental factors *viz.*, temperature and salinity have profound influence on the development and hatching of marine teleosts.

Abundance of eggs and larval stages in relation to the hydrographic features is crucial for the comprehension of the mechanism determining recruitment. Special attention is paid to the stability of the spatial distribution and the interannual and seasonal variability of the abundance index of eggs and larval abundance in relation to environmental features and planktonic predators [36].

Further, the environment also influences biological activities such as spawning and growth. Temperature determines the annual stock at the spawning ground. Fish appears to report to their oceanographic 'climate' rather than to geographically fixed preference period during the spawning season [9].

Studies on eggs and larvae of marine fishes although essential for understanding the dynamics of fish population are hampered by several methodological problems, such as occurrence of eggs and planktonic stages of the different species of fishes showing their seasonal variation in space and time. This might also be related to their spawning period. The heavy rainfall during monsoon considerably reduces the salinity of the estuarine water, which again increases during the post monsoon season. Distribution and abundance of fish eggs and larvae are influenced by hydrographical parameters like temperature, salinity, pH and dissolved oxygen.

Spawning of most marine fishes in tropical waters especially in Indian coastal waters is protracted. Spawning usually begins at the onset of monsoon rains [8]. The previous works on occurrence of finfish eggs and larvae indicate the protracted spawning [37-40].

Present investigation is also indicating protracted spawning behavior of fishes.

Temperature is known to play an important role in determining the fluctuations of ichthyoplankton and their distribution. Bapat [18] has stated that low temperature in Mandapam waters, Hooghly, Maltah estuarine system in the east coast of India, are conductive to growth for many fish eggs and larvae. However in the present study, environmental surface water temperature did not seem to have any direct effect on the distribution of fish eggs although each species of fish prefers to have optimal temperature and time for spawning. Normally eggs are collected in large numbers in increasing temperature and salinity. Salinity of water affects the availability of fish eggs in an estuary [12]. Each species has a certain salinity range. More number of eggs was observed in post monsoon seasons. This corresponds to a decline in salinity from the high salinity of the previous season, stimulating spawning activity in some coastal fishes. Similar observations were made along coleroon estuary and at Parangipettai coastal waters [7,11,12,17].

During the low temperature and salinity conditions the abundance of eggs was low in number but in high temperature and salinity conditions abundance of eggs was more. Perhaps high salinity and temperature conditions may stimulate spawning activity in some coastal fishes as observed presently. This finding is also in agreement with the previous works [11,14,17,30,41].

Of the two years (January 2011 to December 2012) of the present investigation, during the second year the total annual rain fall was lower than the first year. Rainfall and abundance of fish eggs appear to show a reverse relationship as reported in the Maltah estuarine system and from Vellar estuary, east coast of India [17,42]. During the present study the heavy rainfall during monsoon period considerably reduces the salinity of the estuarine and coastal waters so that the abundance of eggs was in less number, which again increases during the post monsoon [7,11,12,17]. Thus the variations observed in the seasonal distribution and abundance of fish eggs in all the stations in the present study may be due to several reasons. Further, the variation in seasonal distribution and abundance of eggs was not similar during the study period of two years (January-2011 to December-2012).

This might also due to the environmental factors such as drainage of water, delay in spawning, predation on eggs and larvae. In addition, it is not unusual that several organisms may show variations from year to year on the magnitude of their population and also in the time of occurrence of maximum and minimum which is slightly earlier or later [43].

The correlation coefficient values between finfish eggs density and environmental parameters for all the stations along Muthupettai waters not significantly correlated, which is in agreement with the past findings [11,44,45]. However, a significant correlation ($P<0.05$) with surface water temperature was observed in station I during the first year along Muthupettai waters and station IV (Tables 4-7) which is supported by the observation [11].

The time and intensity of spawning of fishes may perhaps be controlled by the seasonal cycle of the environmental factors. The physic-chemical parameters recorded from various stations presently observed monthly seasonal and annual variations. Minimum eggs were observed during Monsoon in all the stations along Muthupettai waters. Seasonal distribution of eggs was significant during post monsoon followed by premonsoon, summer and monsoon. Abundant occurrence of eggs during post monsoon may be due to the peak spawning activities of these fishes coinciding the fishing holyday declared by the coastal state government with a view to enhance the breeding and spawning activities of these fishes. Regular sampling of icthyoplankton is essential for locating shoals of adult fishes and their spawning grounds. Ichthyoplankton studies are extensively useful in fishery investigation. Information on fish eggs and larvae of a particular region is useful in understanding the spawning season of fishes of commercial importance. Studies on the early developmental stages of fish allow

Source of Variation	SS	df	MS	F	P-value	F crit
Rows	335531.0625	3	111843.69	22.62194	3.84E-08	2.891564
Columns	576528.2292	11	52411.657	10.60099	5.9E-08	2.093254
Error	163153.1875	33	4944.036			
Total	1075212.479	47				

Table 2: Two-way ANOVA for differences in abundance of fin fish eggs between seasons and stations I to IV along Muthupettai waters for the study period (January-2011 to December-2012).

Source of Variation	SS	df	MS	F	P-value	F crit
Rows	222265.1667	3	74088.389	15.06155	2.4E-06	2.891564
Columns	511608.5	11	46509.864	9.455068	2.26E-07	2.093254
Error	162328.3333	33	4919.0404			
Total	896202	47				

Table 3: Two-way ANOVA for differences in abundance of fin fish eggs between seasons and stations I to IV along Muthupettai waters for the study period (January-2011 to December-2012).

Parameters	Rainfall (mm)	At. Temp. (°C)	Wat. Temp. (°C)	Salinity (%)	pH	DO (ml/l)	Fish eggs
Rainfall (mm)	1.00						
At. Temp. (°C)	-0.65	1.00					
Wat. Temp. (°C)	-0.59	0.55	1.00				
Salinity (%)	-0.53	0.65	0.02	1.00			
pH	-0.29	0.44	0.48	0.09	1.00		
DO (ml/l)	0.10	-0.01	0.38	-0.51	0.23	1.00	
Fish eggs	0.27	-0.51	-0.18	-0.29	-0.28	0.02	1.00

Table 4: Correlation coefficient (r) values between fin fish eggs abundance and physicochemical parameters, at stations I along Muthupettai waters during January-2011 to December-2012.

Parameters	Rainfall (mm)	At. Temp. (°C)	Wat. Temp. (°C)	Salinity (‰)	pH	DO (ml/l)	Fish eggs
Rainfall (mm)	1						
At. Temp. (°C)	-0.67	1					
Wat. Temp. (°C)	0.25	-0.17	1				
Salinity (‰)	-0.75	0.83	-0.31	1			
pH	-0.13	0.10	-0.22	0.072	1		
DO (ml/l)	-0.36	0.41	-0.62	0.40	0.09	1	
Fish eggs	-0.34	0.01	-0.15	0.058	-0.11	0.11	1

Table 5: Correlation coefficient (r) values between fin fish eggs abundance and physicochemical parameters, at stations II along Muthupettai waters during January-2011 to December-2012.

Parameters	Rainfall (mm)	At. Temp. (°C)	Wat. Temp. (°C)	Salinity (‰)	pH	DO (ml/l)	Fish eggs
Rainfall (mm)	1						
At. Temp. (°C)	-0.68	1					
Wat. Temp. (°C)	-0.59	0.82	1				
Salinity (‰)	-0.83	0.80	0.70	1			
pH	-0.65	0.48	0.39	0.72	1		
DO (ml/l)	-0.37	0.46	0.30	0.32	0.14	1	
Fish eggs	-0.26	0.057	0.13	0.026	-0.12	0.28	1

Table 6: Correlation coefficient (r) values between fin fish eggs abundance and physicochemical parameters, at stations III along Muthupettai waters during January-2011 to December-2012.

Parameters	Rainfall (mm)	At. Temp. (°C)	Wat. Temp. (°C)	Salinity (‰)	pH	DO (ml/l)	Fish eggs
Rainfall (mm)	1						
At. Temp. (°C)	-0.66	1					
Wat. Temp. (°C)	-0.55	0.88	1				
Salinity (‰)	-0.74	0.96	0.87	1			
pH	-0.56	0.65	0.54	0.70	1		
DO (ml/l)	0.06	0.41	0.39	0.40	0.23	1	
Fish eggs	-0.31	0.08	0.26	0.07	0.00	-0.54	1

Table 7: Correlation coefficient (r) values between fin fish eggs abundance and physicochemical parameters, at stations IV along Muthupettai waters during January-2011 to December-2012.

us to comprehend the biology of the species besides determining their spawning seasons and to estimate spawning stock abundance. Therefore, study is also an essential prerequisite in undertaking the spawning biomass of target species monitoring, changes in exploitable stocks and yields, forecasting trends of production.

References

1. Rengarajan K, David raj I (1984) On ichthyoplankton of the Cochin Backwaters during spring tides. J Mar Biol Ass India 21: 111-118.
2. Young PC, Leis JM, Hausfeld HF (1986) Seasonal and spatial distribution of fish larvae in waters over the north west continental shelf of Western Australia. Mar Ecol Prog Ser 31:209-222.
3. Matsuura Y (1979) Distribution and abundance of eggs and larval of the Brazilian sardine, Sardinella brasiliensis, during 1974-1975 and 1975-1976 seasons. Bull Jap Soc Fish Oceanogr 34: 1-12.
4. Venkataramanujam K (1975) Life history and feeding habits of Ambassis commersoni (Cuvier) (Ambassidae, Teleosteii). In: Recent Researches in Estuarine Biology, Natara January, R. (Ed.), Hindustan Publ, Corp, New Delhi.
5. Ahlstrom EH, Moser HG (1976) Eggs and larvae of fishes and their role in systematic investigations and in fisheries. Rev Tran Inst Peaches Marit 40: 379-398.
6. Saville A (1964) Estimation of the abundance of a fish stock from egg and larval surveys. Rap Pev Reun Cosn int Explor Mer 153: 164-170.
7. Venkataramanujam K, Ramamoorthi K (1974) Seasonal variation in fish eggs and larvae of Porto Novo coastal waters. Indian J Fish 21: 254 -266.
8. Bensam P (1983) Observation on a few early development stages in some fishes of Porto-Novembero coast, India. Ph.D. thesis, Annamalai University. p 270.
9. Ahlstrom EH (1976) Maintenance of quality in fish eggs and larvae collected during plankton hauls. In: Zooplankton fixation and preservation, Steedman, H. F. (Ed.). UNESCO 313-318.
10. Ganapathi PN, Raju (1961) On the eggs and early development of eels of Waltair coasi HJV'ZOOI SOC India 12: 229-238.
11. Manickasundaram M (1990) Studies on fish eggs and larvae of Coleroon estuary along the southeast coast of India. Ph.D. thesis, Annamalai University.
12. Thangaraja M (1982) Studies on development, distribution and abundance of fish eggs and larvae in the Vellar estuary, Porto-Novo (South India). Ph. D. thesis, Annamalai University.
13. Nair RV (1951) Studies on the life history,bionomics and fishery of the whiter sardine, Kowala coval (Cuv.). Proc Indo-Pac Fish Counc 3rd meet II: 103-118.
14. Siraimeetan P, Marichamy R (1987) Observation on pelagic fish eggs and larvae in the coastal waters of Tuticorin. CMFRI Bull 44: 245-251.
15. Ramaiyan V, Balasubramanian T, Kannupandi T, Ajmal Khan S, Rajagopal S, Lyal PS (2005) Monograph on Eggs and Larvae of Fin and Shell Fish Collected from Parangipettai and Adjacent Waters Along the South East Coast of India. Monograph, Annamalai University, India.
16. Rao KS, Girijavallabhan KG (1973) On the eggs and larvae of an engraulid and two *Carangids* from Madras plankton. Indian J Fish 20: 551-561.
17. Venkataramanujam K (1975) Studies in fish eggs and larvae of Porto-Novo coastal waters. Ph. D. thesis, Annamalai University.
18. Bapat SV (1955) A preliminary study of the pelagic fish eggs and larvae of the Gulf of Mannar and the Palk Bay. Indian J Fish 2: 231-255.
19. Kowtal GV (1967) Occurrence and distribution on pelagic fish eggs and larvae in the Chilka Lake during the years 1964-1965. Indian J Fish 14: 198-214.
20. John MA (1951) Pelagic fish eggs and larvae of the Madras Coast. J Zool Soc India 3: 38-69.
21. Devanesan DW (1943) A brief investigation into the causes of the fluctuations

of the annual fishery of the oil sardine of Malabar (S. longiceps), determination of its age and an account of the discovery of its eggs and spawning grounds. Madras Fish Bull 28: 1-33.

22. Lazarus S (1959) On The Spawning Season And Early Life History Of Oil Sardine *Sardinella longiceps* (Cuvier And Valenciennes) At *Vizhinjam*. VaJ Indian J Fish 6: 343-359.

23. Jones S, Sujansinghani KH (1954) Fish and Fisheries of the Chilka Lake with statistics of the fish catches for the years 1948-50. Indian J Fish 1: 256-344.

24. Vijayaraghavan P (1957) Studies on the fish eggs and larvae of Madras coast. Ph.D. thesis, University of Madras, India.

25. Krishnamurthy K, Prince Jeyaseelan MJ (1981) The early life history of fishes from the Pichavaram mangrove system of India. Rap Pev Reun Cons Inst Expl Mer 187: 416-423.

26. George RM (1988) Ichthyoplankton from the *Vizhinjam* coast. Indian J Fish 35: 258-265.

27. Krishnamurthy K, Prince Jeyaseelan MJ (1983) The Pichavaram (India) mangrove ecosystem. Int J Ecol Envir Sci 9: 79-85.

28. Thangaraja M, Ramamoorthi K (1980) Early life history of *Anadontostoma chacunda* (Ham.-Buch.) from Porto-Novo. Indian J Mar Sci 9: 136-139.

29. Thangaraja M (1984) Laboratory reared Pufferfish, *Arothron hispidus* (Lacepede), eggs and larvae, and subsequent stages from plankton of Vellar estuary, Porto Novo. Indian J Mar Sci 13: 199-201.

30. Koteswarma R (1984) Studies on inshore planktonic fish eggs and larvae off Baptla coast (AndrApriladesh, South India) Ph. D. Thesis, Nagarjuna University, India.

31. Peter KJ (1981) Influence of environmental changes on the distribution of ichthyoplankton in the Bay of Bengal. Rap. -v. Reun Cons int Explor Mer 178: 210-216.

32. Lalithamkbika Devi CB (1993) Seasonal fluctuation in the distribution of eggs and larvae of flat fishes (Pleuronectiformes - Pisces) in the Cochin backwater. J Indian Fish Ass 23: 21-34.

33. Santhakumari V, Saraswathy M (1981) Zooplankton along the Tamil Nadu coast. Mahasagar Bull Nat Inst Oceanogr 14: 289-302.

34. Gea Piroja B (1934) A thesis on a contribution to study of the macroplankton of the Bombay harbor with detailed notes on Hydromedusae, Siphonophore, Sagittae and Stomatopod larvae. M.Sc., thesis, Bombay University.

35. Gajabhiye SN, Govindan K, Desai BN (1982) Distribution of plankton fish eggs and larvae around Bombay waters. Indian J Mar Sci 11: 128-131.

36. Koutsikopoulos C, Lacroix N (1992) Distribution and abundance of sole (Solea solea (L.)) eggs and larvae in the Bay of Biscay between 1986 and 1989. Neth J Sea Res 29: 81-91.

37. Bal DV, Pradhan LB (1945) First progress report on "Investigation of fish eggs and fish larvae from Bombay waters". 1944-45. Govt. Central Press, Bombay.

38. Bal DV, Pradhan LB (1946) Second progress report on "Investigation of fish eggs and larvae from Bombay waters", 1945-46. Govt. Central Press, Bombay.

39. Bal DV, Pradhan LB (1951) Occurrence of fish larvae and post larvae in Bombay waters during 1944-47. J Univ Bombay 20 B: 1-15.

40. Basheeruddin S, Nayar KN (1962) A preliminary study of the juveniles fishes of the coastal waters off Madras city. Indian J Fish 8: 169-188.

41. Flores-Coto C, Barba-Torres F, Sanchery-Robels J (1983) Seasonal diversity, abundance and distribution of ichthyoplankton in Tamiahua Lagoon, Western Gulf of Mexico. Trans Am Fish Soc 112: 247-256.

42. Shetty HPC, Saha SB, Ghosh BB (1963) Observations on the distribution and fluctuation of plankton in the Hooghly - Maltah estuarine system, with notes on their relation to commercial fish landings. Indian J Fish 8: 326-363.

43. Prasad RR (1958) Plankton calendars of the inshore waters at Mandapam, with a note on the productivity of the area. Indian J Fish 5: 170-188.

44. Makhina NV, Dvinina EA (1988) Dynamics of cod abundance at stage of early ontogenesis and assessment of 3 year olds by their eggs and larvae. ICES ELHS (Abstract only).

45. Reis RR, Dean JM (1981) Temporal variation in the utilization of an Intertidal creek by the Bay Anchovy (Anchoa nitchilli). Est 4: 16-23.

17

Jellyfish and Ctenophores in the Environmentally Degraded Limfjorden (Denmark) During 2014 - Species Composition, Population Densities and Predation Impact

Hans Ulrik Riisgård[1]*, Josephine Goldstein[1,2], Kim Lundgreen[1,2] and Florian Lüskow[1]

[1]*Marine Biological Research Centre, University of Southern Denmark, Hindsholmvej 11, DK-5300 Kerteminde, Denmark*

[2]*Max-Planck Odense Center on the Biodemography of Aging and Department of Biology, Campusvej 55, DK-5230 Odense M, Denmark*

Abstract

Species composition, population densities and size of jellyfish and ctenophores were recorded during 5 cruises in the heavily eutrophicated Limfjorden in 2014. No or very few ctenophores (*Pleurobrachia pileus*) and jellyfish (*Aurelia aurita, Cyanea lamarckii*) were recorded in April and June 2014, whereas in August and September numerous small individuals of the invasive ctenophore *Mnemiopsis leidyi* were found on all 4 locations studied, which were strongly reduced in population density during November. *M. leidyi* exerted a notable predation impact, most pronounced in Løgstør Bredning and Skive Fjord in August when the estimated half-lives of zooplankton were 4.8 and 7.3 d, respectively, and in late September, when the half-life in Skive Fjord was only 2.2 d. Severe oxygen depletion in Løgstør Bredning and Skive Fjord between June and September resulted in a release of nutrients. This was followed by a bloom of the dinoflagellate *Noctiluca scintillans* and a subsequent peak in the abundance of copepods which decreased rapidly after the introduction of *M. leidyi* into Limfjorden from the North Sea (between early April and mid-July) to become virtually absent during the rest of the season. This subsequently resulted in starvation and decay of the *M. leidyi* population. The small predatory ctenophore *Beroe gracilis* was recorded on most locations during August and September 2014 but although *B. gracilis* eats small *M. leidyi*, their low number suggested a negligible predation impact on the *M. leidyi* population. Our present understanding of the many biological and environmental factors that control the species composition, abundance and predation impact of jellyfish and ctenophore populations in Limfjorden are discussed.

Keywords: *Mnemiopsis leidyi*; *Beroe gracilis*; *Pleurobrachia pileus*; *Aurelia aurita*; Predation impact; Hydrography; Zooplankton; Limfjorden; Oxygen depletion; Hypoxia

Introduction

Marine areas experiencing jellyfish and ctenophore blooms seem often to be those that reveal the greatest environmental degradation [1,2]. Limfjorden (Denmark) is heavily eutrophicated and large areas suffer from oxygen depletion each summer which has apparently caused a change "from fish to jellyfish" [3]. The abundance and predation impact of jellyfish and ctenophores in Limfjorden have been described in a number of studies conducted since 2003 [3-10] and recently reviewed [11] which was supplemented with observations from 2012 and 2013.

The common jellyfish *Aurelia aurita* is often very abundant in Limfjorden and may exert a considerable predatory impact on zooplankton and fish larvae. However, the population dynamics of *A. aurita* are not only strongly influenced by the high-salinity water brought into Limfjorden from the North Sea, but also by competition with the invasive ctenophore *Mnemiopsis leidyi* that occurred in Limfjorden for the first time in extremely high numbers in 2007 [12]. It has been suggested that Limfjorden may function as an incubator for *M. leidyi* with the potential to further spread *M. leidyi* into the Kattegat and adjacent Danish waters [9,10]. A recent study [11] reported on two bloom events of ctenophores, *Pleurobrachia pileus* and *M. leidyi*, along with their predators (*Beroe* spp.) in Limfjorden in the autumns of 2012 and 2013, when the previously dominating *A. aurita* was absent. Further it was observed that *B. ovata*, which is *M. leidyi*'s native predator, had now occurred as a new introduced species in Limfjorden.

The aim of the present study was to record the seasonal occurrence patterns of jellyfish and ctenophores in Limfjorden during 5 cruises with the Danish Naval Homeguard in 2014 in order to obtain a better understanding of the biological and environmental factors that control their species composition, population densities and predation impact. Our present knowledge on the complex interplay between biological and environmental factors that control the jellyfish and ctenophore populations in Limfjorden is briefly reviewed and discussed, including the importance of recruitment, life cycles, and hydrographic conditions.

Materials and Methods

Study area

Limfjorden is a Danish shallow-water system with a mean water depth of about 4.5 m that connects the North Sea via Thyborøn Kanal in the west to the Kattegat in the east (Figure 1). A dominating west-easterly wind creates an eastward current that brings highly saline North Sea water into Limfjorden, which also receives freshwater inputs from the surrounding land area and this results in a salinity gradient from west to east [13-17]. Limfjorden is heavily eutrophicated and large

***Corresponding author:** Hans Ulrik Riisgård, Marine Biological Research Centre, University of Southern Denmark, Hindsholmvej 11, DK-5300 Kerteminde, Denmark
E-mail: hur@biology.sdu.dk

areas suffer from oxygen depletion each summer [7,13,18], and this has apparently caused the bottom-dwelling fish to disappear and the number of jellyfish and ctenophores to increase [3]. Every year, strong westerly-winds induce about 60 intrusions of North Sea water through Thyborøn Kanal into Limfjorden succeeded by backflow of water into the North Sea on calm days. Annually, about 70 km³ of North Sea water enter Limfjorden, and approximately 65 % flows eastwards into Løgstør Bredning which additionally receives 3 km³ freshwater so that about 8.7 km³ runs eastwards into Kattegat [10]. Inflowing water to Limfjorden mainly comes from the Jutland Coastal Current which flows northwards along the Danish west coast carrying mixed water masses from the English Channel and the southern North Sea [10,19-21].

Jellyfish and ctenophores

Jellyfish and ctenophores were collected with a 2 mm-meshed plankton net (mouth-ring diameter = 1.5 m, mouth area = 1.77 m²) at 4 locations in Limfjorden (Figure 1) during 5 one-day cruises in 2014 with the Danish Naval Homeguard MVH 902 'Manø'. Three hauls per location (haul length ~300 m) were performed at a depth of approximately 1 m at a speed of about 1.5 knots. After the three hauls, the number of medusae and ctenophores in the cod end of the plankton net were counted from sub-samples, and mean population densities were estimated from the densities in the 3 hauls. Further, mean (±SD) sizes were determined to the nearest mm after measuring ctenophore oral-aboral lengths (*L* i.e. distance between mouth and opposite pole) and medusa inter-rhopalia umbrella diameters (*d*) of all, or at least 50 individuals per

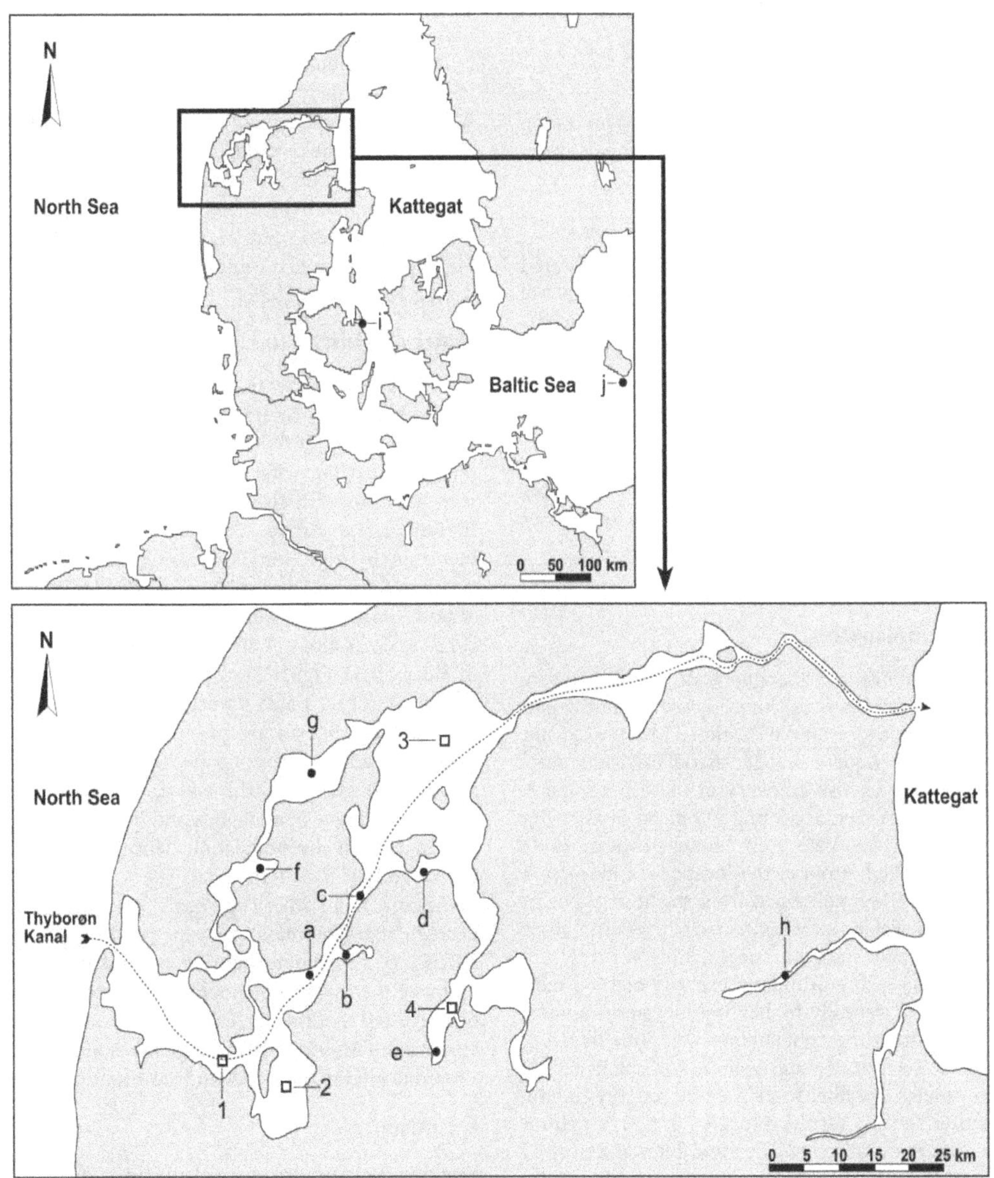

Figure 1: Locations of sampling sites in Limfjorden (Denmark) where jellyfish and ctenophores were collected on cruises during 2014. 1 = Nissum Bredning (56°34'25.80" N, 08°29'41.40" E), 2 = Venø Bugt (56°31'23.40" N, 08°40'27.60" E), 3 = Løgstør Bredning (56°57'09.00" N, 09°03'27.60" E), 4 = Skive Fjord (56°37'15.00" N, 09°05'33.00" E). The path of water exchange between the North Sea in west and Kattegat in the east is indicated by a broken line. Place names: a = Sillerslev, b = Lysen Bredning, c = Glyngøre, d = Branden, e = Skive, f = Dragsted Vig, g = Thisted Bredning, h = Mariager Fjord, i = Kerteminde Fjord/Kertinge Nor, j = Bornholm.

haul. Population clearance rates and predation impact of two species of ctenophores (*Mnemiopsis leidyi*, *Pleurobrachia pileus*) and the moon jellyfish (*Aurelia aurita*) on zooplankton (using copepods as reference) were estimated from recorded population densities using the equations below.

Mnemiopsis leidyi. The following relationship was used for converting oral-aboral length (L, mm) to individual body volume (V, ml ind.$^{-1}$) [8]:

$$V = 0.0226L^{1.72} \quad \text{Eq (1)}$$

The following equation [10,22], was used to estimate the individual clearance rate (Cl_{ind}, l d^{-1}) of *M. leidyi* feeding on copepods as a function of their body volume (V, ml):

$$Cl_{ind} = 2.64V \quad \text{Eq (2)}$$

Pleurobrachia pileus. Clearance rates (Cl_{ind}, l d^{-1}) were estimated from the polar length (L, mm) by using the following equation for ctenophores feeding on copepods [23]:

$$Cl_{ind} = 0.2L^{1.9} \quad \text{Eq (3)}$$

Aurelia aurita. The individual clearance rate (Cl_{ind}, l d^{-1}) of moon jellyfish feeding on copepods was estimated from the mean inter-rhopalia diameter (d, mm) by use of the equation [6]:

$$Cl_{ind} = 0.0073d^{2.1} \quad \text{Eq (4)}$$

The volume-specific population clearance rate of jellyfish and ctenophores (Cl_{pop}, m^3 water filtered by the jellyfish or ctenophore population in one m^3 water per day = m^3 m^{-3} d^{-1}) was estimated as the product of the individual clearance rate (Cl_{ind}, l d^{-1}) and the population density (D, ind. m^{-3}) for each location [10]:

$$Cl_{pop} = Cl_{ind} \times D / 1000 \quad \text{Eq (5)}$$

The time ($t_{1/2}$, d) it takes for a population of jellyfish or ctenophores with known Cl_{pop} to reduce the concentration of prey organisms (copepods) in $V = 1$ m^3 of water by 50 % (i.e. the half-life of prey) was estimated as [10]:

$$t_{1/2} = \ln 2 / Cl_{pop} \quad \text{Eq (6)}$$

Water chemistry and zooplankton

Water chemistry data (salinity, temperature, dissolved oxygen, chlorophyll *a*) and zooplankton biomass from Løgstør Bredning and Skive Fjord for 2014 were obtained from the "National Monitoring and Assessment Programme for the Aquatic and Terrestrial Environments" (NOVANA) conducted by the Danish Ministry of the Environment. All data were collected and analyzed according to official monitoring guidelines [24]. Water chemistry data were obtained from CTD profiles from the surface to the bottom at the locations. Chlorophyll *a* concentrations at 1 m depth were obtained from the absorbance in spectrophotometric analysis of chlorophyll extracted with ethanol from the algae present in filtered water samples. Zooplankton was collected by using a submersible pump with a minimum capacity of 150 l min^{-1}. The water was pumped up vertically by hauling the pump up at a speed of 0.5 m s^{-1}. Collection of the zooplankton was done by using a 60 µm net at the outlet. The filtered water volume was calculated by using a mechanical flow meter. The filter was rinsed thoroughly and the sample preserved in either 2 to 3 % formalin (pH 8.0 to 8.2) or neutral Lugol's solution. The concentrations (ind. m^{-3}) and biomasses (µg C l^{-1}) of zooplankton species and taxonomic groups were subsequently determined.

Results

Jellyfish and ctenophores

Species composition, population densities and size of jellyfish and ctenophores recorded during 5 cruises in Limfjorden in 2014 are listed in Table 1, along with the estimated half-life of zooplankton for the 3 most abundant species. It is seen that only *Mnemiopsis leidyi* exerted a notable predation impact, most pronounced in Løgstør Bredning and Skive Fjord in August, when the estimated half-life of zooplankton (copepods) were 4.8 and 7.3 d, respectively, and in Skive Fjord in late September when the half-life was 2.2 d. The size distribution of *M. leidyi* at the 4 investigated locations is shown in Figure 2. *M. leidyi* was not observed in Limfjorden during the first 2 cruises, 7 April and 3 June, respectively. But on the 3rd cruise (26 August) the ctenophore was found on all 4 locations, with largest body size ($L = 8.6 \pm 4.8$ mm) and lowest population density ($D = 18.5 \pm 19.4$ ind. m^{-3}) in Nissum Bredning in the western part of Limfjorden. The highest density and smallest size were observed in Løgstør Bredning ($D = 158.8 \pm 110.7$ ind. m^{-3}, $L = 4.9 \pm 2.1$ mm) and Skive Fjord ($D = 116.0 \pm 25.6$ ind. m^{-3}, $L = 4.6 \pm 1.9$ mm). This pattern was somewhat changed one month later on the 4th cruise (23 September), and on the 5th cruise (18 November), when the population density of *M. leidyi* on all locations was strongly reduced. Along with the occurrence of *M. leidyi*, the small predatory ctenophore, *Beroe gracilis* was recorded on most sampling locations during the 3rd and 4th cruise, but not on the 5th cruise when only one large ($L = 64$ mm) *Beroe* sp. individual was caught (Table 1). From Figure 3 it appears that *B. gracilis* eats small *M. leidyi*, but their low number suggests that the population predation impact of *Beroe gracilis* exerted on *M. leidyi* during 2014 was negligible.

Water chemistry and zooplankton

Figures 4 and 5 show the temperature, salinity, dissolved oxygen and chlorophyll *a* concentrations in Løgstør Bredning and Skive Fjord during 2014. Figure 6 shows vertical profiles of temperature and salinity on selected dates, with full vertical mixing (Figure 6A) and bottom-near intrusion of high-saline North Sea water (Figure 6B) in Løgstør Bredning, and strong stratification due to combined thermo- and haloclines in Skive Fjord (Figure 6C), respectively. The dissolved oxygen concentration in the near-bottom water decreased to zero several times at both locations between June and September. As indicated by high chl *a* concentrations at the two locations (up to 45 µg l^{-1} in mid-August in Skive Fjord, Figure 5), the severe oxygen depletion had resulted in the release of nutrients (ammonium, phosphate) [7] from the anoxic sediment causing an algal bloom which was subsequently followed by a tremendous bloom of the heterotrophic dinoflagellate *Noctiluca scintillans* (Figure 7A). This was again followed by a pronounced peak in the abundance of copepods and their offspring nauplii (Figure 7B). To judge from the population density of *Mnemiopsis leidyi* and the estimated half-life of copepods ($t_{1/2} = 4.8$ d) in Løgstør Bredning on 26 August 2014 (Table 1) it seems reasonable to suggest that the rapid decrease in the biomass of copepods during late July, and the following absence of zooplankton during the rest of the year was caused by a boom of the invasive ctenophore. As a consequence, it may therefore be suggested that the decrease in population density and the persistent small size of *M. leidyi* during September and November (Table 1) reflect a rapid depletion of prey organisms resulting in starvation and decay.

Discussion

Species composition, population densities and predation impact

The abundance of ctenophores and jellyfish in Limfjorden is characterized by large fluctuations in species composition and population size, and while the yearly development of *Aurelia aurita*

Cruise/Date/Locality	Species	D (ind. m^{-3})	L (mm)	d (mm)	$t_{1/2}$ (d)
Cruise 1, 7 April 2014					
1 Nissum Bredning	*Pleurobrachia pileus*	0.003 ± 0.003	4.1 ± 2.9 (8)		∞
2 Venø Bugt	-				
3 Løgstør Bredning	-				
4 Skive Fjord	-				
Cruise 2, 3 June 2014					
1 Nissum Bredning	*Pleurobrachia pileus*	0.001 ± 0.001	23.0 ± 1.4 (2)		∞
	Aurelia aurita	0.003 ± 0.001		214.8 ± 48.0 (8)	∞
	Cyanea lamarckii	0.003 ± 0.001		66.8 ± 25.3 (8)	
2 Venø Bugt	*Aurelia aurita*	0.001 ± 0.001		212.0 ± 120.2 (2)	∞
3 Løgstør Bredning	*Aurelia aurita*	0.007 ± 0.012		212.5 ± 32.6 (19)	∞
4 Skive Fjord	-				
Cruise 3, 26 Aug 2014					
1 Nissum Bredning	*Mnemiopsis leidyi*	18.5 ± 19.4	8.6 ± 4.8 (50)		15.5
	Beroe gracilis	0.002 ± 0.003	22.0 (1)		
2 Venø Bugt	*Mnemiopsis leidyi*	67.9 ± 50.5	4.5 ± 2.0 (50)		12.9
	Beroe gracilis	0.030 ± 0.019	20.3 ± 5.3 (50)		
3 Løgstør Bredning	*Mnemiopsis leidyi*	158.8 ± 110.7	4.9 ± 2.1 (50)		4.8
	Beroe gracilis	0.002 ± 0.004	13.0 (1)		
4 Skive Fjord	*Mnemiopsis leidyi*	116.0 ± 25.6	4.6 ± 1.9 (50)		7.3
Cruise 4, 23 Sept 2014					
1 Nissum Bredning	*Mnemiopsis leidyi*	38.1 ± 18.1	7.6 ± 2.9 (50)		9.3
2 Venø Bugt	*Mnemiopsis leidyi*	0.2 ± 0.2	6.6 ± 2.0 (50)		∞
	Beroe gracilis	0.001 ± 0.000	20.0 ± 3.6 (3)		
	Aurelia aurita	0.000 ± 0.001		49.0 (1)	∞
	Rhizostoma octopus	0.000 ± 0.001		280.0 (1)	
3 Løgstør Bredning	*Mnemiopsis leidyi*	12.6 ± 5.4	4.3 ± 1.1 (50)		∞
	Beroe gracilis	0.002 ± 0.000	21.5 ± 6.8 (4)		
	Aurelia aurita	0.000 ± 0.001		55.0 (1)	∞
4 Skive Fjord	*Mnemiopsis leidyi*	255.4 ± 116.7	5.8 ± 1.6 (50)		2.2
Cruise 5, 18 Nov 2014					
1 Nissum Bredning	*Mnemiopsis leidyi*	0.5 ± 0.1	15.9 ± 12.7 (50)		∞
	Mnemiopsis leidyi	0.2 ± 0.1	17.8 ± 9.7 (50)		∞
2 Venø Bugt	*Beroe* sp.*	0.000 ± 0.001	64.0 (1)		
3 Løgstør Bredning	*Mnemiopsis leidyi*	6.6 ± 2.9	4.6 ± 1.9 (50)		∞
4 Skive Fjord	*Mnemiopsis leidyi*	19.1 ± 12.2	4.6 ± 1.1 (50)		∞
	Aurelia aurita	0.001 ± 0.002		65.0 (1)	∞

* *Beroe ovata* or *cucumis*

Table 1: Ctenophore and jellyfish species collected during 5 cruises in Limfjorden in 2014. D = mean (± SD) population density; L = mean (± SD, n = number of measured individuals) oral-aboral length of ctenophores; d = mean (± SD, n) inter-rhopalia diameter of medusae; $t_{1/2}$ = estimated half-life of copepods (cf. Eq. 6; $t_{1/2}$ > 3 weeks are indicated by ∞).

populations depends on benthic polyps, incoming water from the North Sea seems to be responsible for a yearly reinvasion by *Mnemiopsis leidyi* and other ctenophore species from the North Sea, and the present study supports the hypothesis that the North Sea serves as a refuge from where *M. leidyi* reinvades Limfjorden [10,11].

No or only very few ctenophores (*Pleurobrachia pileus*) and jellyfish (*Aurelia aurita, Cyanea lamarckii*) were recorded during the first two cruises in April and June 2014, and *Mnemiopsis leidyi* was first recorded on the 3rd cruise in late August (Table 1). However, the present first author (HUR) observed *M. leidyi* at a number of sites in Limfjorden (Sillerslev, Lysen Bredning, Glyngøre, Branden ; Figure 1) between 25 and 28 June 2014. Therefore, *M. leidyi* must have entered Limfjorden with North Sea water through Thyborøn Kanal between 7 April and 25 July. In the period between the 2nd and 3rd cruise, salinity increased in the bottom-near water (from 25.5 to 27.9 psu) in Løgstør Bredning during the period 17 June to 1 July (Figures 5B and 7B). This is also coincident with the observation of the North Sea scyphomedusa *Rhizostoma octopus* which is frequently brought into Limfjorden in the autumn months [11]. The Jutland Coastal Current which carries mixed water masses from the English Channel and the southern North Sea gives rise to seawater inflows into Limfjorden, which also explains the occasional re-introduction of the hydromedusa *Aequorea vitrina* in Limfjorden where it is not able to establish a surviving population [6,8].

Differences in the life-history strategies of jellyfish and ctenophores are important for understanding the often strong variations in species composition and population dynamics. The scyphozoan *Aurelia aurita* has a life cycle which includes a pelagic medusa and a benthic polyp stage. Medusae reproduce sexually and larval development is followed by the disappearance of medusae in late autumn. After settlement, planula larvae metamorphose into polyps to pass the winter. In the early spring, the polyps produce ephyrae that are released into the water column to develop into a new generation of medusae [4,25]. In contrast

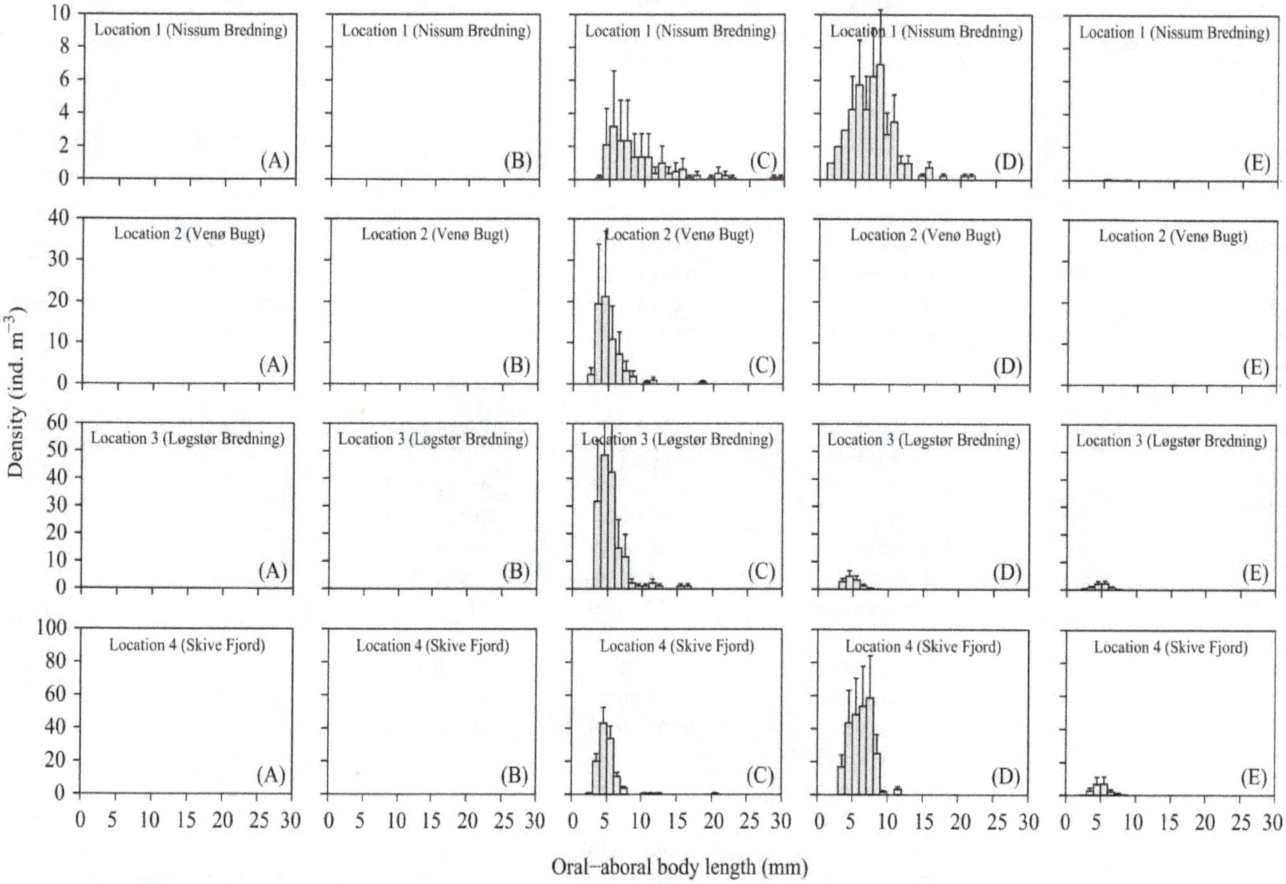

Figure 2: *Mnemiopsis leidyi.* Oral-aboral length distribution at 4 investigated locations in Limfjorden on A) 07 April B) 3 June, C) 26 August, D) 23 September and E) 18 November 2014. Average sizes for each mm size category are displayed (± SD for 3 replicate samples at each location).

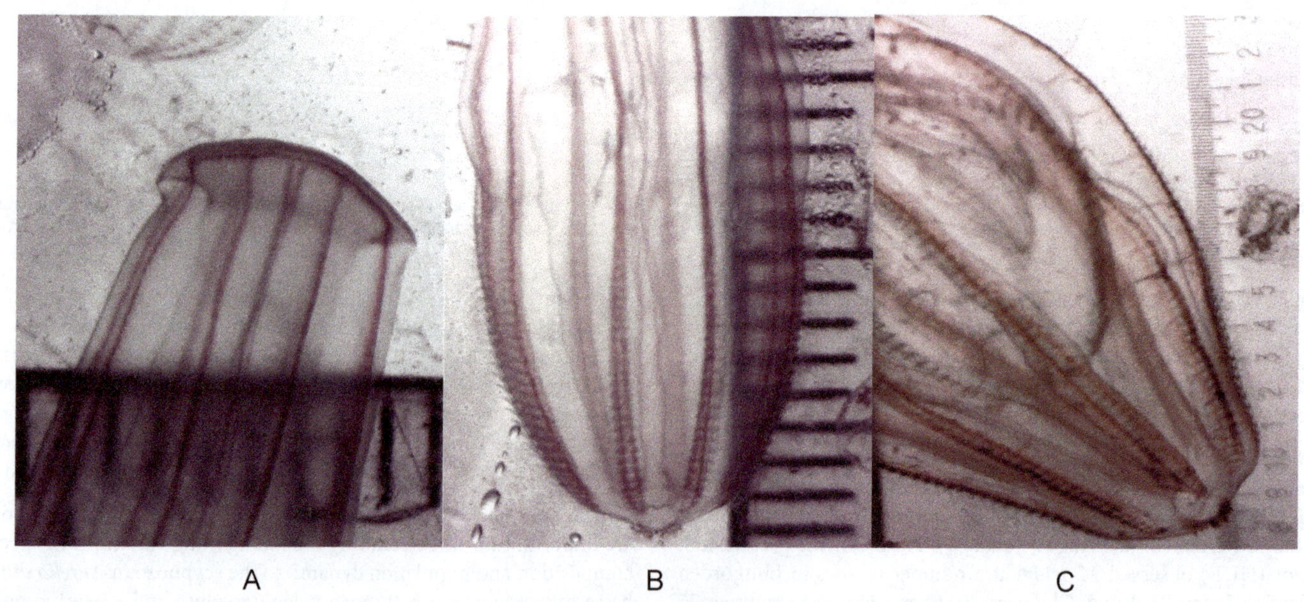

Figure 3: *Beroe gracilis* & *Mnemiopsis leidyi.* Ctenophores captured in net hauls in Limfjorden 26 August 2014. (A) oral end of *B. gracilis*, note a prey organism (*M. leidyi*) close to the left corner; scale: 1.0 mm. (B) aboral end of *B. gracilis*; scale: 1.0 mm. (C) *B. gracilis* with prey organism (*M. leidyi*) in the gastrovascular cavity; scale: 0.1 mm. Photos: Hans Ulrik Riisgård.

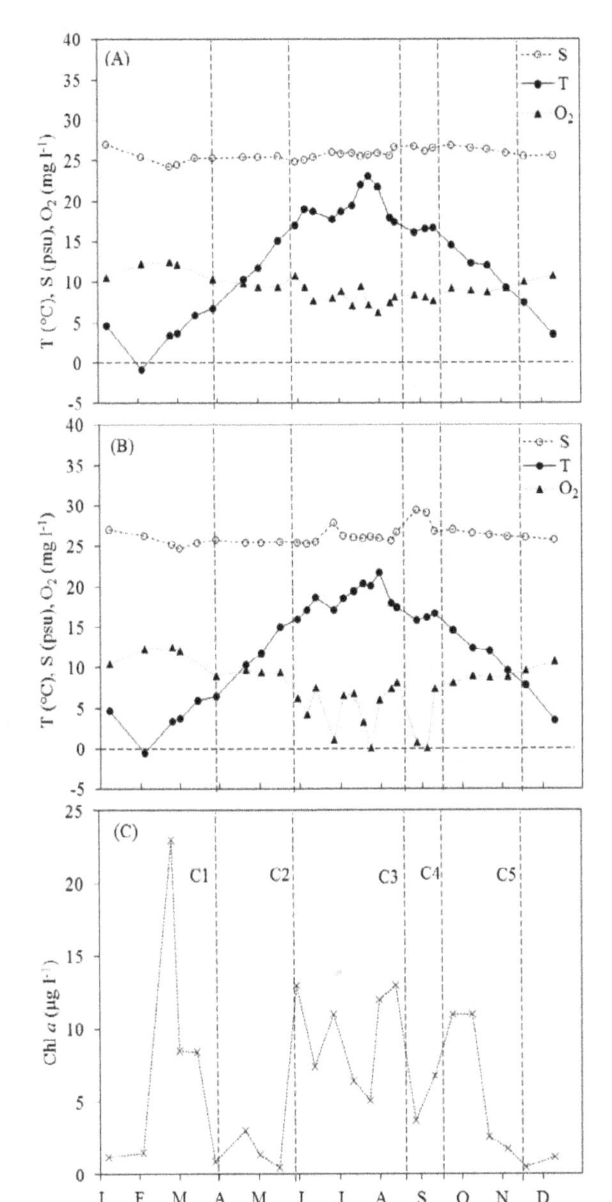

Figure 4: Løgstør Bredning 2014. (A) Temperature (T), salinity (S) and oxygen (O) in the top water layer (0.8 m below the surface), and (B) in the bottom water layer (0.3 to 0.4 m above the bottom). (C) Chlorophyll *a* (Chl *a*, µg l^{-1}) measured at a depth of 1 m. Punctured vertical lines (labelled C1-C5) signify the date of each of the 5 cruises carried out in the present study.

to this, the ctenophore *Mnemiopsis leidyi* relies on a holoplanktonic life cycle. *M. leidyi* is a self-fertilizing hermaphrodite, releasing eggs and sperm in the ambient water where fertilization occurs [26-29]. Thus, high fecundity and rapid generation times during the whole season may explain the ability of *M. leidyi* (and *Pleurobrachia pileus*) to occur in large numbers in Limfjorden [11]. However, unless a major fraction of *A. aurita* ephyrae are washed out of Limfjorden in the early spring, this may also result in a mass occurrence [7] and interspecific competition with *M. leidyi* [9]. The conspicuously few *A. aurita* observed in the present study as well as in the two preceding years [11] indicates that moon jellyfish populations in Limfjorden may currently experience the process of being outcompeted by the more numerous and faster reproducing *M. leidyi*, which could, on the long-term, also lead to a disappearance of local *A. aurita* polyps. However, alternative explanations are also possible, but especially improved understanding of polyp ecology seems necessary for understanding fluctuations in jellyfish numbers [2,30-33].

The abundance of ctenophores and jellyfish in Limfjorden is characterized by often large fluctuations in species composition and population size [11]. While the yearly development of *Aurelia aurita*

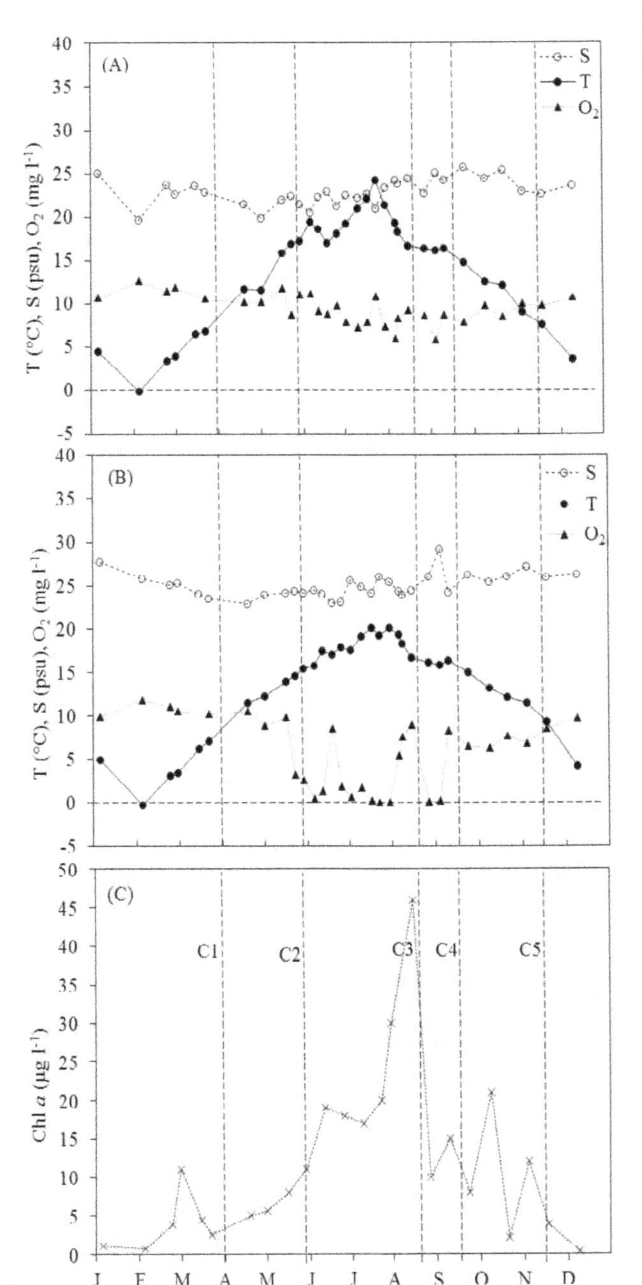

Figure 5: Skive Fjord 2014. (A) Temperature (T), salinity (S) and oxygen (O) in the top water layer (0.8 m below the surface), and (B) in the bottom water layer (0.3 to 0.4 m above the bottom). (C) Chlorophyll *a* (Chl *a*, µg l^{-1}) measured at a depth of 1 m. Punctured vertical lines (labelled C1-C5) signify the date of each of the 5 cruises carried out in the present study.

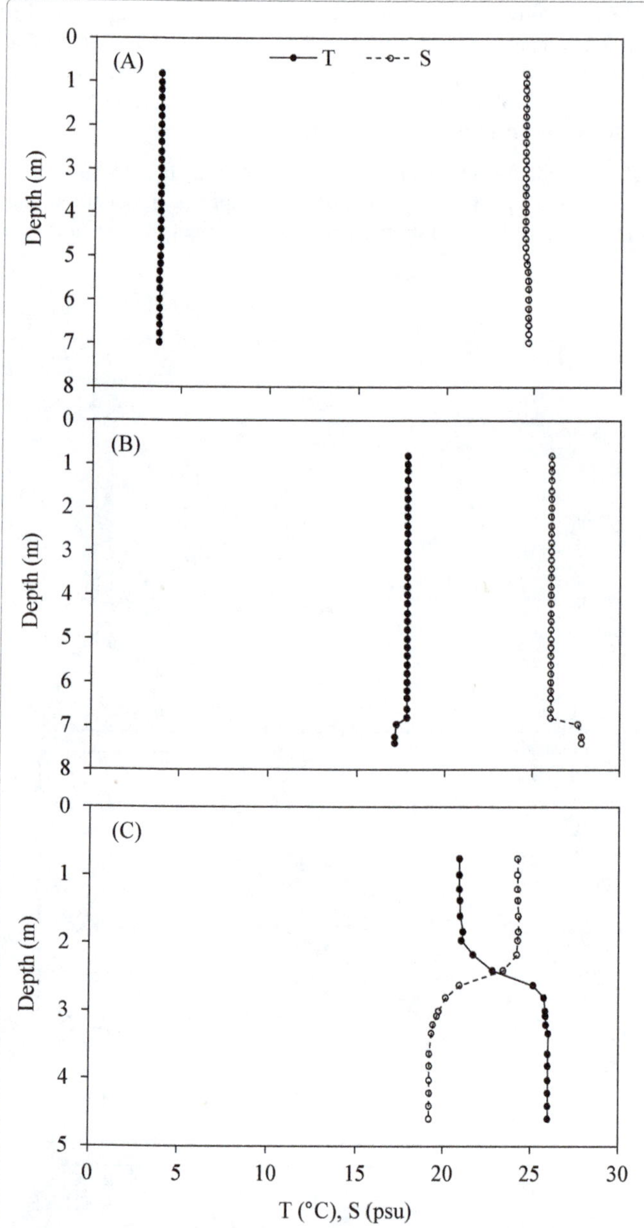

Figure 6: (A) Løgstør Bredning. Vertical profiles of temperature (T, °C) and salinity (S, psu) measured 4 March 2014 which is in between the cruises C1 and C2. (B) Løgstør Bredning. Vertical profiles measured 1 July 2014 (i.e. between cruises C2 and C3 at the time when salinity in the bottom water showed an increase, see Figure 5B). (C) Skive Fjord. Vertical profiles measured 28 July 2014 (i.e. between cruises C2 and C3, a few days before upwelling of anoxic bottom water to the surface as shown in Figure 5).

populations in Limfjorden may largely depend on recruitment via asexual reproduction by local polyps, the incoming water from the North Sea is of decisive importance for the yearly reinvasion by *Mnemiopsis leidyi* and other ctenophore species from the warmer south-western North Sea [34-36]. The present study supports the hypothesis that the North Sea in cold winters serves as a refuge from where *M. leidyi* reinvades Limfjorden, which may function as an incubator for *M. leidyi*, subsequently dispersing into the Kattegat and adjacent waters [10,11]. A number of observations support this hypothesis; thus, first observations of *M. leidyi* outside Limfjorden in 2014 were in Mariager Fjord (8 October, own observation), Kerteminde Fjord (3 October 2014, own observation), Kertinge Nor (7 November 2014, own observation) and the western Baltic Sea near Bornholm (21 November, Marie Stoor-Paulsen). The recent observations of *B. ovata* being a new species in Danish waters [11,37] has not yet caused measurable changes in the number and distribution of *M. leidyi*, and the biological and environmental factors controlling its sporadic occurrence are still poorly understood.

Interplay between biological and environmental factors: our present understanding

Recently, [3] gave an overview of the environmental status of the heavily eutrophicated Limfjorden, including the historical development of nutrient overloading and subsequent oxygen depletion in near-bottom water, and further how the bottom dwelling fish species decreased while the number of jellyfish and ctenophores increased. A case study of Skive Fjord [7] demonstrated the links between primary production, oxygen deficiency, nutrients and jellyfish. Each summer, Skive Fjord suffers from oxygen depletion in the near-bottom water and this causes large amounts of nutrients (phosphate and ammonia) to be released from the anoxic sediment, which subsequently stimulates a phytoplankton bloom, followed by an increase in zooplankton biomass [7] (Figure 2). The same phenomenon was observed in the present study, not only in Skive Fjord but also in Løgstør Bredning, where the chlorophyll *a* concentrations became very high during periods with oxygen depletion (Figures 4 and 5) followed by a bloom of *Noctiluca scintillans* succeeded by an increase in copepod biomass (Figure 7). [7] combined available data on jellyfish with data on dissolved oxygen, nutrients, chlorophyll

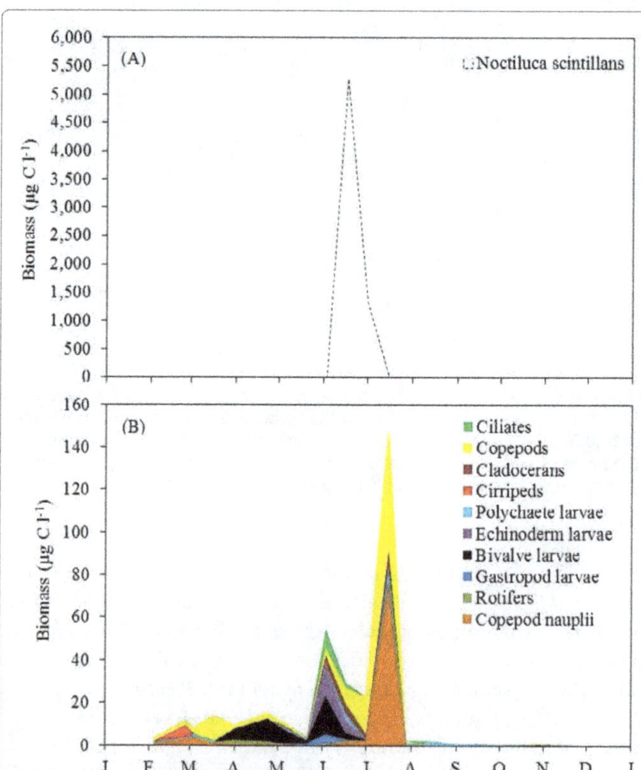

Figure 7: Løgstør Bredning (Figure 1, location 3). (A) June and July 2014 bloom-biomass of the dinoflagellate *Noctiluca scintillans*, (B) zooplankton biomass presented as the contribution of each zooplankton group of total biomass over time.

a, zooplankton, and other data from various studies conducted in Skive Fjord during the period 1996 to 2005 and found that especially severe cases of oxygen depletion apparently take place in years with mass occurrence of jellyfish. Due to their high predatory impact on the zooplankton community, blooming algae are not efficiently grazed on, but settle to the bottom and decay, thus leading to more severe hypoxia.

Mass occurrence of *Mnemiopsis leidyi* was first observed in Limfjorden in 2007, but it probably arrived the previous year [38]. This invasive holoplanktonic ctenophore exerted further predation pressure in addition to the indigenous scyphomedusa *Aurelia aurita*, which frequently caused the zooplankton organisms to become virtually absent, as also observed with some time-lag after the arrival of *M. leidyi* (Figure 7 and Table 1) in the present study. [39] made an integrated ecosystem assessment for Limfjorden over the period from 1984 to 2008 and showed that from 1990 to 1995, the structure shifted from dominance by demersal fish species to that of pelagic fish species (sprat, herring, sticklebacks) and jellyfish. Nutrients (N, P) have been and are still considered as a major bottom-up forcing factor in the system. Since the early 1900s and up to the mid-1980s there was a six-fold increase in nutrients which caused recurrent events of oxygen depletion, but although the loadings have decreased by 33 % (N) and 67 % (P) due to a water action plan decided by the Danish Parliament in 1987 which has so far not resulted in any noticeable improvement in ecosystem state [39,40].

According to [41], the first severe case of oxygen depletion in Limfjorden in 2014 took place in mid-June, including the stratified Skive Fjord (Figure 6C), other inner and central parts and central north-western parts (Thisted Bredning, Dragstrup Vig, Figure 1) where dead fish were observed on the beaches. By the end of June strong westerly winds resulted in mixing of the water column and improved oxygen conditions, but from the beginning of July the water column became stratified resulting in further oxygen depletion soon after. Thus, by the end of July 2014, about 30 % of the total bottom waters in Limfjorden were depleted of oxygen with subsequent release of toxic hydrogen sulphide from the anoxic sediment [13]. In Skive Fjord, off-shore winds in late July and early August pushed the surface water away from the shore, which caused upwelling of near-bottom water containing high concentrations of hydrogen sulphide. Subsequent oxidation made the new surface water white-colored due to precipitation of free sulphur [41]. No systematic studies of the effects of oxygen-depletion on fish, phyto- and zoobenthos have been conducted by the environmental authorities in 2014, and likewise possible effects on jellyfish and ctenophores remain unknown. However, due to the presence of high population densities of *Mnemiopsis leidyi* in Skive Fjord in late August and September (Table 1), it may be suggested that this ctenophore is rather resistant to poor environmental conditions as also emphasized by [22,42] who observed that *M. leidyi* is more tolerant to hypoxia than their prey and competing zooplanktivorous fishes.

From both the present study and earlier accounts [11], it appears that the species composition of jellyfish and ctenophores, their population dynamics and predation impact in Limfjorden is highly unpredictable, depending on the complex interplay between recruitment, life cycles, and hydrographic conditions. Every year since 2007, *M. leidyi* has reinvaded Limfjorden from the North Sea via Thyborøn Kanal with water partly originating from the English Channel [43], the Belgian part of the North Sea [44], and the western Dutch Wadden Sea where it is present the whole year round [35] and recent modelling [46] supports this interpretation. *M. leidyi* has frequently exerted a significant predation impact on the zooplankton biomass in Limfjorden during late summer, but in October and November 2012 it was another ctenophore, *Pleurobrachia pileus*, that controlled the zooplankton in Løgstør Bredning, with $t_{1/2}$ = 1.5 and 2.8 d in October and November, respectively, while only very few *M. leidyi* were observed [11]. However in most years since 2007, *M. leidyi* has thrived extremely well during late summers in Limfjorden, from where the ctenophore with its rapid reproduction may further spread into Kattegat and the inner Danish waters, perhaps even to the western Baltic Sea [47].

In the present study, it was observed that *Beroe gracilis* feed on small (<20 mm) *Mnemiopsis leidyi* (Figure 3) which confirms recent observations by [35,48]. Apparently, *B. gracilis* follows *M. leidyi* from the southern North Sea from where also *Pleurobrachia pileus* comes, but hitherto, *B. gracilis* has been considered a specialized predator on *P. pileus* and new to the inner Danish waters [37]. This phenomenon may be another consequence of the now established self-sustaining populations of *M. leidyi* in the English Channel and southern North Sea coasts.

Herring (*Clupea harengus*) and sprat (*Sprattus sprattus*) are pelagic fish species in Limfjorden and have to date been used for both human consumption and industrial processing for fish meal and oil, but because these fish feed on zooplankton, they may, in certain areas and periods, compete with large numbers of ctenophores and/or jellyfish [11]. The herrings enter Limfjorden from both the North Sea and the Kattegat in late autumn and early winter to spawn in the spring, and the adults leave again in early summer while the juveniles stay (Erik Hoffmann, pers. comm.), but no studies have so far attempted to determine the degree of interspecific competition for zooplankton. The short estimated half-lives of zooplankton caused by *M. leidyi* in August 2014 in Løgstør Bredning and Skive Fjord (Table 1) and the conspicuous reduction of the zooplankton biomass (Figure 7) suggest very poor feeding conditions for juvenile herrings in Limfjorden.

The present study has been focused on the complex interplay between biological and environmental factors that control the jellyfish and ctenophore populations in Limfjorden, and hopefully, the many still unsolved questions may stimulate future research with focus on e.g. how gelatinous predation of zooplankton may reinforce anoxia and further habitat degradation in eutrophicated waters.

Acknowledgements

Thanks to the Danish Ministry of the Environment for providing water chemical data, to skippers Mogens Grimstrup and Henning Bach and their crew aboard the patrol vessel MHV 902 MANØ (Danish Naval Homeguard), to Jan Brandt Wiersma for technical assistance, to Bernd Lüskow for help with the map, and to Jens Würgler Hansen and Ole Secher Tendal for comments on the manuscript. Two researchers, Sabine Holst and Cornelia Jaspers took part in the 3rd and 4th cruise, respectively, and they assisted in counting and measurement of collected jellies.

References

1. Purcell JE, Uye S, Lo WT (2007) Anthropogenic causes of jellyfish blooms and their direct consequences for human: a review. Mar Ecol Prog Ser 350: 153-174
2. Purcell JE (2012) Jellyfish and ctenophore blooms coincide with human proliferations and environmental perturbations. Ann Rev Mar Sci 4: 209-235
3. Riisgård HU, Andersen P, Hoffmann E (2012c) From fish to jellyfish in the eutrophicated Limfjorden (Denmark). Estuar Coasts 35: 701-713
4. Hansson LJ, Moeslund O, Kiørboe T, Riisgård HU (2005) Clearance rates of jellyfish and their potential predation impact on zooplankton and fish larvae in a neritic ecosystem (Limfjorden, Denmark). Mar Ecol Prog Ser 304:117-131

5. Møller LF, Riisgård HU (2007a) Feeding, bioenergetics and growth in the common jellyfish *Aurelia aurita* and two hydromedusae *Sarsia tubulosa* and *Aequorea vitrina*. Mar Ecol Prog Ser 346:167-177

6. Møller LF, Riisgård HU (2007b) Population dynamics, growth and predation impact of the common jellyfish, *Aurelia aurita* and two hydromedusae (*Sarsia tubulosa* and *Aequorea vitrina*) in Limfjorden (Denmark). Mar Ecol Prog Ser 346: 153-165

7. Møller LF, Riisgård HU (2007c) Impact of jellyfish and mussels on algal blooms caused by seasonal oxygen depletion and nutrient release from the sediment in a Danish fjord. J Exp Mar Biol Ecol 351: 92-105

8. Riisgård HU, Bøttiger L, Madsen CV, Purcell JE (2007) Invasive ctenophore *Mnemiopsis leidyi* in Limfjorden (Denmark) in late summer 2007 - assessment of abundance and predation effects. Aquat Invasions 2: 395-401

9. Riisgård HU, Madsen CV, Barth-Jensen C, Purcell JE (2012a) Population dynamics and zooplankton-predation impact of the indigenous scyphozoan *Aurelia aurita* and the invasive ctenophore *Mnemiopsis leidyi* in Limfjorden (Denmark). Aquat Invasions 7: 147-162

10. Riisgård HU, Jaspers C, Serre S, Lundgreen K (2012b) Occurrence, inter-annual variability and zooplankton-predation impact of the invasive ctenophore *Mnemiopsis leidyi* and the native jellyfish *Aurelia aurita* in Limfjorden (Denmark) in 2010 and 2011. BioInvasions Rec 1: 145-159

11. Riisgård HU, Goldstein J (2014) Jellyfish and ctenophores in Limfjorden (Denmark) - mini-review, with recent new observations. J Mar Sci Eng 2: 593-615

12. Riisgård HU (2007) Feeding behaviour of the hydromedusa *Aequorea vitrina*. Scient Mar 71: 395-404

13. Jørgensen BB (1980) Seasonal oxygen depletion in the bottom waters of a Danish fjord and its effect on the benthic community. Oikos 34: 68-76

14. Blanner P (1982) Composition and seasonal variation of the zooplankton in the Limfjord (Denmark) during 1973-1974. Ophelia 21: 1-40

15. Møhlenberg F (1999) Effect of meteorology and nutrient load on oxygen depletion in a Danish micro-tidal estuary. Aquat Ecol 33: 55-64

16. Wiles P, L van Duren LA, Häse C, Larsen J, Simpson JH (2006) Stratification and mixing in the Limfjorden in relation to mussel culture. J Mar Syst 60: 129-143

17. Hofmeister R, Buchard H, Bolding K (2009) A three-dimensional model study on processes of stratification and de-stratification in the Limfjord. Cont Shelf Res 29: 1515-1524

18. Christiansen T, Christensen TJ, Markager S, Petersen JK, Mouritsen LT (2006) Limfjorden i 100 år. Klima, hydrografi, næringsstoftilførsel, bundfauna og fisk i Limfjorden fra 1897 til 2003 Danmarks Miljøundersøgelser. Faglig rapport fra DMU pp: 578

19. Richardson K, Jacobsen T (1990) Jyllandsstrømmen. En transportmekanisme fra tyske bugt til Kattegat. NPOforskning fra Miljøstyreslsen nr. C6 1990 (report in Danish)

20. Aure J, Danielsen D, Svendsen E (1998) The origin of Skagerrak coastal water off Arendal in relation to variations in nutrient concentrations. ICES J Mar Sci 55: 610-619

21. Gyllencreutz R, Backman J, Jakobsson M, Kissel C, Arnold E (2006) Postglacial palaeoceanography in the Skagerrak. The Holocene 16: 973-983

22. Decker MB, Breitburg DL, Purcell JE (2004) Effects of low dissolved oxygen on zooplankton predation by the ctenophore *Mnemiopsis leidyi*. Mar Ecol Prog Ser 280: 163-172

23. Møller LF, Canon JM, Tiselius P (2010) Bioenergetics and growth in the ctenophore *Pleurobrachia pileus*. Hydrobiologia 645: 167-178

24. Novana (2011) Tekniske Anvisninger NOVANA - 2011-2015 (technical instructions for environmental monitoring, in Danish)

25. Lucas CH (2001) Reproduction and life history strategies of the common jellyfish, *Aurelia aurita*, in relation to its ambient environment. Hydrobiologia 451: 229-246

26. Baker LD, Reeve MR (1974) Laboratory culture of the lobate ctenophore *Mnemiopsis mccradyi* with notes on feeding and fecundity. Mar Biol 26: 57-62

27. Shiganova TA, Dumont HJD, Mikaelyan AS, Glazov DM, Bulgakova YV, et al. (2004) Interaction between the invading Ctenophores *Mnemiopsis leidyi* (A. Agassiz) and *Beroe ovata* Mayer 1912, and their influence on the pelagic ecosystem of the northeastern Black Sea. In: Dumont H, Shiganova T, Niermann U, The Aquatic Invasions in the Black, Caspian and Mediterranean Seas. NATO ASI. Environment. Kluwer Academic Publishers, The Netherlands pp: 33-70

28. Costello JH, Bayha KM, Mianzan HW, Shiganova TA, Purcell JE (2012) Transitions of *Mnemiopsis leidyi* (Ctenophora: Lobata) from a native to an exotic species: a review. Hydrobiologia 690: 21-46

29. Jaspers C, Møller LF, Kiørboe T (2015) Reproduction rates under variable food conditions and starvation in *Mnemiopsis leidyi*: significance for the invasion success of a ctenophore. J Plank Res.

30. Condon RH, Decker MB, Purcell JE (2001) Effects of low dissolved oxygen on survival and asexual reproduction of scyphozoan polyps (*Chrysaora quinquecirrha*). Hydrobiologia 451:89-95

31. Boero F, Bouillon J, Gravili C, Miglietta MP, Parsons T, Piraino S (2008) Gelatinous plankton: irregularities rule the world (sometimes). Mar Ecol Prog Ser 356: 299-310

32. Webster CN, Lucas CH (2012) The effects of food and temperature on settlement of *Aurelia aurita* planula larvae and subsequent somatic growth. J Exp Mar Biol Ecol 436: 50-55

33. Hosia A, Falkenhaug T, Naustvoll L-J (2014) Trends in abundance and phenology of *Aurelia aurita* and *Cyanea* spp. at a Skagerrak location, 1992-2011. Mar Ecol Prog Ser 498: 103-115

34. Hamer HH, Malzahn AM, Boersma M (2011) The invasive ctenophore *Mnemiopsis leidyi*: a threat to fish recruitment in the North Sea? J Plank Res 33: 137-144

35. van Walraven L, Langenberg VT, van der Veer HW (2013) Seasonal occurrence of the invasive ctenophore *Mnemiopsis leidyi* in the western Dutch Wadden Sea. J Sea Res 82: 86-92

36. Vansteenbrugge L, Regenmortel TV, Roch MD, Vincx M, Hostens K (2015) Gelatinous zooplankton in the Belgian part of the North Sea and the adjacent Schelde estuary: Spatio-temporal distribution patterns and population dynamics. J Sea Res 97: 28-39

37. Shiganova T, Riisgård HU, Ghabooli S, Tendal OS (2014) First report on *Beroe ovata* in an unusual mixture of ctenophores in Great Belt (Denmark). Aquat Invasions 9: 111-116

38. Tendal OS, Jensen KR, Riisgård HU (2007) Invasive ctenophore *Mnemiopsis leidyi* widely distributed in Danish waters. Aquat Invasions 2: 455-460

39. Tomczak MT, Dinesen GE, Hoffkamm E, Maar M, Støttrup JG (2013) Integrated trend assessment of ecosystem changes in the Limfjord (Denmark): Evidence of a recent regime shift? Estuar Coast Shelf Sci 117: 178-187

40. Krause-Jensen D, Markager S, Dalsgaard T (2011) Benthic and pelagic primary production in different nutrient regimes. Estuar Coasts 35: 527-545

41. Hansen JW, Rytter D, Balsby TJS (2014) Iltsvind i de danske farvande i juli-august 2014. Notat fra DCE - Nationalt Center for Miljø og Energi (Report in Danish with English figure legends and summary, pp: 22

42. Kolesar SE, Breitburg DL, Purcell JE, Decker MB (2010) Effects of hypoxia on *Mnemiopsis leidyi*, ichthyoplankton and copepods: clearance rates and vertical habitat overlap. Mar Ecol Prog Ser 411: 173-188

43. Antajan E, Bastian T, Raud T, Brylinski J-M, Hoffman S. et al, (2014) The invasive ctenophore *Mnemiopsis leidyi* A. Agassiz, 1865 along the English Channel and the North Sea French coasts: another introduction pathway in northern European waters? Aquat Invasions 9: 167-173

44. van Ginderdeuren K, Hostens K, Hoffman S, Vansteenbrugge L, Soenen K, et al. (2012) Distribution of the invasive ctenophore *Mnemiopsis leidyi* in the Belgian part of the North Sea. Aquat Invasions 7: 163-169

45. van der Molen J., van Beek J, Augustine S, Vansteenbrugge L, et al.(2014) Modelling survival and connectivity of *Mnemiopsis leidyi* in the southern North Sea and Scheldt estuaries. Ocean Sci Discuss 10: 1-51

46. David C, Vaz S, Loots C, Antajan E, van der Molen J, Travers-Trolet M (2015) Understanding winter distribution and transport pathways of the invasive ctenophore *Mnemiopsis leidyi* in the North Sea: coupling habitat and dispersal modelling approaches. Biol Invasions 17: 2605-2619"

18

Research of *Ascocotyle (Phagicola) longa* in Heat Treated Fillets of Mullet (*Mugil platanus*)

Marianna Vaz Rodrigues[1]*, Agar Costa Alexandrino de Pérez[2], Thaís Moron Machado[2], Fátima Maria Orisaka[3], Jacqueline Kazue Kurissio[1] and Andrea Lafisca[4]

[1]*Department of Microbiology and Immunology, Biosciences Institute, Univ. Estadual Paulista (UNESP), Distrito de Rubião Júnior s/n, Botucatu, São Paulo, Brazil*
[2]*Reference Unit Laboratory Technology of Seafood–Instituto de Pesca, Agência Paulista de Tecnologia do Agronegócio, Secretaria da Agricultura e Abastecimento, Av Bartolomeu de Gusmão 192, Santos, São Paulo, Brazil*
[3]*Freelance veterinary Distrito de Rubião Júnior s/n, Botucatu, São Paulo, Brazil*
[4]*Veterinary, In-lingua scientific translations and linguistic services, Distrito de Rubião Júnior s/n, Botucatu, São Paulo, Brazil*

Abstract

Seafood can present many biological hazards, such as zoonotic parasites. Among these, *Ascocotyle* (*Phagicola*) *longa* trematode is generally found in mullets (*Mugil platanus*) and is the most common parasite involved in heterophyiosis outbreaks. This research aimed to detect viable metacercariae of *Ascocotyle* (*Phagicola*) *longa* after heating muscle of mullets. The method used was sedimentation followed by microscopy observation. It was found 100% (16/16) of inactivated metacercariae in the analyzed samples. This is the first study involving samples of mullets ready to eat sold directly to consumer. We conclude, consumers must be alerted to the risk of infection by raw mullet eating and proper heating or cooking kills this trematode.

Keywords: Parasite; Food safety; Public health

Introduction

The presence of parasites in marine and freshwater fishes is common and may carry risks, both economic and sanitary [1]. Most of the parasites are found in organs that are discarded during fish processing, some worms may be found in the muscle. In case of consumption of the seafood in an inadequate preparation, consumers may fell ill [2-4]. Among the parasites reported in mullets (*Mugil platanus*), *Ascocotyle* (*Phagicola*) Ransom, 1920 (*Digenea: Heterophyidae*) trematode is very common and can cause disease in human by consumption of parasitized raw seafood [5-7].

Adult *A.* (*Phagicola.*) live in the gut of birds and mammals. Metacercariae develop in mullets tissues [8]. According to Simões et al. [9], mollusks presence is essential for the occurrence of heterophyiosis. Depending on the region studied, particular specie of mollusk is involved with the biological cycle. Simões et al. [9] also reported the presence of the snake *Heleobia australis* as intermediate host for this parasite, increasing the risk of human infection. Even the elevated risk of infection present, this fishborne disease is underestimated due to absence of characteristic clinical signals [10,11]. Heating is the best method for inactivation of these parasites. Coelho [12] recommends heating at 100°C for 60 minutes. Antunes et al. [13] observed ionization with doses of 4.0 kGy gamma rays can also be efficient to kill metacercariae, but this method is not approved by sanitary authorities in some countries. Therefore, this study aimed to detects and identify viable *A. metacercariae* in mullets (*Mugil platanus*) fillet after heat treatment.

Materials and Methods

Sampling

Officers of sanitary police of the State of São Paulo (Brazil) sampled 16 baked fillet of mulltes (*Mugil platanus*) from "mullet festivities" between June and July of 2009 in the following cities: Bertioga, Praia Grande, Santos, and São Vicente in the State of São Paulo (South Eastern Brazil) (Table 1).

In this event, the fish is put in an oven to cook. Normally the product reaches 50-56°C in the centre for 2-3 minutes. The problem is when there are many people because they want the plate fast, and for that reason, the muscle doesn't reach the right temperature. Thereby, if seafood is parasitized by *Ascocotyle*, it can cause illness in the consumer.

Parasitological analysis

To guarantee correct identification, it was taken a piece of muscle of fresh fish and treated by heat of the same animal to equate results. This procedure is essential because heat can cause alteration in the morphology of the parasite, harming their identification.

After sampling, fish were put in plastic bags and destined to the parasitology laboratory of Reference Unit Laboratory Technology of Seafood of the Fishing Institute (*Instituto de Pesca*), in Santos, São Paulo, Brazil, for detection and identification of metacercariae. Five grams of muscle of each fish were taken and submitted to centrifugation with 300 mL of clean, tap water. The content was transferred to the

City	Number of fish sampled
Bertioga	4
Praia Grande	4
Santos	4
São Vicente	4
Total	**16**

Table 1: Number of mullets (*Mugil platanus*) sampled by city during June to July of 2009.

***Corresponding author:** Rodrigues MV, Department of Microbiology and Immunology, Biosciences Institute, Univ. Estadual Paulista (UNESP), Distrito de Rubião Júnior s/n, Botucatu, São Paulo, Brazil, E-mail: mvazrodrigues@gmail.com

conical glass jar remaining 20 minutes for sedimentation. Supernatant was discarded and more 300 mL of potable water were added. After 20 minutes wait, the sediment was collected and put in a slide using Pasteur pipette, to perform microscopic analysis of the sample [14]. Microscopic identification of parasites was performed according to Simões et al. [9].

Statistical analysis

We analyzed the prevalence of parasites. According to the results, samples were divided in two classes: "present" or "absent". The prevalence calculation was performed using R version 2.15.1 software [15].

Results and Discussion

In this study was observed in 100% of samples, the presence of parasites identified as Ascocotyle suggesting a contamination of mullets before cooking. Analysis of parasites suggested these were inactivated (Figure 1).

The high prevalence of metacercariae in mullets we observed in this study had already been described by Hutton [16], Armas de Conroy [17], Almeida-Dias e Woiciechovski [18], Antunes and Almeida Dias [19], Knoff et al. [20], Conceição et al. [21], Oliveira et al. [10], and Santos et al. [22]. The high quantity of studies showing high prevalence of Ascocotyle stresses the importance of the detection of this parasite in the world.

In Brazil, Chieffi et al. [23,24], Antunes and Almeida-Dias [19] illustrated cases of heterophyiosis in the state of São Paulo, probably caused by Ascocotyle (Phagicola). Based on these data, metacercariae detection in the muscle, adequate processing, and consumer awareness are crucial to prevent fishborne disease. Inactivation strategies must be realized to guarantee seafood security, since Santos et al. [22] demonstrated adequate heating importance for the safety to consumers.

This is the first study involving research of Ascocotyle mullets samples ready to eat. "Mullet festivities" attract many consumers and the time of fish cooking varies a lot according to demand. Most of the times, mullets are roasted quickly in high fire, causing external overcooking and internal undercooking. According to Huss et al. [25], inactivation temperature of trematodes is 55°C for 1 minute inside the product.

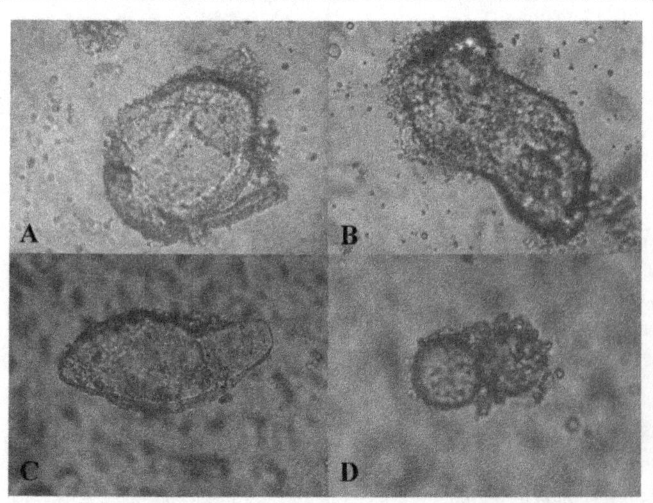

Figure 1: Ascocotyle longa metacercariae inactivated by heating observed in microscopy. 100X. A, B, and C: different larvae forms, D: cysts.

The presence of 100% of inactivated (dead) metacercariae in the samples observed in this study indicates, the temperature of roasting was adequate for parasite inactivation. All samples collected by officers of Sanitary Policy were too roasted, which is not common to observe during these parties, as it was observed by the authors. It is important to control both time and temperature to guarantee the inactivation of metacercariae as described by Huss et al. [25].

Oliveira et al. [10] report, fishes parasitized by Ascocotyle do not present any lesion suggesting any kind of parasitic infection. In 2010, this trematode was included in the Risk Classification of Biological Agents list of Brazil [26].

Sanitary inspection of seafood is not enough to guarantee safety for consumer, once this is based on visual analysis. It is necessary to explain to consumer that raw or undercooked fish eating may carry parasites, such as Ascocotyle which are dangerous to humans.

Acknowledgements

We would like to thank the officers of sanitary police of the State of São Paulo (Brazil) for sampling and help with all the information used in this research.

References

1. Ferrer I (2001) Anisakiosis y otras zoonosis parasitarias transmitidas por consume de pescado. Aquatic 14: 1-21.
2. Rodríguez M (1998) Parásitos de importância en la salud pública. Curso taller: Diagnóstico y control de enfermedades em peces de cultivo. Centro de Investigaciones Pesqueras, Ciudad de la Habana, Cuba.
3. Ubeira FM, Valiñas B, Lorenzo S, Iglesias R, Figueiras A et.al., (2000) Anisaquiosis y alergia. Un estudio epidemiológico en la comunidad Autónoma Gallega. Documentos Técnicos de Salud Pública, Consellería de Sanidade e Serviços Sociais, Xunta de Galicia. Serie B: 102.
4. Lorenzo S (2000) Anisakis y alergia. Imprenta Universitaria, Santiago de Compostela, Tesis.
5. Muller R (2001) Worms and Human Diseases. CABI Publishing, Wallingford.
6. Scholz T, Aguirre-Macedo ML, Salgado-Maldonado G (2001) Trematodes of the family Heterophyidae (Digenea) in Mexico: a review of species and new host and geographical records. J Nat Hist 35: 1733-1772.
7. Fried B, Graczyk TK, Tamang L (2004) Food-borne intestinal trematodiases in humans. Parasitol Res 93: 159-170.
8. Scholz T (1999) Taxonomic study of Ascocotyle (Phagicola) longa Ransom, 1920 (Digenea: Heterophyidae) and related taxa. Syst Parasitol 43: 147-158.
9. Simões SB, Barbosa HS, Santos CP (2010) The life cycle of Ascocotyle (Phagicola) longa (Digenea: Heterophyidae), a causative agent of fish-borne trematodosis. Acta Trop 113: 226-233.
10. Oliveira SA, Blazquez FJH, Antunes SA, Maia AAM (2007) Metacercária de Ascocotyle (Phagicola) longa Ransom, 1920 (Digenea: Heterophyidae), em Mugil platanus, no estuário de Cananéia, SP, Brasil. Ciência Rural 37: 1057-1059.
11. Montejo RD, Yumang AP, Sabay BV (2008) Heterophyidiasis: a re-emerging disease in Davao Region. Epidemiology 19: S165-S166.
12. Coelho MRT (1996) Ação de diferentes métodos de conservação na sobrevivência de metacercárias de Phagicola longus (Ranson, 1920) Price, 1932, parasito de mugilídeos capturados no litoral do Estado do Rio de Janeiro. Niterói. Universidade Federal Fluminense, Faculdade de Veterinária, Dissertation.
13. Antunes SA, Wiendl FM, Almeida Dias ER, Arthur V, Daniotti C (1993) Gamma ionization of Phagicola longa (Trematoda: Heterophyidae) in mugilidae (pisces) in São Paulo, Brazil. Rad Phy Chem 42: 425-428.
14. Coelho MRT, São Clemente SC, Gottshalk S (1997). Ação dos diferentes métodos de conservação na sobrevivência de metacercárias de Phagicola

longus (Ranson, 1920) Price, 1932, Parasito de mugilídeos capturados no litoral do estado do rio de janeiro. Revista Higiene Alimentar. 11: 39-42.

15. Dean CB, Nielsen JD (2007) Generalized linear mixed models: a review and some extensions. Lifetime Data Anal 13: 497-512.

16. Hutton RF (1957) Preliminary notes on trematodes (Heterophydade and Strigeoides) encysted in the heart and flesh of Florida mullet Mugil cephalus L. and Mugil curema Curier & Valencienes. Bulletin of the Dade County Medical Association 2: 2.

17. Armas de Conroy G (1986) Investigaciones sobre la fagicolosis em lisas (Mugilidae) de águas americanas. I. Estúdios taxonômicos de Phagicola sp. (Trematoda: Heterophyidae) em mugílidos sudamericanos. Revista Ibérica de Parasitología 46: 39-46.

18. Almeida-Dias ER, Woiciechovski E (1994) Ocorrência da Phagicola longa (Trematoda: Heterophyidae) em mugilídeos e no homem, em Registro e Canéia, SP. Revista Higiene Alimentar. 8: 43-46.

19. Antunes SA, Almeida Dias ER (1994) Phagicola longa (Trematoda: Heterophyidae) em mugilídeos estocados resfriados e seu consumo cru em São Paulo – SP. Revista Higiene Alimentar 8: 41.

20. Knoff M, Luque JL, Amato JFR (1997) Community ecology of the metazoan parasites of grey mullets, Mugil platanus (Osteichthyes: Mugilidae) from the Littoral of the State of Rio de Janeiro, Brazil. Revista Brasileira de Biologia 57: 441-454.

21. Conceição JCS, São Clemente SC, Matos E (2000). Ocorrência de Phagicola longus (Ransom, 1920) Price, 1932 em tainhas (Mugil sp.) comercializadas em Belém, Estado do Pará. Revista Acadêmica: Ciências Agrárias e Ambientais 33: 97-101.

22. Santos CP, Lopes KC, Costa VS, dos Santos EGN (2013) Fish-borne trematodosis: Potential risk of infection by Ascocotyle (Phagicola) longa (Heterophyidae). Veterinary Parasitology 193: 302-306.

23. Chieffi PP, Leite OH, Dias RM, Torres DM, Mangini AC (1990) Human parasitism by Phagicola sp (Trematoda, Heterophyidae) in Cananéia, São Paulo state, Brazil. Rev Inst Med Trop Sao Paulo 32: 285-288.

24. Chieffi PP, Gorla MC, Torres DM, Dias RM, Mangini AC, et al. (1992) Human infection by Phagicola sp. (Trematoda, Heterophyidae) in the municipality of Registro, São Paulo State, Brazil. J Trop Med Hyg 95: 346-348.

25. Huss HH, Ababouch L, Gram I (2004) Assessment and Management of Seafood Safety and Quality. FAO Fisheries Technical. 444: 60-69.

26. Brasil (2010) Ministério da Saúde. Secretaria de Ciência, Tecnologia e Insumos Estratégicos. Classificação de risco dos agentes biológicos. Normas e Manu.

19

Mouldy Groundnut Cake and Hydrated Sodium Calcium Aluminosilicate in Practical Diet for African Catfish *Clarias gariepinus* (Burchell, 1822)

Comfort Adetutu Adeniji[1]*, Pius Abimbola Okiki[2], Ajani Murano Rasheed[1] and Rasheed Bolaji[1]

[1] *Department of Fisheries, Lagos State University, Lagos, Nigeria*

[2]*Department of Biological Sciences, College of Sciences, Afe Babalola University, Ado Ekiti, Nigeria*

Abstract

The study was designed to detect fungi and quantify aflatoxins B_1 and B_2 microbial load levels and toxin binding efficacy of hydrated sodium calcium aluminosilicate (HSCAS) on growth and nutrient utilization of *Clarias gariepinus* African catfish fingerlings diet formulated with mouldy groundnut cake (MGNC). MGC containing 185.00 ± 7.07 µg/kg and 137.50 ± 10.00 µg/kg of aflatoxins B_1 and B_2 was incorporated into practical diets. Three practical isoproteinous diets were formulated. Diet 1 contained Groundnut Cake (GNC) with 0.20 ± 0.14 cfu microbial load but no apparent mould contamination. Diet 2 contained MGC 4.38 ± 0.40, 1.99 ± 0.01 µg/kg of aflatoxins B_1, B_2 and 2.20 ± 0.14 cfu microbial load. Diet 3 contained MGC, HSCAS; 3.49 ± 0.20, 1.34 ± 0.01 µg/kg and 1.35 ± 0.20 cfu of aflatoxins B_1, B_2 and microbial load, respectively. The diets were fed twice daily to *Clarias gariepinus* fingerlings mean body weight 1.68 ± 0.01g in triplicate of 20 fish each for 12 weeks. Six fungi; *Aspergillus parasiticus, A. flavus, A. niger, A. tamarii, Penicillium citriinum and P. oxalicum* were isolated from MGNC diet. Addition of HSCAS reduced fungi present in diet 3 to *A. flavus, A. tamarii and P. oxalicum*. Also aflatoxins B_1, B_2 and microbial load in the diet and carcass of *Clarias gariepinus* were reduced with HSCAS inclusion in diet 3. Feeding African catfish fingerlings with mould contaminated diet (2) significantly ($P<0.05$) reduced growth performance and feed utilization but had no significant ($P>0.05$) effect on survival of *Clarias gariepinus* fingerlings. The results of this study show that the inclusion of HSCAS reduced the number of fungi, microbial load and aflatoxins B_1 and B_2 in the carcass and diet of African catfish formulated with mouldy GNC. But was not effective in alleviating the growth depression induced by mould contamination of the diet. Hence the need for identification of toxins present in mould contaminated feedstuffs in order to design effective toxins management strategies.

Keywords: Mouldy groundnut cake; HSCAS; Aflatoxin B_1; Aflatoxin B_2; Microbial load; *Clarias gariepinus*; Feed utilization; Growth performance

Introduction

Feed ingredients contamination with toxins may occur anywhere in the supply chain from the field, manufacturing process, transportation to storage. In the tropics conducive environmental condition has majorly been attributed to production and growth of micro-organisms and toxic substances; particularly fungi. This is exacerbated by the cultural practice and lack of adequate regulatory and control system by government for food and feedstuffs contaminant screening in most developing countries in the humid tropics.

Fungi, *Aspergillus flavus* and A. *parasiticus* contamination are mostly responsible for the production of aflatoxins in the humid tropics [1]. They are found when environmental temperature is above 27°C, humidity levels greater than 65% and moisture levels in feedstuff is above 14%. Also suboptimal handling and storage are considered as important factors favouring the growth of aflatoxin-producing fungi [2]. Aflatoxins are the most potent toxic substances occurring naturally. Large yield of aflatoxins are found in feed ingredients with high carbohydrates concentration as found in cereals and to a lesser extent in oil seeds. But in West Africa aflatoxins are of major concern in groundnut. Four major forms of aflatoxins are found in feedstuffs: B_1, B_2, G_1 and G_2. Aflatoxin B_1 is regarded as carcinogenic, most prevalent and toxic of the four. It has significant economic and health implications in both animal and man. Hence efforts have been focused on reducing the impact of aflatoxin B_1 contamination in livestock and fish, with dearth of information on the other aflatoxins.

Recently studies have reported the co-occurrence or co-contamination of substance(s) with more than one aflatoxins [3,4]. This is because in practical situation contaminants are rarely found individually [5]. Especially when environmental condition is conducive, the probability of multiple or co- contamination is always high. Hence there is need to assess contaminants present naturally in feed raw materials particularly in endemic environment like the humid tropics. Sequel to this, there might be need for the identification of fungi present in such substances which are potential toxins producers. This could provide necessary information for designing strategies in the management of these toxins.

Consequently with the recent trend in increase utilization of plant feedstuffs (which are good substrates for aflatoxins) in warm water fish species, there is need to investigate the implications of natural aflatoxins contamination. Also the practice of inclusion of toxin binder like Hydrated Sodium Calcium Aluminosilicate (HSCAS), which is generally routinely added to suspected contaminated (mouldy) feedstuffs in endemic environment [6].

From the foregoing, this study was designed to detect fungi, quantify microbial load and aflatoxins B_1, B_2 of the diets formulated with mouldy groundnut cake. Also the toxin binding efficacy of HSCAS on growth performance and aflatoxins B_1 and B_2 levels in the diets and

***Corresponding author:** Comfort Adetutu Adeniji, Department of Fisheries, Lagos State University, Lagos, Nigeria, E-mail: comfortadeniji@yahoo.com

carcass of the African catfish; *Clarias gariepinus* fingerlings fed diets formulated with mouldy groundnut cake was evaluated.

Materials and Methods

Dietary Design and Feeding

A 50 kg bag of Mouldy Groundnut Cake (MGC) suspected to be naturally contaminated was obtained from a commercial feed ingredients seller at Ojo, Lagos, Nigeria. MGC contained 185.00 ± 7.00 and 137.50 ± 10.00 µg/kg of aflatoxin B_1 and B_2, respectively. Three experimental diets were formulated to contain 35% crude protein each. Diet one contained uncontaminated groundnut cake and other feed ingredients which served as the control, diet two contained contaminated groundnut cake and other feed ingredients while diet three contained contaminated groundnut cake, other feed ingredients and toxin binder (HSCAS) at the manufacturer's recommended inclusion rate of 0.3% of the diet. Percentage gross composition of the experimental diets is shown in Table 1. The feed preparation was done at a reputable commercial fish feed manufacturer: Act livestock consult, Badagry Expressway, Lagos-Nigeria, where all ingredients were weighed, ground into fine powder, thoroughly mixed and pelleted (into 2 mm) with a 350 kg/hr table top pelleting machine. The pellets were crushed into pieces and stored in air tight containers. Fish were fed twice daily at 8.00 and 16.00 hours at the rate of 3% body weight for a period of 12 weeks. Feeding was done by hand and spread evenly across the water surface of each aquarium.

Fish management

The experiment was conducted at the indoor of Lagos State University Hatchery, Ojo, Lagos, Nigeria. Nine (9) plastic aquaria tanks with 80 L water capacity and 0.5 m depth were filled with borehole water to three quarter capacity. The water in the experimental tanks was aerated by electric air pump (Shining model; horsepower 50Hz). *Clarias gariepinus* fingerlings of mean weight of 1.68 ± 0.01 gm were obtained from Lagos State Agricultural Cooperative farm, Ojo, Lagos, Nigeria. The fish were transported early in the morning between the hours of 6.30 and 7.00 from the hatchery and acclimatized for 5 days prior to the start of the experiment. They were randomly distributed to the 9 plastic aquaria at the stocking rate of 60 fingerlings per experimental diet and 20 per replicate making a total of 180 fingerlings. Daily 50% of water in each tank was gently siphoned in exchange for fresh water. This was done to get rid of left over feed and faecal matter.

Ingredients	Diet 1	Diet 2	Diet 3
Maize (CP 9%)	29.50	29.50	29.20
GNC (CP 45%)	21.00	0.00	0.00
MGNC (45%)	0.00	21.00	21.00
SBC (CP 45.30%)	21.00	21.00	21.00
Wheat offal (CP 16%)	8.00	8.00	8.00
Fishmeal (CP 72%)	18.50	18.50	18.50
Soy oil	1.00	1.00	1.00
Premix*	0.50	0.50	0.50
Salt	0.25	0.25	0.25
Vitamin C	0.25	0.25	0.25
HSCAS	0.00	0.00	0.30
Gross energy**	4212.50	4193.80	4207.30
	Proximate composition		
CrudeProtein (N×6.25%)	34.97	34.72	35.00
Fibre (%)	3.71	3.68	3.70
Ether extract (%)	4.41	4.30	4.38
Total ash (%)	4.92	4.95	5.02
Dry matter (%)	92.40	92.18	92.39

*Each kg of diet contained 2,000.000 IU vit A; 4,000.000 IU vitD_3; 2,000.000 vit E; 1,200 mg vit K; 10,000.000 mg vitB_1; 30,000 mg vitB_2; 19,000 mg vitB_6; 1000 mg vit.$B1_2$; 5000 mg Panthotenic acid; 200,000 mg Niacin; 5,000 mg Folic acid; 30 Mn; 40 gm Zn; 40 mg Fe; 4 gm Cu; 5gm I_2; 0.2 mg Co; 600 gm calcium; 400 mg choline chloride; 40 mg biotin; 400,000 mg phosphorus; 100,000 m glysine; 400 gm methionine and 125 IU antioxidant. Gross energy** (KCal kg^{-1}); calorific value of protein 5.65; nitrogen free extract 4.1; lipid 9.45.

Table 1: Percentage composition of the experimental diets.

Fungi isolation and quantification of aflatoxins and microbial loads

Mouldy groundnut cake was subjected to extraction and quantification of aflatoxins B_1 and B_2 before the commencement of the study and other major ingredients were screened against aflatoxins. The coarse ingredients were ground, mixed before samples were collected for analyses. Fungi present in the experimental diets were isolated; the diets were also subjected to extraction and quantification of aflatoxins B_1, B_2 and microbial load. Twenty fingerlings were randomly picked before the start of the experiment. The fingerlings were washed, crushed under sterile condition for microbial load determination. At the end of the experiment 20 fingerlings (divide into two groups) from each treatment were randomly picked for microbial load and aflatoxin B_1 and B_2 determination. The isolation of fungi, extraction and quantification of aflatoxins and microbial load were done at the Institute of Agricultural Research and Training (IAR&T), Ibadan, Oyo-Nigeria [7-10].

Growth and feed utilization

Batch weighing of fish in each replicate aquarium was done at the beginning of the experiment, subsequently biweekly, using a MettleR 20110 top-loading balance. The evaluation of experimental diets for growth and feed utilization was carried out as follows.

Weight gain=Final Body Weight - Initial Body Weight

Specific growth rate (SGR% day^{-1})=($\log_e$FBW-logeIBW)×100

Time

Where, FBW represents final body weight,

IBW represents initial body weight.

Time represents difference in days between final and initial body weight.

$$\text{Protein efficiency ratio (PER)} = \frac{\text{Fish weight gain (kg)}}{\text{Protein fed (kg)}}$$

$$\text{Percentage survival} = \frac{(\text{Initial number of fish stocked} - \text{number of dead fish})}{\text{Initial number of fish stocked}} \times 100$$

Economy analysis

A simple analysis was conducted to assess the cost effectiveness of the experimental diets, only diet cost was used for the calculations with the assumption that all other operating costs remained the same. The average prices of each feedstuff ($) during the study period was used to calculate the amounts required to make the different diets, cost per kilogram of each diet was calculated. The economic conversion ratio (ECR) was determined using this equation: ECR=COST OF DIET×FCR.

Water quality parameters

Daily water temperature was measured using a mercury-in-glass

	Parameters			
Specimen: Diet	AFB$_1$(µg/kg)	AFB$_2$(µg/kg)	Microbial load (cfu)	Fungi isolated
1	ND^c	ND^c	0.20 ± 0.14^c	1*
2	4.38 ± 0.40^a	1.99 ± 0.01^a	2.20 ± 0.14^a	6*
3	3.49 ± 0.20^b	1.34 ± 0.01^b	1.35 ± 0.20^b	3*
Specimen: Carcass				
1	ND^c	ND^c	0.25 ± 0.07^c	ND*
2	2.99 ± 0.02^a	1.05 ± 0.01^a	1.70 ± 0.14^a	ND*
3	2.24 ± 0.01^b	0.89 ± 0.03^b	0.95 ± 0.21^b	ND*
T o x i n absorbed**	0.75 ± 0.02^a	0.16 ± 0.02^b	0.75 ± 0.20^a	ND*

1*= *Aspergillusparasiticus*6*= *A. parasiticus, A. flavus; A. niger, A. tamarii, Penicillumcitriinum and P.oxalicum*3*=*A. parasiticus, A. niger and P. citriinum.* ND=Not detected; ND*=Not Determined. Toxin absorbed**with same superscript in the same row are not significantly (P>0.05) different. Mean with same superscripts in the same column are not significantly different (P>0.05).
a=highest mean, b=intermediate mean and c=lowest mean.

Table 2: Summary of aflatoxins and microbial load in the diets and carcass of African catfish (Clariasgariepinus) fed the experimental diets.

Parameters	Diet 1	Diet 2	Diet 3
Initial weight/fish (g)	1.68 ± 0.01^a	1.68 ± 0.01^a	1.68 ± 0.00^a
Final weight/fish (g)	9.92 ± 1.06^a	5.22 ± 0.64^b	6.12 ± 1.00^b
Weight gain/fish (g)	8.24 ± 1.07^a	3.54 ± 0.64^b	4.44 ± 0.97^b
Specific growth rate	1.58 ± 0.10^a	1.01 ± 0.11^b	1.15 ± 0.14^b
Feed intake/fish (g)	14.64 ± 1.14^a	9.93 ± 0.64^b	10.84 ± 0.96^b
Feed conversion ratio	1.78 ± 0.0^b	2.85 ± 0.41^a	2.50 ± 0.46^a
Protein efficiency ratio	1.61 ± 0.08^a	1.01 ± 0.33^b	1.20 ± 0.13^b
Survival percentage	71.67 ± 3.00^a	74.00 ± 1.01^a	70.00 ± 4.04^a
Economic conversion analysis			
Feed cost/kg ($)	0.79 ± 0.07	0.78 ± 0.06	0.78 ± 0.07
Fish cost/kg ($)	1.41 ± 0.06	2.21 ± 0.06	1.94 ± 0.06

Means with the same superscripts along each row are not significantly different (P>0.05).

Table 3: Growth, feed utilization, survival and economic analysis of *Clariasgariepinus fingerlings* fed the experimental diets.

Parameters	Diet 1	Diet 2	Diet 3	
pH		6.56 ± 0.43	6.49 ± 0.10	6.54 ± 0.32
Temperature (oC)	27.67 ± 0.01	27.40 ± 0.40	27.60 ± 0.20	
DO* (mg/l)	5.77 ± 0.10^a	3.59 ± 0.30^b	3.67 ± 0.50^b	

Means with the same superscripts along each row are not significantly (P>0.05) different. DO* represents dissolved oxygen.

Table 4: Water quality analysis of *Clariasgariepinus* fingerlings fed the experimental diets.

thermometer, hydrogen ions (p^H) concentration measured using p^H meter (Jenway Model 9060) and weekly dissolved oxygen (DO) concentration was measured by oxygen meter (Hanna Model HI-9142).

Observation

The fish were observed for behavioural and physical abnormalities that could suggest possibility of disease situation.

Statistical and chemical analyses

Feed proximate composition was done according to the method of Association of Analytical Chemists [11]. Data collected from feeding trial were subjected to one-way analysis of variance (ANOVA). Significance difference in means was evaluated by Duncan's Multiple Range Test using SPSS for windows (version 11). Values are expressed as Means ± SD.

Results

Fungi, aflatoxins content and microbial load of the experimental materials:

The results of the fungi isolated, aflatoxins and microbial load of the experimental diets, fish and carcass are presented in Table 2. Summary of growth performance characteristics and water quality results are presented in Tables 3 and 4, respectively.

Six fungi from two genera comprising of *Aspergillus* and *Penicillum* were isolated from the three diets. *Aspergillus parasiticus* was found in all the experimental diets. In addition diet 2 contained five other fungi; *A. niger, A. flavus, A. tamarii, Penicillum citriinum* and *P. oxalicum.* Inclusion of HCSAS to diet 3 reduced the number of fungi isolated to two; *A. niger* and *P. citriinum.*

Aflatoxins B_1 and B_2 were not detected in the diet and carcass of African catfish fed diet 1 (uncontaminated). Generally addition of HCSAS significantly (P<0.05) reduced the concentrations of aflatoxins B_1, B_2 and microbial load in the diets and the carcass of African catfish (*Clarias gariepinus*) in this study.

Feed utilization, growth and economic analyses:

Inclusion of mouldy groundnut cake in the diet of *Clarias gariepinus* had no significant (P>0.05) effect besides survival, but significantly (P<0.05) reduced final weight, weight gain, specific growth rate, feed intake, feed conversion ratio and protein efficiency ratio. However, addition of HCSAS to diet 3 slightly but not significantly (P>0.05) improved all growth and feed utilization parameters assessed.

Economic analysis showed that mould contaminated diet (2) had the least cost of production ($ 0.791), but most expensive ($ 2.21) for African catfish to attain a kilogram body weight when fed with this diet.

Water quality analysis:

Inclusion of mouldy groundnut cake to the diet of *Clarias gariepinus* significantly (P<0.05) reduced dissolved oxygen concentration of catfish fed the contaminated diet.

Observation:

No external changes or unusual behaviour was observed in any of the fish fed the experimental diets throughout the study. Fish in all experimental diets appeared healthy and normal throughout the period of study.

Discussion

There is dearth of information on natural aflatoxins contamination of feedstuffs and diets of the African catfish in spite of their prevalence and the economic importance of the fish in this part of the world. The results of fungi isolated in the three experimental diets showed the presence of two of the three fungi genera (*Aspergillus, Penicillum and fusarium*) associated with mycotoxins production of agriculture and human health significance. The presence of *Aspergillus parasiticus* in the three diets suggests potentials for the production of the four forms of aflatoxins. Although aflatoxins B_1 and B_2 were not detected in diet 1; which may be due to insufficient availability of conditions necessary for their production in this study [12,13]. The co- occurrence of *A. flavus, A. niger and Penicillium* in the contaminated diet could probably encourage the production of ochratoxins and cyclopiazonic acid (CPA), which have been reported to be commonly found with aflatoxins [13-15]. Also possibilities exist for the production of citrinin, secalonic and oxalic acid *by P. citrinin* and *P. oxalicum* [16,17]. Their co-occurrence may encourage additive, synergy and or antagonistic reactions among these toxins. The addition of HSCAS (diet 3) reduced the number of fungi isolated, however potentials exit for the production of all toxins listed against MGC diet, except secalonic acid which is produced by *P. oxalicum* [18]. The increased microbial load observed in the mould contaminated diet could have resulted from the high number of fungi present in the MGC diet compared to other diets. Therefore from the fungi isolated, there might be need to investigate the presence of other mycotoxins when mouldy groundnut cake is used in diet formulation.

The reduced levels of aflatoxins B_1 and B_2 in HSCAS diet further attest to the ability of HSCAS in reducing aflatoxins levels particularly B_1 [19,20]. The reduced aflatoxins levels may have contributed to lower microbial load observed in the diet and carcass of African catfish fingerlings fed diet containing HSCAS. It is however imperative to note that the level of aflatoxin B_1 (2.24 ± 0.01) found in fish carcass was more than 0.05 µg/kg [21]. Consequently, the addition of HSCAS alone may not be effective in combating mould contamination in the diets of African catfish.

The feed utilization parameters measured showed that African catfish fed mouldy GNC diets had lower weight gain compared to uncontaminated diet. The reduced weight gain was slightly but not significantly improved by the addition of HSCAS to the diet. This is in agreement with the findings [20,22,23]. The lower weight attained may have emanated from low feed consumed by fingerlings on these diets, occasioned by mould contamination which is known to have negative effect on feed taste and nutritive quality [6,24,25]. The low feed intake and nutritive quality may have contributed to the lower growth and feed utilization; feed conversion and protein efficiency ratio observed with the aflatoxins contaminated diets. However, the results of this study disagree with the findings of [26,27] who reported no significant change in growth when tilapia and channel catfish were fed 250 and 2150 ppb purified aflatoxin B_1. The discrepancy might be due to the different species of fish, as well as the nature of toxins used (purified toxin was used in their study as against natural toxins present in feedstuff used in this study). This is besides the fact that more than one toxin was assessed in this this study. These might have aggravated the observed adverse effects. Also the inability of HSCAS to significantly improve the growth, feed utilization and economy of production is in agreement with the findings of [20,28,29] who reported ineffectiveness of HSCAS in diets containing multiple mycotoxins. Consequently supporting the need to determine fungi and all toxins present when mouldy feedstuffs are used in diet formulation. The low dissolved oxygen recorded with contaminated diets could have arisen from the quality of the feed fed. This assumption is based on the fact that aflatoxin is known to reduce nutritive quality of feed [26,30]. The reduced nutritive quality could have reduced fish's appetite resulting in reduced feed intake and increase feed or solid suspension in the culture medium, hence the reduced dissolved oxygen. However, the low dissolved oxygen values were within tolerable levels for catfish production.

Conclusion

The results of this study showed the co-occurrence of six fungi; *A. parasiticus, A. niger A. flavus, A. tamarii, P. citriinum and oxalicum* in diets formulated with mouldy GNC. These fungi have potentials for production of the four forms of aflatoxins, ochratoxins, CPA, citrinin, secalonic and oxalic acid. The co-occurrence of aflatoxins B_1 and B_2 evaluated in this study adversely affected growth, feed utilization and resulted in deleterious impact on production cost of African catfish (*Clarias gariepinus*) fingerlings. The addition of HSCAS slightly improved the adverse effects of the contamination with the elimination of potential for secalonic acid production. However, aflatoxin B_1 levels in the carcass of African catfish; *Clarias gariepinus* is beyond the 21 acceptable level. Consequently, there might be need for identification and quantification of all toxins present in diets formulated with mould contaminated feedstuff (in this study MGC) to proffer needed strategies for the toxins management in catfish diets.

References

1. Binder ME, Tan LM, Chi LJ, Handle J, Richard J (2007) Worldwide occurrence of mycotoxins in commodities, feeds and feed ingredients. Animal Feed Science Technology 137: 265-282.
2. Chulze SN (2010) Strategies to reduce mycotoxin levels in maize during storage: A review. Food Addit Contam Part A Chem Anal Control Expo Risk Assess 27: 651-657.
3. Ritchie JC (2007) Aflatoxin: In Molecules of Death (2nd edn). Imperial College Press, London.
4. Manning BB (2010) Mycotoxins in aquaculture feeds. SRAC Publication.
5. Braicu C, Berindan-neagoe I, Chedea VS, Balacescu L, Brie I et al (2010) Individual and combined cytotoxic effects of the major four aflatoxins in different in vitro stabilized system. Journal of Food Biochemistry 34: 1079-1090.
6. Ayasha A, Rahman M, Hasan M (2010) Effects of aflatoxin B_1 on growth and bioaccumulation in common carp fingerlings in Bangladesh. Asian-Pacific Journal of Rural Development 20: 1-13.
7. Rodricks JV, Stoloff L (1997) Determination of concentration and purity of aflatoxin standards. Journal of Association Official and Analytical Chemists 53: 92-95.
8. Hara S, Fennel D, Hesseltine CW (1974) Aflatoxin producing strains of *Aspergillus flavus* detected by fluorescence of agar medium under ultra violet light. Appl Microbiol 27: 1118-1123.
9. Robertson JA, Pons WA, Godblatt LA (1967) Preparation of aflatoxin and determination of their ultra violet and fluorescent characteristics. J Agric Food Chem 15: 798-801.
10. Lee PS (2009) Quantitation of Microorganisms: Practical Handbook in Microbiology. 2ndEdn CRC Press, Taylor & Francis Group, U.S.
11. AOAC (1995) Official Method of Analysis 16th edition. Association of Official Analytical Chemists, Arlington, VA.
12. Manning BB, Li MH, Robinson EH (2005) Aflatoxins from moldy corn cause no reduction in channel catfish *Ictalurus punctatus* performance. Journal of World Aquaculture 36: 59-67.
13. Dorner JW (1983) Production of cyclopiazonic acid by *A. tamarii*. Appl Environ Microbiol 46: 1435-1437.
14. Balachandran C, Parthasarathy KR (1996) Occurrence of CPA in feeds and feedstuffs in Tamil Nadu, India Mycopathologia 133: 159-162.
15. Fraga ME, Curvello F, Gatti MJ, Cavaglieri LR, Dalcero CA et al. (2007) Potential Aflatoxin and Ochratoxina production by *Aspergillus* species in Poultry feed processing. Vet Res Commun 31: 343-353.
16. Ciegler A, Hayew AW, Vesonder RF (1980) Production and biological activity of secalonic acid D. Appl Environ Microbiol 39: 285-287.
17. Zaied C, Zouaoui N, Bacha H, Abid S (2012) Natural occurrence of citrinin in Tunisian wheat grains. Food Control 28: 106-109.
18. Amadi JE, Adeniyi DO (2009) Mycotoxin production by fungi isolated from stored grains. African Journal of Biotechnology 8: 1219-1221.
19. Arunlertaree C, Sonngam L, Hutachareon R (2007) Vermiculite and HSCAS as the agent of aflatoxinB1 absorption for black tiger shrimp diet. Environment and Natural Resources Journal 5: 50-58.
20. Agouz HM, Anwer W (2010) Effect of biogen and myco-ad on growth performance of common carp (*Cyprinuscarpio*) fed a mycotoxin contaminated aquafeed. Journal of Fisheries and Aquatic Science 6: 334-345.
21. Aflatoxin contamination in foods and feeds in the Philippines. (2004) FAO/WHO Regional Conference on food safety for Asia and Pacific, Malaysia.

22. Oluwafemi F, Dahunsi O (2009) Performance of catfish fed different doses of aflatoxin in the diet. Journal of Natural Sciences, Engineering & Technology 8: 15-24.

23. Lopes PRS, Povey JLOF, Enke DBS, Mallamnn CA, Kich HA et al (2009) Use of adsorbent in diets containing aflatoxin for silver catfish fingerlings. R Bras Zootec 38: 589-595.

24. Giambrone JJ, DeinerUl, Davis ND, Panagola AS, Hoerr J (1985) Effects of purified aflatoxin on turkeys. Poultry Science 64:859-865.

25. Lim HA, Ng WK, Lim SL, Ibrahim CO (2001) Contamination of palm kernel meal with *Aspergillus flavus* affects its nutritive value in pelleted feed for tilapia *Oreochromis mossambicus*. Aquaculture Research 32: 895-905.

26. Tuan NA, Grizzle JM, Lovell RT, Manning BB, Rottinghaus GE (2002) Growth and hepatic lesions of Nile tilapia (*Oreochromis niloticus*) fed diets containing aflatoxin B1. Aquaculture 212: 311-319.

27. Jantrarotai W, Lovell RT (1990) Sub chronic toxicity of dietary aflatoxin B1 to channel catfish. Journal of Aquatic Animal Health 2: 248-254.

28. Garcia AR, Avica E, Rosiles R, Petrone VM (2003) Evaluation of two mycotoxin binders to reduce toxicity of broiler diets containing ochratoxin A and T-2 toxin contaminated grain. Avian Dis 47: 691-699.

29. Kolosova A, Stroka J (2012) Evaluation of the effect of mycotoxin binders in animal feed on the analytical performance of standardized methods for the determination of mycotoxins in feed. Food Addit Contam Part A Chem Anal Control Expo Risk Assess 29: 1959-1971.

30. Akande KE, Abubakar MM, Adegbola TA, Bogoro SE (2006) Nutritional and health implications of aflatoxins in animal feeds: A review. Pakistan Journal of Nutrition 5: 398-403.

20

Potential Use of Mussel Farms as Multitrophic On-growth Sites for American Lobster, *Homarus americanus* (Milne Edwards)

Guoqiang Wang and Iain J McGaw*

Department of Ocean Sciences, 0 Marine Lab Road, Memorial University, St John's, NL A1C 5S7, Canada

Abstract

Mussel (*Mytilus edulis*) farms in Newfoundland, Canada were investigated as potential sites to hold adult lobsters *Homarus americanus* in inshore benthic cages. The goals of this project were to determine if lobsters can be maintained for prolonged periods in cages and survive and grow by feeding on mussels dropping-off culture lines. The effects of biotic and abiotic factors on the moulting, growth rates and serum protein concentrations were monitored at regular intervals in both the field and the lab over 6 months. Although survival rates were high under mussel lines, the moulting rate was low and analysis of serum protein concentration showed they were in a poorer condition than fed lobsters in lab experiments. In the laboratory diet type, temperature, feeding frequency and compartment size were manipulated to determine possible factors influencing survival and growth of the lobsters in the field. In the lab, moulting was highest at 15°C and survival lowest at 5°C; lobsters fed a mixed versus a mussel only diet were healthier. In a separate lab experiment, lobsters that were fed twice weekly attained a larger size at post-moult than those fed once per month. However, feeding frequency did not affect survival or the number of animals moulting. The lab experiments suggested that the combination of low temperature and infrequent food input was the cause of the low moulting rate and overall quality of the lobsters in the field. This project showed although lobsters can be stored in benthic cages in the field for up to 6 months, relying on mussel drop-off alone is limited, and lobsters may need supplemental feeding in order to produce a larger, higher quality product for market. Initial results also suggest the promise of incorporating lobsters into a multitrophic aquaculture system as a means to remove moribund mussels underneath culture lines.

Keywords: Aquaculture; Blue mussel; Growth; Lobster; Moulting; Serum protein; Benthic cages; *Homarus americanus*

Introduction

The American lobster, *Homarus americanus* (Milne-Edwards) supports is a multi-billion dollar fishery in Atlantic Canada and New England. Generally it takes between 4 and 12 years for *H. americanus* to reach marketable size [1]. Because of this long growth time the fishery is highly regulated. In Atlantic Canada, the lobster fishing season varies in duration among the lobster fishing areas, ranging from 2 to 6 months [2]. On the island of Newfoundland, the lobster fishing season is comparatively short compared with the rest of Canada and New England, starting between April and May and closing between June and July. Because the lobster fishing seasons are regulated, the commercial trade of adult American lobsters has largely focused on development of holding methods so that live lobsters are available year round. For example, most of the lobster marketing companies in Nova Scotia and New Brunswick hold lobsters in indoor tanks or outdoor impoundments. In the indoor facilities the lobsters are held at temperatures of 1-3°C. This low temperature reduces the lobster's metabolic rate and maintains them in the intermoult (hard shell) stage, allowing the animals to be held for several months with minimal loss of product. The lobsters are not fed during this time and in order to combat the effects of starvation only lobsters with high serum protein concentrations can be stored in this way (Stewart Lamont, Tangier Lobster Company Limited, pers. comm.). However, the maintenance and logistics required for these holding methods are expensive, and cannot be employed on a small scale or in remote locations.

Although there have been efforts to raise and release juvenile clawed lobsters to repopulate areas [3,4] and there has been some work on on-growth of juvenile clawed lobsters in the field [5-8] they have had limited success. Therefore, the harvest of both *H. americanus* and its European counterpart, *Homarus gammarus* (Linnaeus) primarily remain a wild capture fishery. In contrast there has been much more research directed towards the potential for aquaculture and on-growth in spiny lobsters (Genus *Jasus* and *Panulirus*) [9]. These can either be cultured all the way from the larval pueruli stage [10-12], or sub-adults can be "fattened" for market [13,14]. The animals are held in cages, and although feeding with commercial pellet meals has met with some success, the highest growth and survival rates have been obtained when the lobsters are fed fresh mussels [10,15,16].

The high survival and growth of spiny lobsters on mussel flesh is echoed by the fact that populations of *H. americanus* have been enriched below or near commercial blue mussel (*Mytilus edulis*, Linnaeus) operations. The anchor line buoys provide shelter for the animals and the mussels dropping off the culture lines may be a potential food source for these lobsters [17-20]. Although the presence of bivalve farms has been reported to enrich lobster populations, the rapid expansion of this sector in recent years has led to concerns about its sustainability and in particular the problems with the input of excess nutrients into the environment and the impacts on local fauna [21-23].

In a pilot study, lobsters maintained for 6 months in cages in 6-8 m of water and fed twice weekly by hand had an 85% survival rate. 40% of

***Corresponding author:** Iain J McGaw, Department of Ocean Sciences, 0 Marine Lab Road, Memorial University, St John's, NL A1C 5S7, Canada
E-mail: ijmcgaw@mun.ca

the animals moulted and increased their body mass by approximately 35% [24]. However, the logistics and costs associated with hand feeding precluded the development of this method on a large scale. The goal of the present study was to investigate an alternative method for storage to determine if lobsters can survive and grow when held for extended periods. Lobsters were held in the field under blue mussel farms with the idea that mussels dropping off the culture lines could supply a food resource for lobster growth. In turn the lobsters could help remove moribund mussels that would otherwise rot and stagnate on the bottom. Because of the logistics associated with constant monitoring of animals and environmental conditions in the field, experiments were also conducted in the laboratory. Using the current literature, feedback from the mussel growers, and diver observations, the potential variables that lobsters may experience in the field such as the temperature change, the frequency of mussel drop-off and type of food items reaching the cages (mussels only or mussels with supplemental items) were manipulated in the laboratory. These experiments allowed us to more accurately determine how these variables potentially affected the moulting, survival and health of the cage-held lobsters in the field.

Methods

Housing and management

Three series of experiments were carried out, two were performed in the laboratory (Department of Ocean Sciences, Memorial University) where factors could be manipulated and a field experiment was conducted at Sunrise Fish Farms (Triton, NL). The animals used in the laboratory experiments were purchased from Clearwater Seafood, Nova Scotia, and both male and female intermoult animals were used. Due to permitting regulations, the lobsters used in the field experiments were purchased from local harvesters at Triton and only male intermoult lobsters were used. All the experimental animals (both in the lab and field) were held in individual compartments in plastic coated 2.5 cm wire mesh cages during the 6 month experimental period. The cages measured 1.20 × 0.90 × 0.30 m in depth either with 24 separate compartments, individually measuring 0.30 × 0.15 × 0.30 m in depth, or with 12 larger compartments, individually measuring 0.45 × 0.20 × 0.30 m in depth, with 2 cages held side by side (total n=24). Each individual lobster, isolated from other lobsters in separate compartments of the cage acted as a replicate. This cage design and experimental set-up was chosen as it represented the exact protocol for lobster on-growth in Newfoundland that would be used by Jerseyman Island Fisheries Ltd [24].

Sampling methods

The lobsters in the two laboratory experiments were checked every other day and any mortalities were recorded and removed, at the same time any lobsters undergoing moulting were noted. The following parameters were measured once every 2 months in the laboratory and once every 3 months in the field.

Lobster growth was measured by recording body mass and carapace length. For body mass the lobsters were removed from the cages and the water was allowed to drain from the bronchial chambers for 3 to 5 mins. The animals were then wiped dry and measured to the nearest 0.1 g. The carapace length was measured along the dorsal line between the eye socket and the posterior margin of the carapace.

The serum protein concentration is a good indicator of quality (meat content and health) and physiological condition in lobsters. It was measured by withdrawing a 500 μl sample of haemolymph from arthrodial membrane at the base of the fourth walking legs. This sample was then injected onto the sample well of a pre-calibrated Brix/RI-Chek Digital Pocket Refractometer (Reichert Analytical Instruments, Depew, NY, USA). The time between withdrawal of the haemolymph and processing of the sample did not exceed 90 seconds. The total serum protein concentration was then calculated from the RI as outlined in Wang and McGaw [25].

Experimental protocols

Field experiment: Cage location and compartment size: The objective of the field experiment was to assess the input of blue mussels as a food source as well as the effect of compartment size on lobster survival, growth, and health status. In line with the Department of Fisheries and Oceans permitting requirements, all the lobsters used in the field experiments were males and had to be purchased from local harvesters; they had a body mass (mean ± SD) of 601.0 ± 92.8 g. The lobsters (n=192) were held in benthic cages (10 to 13 m depth) at Sunrise Fish Farms near Triton, NL (N49° 29' 03.03', W55° 45' 03.58'). Half of the lobsters were maintained in cages with regular sized compartments, while the other half were kept in cages with the larger compartments. Each series of cages was strapped together to ensure they would remain in the same location. Temperature loggers (iBCod, type G, Ste-Juline, QC, Canada) were attached to the cages and recorded the temperature every 4 h during the 6 months (June to December 2013) experimental period. Half of the experimental animals (n=48 in regular compartments, 48 in large compartments) were placed directly under the mussel culture lines of Sunrise Fish Farms. The idea being that they would be able to feed on mussels dropping off the culture lines. The remaining cages (n=96 lobsters) were set approximately 15-25 m away from the mussel farm where there was no evidence of mussel drop-off from the culture lines.

Lab experiment 1: Temperature and diet type: The first lab experiment was designed to test the effects of temperature and diet type on survival and growth of cage-held lobsters. The lobsters were held in 3000 L flow-through seawater tanks maintained at either 5, 10 or 15°C, representing typical temperature ranges experienced by lobsters in the wild [26,27]. The temperature in each tank was checked daily; the temperature in the 10°C and 15°C tanks varied ± 1°C, while the 5°C tanks fluctuated ± 2°C, during the 6 month experimental period. Each tank was equipped with air diffuser stones which maintained the water oxygen content between 92% and 98% saturation. The photoperiod was maintained on a 12L:12D cycle. Approximately equal number of lobsters (intermoult stage) of both sexes, with a body mass (mean ± SD) of 548.8 ± 53.3 g were purchased from Clearwater Seafood, NS. The lobsters were acclimated to laboratory conditions (and fed) for one month before the experiment commenced. The experiment was carried out between January and June 2013, 48 lobsters were held at each temperature (5°C,10°C,15°C) with 24 lobsters per cage, isolated in separate (small size) compartments; each cage (total of 6 cages) was housed in a separate tank. Each individual animal was fed to satiation twice weekly; one group of lobsters (n=24) in each temperature regime was fed a mixed diet (diet changed weekly-shrimp, squid, fish, blue mussel, scallop mantle or crab); while the other group (n=24) was only fed blue mussels. The tanks were cleaned at the end of each week, when any uneaten food and mortalities were removed.

The identification of the different moult stages of adult lobster was carried out by staging the developmental morphology of the pleopods following the methods outlined in reference [28]. In the present study, lobster pleopods were sampled at 1 month intervals and each sample was immediately photographed using Infinity Capture Imaging Software at 40X magnification. Haemolymph samples were collected weekly for lobsters in premoult and those that underwent moulting and the serum protein concentration was measured immediately; haemolymph samples continued to be collected from the moulted lobsters until the experiment was terminated.

Lab experiment 2: Feeding frequency and compartment size: In the second laboratory experiment the effect of feeding frequency and cage compartment size was investigated. The animals were maintained in cages (6 cages total) in the laboratory from June to December 2013 in a 45,000 l flow through tank, using ambient aerated sea water pumped from 5 m depth in Logy Bay, Newfoundland, Canada. Temperature data loggers (iBCod, type G, Ste-Juline, QC, Canada) were attached to cages and recorded water temperature every 4 h. Ninety six intermoult male and female lobsters with a body mass (mean ± SD) of 363.0 ± 59.3 g were purchased from Clearwater Seafoods. Half of the lobsters (n=48) were held in cages with 24 individual compartments of 0.20 × 0.30 × 0.30 m in depth, while the other half were held in cages with 12 larger compartments of 0.45 × 0.20 × 0.30 m in depth. Since a mussel only diet proved to be restrictive (lab experiment 1) and divers also noted other prey items in the vicinity of the cages each individual lobster was fed a mixed diet comprising of approximately 50% blue mussels supplemented with fish, squid, crab and scallop mantle. Following discussion with the mussel growers on potential timing and rates of drop-off, one group (n=24 lobsters per both compartment sizes) was fed to satiation twice weekly, while the remaining lobsters were fed once per month. To avoid fouling of the water, any remaining feed was removed at the end of each week.

Analytical Methods

Statistical analyses of lobsters mortality and moulting rates in various conditions, were analyzed using Kaplan-Meier survival curves (Prism v5.0, GraphPad Software Inc., La Jolla. CA). The Kaplan-Meier survival curves plot fractional survival/moulting (Y) as a function of time (X). It can be used to analyse the time to any event (usually death/moult) that can only happen once. The data from the survival curves was then compared using a Mantel-Cox log rank test which examines the actual amount of events in relation to expected number of events and gives a Chi-squared statistic.

Changes in serum protein concentration as a function of diet type (Figure 1) were analysed with linear regression analyses using Prism (GraphPad Software Inc., La Jolla. CA). Growth (body mass and carapace length increments) and serum protein concentration were analysed using either ANOVA or repeated measures ANOVA (Sigma Stat). If significant differences were obtained they were further analysed with Tukey post-hoc tests. Statistical significance in all tests was accepted at the $P<0.05$ level. All the data is presented as the mean value ± the standard deviation.

Results

Field experiment: Cage location and compartment size

The benthic water temperature (10-13 m) at Triton increased from 3.5°C in late June reaching between 5°C and 7°C, during the period between mid July to early November. Thereafter the water temperature dropped to around 2°C by mid-December (Figure 2). The temperatures recorded on cages under the mussel lines and those situated away from the mussel lines were almost identical to one another.

The lobsters held in cages in the field exhibited a low mortality rate, ranging from 2.1% to 4.2% (Table 1) and the majority of these mortalities occurred in the last 3 months (mid-September to mid-December). The moulting rate was comparatively low and ranged between 10.4% to 18.8% in all treatment groups (Table 1). Statistical analysis could not be performed on this data, because the limited inspection at the field sites (once per 3 months) did not allow accurate assessment of the exact time of each individual mortality/moulting event.

The moulted lobsters in cages on the open sea bottom exhibited an increase in body mass and carapace length of 22.32% ± 7.74% and 6.45% ± 1.53% respectively, while for those situated under mussel lines body mass and carapace length increased by mean levels of 20.33% ± 4.70% and 7.04% ± 2.26%, respectively. There was no significant effect of cage location or compartment size on these values (Figure 3) (Two way ANOVA, location, $F=0.58$, $P=0.459$; compartment size, $F=0.12$, $P=0.738$). Non-moulted lobsters were able to increase overall body mass during the 6 month period, nevertheless this increase was low, ranging between 0.39% and 1.92%, for both cage locations. The non-moulted lobsters in cages on the open seabed had a significantly higher

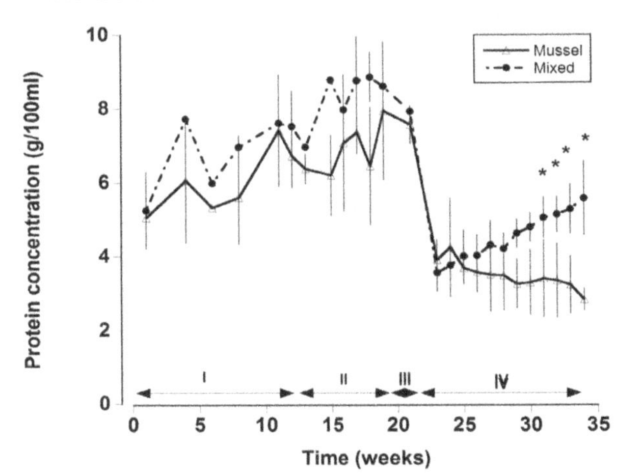

Figure 1: The effect of diet type on serum protein concentration of adult lobsters maintained for 6 months at 5°C in the lab. The different moult stages are indicated, where I is intermoult, II is proecdysis, III is ecdysis, and IV is postmoult. The arrow indicates the time of moult. The data represents the mean ± SD of 14 lobsters fed mixed diet (reproduced from Wang and McGaw [25]) and 19 lobsters fed the blue mussel diet. Asterisks indicate significant differences between treatments.

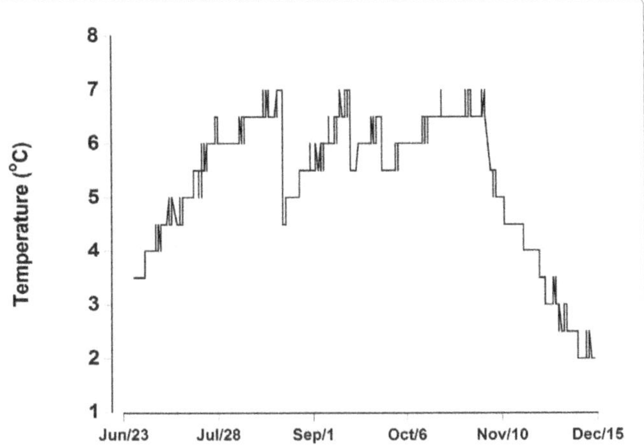

Figure 2: Temperature range (°C) from late June to mid December (2013) in the Triton area on the north-western coast of Newfoundland. Data loggers were attached to the cages which were set in 10-13 m of water and recorded the temperature every 4 h.

Cage Location/ Compartment Size	Parameter	Number	Cumulative Mortality and Moulting Over 6 Months (Number and Percent)		
			June	September	December
Under Mussel Line (Large)	Mortality Moulting	48	0 (0) 0 (0)	1 (2.1%) 1 (2.1%)	1 (2.1%) 6 (12.5%)
Under Mussel Line (Small)	Mortality Moulting	48	0 (0) 0 (0)	1 (2.1%) 3 (6.3%)	2 (4.2%) 5 (10.4%)
Away Mussel Line (Large)	Mortality Moulting	48	0 (0) 0 (0)	0 (0) 9 (18.8%)	2 (4.2%) 9 (18.8%)
Away Mussel Line (Small)	Mortality Moulting	48	0 (0) 0 (0)	0 (0) 4 (8.3%)	2 (4.2%) 5 (10.4%)

Table 1: Mortality and moulting rates of adult lobsters held in the field at Triton, NL. The animals were either held under mussel lines, or on the open sea bed and in two different compartment sizes and measurements taken at the start of the experiment (June) and at 3 month periods thereafter (September, December). Data is shown as cumulative numbers and these are expressed as a percentage of the total number in parentheses.

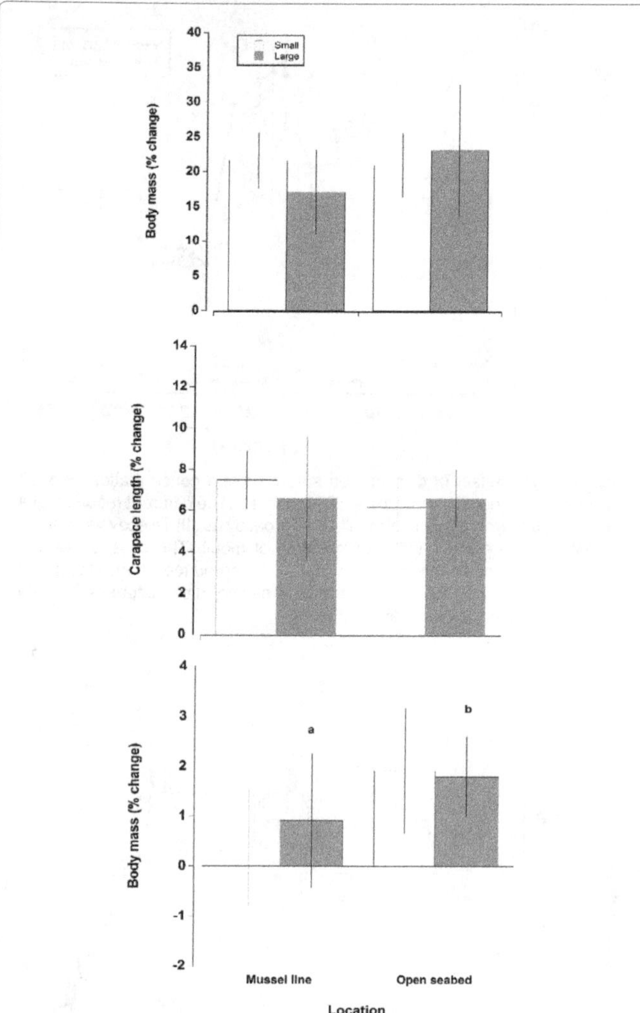

Figure 3: Effect of cage location and compartment size on-growth of adult lobsters maintained for 6 months in benthic cages near Triton, NL. A) Percent increase in body mass for moulted lobsters (n=6-8); B) Percent increase in carapace length for moulted lobsters (n=6-8); C) Percent increase in body mass for non-moulted lobsters (n=32-38). Data are expressed as mean ± SD. Different lowercase letters indicate significant differences (P<0.05) between the 2 cage locations for large compartments; different capital letters indicate significant differences between the 2 locations for animals maintained in small compartments.

increase in body mass (1.85% ± 1.15%), than those maintained under mussel lines (0.67% ± 1.29%) (Two way ANOVA, F=32.23, P<0.001). The compartment size did not have a significant impact on change in body mass of non-moulted lobsters (Two way ANOVA, F=0.92, P=0.34).

The serum protein levels decreased significantly over the 6 month study for both lobsters under mussel lines and on the open seabed (Table 2) (Two way RM ANOVA, F=279.3, P<0.001). The decrease in final serum protein levels was more pronounced in lobsters maintained on the open seabed (6.00 ± 1.39 to 2.35 ± 1.02 g/100 ml) compared with those held under mussel lines (5.83 ± 1.02 to 3.25 ± 0.911 g/100 ml) (Tukey post-hoc test, P<0.001). The compartment size also had a significant effect on the final serum protein concentration, but only for the lobsters ranched underneath mussel lines (Tukey post-hoc test, P<0.01; Table 2). The location of the compartment in the cage also had an effect on serum protein levels (Figure 4). For example, in the 24 compartment cages the 4 corner compartments had a large surface area that could potentially come into direct contact with organic material,

Sea Cage Location	Compartment Size	Serum Protein (g/100 ml) Over 6 Months Mean ± SD		
		June	September	December
Under Mussel Line	Large	5.86 ± 1.11	3.86 ± 0.92	3.50 ± 0.92^{aA}
Under Mussel Line	Small	5.73 ± 0.90 (7.12 ± 1.68, n=3)	3.60 ± 0.94 (2.20 ± 0.511, n=3)	2.98 ± 0.86^{bA} (1.76 ± 0.57, n=3)^{A}
Away Mussel Line	Large	6.05 ± 1.39	2.65 ± 1.01	2.22 ± 0.88^{aB}
Away Mussel Line	Small	5.96 ± 1.41 (8.33 ± 1.08, n=11)	2.95 ± 1.18 (1.99 ± 0.52, n=11)	2.47 ± 1.13^{aB} (1.46 ± 0.17, n=11)^{A}

Table 2: Serum protein concentration of adult lobsters at Triton area, at 0, 3 and 6 months (June-December). The animals were held under mussel lines or on the open seabed in 2 different compartment sizes. The data represent the mean ± SD of 46-48 individual lobsters at each time point. Different lowercase letters indicate significant differences (*P*<0.05) between the 2 compartment sizes in the same location; different capital letters indicate significant differences (*P*<0.05) between the 2 sea cage locations for the same compartment size. Data in parentheses are from molted lobsters.

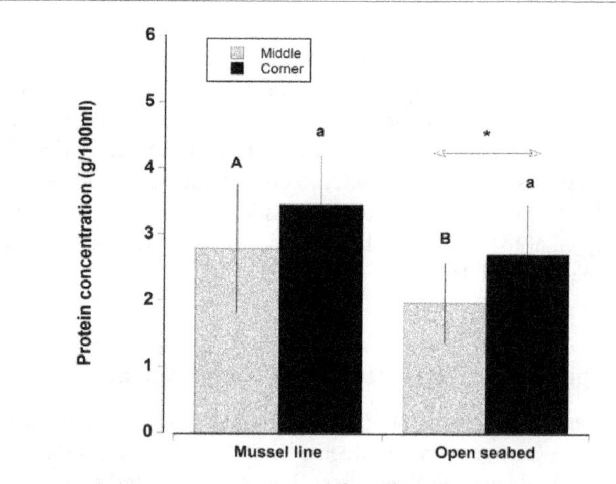

Figure 4: Effect of compartment location in the cage (corner or middle) and sea cage location on lobster serum protein concentrations (g/100 ml). Data are expressed as mean ± SD of 7-9 individuals. Different lowercases indicate significant differences (P<0.05) between the middle and corner compartments; different capital letters indicate significant differences (P<0.05) between the middle and corner locations for compartments with a small surface area; an asterisk indicates a significant difference within group.

while the lobsters in the middle 6 compartments were surrounded by animals in other compartments therefore had a much lower surface area (1 upper and 1 lower) directly in contact with the environment. On the open seabed, the lobsters held in corner compartments had significantly higher serum protein levels (2.70 ± 0.74 g/100 ml) than those held in the middle compartments (1.94 ± 0.56 g/100 ml) (Two way ANOVA, F=6.1, P<0.05). Although a similar trend was observed for lobsters held under the mussel lines (corner=3.45 ± 0.73 g/100 ml, middle=2.93 ± 0.9 g/100 ml), this proved to be statistically insignificant (Tukey post-hoc test, P=0.231).

Only a few lobsters moulted and because the cages were only checked every 3 months, the data for moulted lobsters was limited (shown in parantheses in Table 2). The general trend was that serum protein levels dropped after moulting, and serum protein levels (at both locations) continued to decrease thereafter. This decrease appeared to be more pronounced for lobsters settled on the open seabed (8.84 ± 1.08 to 1.46 ± 0.17 g/100 ml), than for lobsters settled under mussel lines, (7.12 ± 1.68 to 1.76 ± 0.57), however, this difference proved to be statistically insignificant (Student t test, T=1.62, P=0.131).

Lab experiment 1: Temperature and diet type

Temperature had a significant effect on lobster survival rate (Table 3) (Mantel-Cox Test, Chi square=53.49, P<0.001). The lowest mortality occurred at 10°C (4.2%-12.5%), and at this temperature the diet did not have a significant effect on survival rate (Mantel-Cox Test, Chi square=1.09, P=0.297); these mortalities only occurred during the final 2 months of the experiment. The highest mortality rate (79.2%) was recorded at 5°C for lobsters fed the mussel only diet; this mortality rate was significantly higher than the 50% rate measured for the group fed a mixed diet at 5°C (Mantel-Cox Test, Chi square=4.11, P<0.05). The mortality rate was also relatively high (50%) for lobsters in 15°C fed on the mussel diet and this was significantly higher than its counterparts (12.5%) fed a mixed diet (Mantel-Cox Test, Chi square=6.58, P<0.05). The mortalities in the 15°C mussel diet nearly all occurred during the final month of the study and all these had recently moulted. In contrast, very few mortalities were observed in post-moult lobsters fed a mixed diet at 15°C.

The experimental temperature also influenced the incidence of moulting (Table 3) (Mantel-Cox Test, Chi square=71.70, P<0.001). In 5°C, only 2 lobsters moulted during the 6 month experimental period, these occurred at beginning of the study and both of these animals were maintained on a mixed diet. There was a significant effect of diet on moulting at 10°C (Mantel-Cox Test, Chi square=7.65, P<0.01). Moulting rates were similar (and low) between the 2 groups during the first 5 months; there was a substantial increase in the moulting rate for the mixed diet lobsters in the final month, but none of the animals fed the mussel only diet moulted at this time. The highest moulting rates occurred in lobsters maintained at 15°C with most of the animals undergoing this process during the final three months of the study (66.7% for mixed diet and 83.3% for mussel diet).

The limited amount of data for the 5°C treatment and 10°C mussel diet precluded statistical analysis on all combinations. Analysis of the remaining data showed no significant effect of temperature or diet on growth (Two way ANOVA, F=0.21, P=0.818). Following moulting lobsters increased in body mass by 29.32% ± 7.21% while an increase of carapace length of 8.89% ± 1.36% was observed (Figures 5A and 5B). There were only slight increases (1.3%-2.5%) in body mass for non-moulted lobsters (Figure 5C) and there was no consistent trend in change in body mass as a function of temperature or diet in these animals.

Serum protein concentration was used as an indicator of the physiological and nutritional status of the animal. Temperature and diet had a significant effect on the final serum protein concentration (Table 4). Serum protein concentration increased with increasing temperature in non-moulted lobsters fed both diets after 4 months (2 way RM ANOVA, F=105.1, P<0.001). Lobsters in 5°C and 15°C fed a mixed diet had higher serum protein levels than those fed the mussel diet, but there was no effect of diet at 10°C (Two way ANOVA; diet, F=5.96, P<0.05; interaction, diet and temperature, F=4.25, P<0.05) in the final measurement of serum protein concentration of non-moulted animals (5°C and 10°C in the 6th month; 15°C in the 4th month).

In addition to temperature and diet type, the lobster serum protein concentration changed over time for both diet types (Two way RM

Temperature and Diet	Parameter	Number	Monthly Cumulative Mortality and Moulting Over 6 Months					
			February	March	April	May	June	July
5°C Mussel	Mortality Moulting	24	1 (4.2%) 0 (0)	2 (8.3%) 0 (0)	6 (25.0%) 0 (0)	9 (37.5%) 0 (0)	13 (54.2%) 0 (0)	19 (79.2%) ^{aA} 0 (0) ^{aB}
5°C Mixed	Mortality Moulting	24	0 (0) 1 (4.2%)	1 (4.2%) 2 (8.3%)	2 (8.3%) 2 (8.3%)	3 (12.5%) 2 (8.3%)	8 (33.3%) 2 (8.3%)	12 (50.0%) ^{bA} 2 (8.3%) ^{aC}
10°C Mussel	Mortality Moulting	24	0 (0) 0 (0)	0 (0) 1 (4.2%)	0 (0) 1 (4.2%)	0 (0) 1 (4.2%)	1 (4.2%) 1 (4.2%)	1 (4.2%) ^{aC} 1 (4.2%) ^{bB}
10°C Mixed	Mortality Moulting	24	0 (0) 0 (0)	0 (0) 1 (4.2%)	0 (0) 1 (4.2%)	0 (0) 2 (8.3%)	2 (8.3%) 2 (8.3%)	3 (12.5%) ^{aB} 9 (37.5%) ^{aB}
15°C Mussel	Mortality Moulting	24	0 (0) 0 (0)	1 (4.2%) 0 (0)	1 (4.2%) 3 (12.5%)	1 (4.2%) 11 (45.8%)	2 (8.3%) 15 (62.5%)	12 (50.0%) ^{aB} 20 (83.3%) ^{aA}
15°C Mixed	Mortality Moulting	24	0 (0) 0 (0)	0 (0) 0 (0)	3 (12.5%) 2 (8.3%)	3 (12.5%) 7 (29.2%)	3 (12.5%) 11 (45.8%)	3 (12.5%) ^{bB} 16 (66.7%) ^{aA}

Table 3: Mortality and moulting rates of adult lobsters maintained in the lab at 3 temperatures (5, 10, 15°C) and fed 2 diet types (mussel and mixed) over a 6-month period in the lab. The monthly recorded data are shown as cumulative numbers and these are expressed as a percentage of the total number in parentheses. Different lowercase letters indicate significant differences (*P*<0.05) between the 2 diets in the same temperature; different capital letters indicate significant differences (*P*<0.05) among the 3 temperatures for the same diet.

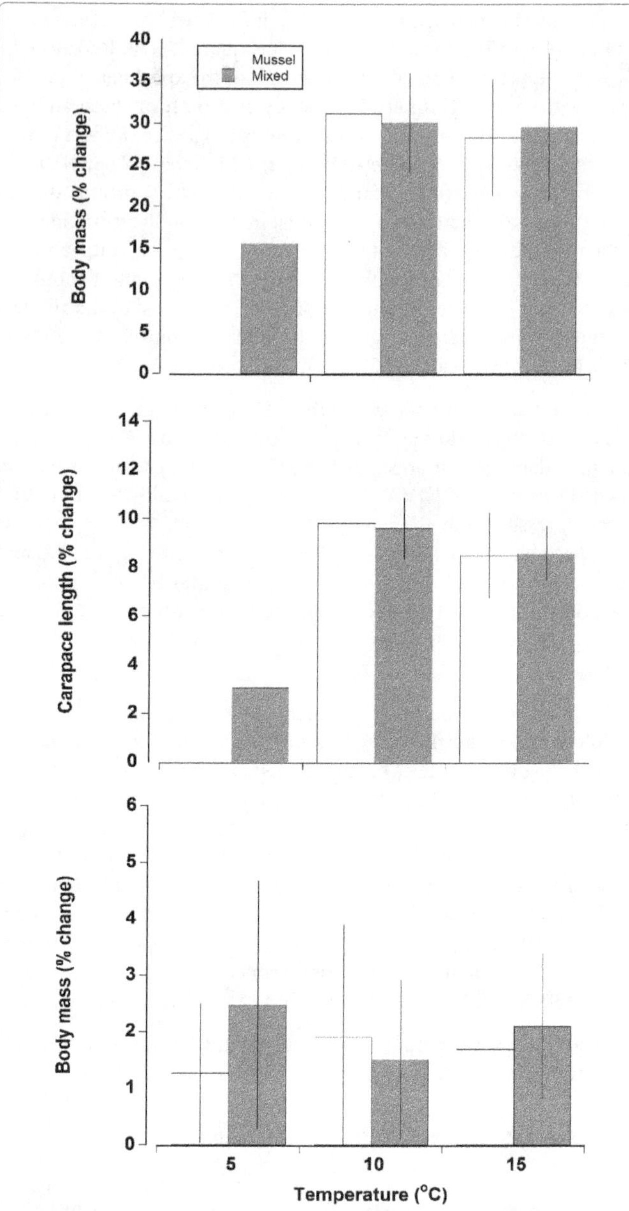

Figure 5: Effect of temperature and diet on growth of adult lobsters maintained in the lab. The data represent mean ± SD of the changes observed at the end of a 6 month experimental period. A) Percent increase in body mass for moulted lobsters (n=9-18); B) Percent increase in carapace length for moulted lobsters (n=9-18); C) percent increase in body mass for non-moulted lobsters (n=11-21).

ANOVA, time, F=15.19, P<0.001; interaction of temperature and time, F=16.63, P<0.001). At 5°C, the serum protein concentration was maintained at stable levels during the first 4 months, but increased significantly during the last 2 months of the study (Tukey post-hoc test, P<0.05). At 10°C, the serum protein concentration increased significantly at each two month sampling period, reaching its highest level at the end of the 6 month experimental period (Tukey post-hoc test, P<0.001). Serum protein concentrations at 15°C also increased significantly during the first 4 months, thereafter a significant decrease in serum protein concentration occurred. This was due to lower serum protein levels measured in post-moulted lobsters (Tukey post-hoc test, P<0.001).

Haemolymph samples were collected at weekly intervals in both pre and post-moult animals in 15°C (Figure 1). Serum protein concentrations (both diet types) increased steadily during intermoult and early proecdysis, reaching a peak in late proecdysis. There was a trend for the lobsters fed a mixed diet to exhibit higher serum protein levels than those fed the mussel diet, however, this difference proved to be statistically insignificant (Linear regression, F=2.98, P=0.088). The majority of the lobsters (n=33) moulted during the 22nd week. Following moulting, serum protein levels dropped to their lowest levels of between 3.55 g and 3.91 g/100 ml. Thereafter there was a significant effect of diet type on serum protein levels. Serum protein concentration slowly increased over the following 12 weeks in lobsters fed a mixed diet (Two way RM ANOVA, F=6.6, P<0.001). In contrast, serum protein levels of lobsters fed a mussel only diet declined steadily reaching levels that were significantly lower than those of the mixed diet animals after 30 weeks (Two way RM ANOVA, diet type, F=56.0, P<0.001; time, F=28.34, P<0.001; interaction, F=2.14, P<.01). Eighty nine percent (n=19) of post-moulted lobsters fed on a mussel diet died, while only 7% of post-moulted lobsters (n=14) fed a mixed diet died during the same time period.

Lab experiment 2: Feeding frequency and compartment size

There was a significant variation in ambient water temperature during the 6 month experimental period (Figure 6). Water temperature increased from around 5°C at the start of the experiment in early June, reaching maximal levels of approximately 15.5°C at the end of July. The temperature remained steady for around 3 months, after which the water temperature decreased to 10°C in early October. There was a further decrease from late October onwards, reaching the lowest measured temperature of 2°C in mid-December.

The mortality rate ranged between 12.5% and 37.5% and a large proportion of these mortalities occurred in last 60 days in all treatments (Table 5). There was no significant effect of feeding frequency or compartment size on mortality rates (Mantel-Cox Test, Chi square=6.588, P=0.086). Feeding frequency and compartment size did not have any significant effect on moulting rate, which ranged from 20.8% to 37.5% among the different treatments (Table 5) (Mantel-Cox Test, Chi square=0.09, P=0.761). Moulting started in mid-August, peaked during September to October and declined substantially during November and December.

Feeding frequency and to a lesser degree compartment size did have a significant effect on growth of moulted lobsters (Figures 7A-7C).

Temperature	Diet	Serum Protein (g/100ml) over 6 Months Mean ± SD			
		January	March	May	July
5°C	Mussel	4.62 ±0.75	4.99 ± 0.88	4.85 ± 0.80^{aC}	4.98 ± 1.44^{bB}
5°C	Mixed	4.62 ± 1.76	4.96 ± 1.24	5.67 ±1.10^{aC}	6.32 ± 0.85^{aB}
10°C	Mussel	3.97 ± 1.18	5.55 ± 1.03	6.77 ± 1.00^{aA}	7.92 ± 1.05^{aA}
10°C	Mixed	5.12 ± 0.99	5.89 ± 1.06	6.69 ± 0.75^{aB}	7.65 ± 0.78^{aA}
15°C	Mussel	5.20 ± 0.90	6.37 ± 1.25	7.73 ± 1.85^{bAB}	*3.47 ± 0.48^{aC}
15°C	Mixed	5.22 ± 1.26	6.91 ± 0.79	8.35 ± 0.85^{aA}	*4.16 ± 1.83^{aC}

Table 4: Serum protein concentrations (g/100 ml) in 3 different temperature and 2 diet treatments. Samples were taken at 2 month intervals between January and July, data represents the mean + SD of 10-24 individuals at each time point. For the 4th and 6th month, different lowercase letters indicate significant differences (*P*<0.05) between the 2 diets in the same temperature regime; different capital letters indicate significant differences (*P*<0.05) among 3 temperatures in the same feed type condition. At 15, the data for months 0, 2 and 4 were from non-molt lobsters; at 6th month, all previously sampled lobsters had molted. * indicates serum protein values of post-molted lobsters.

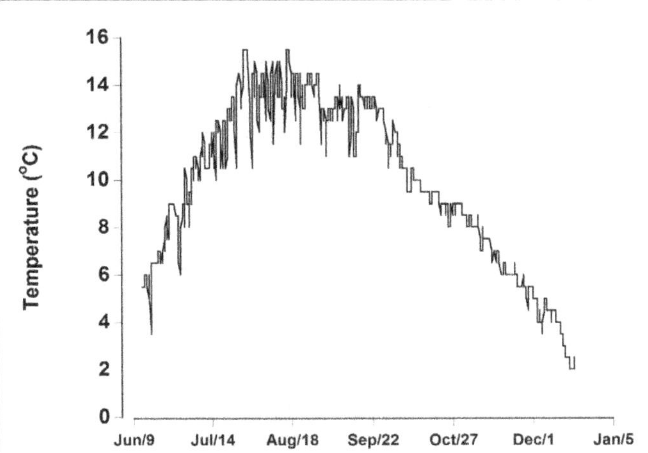

Figure 6: Temperature range (°C) from mid June to mid December (2013) in experimental tanks in the lab. The seawater was pumped directly from approximately 5 m depth in Logy Bay, St. John's, Newfoundland. Data was collected every 4 h using iBCod data tags (type G, Ste-Juline, QC, Canada) affixed to the cages.

Lobsters fed in the high feeding frequency treatment had a significantly higher increment of both body mass and carapace length (37.07% ± 10.94% and 10.03% ± 2.07%, respectively) compared with lobsters in the low feeding frequency treatment (20.49% ± 7.39% and 6.85% ± 1.84%, respectively) in both compartment sizes (Two way ANOVA, F=22.11, P<0.001; F=13.03, P<0.01 for body mass and carapace length respectively). Lobsters maintained in large compartments (both frequent and infrequent feeding) exhibited a trend of a larger increment in body mass but this was only statistically significant for body mass in the low feeding frequency treatment (Two way ANOVA, F=7.61, P<0.05). There was also a change in body mass for non-moulted lobsters. Lobsters in the low feeding frequency could not maintain their body mass and it decreased on average by 1.9% during the 6 month trial. Lobsters in high feeding frequency treatment maintained their body mass with a mean increase of 0.5%. However, the effect of feeding frequency on non-moulted lobster growth was only statistically significant in the small compartment sizes (Two way ANOVA, feeding frequency, F=20.91, P<0.001).

The serum protein concentration of the lobsters was significantly impacted by feeding frequency, but was not affected by the compartment size (Table 6) (Two way RM ANOVA, feeding frequency, F=54.81, P<0.001; compartment size, F=1.89, P=0.175). Non-moulted lobsters with a high feeding frequency exhibited an increase in serum protein levels during the 6 month period. The highest levels were measured in mid-October and although serum protein levels decreased slightly in the last 2 months, they were still significantly higher (7.08 g ± 2.06 g/100 ml) than levels measured at the start of the experiment (5.26 g ± 0.92 g/100 ml) (Student t test, T=3.533, P<0.01). In contrast, the lobsters fed once per month were unable to maintain serum protein levels and they declined significantly from initial levels of 5.58 g ± 0.94 g/100 ml, reaching their lowest levels of 3.79 g ± 1.09 g/100 ml at the end of the experimental period (Student t test, T=9.54, P<0.001).

Because of a large difference in the timing of the moult for individual lobsters and the close relationship between time after moulting and serum protein levels (Figure 1), there were not enough replicates to perform reliable statistically analysis. Nevertheless, the trend for moulted lobsters was consistent with the non-moulted lobsters. In the low feeding frequency group, the serum protein concentration of moulted lobsters decreased from 3.7 g ± 1.15 g/100 ml down to 1.84 g ± 0.25 g/100 ml (n=2) at the end of the experimental period. In the high feeding frequency group, the moulted lobster's serum protein level increased from 3.63 g ± 0.21 g/100 ml, reaching levels as high as 6.03 g ± 1.03 g/100 ml (n=3) after 2 months.

Discussion

Survival

The survival rates of adult lobsters stored in benthic cages under mussel farms were very high (>95%). In commercial holding facilities and during live transport, chilling coma (<1°C) is used to enhance survival rates [29,30]. The water temperature in the Triton area varied between 2°C-7°C, thus this relatively low temperature may have enhanced survival. However, the results of the laboratory experiments did not fully support this assumption. Lobsters held in the lab at 5°C exhibited a high mortality rate (>50%), especially those that were fed the mussel only diet. In the lab the haemolymph protein concentration decreased with decreasing water temperature (Table 3). The haemolymph contains important proteins which are involved in the immune response [31,32] and in lobsters, the rate of phagocytosis is positively related to temperature [33]. The reasons for the higher mortality in the lab are unclear. Although the experimental tanks were supplied with a constant flow of seawater and were cleaned

Feed Frequency/ Compartment Size	Parameter	Number	Cumulative Mortality and Moulting (Number and Percent) Over 6 Months					
			July	August	September	October	November	December
1 Feeding/Month Large	Mortality	24	0 (0)	0 (0)	0 (0)	0 (0)	1 (4.2%)	3 (12.5%)
	Moulting		0 (0)	2 (8.3%)	4 (16.7%)	4 (16.7%)	5 (20.8%)	5 (20.8%)
1 Feeding/Month Small	Mortality	24	0 (0)	1 (4.2%)	1 (4.2%)	1 (4.2%)	2 (8.3%)	3 (12.5%)
	Moulting		0 (0)	1 (4.2%)	4 (16.7%)	5 (20.8%)	8 (33.3%)	9 (37.5%)
2 Feeding/Week Large	Mortality	24	1 (4.2%)	3 (12.5%)	3 (12.5%)	3 (12.5%)	5 (20.8%)	9 (37.5%)
	Moulting		0 (0)	2 (8.3%)	3 (12.5%)	5 (20.8%)	6 (25%)	7 (29.2%)
2 Feeding/Week Small	Mortality	24	0 (0)	1 (4.2%)	1 (4.2%)	1 (4.2%)	3 (12.5%)	5 (20.8%)
	Moulting		0 (0)	1 (4.2%)	2 (8.3%)	2 (8.3%)	4 (16.7%)	6 (25%)

Table 5: Mortality and moulting rates of adult lobsters held in the lab from June to December 2013 on an ambient temperature cycle. The animals were held in 2 different compartment sizes and fed either twice per week or once per month. Monthly data are shown as cumulative numbers and these are expressed as a percentage of the total number in parentheses.

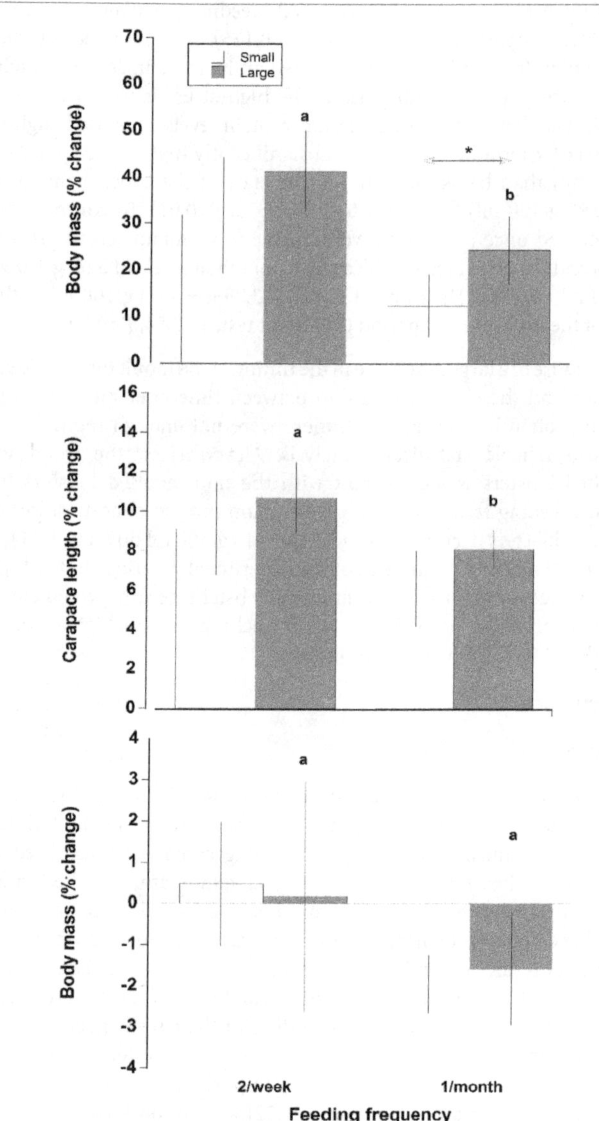

Figure 7: Effect of feeding frequency and compartment size on-growth of adult lobsters maintained in the lab for a 6-month period in the lab. A) Percent change in body mass for moulted lobsters (n=6-8); B) Percent increase in carapace length for moulted lobsters (n=6-8); C) Percent increase in body mass for non-moulted lobsters (n=13-16). Data are expressed as mean ± SD. Different lowercase letters indicate significant differences (P<0.05) between the 2 feeding frequencies in the large compartment size; different capital letters indicate significant differences (P<0.05) between the 2 feeding frequencies in small compartment size. An asterisk indicates significant difference as a function of compartment size at any feeding frequency treatment.

weekly, there is the potential "wall effect" where bacterial build-up occurs on flat surfaces in these semi-enclosed laboratory systems [34-36]. Since serum protein concentrations were low in the animals maintained in 5°C, it would suggest that they may have compromised defense mechanisms, leaving them more vulnerable to infection from pathogens. Rao et al. also report higher mortality rates in tank versus cage-held spiny lobsters *Panulirus homarus* (Linnaeus) and suggest that this may be due to higher levels of stress in tank held animals [37]. This highlights some of the potential problems of extrapolating responses in the lab with those in the field.

There was also a high mortality rate in the lab at 15°C, but this was primarily for post moulted lobsters. Although mortality rates increase during moulting, and recently moulted lobsters are more physiologically sensitive and vulnerable [26,38,39], this was not the case here. The mortalities primarily occurred between 31 and 80 days after moulting and nearly all of them were lobsters fed the mussel diet at 15°C. The drop in serum protein levels in the lobsters fed mussels indicated that this diet was not sufficient to maintain health [25]. Amino acids such as asparagine, alanine and glutamic acid are deficient in mussels [40,41]. Astaxanthin is also lacking in the mussel diet and this plays an important role in immunocompetence and stress tolerance in crustaceans [42,43]. Post-moulted lobsters also required a higher levels of calcium intake for hardening of the shell [44]. This suggests the mussel only diet was not sufficient to provide nutrients for post-moult processes such as laying down muscle and hardening of the carapace and for dealing with increased pathogen loads in the experimental tanks.

The effects of a restricted diet (mussel only) diet on post-moult survival could be a potential concern when holding lobsters under mussel farms. The lobsters held under culture lines were likely feeding on mussels because large numbers of empty mussel shells were found in and around the cages when they were retrieved by the divers. However, the remains of gastropods, sea urchins and sea stars were also found inside the cages and green algae were growing on the cages. As lobsters are omnivorous, it was likely they were also feeding opportunistically on animals that entered the cages and therefore were not feeding exclusively on mussels, but rather getting a broad range of nutrients in their diet.

Moulting

The low moulting rate in Triton area (13%) was probably due to the cooler water temperatures (2°C to 7°C). In the lab, moulting rate was also very low in 5°C (4%) and most animals remained in the intermoult stage throughout the 6 month experimental period. The European lobster *H. gammarus* moults when the temperature reaches between 12°C-14°C [45]. The present results suggest that for *H. americanus*, rather than needing to be exposed to a certain temperature to induce

Feed Frequency	Compartment Size	Serum Protein (g/100 ml) Over 6 Months Mean ± SD			
		June	August	October	December
1 Feeding/Month	Large	5.63 ± 0.97	4.86 ± 1.10	4.13 ± 1.05	3.81 ± 1.11^{aB}
1 Feeding/Month	Small	5.52 ± 0.94 (8.75 ± 1.37, n=2)	4.74 ± 1.06 (3.70 ± 1.14, n=2)	4.08 ± 1.14 (2.07 ± 0.58, n=2)	3.76 ± 1.12^{aB} (1.84 ± 0.25, n=2)^{B}
2 Feeding/Week	Large	5.40 ± 0.36	5.25 ± 0.67	7.46 ± 2.20	6.24 ± 2.33^{Aa}
2 Feeding/Week	Small	5.17 ± 1.13 (8.68 ± 0.35, n=3)	5.90 ± 1.04 (3.63 ± 0.21, n=3)	7.65 ± 1.82 (6.03 ± 1.03, n=3)	7.56 ± 1.80^{aA} (5.84 ± 0.49, n=3)^{A}

Table 6: Serum protein concentration of adult lobsters held in the lab in two compartment sizes and on 2 feeding schedules over a 6 month period on an ambient temperature cycle. The data represents the mean + SD of 15-24 individuals at each time point. Different lowercase letters indicate significant differences (*P*<0.05) between the 2 compartment sizes in the same feeding frequency conditions; different capital letters indicate significant differences (*P*<0.05) between 2 feeding frequencies for the same compartment size. Data in parentheses are from molted lobsters.

moulting, the lobsters might need to be exposed for a number of degree days. The growing degree day (thermal integral) is used as a reliable predictor of growth and development in fish species and this likely also applies to crustacean moulting and growth [46].

The low moulting rate observed at Triton could also be due to food limitation since changes in food abundance impact moulting frequency in larval and juvenile stages of clawed, rock and spiny lobsters [47-49]. Crustaceans can refrain from moulting during starvation in order to save energy to maintain basal metabolic functions [50]. However, this did not appear to be the case here for adult lobsters. Lobsters in the lab fed once per month had similar moulting rates to those fed twice weekly. In addition, lobsters maintained in the laboratory at 5°C and fed had similar low moulting rates to animals at Triton where food input was limited. Crustaceans expend energy at moult and the hepatopancreas functions as a major source of energy during moulting [51]. Even though the hepatopancreas was smaller in infrequently fed animals and those at Triton it suggests that the lipid and glyceride stores would still have been sufficient to facilitate moulting [25].

Growth

The growth increment of moulted lobsters at Triton was lower than those maintained in the lab. The cooler temperatures suppress the lobster's metabolism [52,53], and subsequently the lobsters would have consumed less food. In support of this, post moult lobsters fed frequently in the lab were significantly larger than infrequently fed lobsters. Lobsters with access to enough food would have enough reserves to lay down more muscle tissue and have energy adequate energy reserves to produce larger organs.

The effect of feeding frequency on-growth of non-moulted lobsters was somewhat different. Although one feeding per month was adequate to keep the lobsters alive, the mass of non-moulted lobsters in the lab experiment tended to decrease. In contrast, the lobsters at Triton (without artificial feeding) were able to maintain or even increase their body mass slightly. The lower temperatures at Triton probably slowed the lobsters metabolic rate and use of stored nutrients [27,54]. In spite of this, the lobsters at Triton had a more pronounced decrease in hepatopancreas size and edible meat content than those fed once per month in the lab [25]. During starvation, crustaceans metabolize their tissues, resulting in a decrease in organ mass [55,56]. One possible explanation for the starved lobsters at Triton area maintaining or even slightly increasing (1% to 2%) their body mass is that an increased water uptake would compensate for the decrease in organ mass. The body mass of white shrimp *Litopenaeus vannamei* (Boone) and king crabs *Lithodes santolla* (Molina) also remains constant during short-term starvation and is likely due to an increased water content in the body [50,57,58].

There was no effect of doubling cage size on growth of adult lobsters in the field or in the lab. In contrast, juvenile *H. americanus* respond to an increase in container size with a significant increase in carapace length and body mass [8,59,60]. The cages used by Beal and Protopopescu's were large enough for juvenile lobsters to freely move around and the large surface area for settling organisms supplied plenty of food for the lobsters suggesting that in the current experiments a much larger increase in cage size relative to adult body size would be required in order to have any discernible effect [8].

Health and physiological condition

The serum protein concentration is a rapid and effective way of determining the quality and physiological condition of lobsters. There is a strong positive correlation between serum protein concentration and hepatopancreas size, heart size and edible meat content and a negative correlation with moisture content of the hepatopancreas and muscle tissue [25].

The final serum protein concentrations of the lobsters at Triton area were lower than any of the lab treatments, although the colder water temperatures may have contributed, the decrease would primarily be related to the lower food input, because lobsters maintained in the lab at 5°C and fed regularly exhibited an increase in serum protein concentration. In support of this assumption, the decrease in serum protein concentration was greater for lobsters set on open sea bottoms where they would not get the input of mussels. Interestingly, lobsters held in corner compartments had higher serum protein concentrations than lobsters held in the centre compartments. The corner compartments had a larger surface area in direct contact area with the surrounding environment, allowing more surface area to forage and these lobsters would be the first to come into contact with any organic material that drifted into the cages.

The slow increase in serum protein during post moult represents body tissue growth, which replaces the water [55,61]. Serum protein concentration declined in post-moulted lobsters in 15°C fed mussels, but increased in those fed the mixed diet. Post-moulted lobsters appeared to have a poor appetite for the mussel diet, while those fed the mixed diet continued feeding. Low serum protein concentration from the mussel diet could also be attributed the lower energetic content of molluscs when compared with other benthic invertebrates [40]. This suggests that although mussels are readily eaten by lobsters [62] and are a good source of calcium for exoskeleton hardening, they lack all the essential nutrients needed for survival [40,42,63]. This may be an important consideration when attempting to hold lobsters under mussel farms and additional feeding may be required for post-moulted lobsters if they are destined for market.

Conclusion

Inshore benthic storage cages could be useful in remote areas with short fishing seasons, enabling harvesters to hold lobsters and release them when market price dictates. Survival rates in the field will likely be high; the deeper cold water reduces a lobsters metabolism and need for food, thus extending their storage time [52,53]. After 3 months serum protein levels were still relatively high, indicating a healthy, quality product [25]. Taste tests showed that although people could discern a difference between the cage-held lobsters from Triton and store bought lobsters, there was no preference for either type [64]. Nevertheless, the benthic cage method may be limited for longer term storage (>3 months) and on-growth. Although cold water enhances survival [52], it reduces moulting (growth rate) and overall quality (edible meat content). After 6 months of storage the lobsters had a low serum protein concentration and were more susceptible to the effects of emersion during transport to market [24,25]. Longer term storage success could be remedied with supplemental feeding of the lobsters similar to that employed with on-growth of spiny lobsters [14]. However, the number of mussel farm sites in Newfoundland that are suitable for on-growth may be limited because the mussels lines are typically situated in deep, colder water (Laura Halfyard, Sunrise Fish Farms pers comm). Results from the laboratory experiments suggest that warmer shallow water sites typical of those found in Prince Edward Island or New Brunswick mussel aquaculture operations would be most effective at promoting moulting, and size at moult could be enhanced with supplementary feeding.

Multitrophic integrated aquaculture has typically focused on the use of bivalves and seaweeds to remove particulate and dissolved organic material around finfish farms [65-67]. There have also been recent advances in the use of sea cucumbers and sea urchins as potential vectors to control benthic deposits [68-70]. The present study showed that lobsters will survive in the vicinity of mussel farms, and when divers retrieved the cages they were full of broken mussel shells; both these factors suggest lobsters have the potential to be incorporated into a multitrophic system. Due to logistics associated with the remote location of the site, we were unable to fully assess the amount and frequency of mussel drop-off into the cages. Future work will be aimed at quantifying the amount of mussel drop-off, as well as the number of lobsters required to significantly impact the removal of benthic mussel deposits.

Acknowledgement

This research was supported by a NSERC Discovery Grant and a grant from the Canadian Centre for Fisheries Innovation (Memorial University) to IJM. We thank the staff from the workshop, dive team and the Joe Brown Aquatic Research Building (JBARB) (Ocean Sciences, Memorial University) for assistance with experiments. We thank the Halfyard family and the staff of Sunrise Fish Farms for access to their mussel farms. We also thank Stewart Lamont of Tangier Lobster, NS for providing helpful discussion on holding conditions for commercial lobsters.

References

1. Copper RA (1977) Growth of deep water American lobsters (Homarus americanus) from the New England continental shelf. CSIRO Division of Fisheries and Oceanography (Australia) Circulation 7: 27-28.
2. Department of Fisheries and Oceans Canada (2011) Lobster.
3. Bannister RCA, Addison JT (1998) Enhancing lobster stocks: a review of recent European methods, results, and future prospects. Bulletin of Marine Science 62: 369-387.
4. Nicosia F, Lavalli K (1999) Homarid lobster hatcheries: Their history and role in research, management, and aquaculture. Mari Fish Review 61: 1-57.
5. Beal BF, Mercer JP, O'Conghaile A (2002) Survival and growth of hatchery-reared individuals of the European lobster, Homarus gammarus (L.), in field-based nursery cages on the Irish west coast. Aquacul 210: 137-157.
6. Benavente GP, Uglem I, Browne R, Balsa CM (2010) Culture of juvenile European lobster (Homarus gammarus L.) in submerged cages. Aquacul Intern 18: 1177-1189.
7. Beal BF (2012) Ocean-based nurseries for cultured lobster (Homarus americanus Milne Edwards) postlarvae: initial field experiments off the coast of eastern Maine to examine effects of habitat and container type on growth and survival. Jour of Shellfi Rese 31: 167-176.
8. Beal BF, Protopopescu GC (2012) Ocean-based nurseries for cultured lobster (Homarus americanus Milne Edwards) postlarvae: field experiments off the coast of eastern Maine to examine effects of flow and container size on growth and survival. Jour of Shellfi Rese 31: 177-193.
9. Booth JD, Kittaka J (2006) Spiny lobster growout. In Spiny lobsters fisheries and culture. Eds Phillips BF, Kittaka, J Blackwell Science pp: 556-585.
10. Jeffs AG, James P (2001) Sea-cage culture of the spiny lobster Jasus edwardsii in New Zealand. Mari and Freshw Resea 52: 1419-1424.
11. Johnston D, Melville-Smith R, Hendriks B, Maguire GB, Phillips B (2006) Stocking density and shelter type for the optimal growth and survival of western rock lobster Panulirus cygnus (George). Aquacul 260: 114-127.
12. Rogers PP, Barnard R, Johnston M (2010) Lobster aquaculture a commercial reality: a review. Jour of the Mari Biolog Assoc of India 52: 327-335.
13. Lorkin M, Geddes M, Bryars S, Leech M, Musgrove R, et al. (1999) Sea-based live holding of the southern rock lobster, Jasus edwardsii: a pilot study on long term holding feeding. SARDI Research Report Series 46: 22.
14. Bryars SR, Geddes MC (2005) Effects of diet on the growth, survival, and condition of sea-caged adult southern rock lobster, Jasus edwardsii. New Zealand Journ of Mari and Fresh wat Resea 39: 251-262.
15. Crear CJ, Thomas CW, Hart PR, Carter CG (2000) Growth of juvenile southern rock lobsters, Jasus edwardsii, is influenced by diet and temperature, whilst survival is influenced by diet and tank environment. Aquacul 190: 169-182.
16. Simon CJ, James PJ (2007) The effects of different holding systems and diets on the performance of spiny lobster juveniles Jasus edwardsii (Hutton, 1875). Aquacul 266: 166-178.
17. Clynick BG, McKindsey CW, Archambault P (2008) Distribution and productivity of fish and macroinvertebrates in mussel aquaculture sites in the Magdalen Islands (Québec, Canada). Aquacul 283: 203-210.
18. D'Amours O, Archambault P, McKindsey CW, Johnson LE (2008) Local enhancement of epibenthic macrofauna by aquaculture activities. Marine Ecology Progress Series 371: 73-84.
19. McKindsey CW, Archambault P, Callier MD, Olivier F (2011) Influence of suspended and off-bottom mussel culture on the sea bottom and benthic habitats: a review. Canad Journ of Zool 89: 622-646.
20. Drouin A, Archambault P, Clynick B, Richer K, McKindsey CW (2015) Influence of mussel aquaculture on the distribution of vagile benthic macrofauna in îles de la Madeleine, Eastern Canada. Aquacul Environ Inter 6: 175-183.
21. Callier MD, McKindsey CW, Desrosiers G (2008) Evaluation of indicators used to detect mussel farm influence on the benthos: two case studies in the Magdalen Islands, Eastern Canada. Aquacul 278: 77-88.
22. Cranford PJ, Kamermans P, Krause G, Mazurié J, Buck BH (2012) An ecosystem-based approach and management framework for the integrated evaluation of bivalve aquaculture impacts. Aquacul and Environ Interact 2: 193-213.
23. Gallardi D (2014) Effects of bivalve farming on the environment and their possible mitigation: A review. Fisher and Aquacul Journ 5: 1-8.
24. Wang GQ (2015) Storage and on-growth of adult lobsters Homarus americanus in inshore benthic cages. MSc Thesis. Memorial University pp: 171.
25. Wang G, McGaw IJ (2014) Use of serum protein concentration as an indicator of quality and physiological condition in the lobster, Homarus americanus (Milne-Edwards, 1837). Journ of Shellfish Resea 33: 1-9.
26. McLeese DW (1956) Effects of temperature, salinity and oxygen on the survival of the American lobster. Journ of the Fish Rese Boar of Canada 13: 247-272.
27. Stewart JE, Horner GW, Arie B (1972) Effects of temperature, food, and starvation on several physiological parameters of the lobster Homarus americanus. Jour of the Fish Rese Boar of Canada. 29: 439-442.
28. Aiken DE (1973) Proecdysis, setal development, and molt prediction in the American lobster (Homarus americanus). Journal of the Fish Rese Board of Canada 30: 1337-1344.
29. Lorenzon S, Giulianini PG, Martinis M, Ferrero EA (2007) Stress effect of different temperatures and air exposure during transport on physiological profiles in the American lobster Homarus americanus. Comparative Biochemistry and Physiology 147: 94-102.
30. Lorenzon S, Giulianini PG, Libralato S, Martinis M, Ferrero EA (2008) Stress effect of two different transport systems on the physiological profiles of the crab Cancer pagurus. Aquaculture 278: 156-163.
31. Le Moullac G, Le Groumellec M, Ansquer D, Froissard S, Levy P, et al. (1997) Haematological and phenoloxidase activity changes in the shrimp Penaeus stylitrostris in relation with the moult cycle: protection against vibriosis. Fish and Shel Immuno 7: 227-234.
32. Vargas-Albores F, Jiménez-Vega F, Yepiz-Plascencia GM (1997) Purification and comparison of ß-1,3-glucan binding protein from the white shrimp (Penaeus vannamei). Comparative Biochemistry and Physiology 116: 453-458.
33. Paterson WD, Stewart JE (1974) In vitro phagocytosis by hemocytes of the American lobster (Homarus americanus). Jour of the Fish Rese Board of Canada 31: 1051-1056.
34. Zobell CE (1943) The Effect of Solid Surfaces upon Bacterial Activity. J Bacteriol 46: 39-56.
35. Eilers H, Pernthaler J, Amann R (2000) Succession of pelagic marine bacteria during enrichment: a close look at cultivation-induced shifts. Appl Environ Microbi 66: 4634-4640.
36. Baltar F, Lindh MV, Parparov A, Berman T, Pinhassi J (2012) Prokaryotic community structure and respiration during long term incubations. Microscopy Open 1: 214-224.

37. Rao GS, George RM, Anil MK, Saleela KN, Jasmine S, et al. (2010) Cage culture of the spiny lobster Panulirus homarus (Linnaeus) at Vizhinjam, Trivandrum along the south-west coast of India. Indian Journ of Fishe 57: 23-29.

38. Mykles DL (1980) The mechanism of fluid absorption at ecdysis in the American lobster, Homarus americanus. Jour of Experi Biol 84: 89-102.

39. Bowser PR, Rosemark R (1981) Mortalities of cultured lobsters, Homarus, associated with molt death syndrome. Aquaculture 23: 11-18.

40. Brawn VM, Peer DL, Bentley RJ (1968) Caloric content of the standing crop of benthic and epibenthic invertebrates of St. Margaret's Bay, Nova Scotia. Jour of the Fishe Resea Boar of Canada 25: 1803-1811.

41. Mente EI (2010) Survival, food consumption and growth of Norway lobster (Nephrops norvegicus) kept in laboratory conditions. Integr Zool 5: 256-263.

42. Barclay MC, Irvin SJ, Williams KC, Smith DM (2006) Comparison of diets for the tropical spiny lobster Panulirus ornatus: astaxanthin-supplemented feeds and mussel flesh. Aquac Nutr 12: 117-125.

43. Chien YH, Pan CH, Hunter B (2003) The resistance to physical stresses by Penaeus monodon juveniles fed diets supplemented with astaxanthin. Aquaculture 216: 177-191.

44. Donahue DW, Bayer RC, Riley JG (1998) Effects of diet on weight gain and shell hardness in new shell American lobster Homarus americanus. Journ of Appl Aquac 8: 79-85.

45. Schmalenbach I, Buchholz F (2013) Effects of temperature on the moulting and locomotor activity of hatchery raised juvenile lobsters (Homarus gammarus) at Helgoland (North Sea). Mari Biol Resea 9: 19-26.

46. Neuheimer AB, Taggart CT (2007) The growing degree-day and fish size-at-age: the overlooked metric. Canad Journ of Fishe and Aqua Sci 64: 375-385.

47. Templeman W (1936) The influence of temperature, salinity, light and food conditions on the survival and growth of the larvae of the lobster (Homarus americanus). Journ of the Fishe Rese Board of Canada 2: 485-497.

48. Chittleborough RG (1975) Environmental factors affecting growth and survival of juvenile western rock lobsters Panulirus longipes (Milne-Edwards). Austra Jour of Mari and Freshwa Resc 26: 177-196.

49. Vijayakumaran M, Radhakrishnan EV (1986) Effects of food density on feeding and moulting of phyllosoma larvae of the spiny lobster, Panulirus homarus (Linnaeus). Proceedings of the Symposium of Coastal Aquaculture 4: 1281-1285.

50. Comoglio L, Gaxiola G, Roque A, Cuzon G, Amin O (2004) The effect of starvation on refeeding, digestive enzyme activity, oxygen consumption, and ammonia excretion in juvenile white shrimp Litopenaeus vannamei. Jour of Shellf Resea 23: 243-249.

51. Read GH, Caulton MS (1980) Changes in mass and chemical composition during the moult cycle and ovarian development in immature and mature Penaeus indicus Milne Edwards. Comparative Biochemistry and Physiology 66: 431-437.

52. Nelson SG, Armstrong DA, Knight AW, Li HW (1977) The effects of temperature and salinity on the metabolic rate of juvenile Macrobrachium rosenbergii (Crustacea: Palaemonidae). Compar Biochem and Physio 56: 533-537.

53. Childress JJ, Cowles DL, Favuzzi JA, Mickel TJ (1990) Metabolic rates of benthic deep-sea decapod crustaceans decline with increasing depth primarily due to the decline in temperature. Deep Sea Research. Oceanogr Rese Papers 37: 929-949.

54. Stewart JE, Cornick JW, Foley DM, Li MF, Bishop CM (1967) Muscle weight relationship to serum proteins, hemocytes, and hepatopancreas in the lobster, Homarus americanus. Jour of the Fishe Resea Board of Canada 24: 2339-2354.

55. Dall W, Smith DM (1987) Changes in protein-bound and free amino acids in the muscle of the tiger prawn Penaeus esculentus during starvation. Mari Bio 95: 509-520.

56. Depledge MH, Bjerregaard P (1989) Haemolymph protein composition and copper levels in decapod crustaceans. Helgoländer Meeresunters 43: 207-223.

57. Comoglio L, Goldsmit J, Amin O (2008) Starvation effects on physiological parameters and biochemical composition of the hepatopancreas of the southern king crab Lithodes santolla (Molina, 1782). Revista de Biología Marina y Oceanografía 43: 345-353.

58. D'Agaro E, Sabbioni V, Messina M, Tibaldi E, Bongiorno T, et al. (2014) Effect of confinement on stress parameters in the American lobster (Homarus americanus). Itali Jour of Anim Sci 13: 891-896.

59. Shleser RA (1974) The effects of feeding frequency and space on the growth of the American lobster, Homarus americanus. World Mariculture Society 5: 149-155.

60. Aiken DE, Waddy SL (1978) Space, density and growth of the lobster (Homarus americanus). World Mariculture Society 9: 459-467.

61. Oliver MD, MacDiarmid AB (2002) Blood refractive index and ratio of weight to carapace length as indices of nutritional condition in juvenile rock lobsters (Jasus edwardsii). Marine and Freshwater Research 52: 1395-1400.

62. Ennis GP (1973) Food, feeding, and condition of lobster, Homarus americanus, throughout the seasonal cycle in Bonavista Bay, Newfoundland. Jour of the Fishe Rese Board of Canada 30: 1905-1909.

63. Smith DM, Williams KC, Irvin SJ (2005) Response of the tropical spiny lobster Panulirus ornatus to protein content of pelleted feed and to diet of mussel flesh. Aquacul Nutri 11: 209-217.

64. Thompson M (2014) Sensory evaluation of lobster held in a mussel farm environment. Project report, Centre for Aquaculture and Seafood Development, Marine Institute, Memorial University pp: 1813.

65. Barrington K, Chopin T, Robinson S (2009) Integrated multi-trophic aquaculture (IMTA) in marine temperate waters. Integrated mariculture: a global review. FAO Fisheries and Aquaculture Technical Paper 529: 7-46.

66. Chopin T, MacDonald B, Robinson S, Cross S, Pearce C, et al. (2013) The Canadian integrated multi-trophic aquaculture network (CIMTAN)-A network for a new era of ecosystem responsible aquaculture. Fisheries 38: 297-308.

67. Chopin T (2015) Marine aquaculture in Canada: Well-established monocultures of finfish and shellfish and an emerging integrated multi-trophic aquaculture (IMTA) approach including seaweeds, other invertebrates, and microbial communities. Fisheries 40: 28-31.

68. Hannah L, Pearce CM, Cross SF (2013) Growth and survival of California sea cucumbers (Parastichopus californicus) cultivated with sablefish (Anoplopoma fimbria) at an integrated multi-trophic aquaculture site. Aquaculture 407: 34-42.

69. Orr LC, Curtis DL, Cross SF, Gurney-Smith H, Shanks A, et al. (2014) Ingestion rate, absorption efficiency, oxygen consumption, and faecal production in green sea urchins (Strongylocentrotus droebachiensis) fed waste from sablefish (Anoplopoma fimbria) culture. Aquaculture 423: 184-192.

70. Yu Z, Zhoub Y, Yang H, Ma Y, Hu C (2014) Survival, growth, food availability and assimilation efficiency of the sea cucumber Apostichopus japonicus bottom-cultured under a fish farm in southern China. Aquaculture 426: 238-248.

Longitudinal Exploitation of the Transversal Gradient of *Oreochromis niloticus* (Linné , 1758) Gill System by Four Monogenean Species at Melen Fish Station (Yaounde, Cameroon)

Tombi Jeannette*, Akoumba John Francis, Mieguim Ngninpogni Dominique and Bilong Bilong Charles Felix

Department of Animal Biology and Physiology, University of Yaounde I, Yaounde, Cameroon

Abstract

From February 2012 to February 2013, 406 fish were caught in the Melen fish station, then fixed in 10% formalin solution and taken to the laboratory to be examined. Mounting of monogeneans carried out under binocular magnifying glass; the determination of the various species was further done using the optical microscope. The colonization of the four pairs of gill arches by *C. thurstonae* occurred in the anterio-posterior direction. The other three species showed non-specific model of occupation of the transversal gradient. These different patterns have undergone permanent modifications. The results obtained in this study could be explained based on the heterogeneity of the gill system, the ventilation of current flow, the model of gills colonization by the oncomiracidiums of the Monopisthocotylea. The low diversity values obtained between seasons indicate that this period is harmful for the parasites studied. The species *C. halli* has exploited the resource space better in all. The populating of arches II and III were the best organized.

Keywords: Distribution; Monopisthocotylea; Fish; Arches; Months

Introduction

Modern ecology has highly emphasized on the importance of parasites as study models of structure and organization of communities [1]. In most of these works, species diversity was often used to measure the composition of species in a well defined ecosystem or in one of its components. However, the study of diversity could be more revealing if carried out in a spatio-temporal context as the relative abundance of species changes within the same community [2-4]. Moreover, some research works showed not only the heterogeneity of fish gill biotope but also its high complexity with time [5-7]. These authors suggested that, spatio-temporal variability should be taken into consideration for a better understanding of population dynamics and the structure of fish parasites communities. For Hawkins, understanding the determinants of spatial variation in species diversity remains a fundamental problem in modern ecology [8].

The study of niche amplitude and overlap permits not only to know the modalities of community organization, but to also understand the extent of exploitation of the same resource [9]. Such works were carried out in Finland, Japan and Cameroon on monogenean gill parasites of different fish species [6,10,11]. It appears therefore that, the consideration of the diversity indexes, the amplitude and the niche overlap might be more promising and complementary for the analysis of population dynamics, enhancing knowledge of the biology of fish gill monogenean parasites guild. The present investigation carried out basically with ecological indexes was aimed to study the monthly colonization of the various gill arches of *Oreochromis niloticus* (Linné, 1758) by four monopisthocotylea monogenean species.

Materials and Methods

The host fish were caught from the Melen fish station located at 3°52'N and 11°31' E. This station consists of 17 interconnected ponds. It is found in the valley which is limited in the North by the campus of University of Yaounde I, in the East by Obili quarter, in the West by the University Hospital Center of Yaounde and in the South by the Cameroon National Herbarium. Samples were collected in the largest pond in which the density of fish was highest. *Oreochromis niloticus* farming is the predominant activity in this fish station. The vegetation is less diversified. The subequatorial climate in this area is characterized by four well known seasons : the short dry season (from July to August); the long rainy season (from September to mid-November); the long dry season (from mid-November to mid-March) and then a short rainy season (from mid-March to June) [12].

A total of 406 fish were examined with total body length from 25 to 162 mm and weight from 0.3 to 79.6 g. Each month, 30 to 33 fish were caught using gill-nets from February 2012 to February 2013. Once the fish were caught after draining the pond, they were immediately fixed in 10% formalin solution. In the laboratory, the gill arches of each fish side numbered from 1 to 4 in the anterio-posterior direction were removed and each placed in a Petri dish containing tap water for subsequent examination using a stereomicroscope, Wild Heebrugg M50. Parasites observed on gill filaments were colored with eosin and examined under the optical microscope, Olympus CH2, in order to determine the parasite species based on the sclerified parts of the parasites.

Shannon-Weaver (H′) index and Pielou equitability (E) index were used to study the monthly variation of parasite diversity of monogeneans collected. These indexes are calculated as follows: $H' = -Pi\sum_{i=1}^{s}\log_2 Pi$ and $E=\frac{H'}{\log_2 S}$ where for a given gill arch, Pi represents the relative abundance of parasite species i and S corresponds to the specific richness. The monthly evolution of niche amplitude of each parasite species was calculated using the standardized index of Levins

***Corresponding author:** Tombi Jeannette, Faculty of Science, Department of Animal Biology and Physiology, University of Yaoundé I, PO-Box: 812, Yaoundé, Cameroon, E-mail: tombijeannette2007@yahoo.fr

as follows $B=\frac{1}{\sum_{i=1}^{S} Pj^2}$ where Pj is the relative abundance of a given parasite species on the gill arch j. The mean monthly abundances of various arches were compared using Kruskall-Wallis test followed by Student t test when necessary. Data analysis was done using excel software and statistix version 2.0.

Results

Colonization of the four gill arches by *Cichlidogyrus thurstonae* was mostly done in an anterio-posterior direction without equipartition (Figure 1). This pattern was perturbed in the months of May 2012, when the median arches II and III were more colonized than arch I; July 2012 during which equipartition was observed and August 2012 when arch I was statistically less parasitized. Gill exploitation by *C. halli* was very unstable and varied from one month to another. Generally, the parasite load of this helminth on arch IV was lower (Figure 2) with no significant difference. During the months of March, June and September, the colonization gradient occurred in an anterio-posterior direction. This pattern was slightly modified in April, October 2012 and February 2013. Thereafter, the colonization showed no fixe pattern except in July 2012 where equipartition was observed. Similarly, the occupation models of gills by *C. tilapiae* varied from one month to another and latter stabilized in December 2012 and January 2013. In most cases, the parasite load of this species was more important on one of the median gills. This parasite species disappeared from arch II in August and from arch IV in March 2012 (Figure 3). No equipartition was observed during the study period. The colonization of transversal gradient by *Scutogyrus longicornis* was equally very perturbed and an equipartition was observed in February and November 2012. *S. longicornis* was not observed on the four pairs of gill arches in March 2012 and on the last three pairs in February 2013 (Figure 4). In June, colonization occurred in an anterio-posterior direction and then was unfixed.

The specific diversity of each gill arch varied from one month to another. The monthly values of Shannon-Weaver were found between 0.76 and 1.53; 1.1 and 1.78; 1.21 and 1.79; 0.44 and 1.78 respectively for arches I, II, III and IV. Thus, this diversity varied in this way: AIV>AIII>AII>AI. The most posterior arch indicated the highest monthly diversity in most cases. The smallest values were observed in September (arch I), June (arches II and IV) and March (arch IV) of 2012.

The equitability profile of each gill arch was similar to that of Shannon-Weaver index. The values of this parameter varied between 0.38 and 0.86 (arch I); 0.54 and 0.90 (arch II); 0.50 and 0.85 (arch III) and 0.44 and 0.89 (arch IV) and the monthly equitability values of the various gill arches were less than 0.80 in 21.15% of cases.

The variation in niche amplitude revealed that, *C. halli* is the monogenean that used the spatial resource better. This parasite

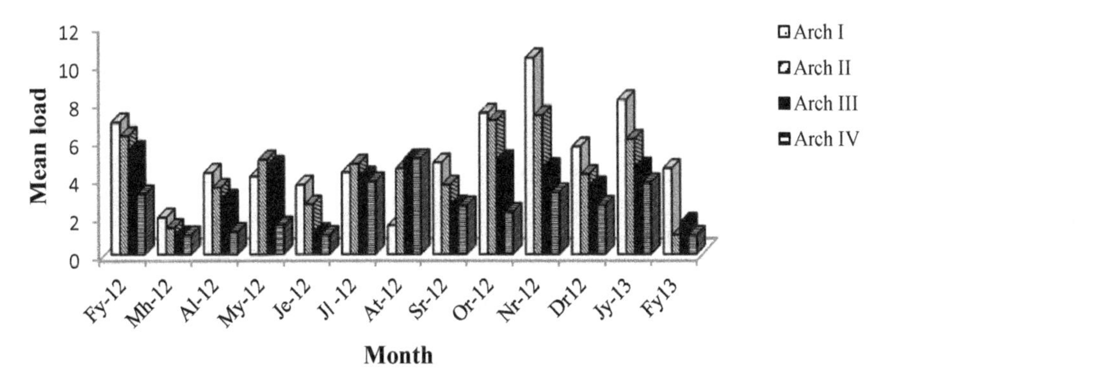

Figure 1: Longitudinal exploitation of the transversal gradient by *Cichlidogyrus thurstonae*. January: Jy; February: Fy; March: Mh; April: Al; May: My; June: Je; July: Jl; August: At; September: Sr; October: Or; November: Nr; December: Dr.

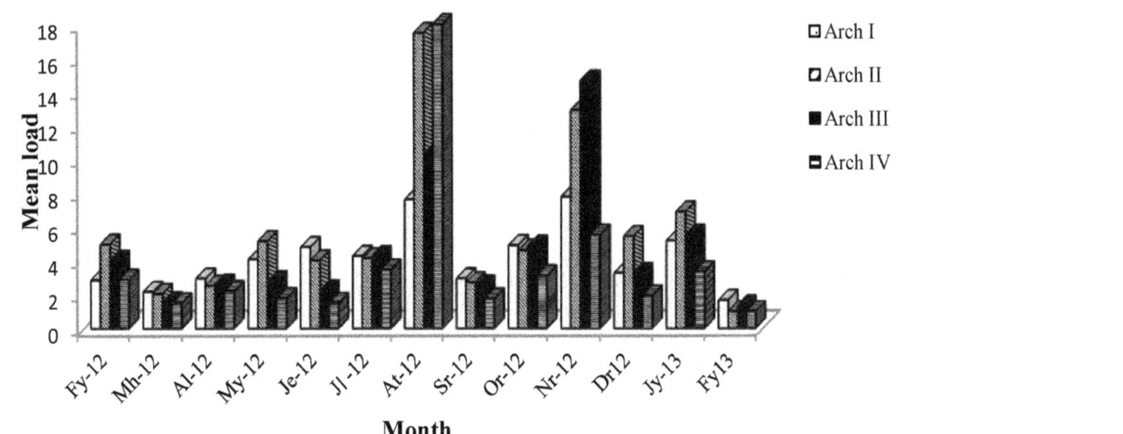

Figure 2: Longitudinal exploitation of the transversal gradient by *Cichlidogyrus halli*. January: Jy; February: Fy; March: Mh; April: Al; May: My; June: Je; July: Jl; August: At; September: Sr; October: Or; November: Nr; December: Dr.

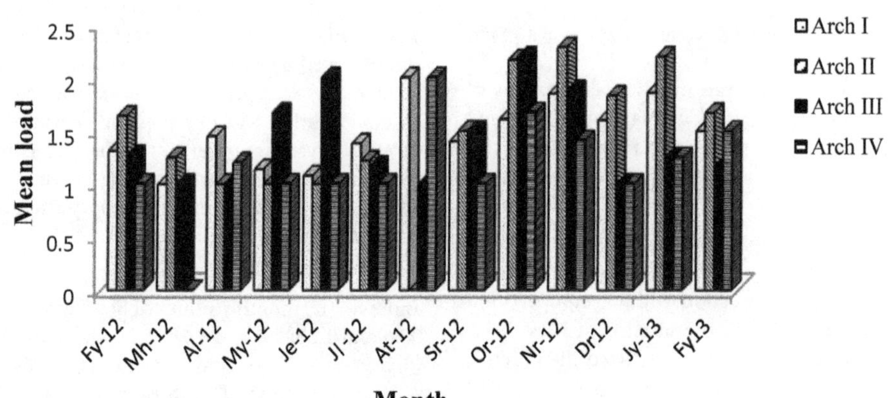

Figure 3: Longitudinal exploitation of the transversal gradient by *Cichlidogyrus tilapiae*. January: Jy; February: Fy; March: Mh; April: Al; May: My; June: Je; July: Jl; August: At; September: Sr; October: Or; November: Nr; December: Dr.

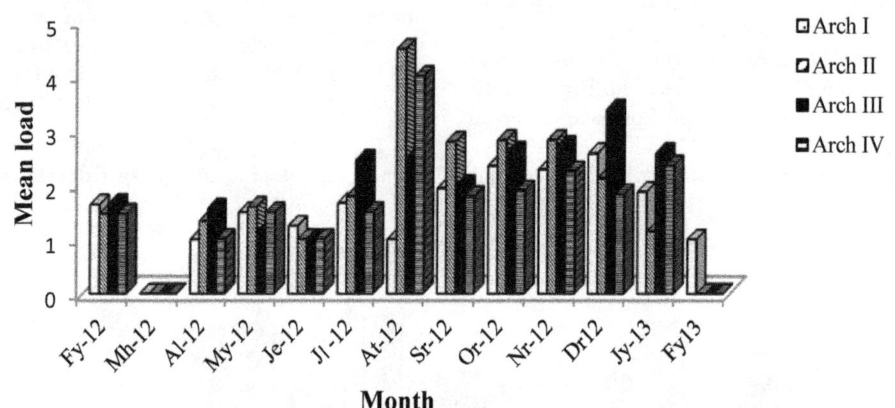

Figure 4: Longitudinal exploitation of the transversal gradient by *Scutogyrus longicornis*. January: Jy; February: Fy; March: Mh; April: Al; May: My; June: Je; July: Jl; August: At; September: Sr; October: Or; November: Nr; December: Dr.

exploited simultaneously the four gill arches of *O. niloticus* during the months of February-May, July-August 2012 and January 2013. While the species, *C. tilapiae,* occupied most of the space in June, October and November 2012. The best niche overlap of *S. longicornis* was observed in February, September and October 2012 while that of *C. thurstonae* was the most restricted.

Discussion

During the study period, *C. thurstonae* showed a precise model of gill occupation. Overall, the colonization of gills by this helminth occurred in the anterio-posterior direction with some perturbations. Various specific models of transversal gradient exploitation by some monogenean species have previously been mentioned. The gill arches colonization of *Hemichromis fasciatus* by *Onchobdella voltensis* and *C. longicornis* was done in the anterio-posterior direction and by equipartition respectively [13]. Also, the gill occupation model of *Barbus martorelli* by *Dactylogyrus bopeleti* and *D. insolitus* was an equipartition [14]. In this study, the observation of a precise model of gill occupation by *C. thurstonae* might indicate a low impact of environmental conditions on this species.

Although gill arches colonization by *C. tilapiae* was perturbed, its monthly number in most cases was higher in one of the median arches and lower on gill IV. The works of Paling, considered as a reference for Teleosteans has shown that the water flux through the gills presents the following gradient: II>III>I>IV [15]. These works allow us to understand firstly that arches II and III which are larger and more ventilated harbor more specimens of *C. tilapiae,* and secondly that, arch IV which is the smallest and less ventilated harbor less specimens of this species.

Cichlidogyrus halli and *S. longicornis* showed no precise colonization model of transversal gradient. In fact, the spatial structure of each of these species was perpetually modified. These two species could be more sensitive to variations in environmental conditions. However, *S. longicornis* which disappeared completely from its host in March 2012 and from arches II, III, IV in February 2013 could be the most sensitive to perturbations. This observation falls in line with the conclusion made by Dogiel which shows that, some fish parasites react highly even to the least physico-chemical modifications of the environment which is considered less perturbed [16]. In August 2012, *C. halli* was more abundant on arch IV which is the most posterior, the smallest and the less ventilated. This observation was also made with *Birgiellus kellensis* and *Quadriacanthus* sp, gill parasites of *Clarias camerounensis* [17]. According to these authors, the present model can be explained based on the way in which oncomiracidiums of monopisthocotylea colonize fish gills. They first fixe on the body of the host and latter migrate towards the gill in the posterior-anterior direction.

Whatever the gill arch, lower diversity values were observed between close seasons. Therefore, seasonal fluctuations have an impact on the colonization of the various gill arches of *O. niloticus* of Melen fish station. A seasonal variation of the specific diversity in the community of monogenean gill parasites of *Rutilus rutilus* was equally observed in Finland [10]. The same phenomenon was noted in monogenean gill parasites of *H. fasciatus* in Cameroon [13]. Diversity variation is often explained by the reduction of specific richness during some periods of the year, by pollution or by eutrophication [10,13,18].

Specific diversity calculated by Shannon-Weaver index is a measure of the degree of the community organization; low diversity is synonymous to good organization while high diversity reveals poor organization [7]. It appears therefore that, the communities of the two anterior arches were more organized with time compared to those of the two posterior ones. Monthly equitability of each arch was less than 1 in most cases, implying that monthly arch diversity did not rich the maximal value. Niche amplitude of *C. halli* appeared to be higher of that of the three other species. Thus, the sensitivity of monogeneans to environmental modifications varies with each species. For Rohde these variations could be more or less perceived according to the selected spatial dimension [19]. This author adds that, it is possible to observe unlike infracommunities between individuals of a same host belonging to two closer localities.

References

1. Mackensie K, Abaunza P (2005) Parasites as biological tags. In: Cadrin SX, Friedland KD, Waldman JR, Stock identification methods: Applications in fishery sciences. Elsevier Academic press, London pp: 221-225.
2. Begon M, Harper JL, Townsend CR (1996) Ecology: Individuals, population and communities. Oxford: Blackwell Science.
3. Sasal P, Morand S, Guégan JF (1997) Determinants of parasite species richness in Méditerranean marine fishes. Marine Ecology Progress Series 149: 61-71.
4. Ternengo S, Levron C, Mouillot D, Marchand B (2009) Site influence in parasite distribution from fishes of the Bonifacio Strait Marine Reserve (Corsica Island, Mediterranean Sea). Parasitology Research 104: 1279-1287.
5. Bilong Bilong CF, Le Pommelet E, Silan P (1999) The gills of Hemichromis fasciatus Peters, 1858 (Teleostei, Cichlidae), a biotope for ectoparasites: structure, heterogeneity and growth model. Ecologie 30: 125-130.
6. Bilong Bilong CF, Tombi J (2004) Heterogeneity of Barbus martorelli (Pisces Cyprinidae) gill system and growth model. Journal of Cameroon Academy of Sciences 4: 211-218.
7. Caltran H, Silan P (1996) Gill filaments of Liza ramada, a biotope for ectoparasites: Surface area acquisition using image analysis and growth models. Journal of Fish Biology 49: 1267-1279.
8. Hawkins BA (2004) Are we making progress toward understanding the global diversity gradient? Basic and Applied Ecology 5: 1-3.
9. Barbault R (1992) Ecology of population. Structure, dynamics and evolution. Masson, Paris.
10. Koskivaara M, Valtonen ET (1992) Dactylogyrus (Monogenea) communities on the gills of roach in three lakes in central Finland. Parasitology 104: 263-272.
11. Simkova A, Gelnar M, Morand S (2001) Order and desorder in ectoparasite communities: The case of congeneric gill monogeneans (Dactylogyrus spp). International Journal for Parasitology 31: 1205-1210.
12. Moby Etia P (1979) Climate In: Atlas of the United Republic of Cameroon. Laclavere, G. Paris pp: 16-19.
13. Bilong Bilong CF, Atyame Ntem CM, Njine T (2004) Structure of the guild of monogenean gill parasites of fish Hemichromis fasciatus in Yaounde municipal lake. Journal of Cameroon Academy of Sciences 4: 33-40.
14. Tombi J, Bilong Bilong CF (2013) Gill arch occupation models of parasite communities of Barbus martorelli (Teleostean: Cyprinidae). International Journal of Fisheries and Aquaculture 5: 215-220.
15. Paling E (1968) A method of estimating the relative volumes of water flowing over the different gills of a freshwater fish. Journal of Experimental Biology 48: 533-544.
16. Dogiel VA (1961) Ecology of the parasites of freshwater fishes. In: Dogiel VA, Petrushevski GK and Polyanski Yu I, Parasitology of fishes, Edinburg.
17. Nack J, Tombi J, Bitja Nyom A, Bilong Bilong CF (2010) Site selection of two Dactylogyridae monogenean gill parasites of Clarias camerunensis: evidence on the Monopisthocothylea mode of infestation. Journal of Applied Biosciences 33: 2076-2083.
18. Bagge AM, Valtonen ET (1996) Experimental study on the influence of paper and pulp mill effluent on the gill parasite communities of roach (Rutilus rutilus). Parasitology 112: 499-508.
19. Rohde K (1993) Ecology of marine parasites. An introduction to marine parasitology. Cab international Wallingford. Oxon.

Retention of Fillet Coloration in Rainbow Trout After Dietary Astaxanthin Cessation

Katherine R Brown[1], Michael E Barnes[1*], Timothy M Parker[1] and Brian Fletcher[2]

[1]*McNenny State Fish Hatchery, South Dakota Department of Game, Fish and Parks, 19619 Trout Loop, Spearfish, South Dakota-57783, USA*
[2]*Cleghorn Springs State Fish Hatchery, South Dakota Department of Game, Fish and Parks 4725 Jackson Boulevard, Rapid City, South Dakota-57702, USA*

Abstract

This study was conducted to determine the retention time of rainbow trout (*Oncorhynchus mykiss*) fillet coloration, as indicated by hue, chroma, and entire color index (ECI) values, after the cessation of dietary astaxanthin. After 12 weeks of receiving a diet containing astaxanthin, rainbow trout [mean (SD) length 240 (26) mm, weight 194 (65) g] were either switched to a non-astaxanthin diet or continued to receive the same astaxanthin diet for 55 d. In addition, a control group was fed a non-astaxanthin diet for the entire duration of the study. Digital coloration measurements (L*, a*, and b*) were recorded on fillets and used to calculate hue, chroma, and ECI values. Hue, chroma, and ECI did not significantly change within any of the treatments over the course of the study. Hue, chroma, and ECI were also not significantly different between the fillets of those fish that either continued to receive astaxanthin-supplemented feed compared to those that were switched from an astaxanthin diet to diet free of astaxanthin. However, ECI, hue, and chroma were significantly different in fillets from the fish that had never received dietary astaxanthin compared to fillets from the other two treatments. These results indicate that catchable-sized rainbow trout stocked into recreational fishing waters with minimal natural feed should retain fillet coloration for at least 55 d after stocking, making astaxanthin supplementation in recreational hatcheries a viable tool to improve angler satisfaction.

Keywords: Rainbow trout; Astaxanthin; *Oncorhynchus mykiss*; Coloration

Introduction

Consumer and angler satisfaction is of great importance for both commercial and recreational aquaculture. The work of Forsberg and Guttormsen stated that with salmon, the appearance of wild characteristics is the most important criteria of perceived quality by consumers [1]. Fillet coloration is particularly important. In rainbow trout (*Oncorhynchus mykiss*), fillets containing red pigmentation are deemed much more desirable than non-pigmented, white fillets [2]. A 2009 study also found a positive relationship between color and consumer demand in snapper (*Pagrus auratus*) [3]. In recreational fisheries, Simpson reported that anglers overwhelmingly preferred more brightly colored rainbow trout, as well as more red-colored trout fillets [4].

The red fillet color preferred by both consumers and recreational angler's results primarily from the consumption of astaxanthin, a carotenoid pigment that fish cannot synthesize de novo [5]. Astaxanthin occurs in many prey items consumed by wild fish, while in cultured fish this pigment must be provided in the artificial diets [5]. Including the synthetic or natural forms of astaxanthin in formulated feeds increases production costs, but this is typically more than compensated by increases in pricing power [1], or in the case of natural resource agencies, by increases in angler satisfaction [4]. However, anglers in the Simpson survey were not told that such coloration in hatchery fish would require the addition of supplemental astaxanthin [4].

The time and amount of astaxanthin supplementation required to produce red-colored fillets in salmonids is well documented [1,6-8]. However, unlike trout produced for direct human consumption by commercial aquaculture which are slaughtered at peak coloration, trout produced for stocking into water bodies by recreational hatcheries may be harvested months after stocking [9].

A 1998 study reported few differences in the color of frozen fillets from Atlantic salmon Salmo salar starved for up to 86 days after being fed a diet containing astaxanthin [10]. Except for the very short duration experiment of López-Luna, we are not aware of any studies that have examined long-term coloration depletion rates and subsequent changes in fillet coloration following the cessation of feeding dietary astaxanthin in rainbow trout [11]. This information is needed by resource agencies and other organizations to determine the effectiveness and utility of stocking of catchable-sized rainbow trout with red-colored fillets due to supplemental astaxanthin. Thus, the objective of this study was to examine the long-term effects on rainbow trout fillet coloration after the removal of supplemental dietary astaxanthin.

Methods

This study was conducted at Cleghorn Springs State Fish Hatchery, Rapid City, South Dakota, USA, using degassed and aerated spring water (11°C, total hardness as $CaCO_3$, 36 mg L^{-1}; alkalinity as $CaCO_3$, 210 mg L^{-1}; p^H-7.6; total dissolved solids, 390 mg L^{-1}). Four circular production tanks (6.1 m diameter, 0.9 m operating depth), each containing approximately 5,500 Erwin-Arlee strain rainbow trout [mean (SD) length 240 (26) mm, weight 194 (65) g] were initially used in this experiment. Beginning on June 4, 2008, two of the tanks were fed a 4.5 mm classic trout diet (Skretting, Tooele, Utah, USA), while

***Corresponding author:** Michael E Barnes, McNenny State Fish Hatchery, South Dakota Department of Game, Fish and Parks, 19619 Trout Loop, Spearfish, South Dakota-57783, USA, E-mail: mike.barnes@state.sd.us

the other two tanks were fed the same diet with the inclusion of 40 mg/L synthetic astaxanthin. Feeding rates were based on a projected growth rate of 0.065 cm/day and an anticipated feed conversion ratio of 1.1. Because the fish were part of normal hatchery production, small numbers of fish were periodically removed from each tank for stocking into recreational fishing waters. As the number of fish in each tank changed, feeding amounts were adjusted correspondingly.

Weekly measurements of fillet color were obtained from fish euthanized by a lethal dose of 250 mg/L MS-222 using a HunterLab MiniScan XE Plus Colorimeter (HunterLab, Reston, Virginia, USA). Measurements occurred mid-fillet, below the dorsal fin and above the lateral line. Color was measured from whole fillets, rather than ground tissue, because this coloration would be most evident to the recreational angler. The colorimeter measured color along the CIELAB color scale (CIE 1976), which includes L*, a*, and b* values. The a* score indicates red-green chromaticness and b* indicates yellow-blue chromaticness [7]. L* indicates white-to-black values [12]. At 12 weeks, after which a* values were no longer increasing in the tanks receiving supplemental astaxanthin, one-half of these fish were switched to a non-astaxanthin diet and moved into two additional tanks. Fillet color data was then collected at approximately 14 day intervals for the next 55 days. The focus of this study was solely on changes in coloration, not growth or any other rearing variable. Thus, only coloration data was collected.

The L*, a*, and b* values were used to calculate Hue, Chroma, and the Total Color Index using the following formulas:

$$Hue = \arctan\left(b^*/a^*\right)$$

$$Chroma = \left[\left(a^*2 + b^*2\right)1/2\right]$$

$$Entire\ Color\ Index\left(ECI_i\right) = C_i * \cos\left(H_i - H_{mean}\right),$$

Where $H_{mean} = mean\ hue$ and C_i, H_i =the chroma and hue values for each measurement [13].

Statistical analysis was done using Minitab 16 software (Minitab Inc., State College, Pennsylvania, USA). Color depletion data was analyzed using multiple regression models. Models were done using the overall means of the treatments at each measurement to determine significance in changes of color variables. Statistical significance was predetermined at $P<0.05$.

Results

Hue, chroma, and ECI did not significantly change in any of the treatments over the course of the study (Figures 1-3). Hue, chroma, and ECI were also not significantly different between the fillets of those fish that either continued to receive astaxanthin-supplemented feed (Color) compared to those that were switched from an astaxanthin diet to diet free of astaxanthin (Depletion). However, ECI, hue, and chroma were significantly different in fillets from the fish that had never received dietary astaxanthin (No Color), compared to fillets from both Color and Depletion treatment fish.

Best fitted models for each color variable are as follows:

a) **Hue:** No Color=-27.75-0.00073 (Date); Color and Depletion=-27.9-0.00073 (Date) R^2=60.1, d.f=2,10.

b) **Chroma:** No Color=261.15-0.0062 (Date); Color and Depletion=268-0.0062 (Date) R^2=91.8, d.f=2,10.

c) **ECI:** No Color=258.18-0.0061(Date); Color and Depletion=263-0.0061 (Date) R^2=91.8, d.f=2,10.

Discussion

These results indicate that, at least for the 55 d duration of the study, there was no loss in fillet coloration after the cessation of feeding dietary astaxanthin. This timing is important in recreational fisheries sustained by the stocking of catchable-sized rainbow trout into waters with minimal natural feed. In a study conducted by Barnes, most of the rainbow trout stocked into two public fishing lakes were caught within the first few weeks after stocking, and nearly all of the fish were harvested within the first 55 days [9]. Thus, natural resource agencies can feed supplemental astaxanthin to catchable rainbow trout without concern that the internal and external benefits of pigmentation will disappear before the fish are caught, thereby meeting angler preferences for more colorful fish and fillets [4].

The results from this study also have application to commercial trout hatcheries, where fillet coloration influences customer perceptions of quality, customer demand, and pricing power [1-3,13,14]. Rather than incur the expense of feeding supplemental astaxanthin until the trout

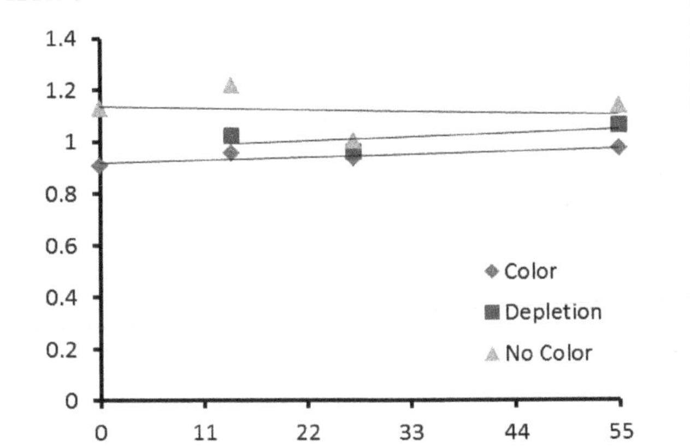

Figure 1: Mean hue values from fillets of rainbow trout subjected to one of three treatments:
a) Fed feed void of supplemental astaxanthin (no color),
b) Fed non-astaxanthin supplemented feed (color), or
c) Fed feed supplemented with astaxanthin for 12 weeks prior to the start of the experiment and then switched to non-astaxanthin feed (depletion).

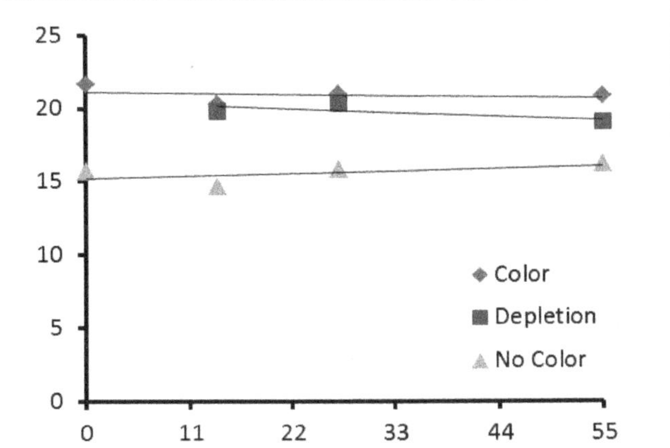

Figure 2: Mean chroma values from fillets of rainbow trout subjected to one of three treatments:
a) Fed feed void of supplemental astaxanthin (no color),
b) Fed non-astaxanthin supplemented feed (color), or
c) Fed feed supplemented with astaxanthin for 12 weeks prior to the start of the experiment and then switched to non-astaxanthin feed (depletion).

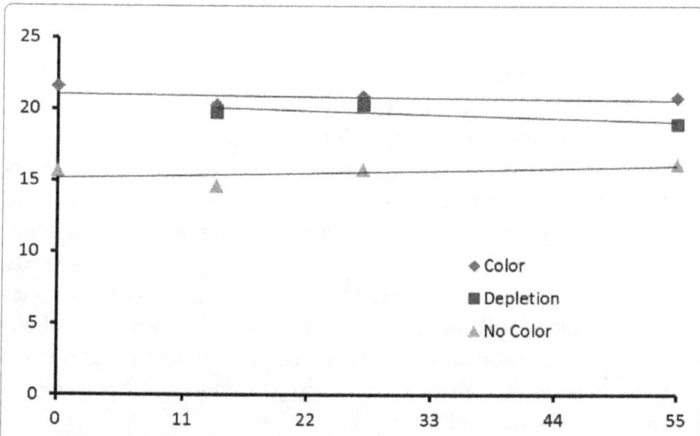

Figure 3: Mean Entire Color Index values from fillets of rainbow trout subjected to one of three treatments:
a) Fed feed void of supplemental astaxanthin (no color),
b) Fed non-astaxanthin supplemented feed (color), or
c) Fed feed supplemented with astaxanthin for 12 weeks prior to the start of the experiment and then switched to non-astaxanthin feed (depletion).

are ready for market, producers could discontinue feeding astaxanthin at 55 d pre-harvest and still maintain the fillet color so desired by consumers. With astaxanthin accounting for as much as 10%-15% of total feed costs, the cessation of feeding astaxanthin for nearly two months would produce considerable cost-savings [3].

López-Luna noted no difference in fillet L*, a*, and b* values of rainbow trout previously fed astaxanthin after three days of starvation at 11.5°C (34.1 degree-days, whereby one degree-day equals 24 hours at 1°C) [11]. Our study extends this timeframe out to 605 degree-days (55 days at 11°C). In frozen fillets from Atlantic salmon starved for up to 86 days after being fed a diet containing astaxanthin, no differences in hue or red (a*) values were observed by Einen and Thomassen [10]. The lack of decrease in fillet coloration over 55 days likely reflects the ability of rainbow trout muscle to store relatively large amounts of astanxanthin [15].

The color values are well within the range reported for rainbow trout in other studies involving astaxanthin [6,11,12,16,17]. Although the results from this study are likely applicable to other strains of rainbow trout, the influence of other dietary factors on astaxanthin absorption and ultimately fillet coloration could potentially influence the results if different diets are used [18].

This study only examined the loss of coloration in trout fillets. It does not address any potential issues that recreational anglers or consumers may have about the use of synthetic astaxanthin in hatchery diets [19,20].

Acknowledgement

We thank Greg Simpson, Will Sayler, John Carriero, and Joseph Barnes for their assistance with this study.

References

1. Forsberg O, Guttormsen A (2006) Modelling optimal dietary pigmentation strategies in farmed Atlantic salmon: Application of mixed-integer non-linear mathematical programming techniques. Aquaculture 261: 118-124.
2. Folkstead A, Wold JP, Rorvik K (2008) Rapid and non-invasive measurements of fat and pigment concentrations in live and slaughtered Atlantic salmon (Salmo salar L.). Aquaculture 280: 129-135.
3. Doolan BJ, Booth MA, Allan GL (2009) Effects of dietary astaxanthin concentration and feeding period on the skin pigmentation of Australian snapper Pagrus auratus. Aquaculture Res 40: 60-68.
4. Simpson G (2009) Angler Use and Harvest Survey of Center Lake, South Dakota. South Dakota Department of Game, Fish and Parks Report, Pierre, South Dakota pp: 4-10.
5. Choubert G, Blanc JM (1997) Colour measurements, using the CIELCH colour space, of muscle of rainbow trout, Oncorhynchus mykiss (Walbaum), fed astaxanthin: effects of family, ploidy, sex, and location of reading. Aquaculture Res 28: 15-22.
6. Nickel DC, Bromage NR (1998) The effect of timing and duration of feeding astaxanthin on the development and variation of fillet colour and efficiency of pigmentation in rainbow trout (Oncorhynchus mykiss). Aquaculture 169: 233-246.
7. Smith B, Hardy RW, Torrissen OJ (1992) Synthetic astaxanthin deposition in pan-sized coho salmon (Oncorhynchus kisutch). Aquaculture 104: 105-119.
8. Storebakken T, Choubert G (1991) Flesh pigmentation of rainbow trout fed astaxanthin or canthaxanthin at different feeding rates in freshwater and saltwater. Aquaculture 95: 289-295.
9. Barnes ME, Simpson G, Durben DJ (2009) Post stocking Harvest of Catchable-Sized Rainbow Trout Enhanced by Dietary Supplementation with a Fully Fermented Commercial Yeast Culture during Hatchery Rearing. NA J Fish Manage 29: 1287-1295.
10. Einen O, Thomasson MS (1998) Starvation prior to slaughter in Atlantic salmon (Salmo salar) II. White muscle composition and evaluation of freshness, texture and colour characteristics in raw and cooked fillets. Aquaculture 169: 37-53.
11. López-luna J, Torrent F, Villarroel M (2014) Fasting up to 34°C days in rainbow trout, Oncorhynchus mykiss, has little effect on flesh quality. Aquaculture 420: 63-70.
12. Choubert G (2010) Response of rainbow trout (Oncorhynchus mykiss) to varying dietary astaxanthin/canthaxanthin ratio: colour and carotenoid retention of the muscle. Aquaculture Nutrition 16: 528-535.
13. Pavlidis M, Papandroulakis N, Divanach P (2006) A method for the comparison of chromaticity parameters in fish skin: Preliminary results for coloration pattern of red skin Sparidae. Aquaculture 258: 211-219.
14. Doolan BJ, Booth MA, Jones PL, Allan GL (2007) Effect of cage colour and light environment on the skin colour of Australian snapper Pagrus auratus (Bloch & Schneider, 1801). Aquaculture Res 38: 1395-1403.
15. Choubert G (1985) Effects of starvation and feeding on canthaxanthin depletion in the muscle of rainbow trout (Salmo gairdneri Rich). Aquaculture 46: 293-298.
16. Choubert G, Mendes-Pinto MM, Morais R (2006) Pigmenting efficacy of astaxanthin fed to rainbow trout Oncorhynchus mykiss: Effect of dietary astaxanthin and lipid sources. Aquaculture 257: 429-436.
17. Choubert G, Cravedi JP, Laurentie M (2009) Effect of alternative distribution of astaxanthin on rainbow trout (Oncorhychus mykiss) muscle pigmentation. Aquaculture 286: 100-104.
18. Chimsung N, Tantikitti C, Milley JE, Trichet VV, Lall SP (2014) Effects of various dietary factors on astaxanthin absorption in Atlantic salmon (Salmo salar). Aquaculture Res 45: 1611-1620.
19. Olesen I, Alfnes F, Røra MB, Kolstad K (2010) Eliciting consumer's willingness to pay for organic and welfare-labelled salmon in a non-hypothetical choice experiment. Livestock Sci 127: 218-226.
20. Steine G, Alfnes F, Rør MB (2005) The effect of colour on consumer WTP for farmed salmon. Marine Resource Econ 20: 211-219.

Present Yield Status, Percentage Composition and Seasonal Abundance of Shark in Two Geographically Important Zones of Bangladesh

Monjurul Hasan Md*[1], Bhakta Supratim Sarker[1], Mahabubur Rahman[1], Shamsul Alam Patwary Md[1], Jahangir Sarker Md[1], Shahriar Nazrul KM[2] and Mohammed Rashed Parvej[3]

[1]*Department of Fisheries and Marine Science, Noakhali Science and Technology University, Sonapur, Noakhali, Bangladesh*
[2]*Department of Fisheries, Ministry of Fisheries and Livestock, Bangladesh*
[3]*Bangladesh Fisheries Research Institute, Mymensingh, Bangladesh*

Abstract

A study was carried out during January-December (2014) in two selected shark landing centers; BFDC Fish harbor, Cox's Bazar and Fishery Ghat fish landing center, Chittagong situated at the North-eastern part of the Bay of Bengal. Data were collected through semi-structured interview, case study, frequent visit to the informants found in and between the trade channels. A total of 9 shark species belonging to 3 families (Carcharhinidae, Hemiscylliidae and Sphyrnidae) were recorded. It was found that sharks were exploited mostly at small sizes (45%) in those landing centers. The highest and lowest yield were found in the month of January and July respectively at Chittagong whereas November and July at Cox's Bazar. The highest and lowest landed number was found in November and July respectively at both the landing centers. Dog shark was the most dominant species followed by Hammerhead shark in terms of yield and landed number at Chittagong and Cox's Bazar contributed 55.794 MT (60% 90) and 17.675 MT (19%) among the total yield and 174,877 (83%) and 25,733(12%) landed number respectively. Yield and landed number of other species contribution altogether were only 21% and 5% respectively of total. Total yield was found 6 folds in Cox's Bazar than that of Chittagong. Abundance reveals that the highest catches of shark were found during October to December (42%) and the lowest catches during January to March (16%). Yield of shark was found to be declining than the previous years and a clear deviation of seasonal abundance is also occurring. As there is no gear size limitation or seasonal restriction in the Fish Act, small sized sharks were found to be caught mostly in those landing centers which may also pose a threat to shark species composition in the Bay of Bengal region, Bangladesh.

Keywords: Shark yield; Bay of Bengal; Percentage contribution; Seasonal abundance; Fish act

Introduction

Sharks are a highly diverse group of fish that evolved over 400 million years ago. These are predominantly marine, oceanic and are widely distributed in the tropical, subtropical and temperate waters of the seas around the world [1]. More than 60% catches were reported from central (tropical) regions, in particular from the Indian Ocean (26%) followed by Western central pacific (14%) and the Eastern Central Atlantic (10%). Total 26 top shark-fishing countries were responsible for 84% of global shark catches [2]. In the Indian Ocean deep sea, there were species of shark-like fishes including 8 orders, 23 families and 46 genera [3]. At least 171 species of elasmobranches, representing 68 genera and 34 families, were recorded from fresh or estuarine waters [4]. In Bay of Bengal there are 11 species of sharks identified [5]. The Bay of Bengal is one of the most heavily fished regions in world's ocean for shark. The major shark hunting grounds of Bangladesh include the coastal waters of Kuakata, Sonar Char, Ruper Char, Fatrar Char, Char Gongmoti and Dublar Char in Patuakhali and Ashar Char, Patharghata Barguna, the Sunderbans, Sandwip, Kutubdia, Moheshkhali, Cox's Bazar and Teknaf [6]. Sharks are harvested as target species mainly by shark net (modified gill net) and hooks and lines and as a by catch in other commercial fishing. A large numbers of small sized juveniles or new born sharks and rays are caught by shrimp and fish trawlers which were not recorded or reported, for small size and low market value and discarded it as a trash [7]. Some 50% of the estimated global catch of chondrichthyans is taken as by-catch, does not appear in official fishery statistics, and is almost much unmanaged. When taken as by-catch, they are often subjected to high fishing mortality. Consequently, some skates, sawfish, and deep-water dogfish have been virtually extirpated from large regions [8]. Recently, number of shark fishing boats, fishing days and export trade have been expanding rapidly in Bangladesh which gives some cause of alarm. Moreover, catch of small sized or juvenile sharks has increased with the decrease of large size sharks reminding us that the stock may be undergoing overexploitation [9]. Some shark species were found frequently in a season which is not found now as before some has entered into the IUCN threatened and endangered list which gives concern about the decrease of species composition because of overexploitation or illegal fishing activities.

In this present study, effort had been made to determine the shark species composition and percentage contribution from January, 2014 to December, 2014 in BFDC Fish harbor, Cox's Bazar and Fishery Ghat fish landing center, Chittagong. The main objective of this study was to find out the landing trends and seasonal abundance of shark from the sustainable yield and conservation point of view in the Bay of Bengal region. It is expected that the statistical interpretation would rightly focus on the status of the shark fisheries and contribute towards any national management plan for shark fishery of Bangladesh in the Bay of Bengal region.

***Corresponding author:** Monjurul Hasan Md, Department of Fisheries and Marine Science, Noakhali Science and Technology University, Sonapur, Noakhali, Bangladesh, E-mail: mhshihab.hasan@gmail.com

Materials and Methods

Study site and duration

The field study was conducted from January-December(2014) in two selected shark landing centers, BFDC Fish harbor, Cox's Bazar and Fishery Ghat fish landing center, Chittagong district situated at the North-eastern part of the Bay of Bengal. These two study areas were selected to cover most of the landing centers, retail, and wholesale markets of shark from the Bay of Bengal of Bangladesh region.

Data collection method

Information regarding harvesting procedure, trip duration per month, auction procedure, harvesting gears and vehicles were collected at both landing stations through formal face to face interview of boat owners/divers of commercial fishing vessels, fishermen, retailers and buyers. Collected information was verified by the key informants. Photos were captured by a digital camera. The length and weight of the fishes were measured directly by using measuring tape and balance. Missing information were collected through the phone call and verified by the respective officers.

Month wise data regarding total yield and landed numbers of each shark species, number of fishing days were collected from Fishery officials of the Marine Fishery Wing (Department of Fisheries), Cox's Bazar and the Marine Fisheries Survey Management Unit, Chittagong.

Statistical analysis

The species wise weight was measured in kilogram and then it was converted into metric tons (MT). Statistical software MS Excel (version 2013) was used for data analyzing.

Results

Species composition

A total 9 species of sharks recorded from this study which was Dog shark (*Scoliodon laticaudus*), Hammerhead shark (*Sphyrna lewini*), Milk shark (*Rhizoprionodon acutus*), Tiger shark (*Galeocerdo cuvier*), Silky shark (*Carcharhinus falciformis*), Ridge back cat shark (*Chiloscyllium indicum*), Black tip reef shark (*Carcharhinus melanopterus*), Bull shark (*Carcharhinus leucas*) and Spot tail shark (*Carcharhinus sorrah*).

Size abundance

Major findings from this study were that the highest mean length was found in Bull shark (151.67 ± 7.92 cm) followed by Black tip reef shark (116.25 ± 2.27 cm) and Silky shark (102.19 ± 6.29 cm) whereas the lean was found in Ridge back cat shark (46.02 ± 2.98 cm) (Table 1). The highest mean weight was found in Bull shark (29.20 ± 4.34 kg) and the lean in Ridge back cat shark (0.36 ± 0.11 kg) (Table 1). Big sized sharks were caught rarely in Chittagong only.

Sharks were mostly caught at small sizes (45%) as shown in (Figure 1) Large sized sharks were caught in a very small amount (22%).

Harvesting depth, vehicles and gears

Sharks were harvested by the Chittagong and Cox's Bazar, Kutubdia, Moheshkhali, Cox's Bazar and Teknaf coasts. The usual harvesting depth was 10-50 m. Shark nets were widely used mainly for harvesting target species of shark like Dog shark, Milk shark, and Hammerhead shark. Sharks were exploited as by catch of Lakkha net in both Chittagong and Cox's Bazar. Gill nets, Set net bag, Trammel net were also used for exploitation. Anchors as well as hooks and lines were found to be used only in Cox's Bazar. These were operated by wooden mechanized boats.

Shark species	Mean length (cm) ± SD	Mean weight (kg) ±SD
Silky shark	102.19 ±6.29	9.25 ± 3.12
Dog shark	50.87 ±2.18	0.60 ± 0.19
Spot tail shark	55.75 ± 2.20	0.91 ± 0.27
Black tip reef shark	116.25 ± 2.27	4.53 ± 0.92
Bull shark	151.67 ±7.92	29.20 ± 4.34
Tiger shark	70.21 ±2.13	4.14 ± 1.11
Hammerhead shark	70.36 ±1.68	2.30 ± 0.54
Ridge back cat shark	46.02 ±2.98	0.36 ± 0.11
Milk shark	58.67 ±5.00	0.85 ± 0.25

*SD=Standard Deviation.

Table 1: Species wise mean length and weight (±SD) of shark.

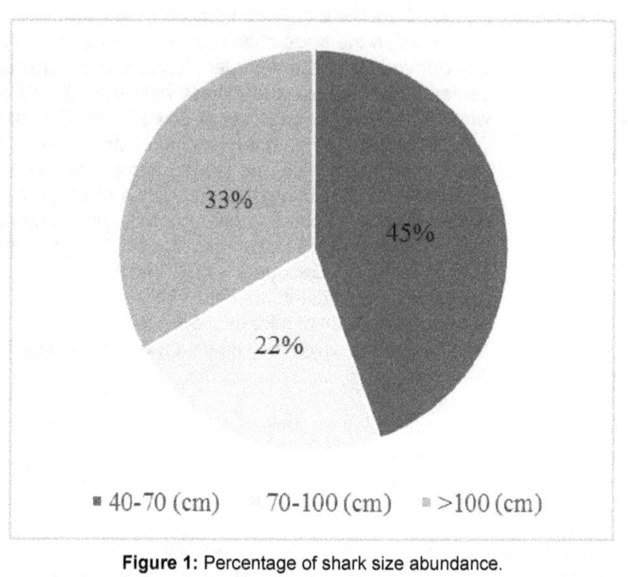

Figure 1: Percentage of shark size abundance.

Total yield and landed number

During the study period, the highest yield was recorded 2.37 MT in January and landed number was 5,200 in November in Chittagong. Total yield and landed number were found 12.824 MT and 23,245 respectively.

In Cox's Bazar, the yield and landed number was found to be the highest in November and no catch was recorded in July. Total yield and landed number were found 80.34 MT and 188,407 respectively (Figures 2 and 3).

According to the landing data, Dog shark was found to be the most dominant species followed by Hammerhead shark in terms of yield in both Chittagong and Cox's Bazar. Dog shark yield was found 5.696 and 50.098 MT in Chittagong and Cox's Bazar respectively. The lowest yield was found in Black tip reef shark (1.807 MT). Milk shark was found to be landed only in Chittagong whereas Spot tail shark in Cox's Bazar. Total yield was found much higher in Cox's Bazar (80.34 MT) than Chittagong (12.824 MT) (Figure 4).

The landed number of Dog shark was found to be the most dominant followed by Hammerhead shark in both Chittagong and Cox's Bazar than the other shark species (Figure 5). During the study period, the highest landed number was found in Dog shark species (153,696) in Cox's Bazar and the lean in Ridge back cat shark (9) in Chittagong.

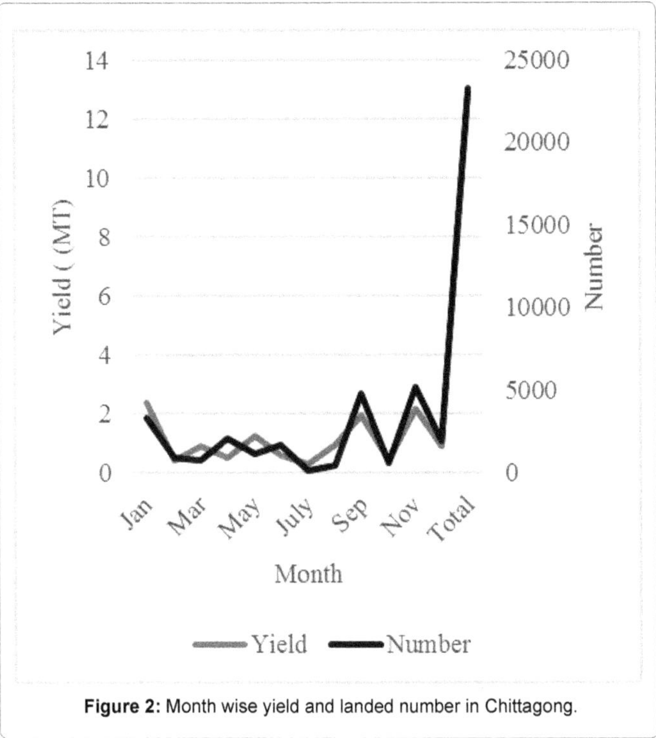

Figure 2: Month wise yield and landed number in Chittagong.

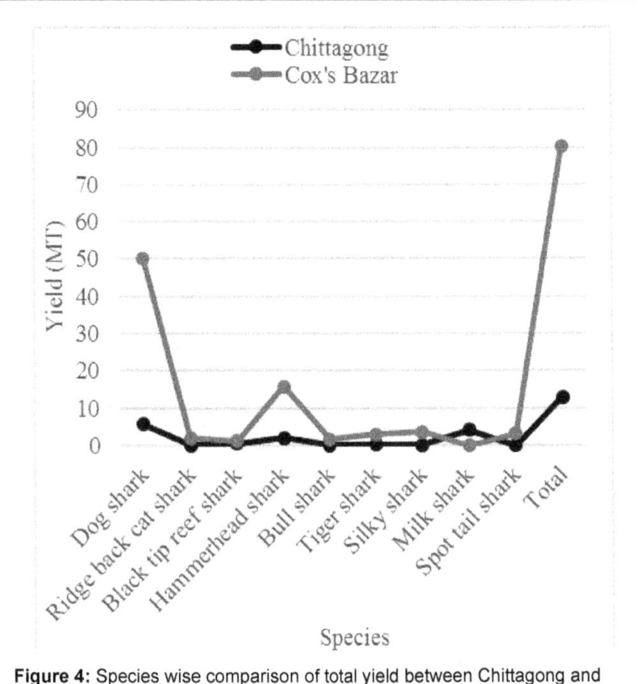

Figure 4: Species wise comparison of total yield between Chittagong and Cox's Bazar.

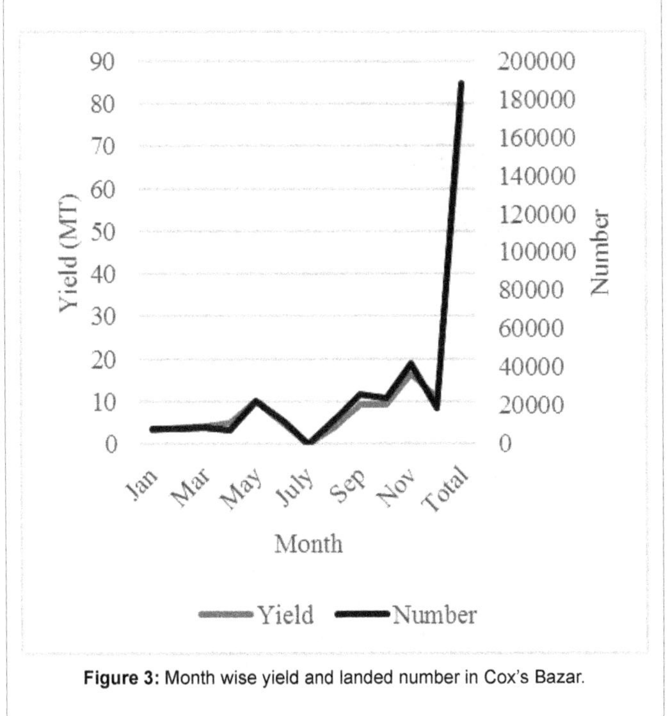

Figure 3: Month wise yield and landed number in Cox's Bazar.

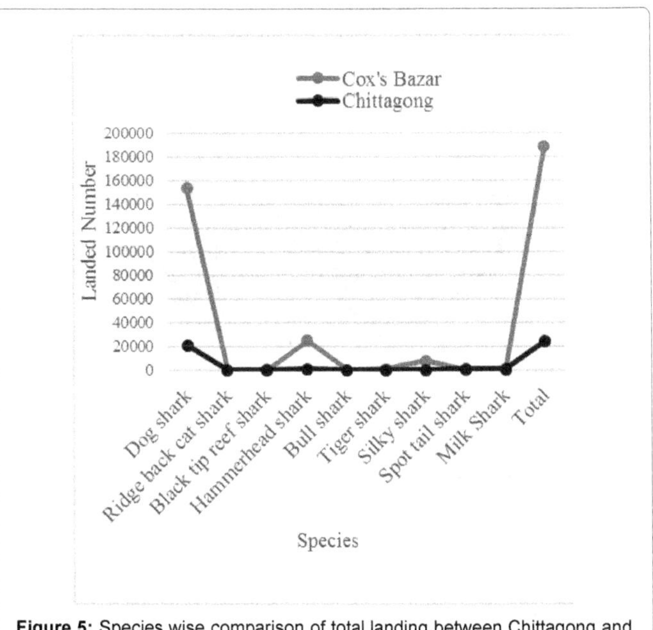

Figure 5: Species wise comparison of total landing between Chittagong and Cox's Bazar.

Percentage contribution

Among the recorded shark species, greatest contribution came from the Dog shark (60%) in total yield followed by Hammerhead shark (19%) in both landing centers. Yield of other shark species altogether was found only 21% (Figure 6).

The highest percentage contributor shark species in total landed number was found in Dog shark (83%) followed by the Hammerhead shark (12%). Landed number of other shark species altogether was found only 5% (Figure 7).

Cox's Bazar contributes the highest shark yield (86%) between the two selected shark-landing centers (Figure 8). Total yield was found 6 folds in Cox's Bazar than that of Chittagong.

From the seasonal abundance data the highest abundance of shark was found during October to December (42%) and the lowest during

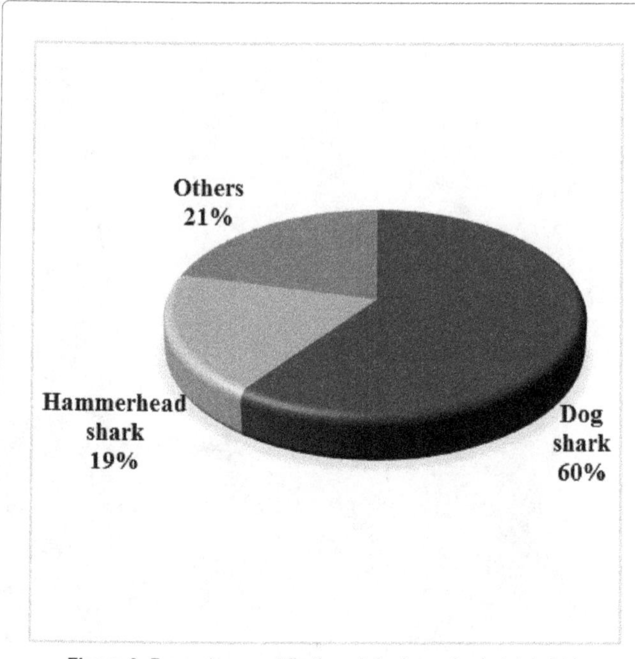

Figure 6: Percentage contribution of shark species in total yield.

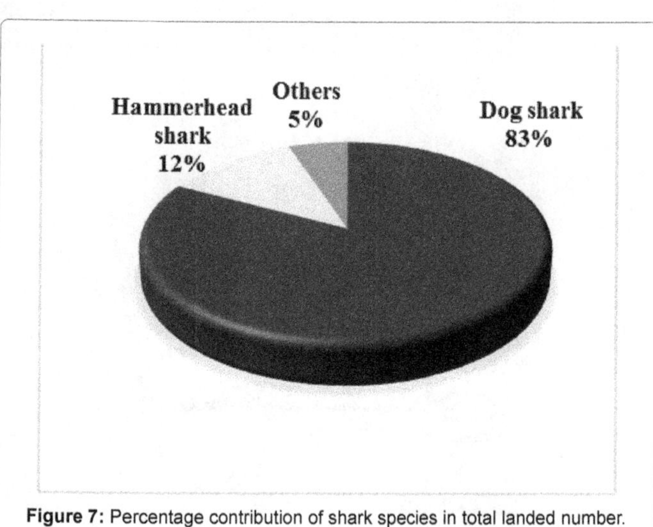

Figure 7: Percentage contribution of shark species in total landed number.

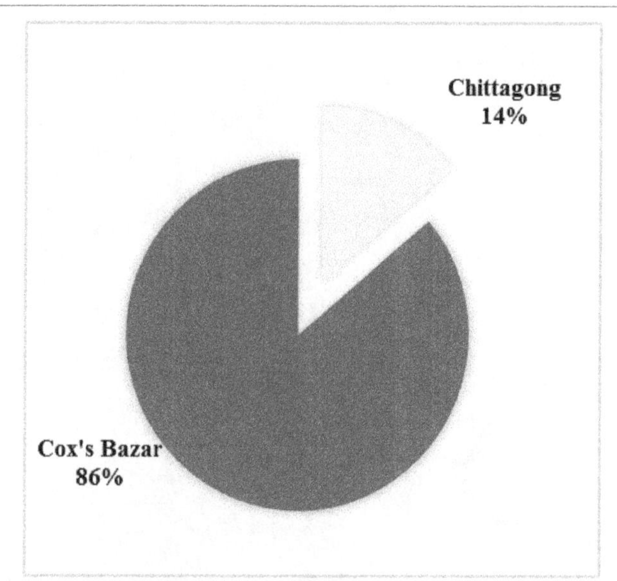

Figure 8: Comparison of shark catch contribution percentage between Chittagong and Cox's Bazar.

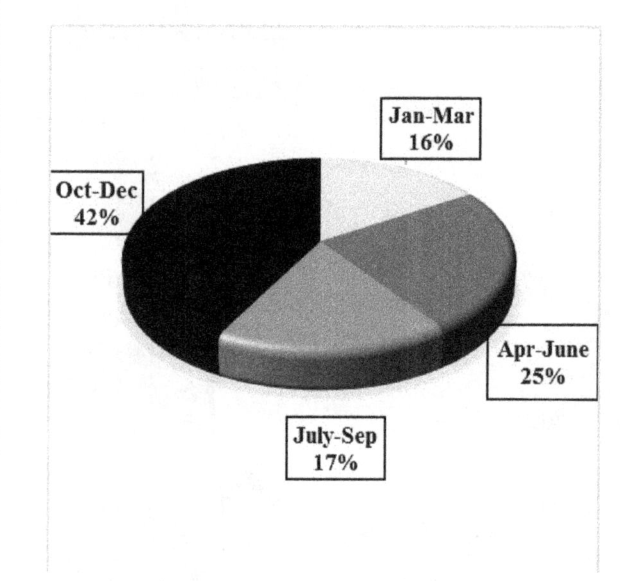

Figure 9: Seasonal abundance of shark in Chittagong and Cox's Bazar.

January-March (16%) (Figure 9).

Fishing season

The total harvesting and fishing days from both Chittagong and Cox's Bazar were 93.164 MT and 193 days respectively. The highest harvesting was found in the month of November (18.56 MT). In the month of July, no fishing was done in Cox's Bazar (Figure 10).

Discussion

The number of shark species in Bangladesh reported by different authors varies in different times [10] recorded 10 species of sharks belonging to 3 families. According to [11] the total number is 56, while [1,7,12-15] mentioned the number as 51, 22, 36, 21, 63 and 56 respectively. There is a clear indication of declining shark species composition from the previous years.

In the present study, dominance of smaller size shark was observed as maximum harvesting came from Dog shark. The mean size and mean weight of Dog shark was found 50.87 ± 2.18 cm and 0.60 ± 0.19 kg which was different to the observation of [16] having average size between 50-52 cm in total length and 0.15-0.2 kg in weight. It might create threats to Dog shark population due to overexploitation in the near future. Since, larger size sharks are mostly common in offshore water which is beyond the reach of our artisanal fishermen, it is more likely that sharks more than 100 kg are less common in the catch.

Dog shark was found to be the top listed shark in respect to the total landing 55.794 MT and contribution (60%) to overall catch at both landing centers which agree with the work of [17] who showed that the highest landing (134 MT) and contribution (76%) to total catch for the whole sampling period was found from Dog shark [1] found that

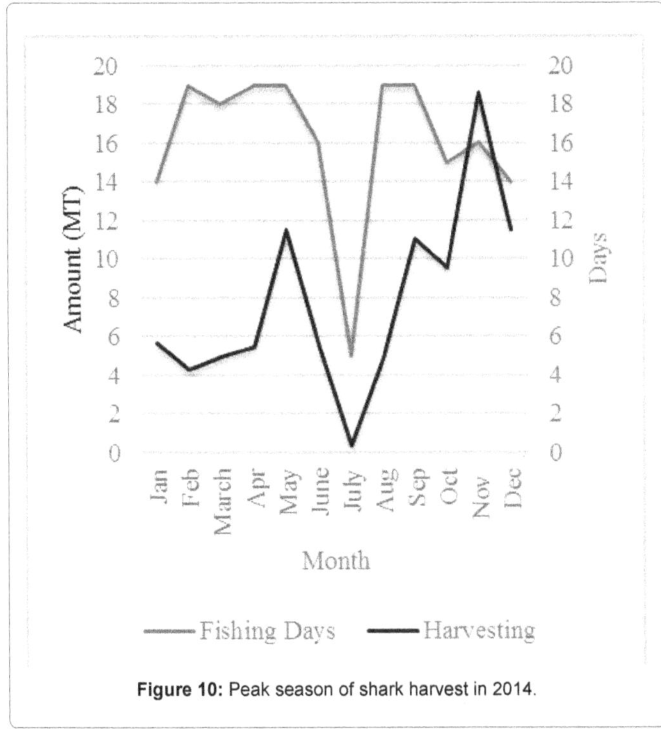

Figure 10: Peak season of shark harvest in 2014.

mostly small sized sharks were caught because of gear limitations. For a number of species shifts in length compositions to smaller sizes have been attributed to over exploitation [18]. Further study need to be done on species size distribution through time and area to understand the issue of overexploitation.

In the present investigation, it was found that sharks were mainly caught by Hooks and lines, Gill nets, Set net bag, Trammel net and Shark nets [16] noted in their study that shark target fishing has been developed for the last 5-10 years mainly by using hooks and lines during winter months. In their study they included that during 2007-2008 about 53% of total shark landings were caught by gill net (shark net) followed by hook and lines (34%) and trammel net (8%) and the minimal catch was from set bag net (5%). They also predicted that most sharks were caught 10-40 meter depth as major landings came from mechanized boats. This was one of major causes of smaller size shark catches.

Total yield and landed number in the year of 2014 was 93.16 MT and 211,652 respectively from two landing centers which was lower comparing with 136.45 MT yield and 449,133 landed number found by [17]. Catch records clearly reflect declining trends. The highest and the lowest landing were in the month of July and November (2014) respectively which was different from the study of [16] who showed the highest landing was found in the month of October 2011 and lowest in the month of January 2012.

During the study, it was observed that some shark species were abundant and some were less abundant in the fish landing centers in Chittagong and Cox's Bazar districts [16] found the most common and widely distributed two shark species were *Scoliodon laticaudus* and *Sphyrna lewini* which were also common in the present study. They also found five relatively common species which were *Rhizoprionodon acutus, Chiloscyllium indicum, Galeocerdo cuvier, Carcharhinus melanopterus,* and *Carcharhinus falciformis* but in the present study *Rhizoprionodon acutus, Galeocerdo cuvier*, and *Carcharhinus falciformis* were relatively common though not frequently caught. They showed rarely found species were *Rhizoprionodon oligolinx, Carcharhinus leucas and Carcharhinus sorrah though Chiloscyllium indicum, Carcharhinus leucas, Carcharhinus melanopterus, Carcharhinus sorrah* were rare in the present study and *Rhizoprionodon oligolinx* was totally uncommon [17] stated that the least common species in the catch gives cause for some concern and requires investigation on their population status. They also noted that changes in species contribution takes long time and require many years' data to draw conclusion and there may be other factors like changes in fishing effort and fishing practices having significant influence on catch.

There were some shark species that were available at least 5-10 years ago, but now they are not available and some are vulnerable. Totally absent Milk shark in Cox's Bazar and spot tail shark in Chittagong might be due to their late maturity, highly mobile and migratory characters; and harvesting of small sizes did not possible to renew the stock quickly like other shark species.

From the study, Shark abundance was found to be the highest during October to December, and lowest during January-March whereas [6] found the peak during January-March and lowest during July-September in their study during (2003-2013) conducted in those landing centers. There was a clear indication of change in shark seasonal abundance which might be due to the large exploitation of Dog shark during October to December than the previous years. Besides, many species of sharks and rays are highly seasonal and erratic in their occurrence i.e., vary over geographical locations, therefore, country wide and regional catch record is very important to track changes in elasmobranches diversity [16].

In the present study, catch compositions varied from month to month of the sampling year [7]. Stated that shark fishing was done throughout the year but the main season was November to March and a peak was found in June, in the present study it was found October to December and a peak in November. Similar study of [9] showed that the main shark fishing season was November to March and a peak was found in July where in the present study no fishing was found to be done in July due to heavy rainfall.

Increased number of active fishing days and new fishing techniques indicated that all shark species were in high fishing pressure. Many shark populations have declined where they were once common due to increased human pressure.

Conclusion

Sharks were mainly harvested by shark nets and hooks and lines as target species. But as by catch some new borne juveniles harvested by shrimp and fish trawler were discarded as a trash for very small size and low market value and not recorded. Catch record clearly reflects declining trends and bulk of the catch were small sized due to overfishing and lack of gear size limitation. A clear deviation in shark seasonal abundance was observed. Decreasing in shark species composition was also found which might pose serious threats to shark population and total yield. Steps should be taken to ensure maximum sustainable yield and conservation through the inclusion of shark in the Fish Act restricting overexploitation or illegal exploitation.

Acknowledgements

The author expresses his gratitude towards Bikram Jit Roy, Scientific Officer, Marine Fisheries Survey Management Unit, Chittagong, Bangladesh for providing necessary data. Thanks to Roaim Ahmed Hridoy for writing assistance.

References

1. Hoq ME, Haroon MKY, Karim E (2014) Shark fisheries status and management approach in the Bay of Bengal, Bangladesh 233-246.
2. FAO (2014) The State of World Fisheries and Aquaculture, Rome p. 223.
3. Ebert DA, Stehmann M (2013) Sharks, Batoids and Chimaeras of the North Atlantic.
4. Martin R, Aidan (2005) Conservation of freshwater and euryhaline elasmobranchs: A review, J Mar Biol Ass UK 85: 1049-1073.
5. Roy, Bikram Jit, Nripendra Kumar Singha, Hasan Ali SM, Gaziur Rhaman Md (2012) Availability of Vulnerable Elasmobranches in the Marine Water of Bangladesh. Bangladesh J Zool 40: 221-229.
6. Roy, Bikram Jit, Nripendra Kumar Singha, Gaziur Rahman Md, Fukrul Alam Md (2015) In the Bay of Bengal of Bangladesh Region Shark Fisheries Exploitation, Trade, Conservation and Management. International Journal of Comprehensive Research in Biological Sciences 2: 54-65.
7. Roy, Bikram Jit, Dey MP, Alam MF, Nripendra Kumar Singha (2007) Status of shark fishing in the marine water of Bangladesh. UNEP/CMS/MS/Inf 10.17.
8. Stevens JD, Bonfil R, Dulvy NK, Walker PA (2000) The effects of fishing on sharks, rays and chimaeras (chondrichthyans) and the implications for marine ecosystems. ICES Journal of Marine Science 57: 476-494.
9. Haldar GC (2010) National plan of action for shark fisheries in Bangladesh 75-89.
10. Roy, Bikram Jit, Hasan Ali SM, Nripendra Kumar Singha, Gaziur Rahman Md, et al. (2014a) Sharks and Rays Fisheries of the Bay of Bengal at the landing centers of Chittagong and Cox's Bazar, Bangladesh. World J Biol Med Science 1: 1-14.
11. IUCN (2000) Red book of threatened fishes of Bangladesh, IUCN Bangladesh Country office, Dhaka, Bangladesh p. 116.
12. Rahman AKA, SMH Kabir, Ahmad M, Ahmed ATA, Ahmed ZU, et al. (2009) Encyclopedia of Flora and Fauna of Bangladesh. Marine Fishes, Asiatic Society of Bangladesh, Dhaka 24: 485.
13. Quddus MMA, Sarker MN, Banarjee AK (1988) Studies on the Chondrichthyes fauna (sharks, skates and rays) of the Bay of Bengal. J NOAMI 5: 19-39.
14. DAY F (1878) The fishes of India being a natural history of the fishes known to inhabit the Seas and Fresh water of India, Burma and Ceylon. New Delhi: 730-740.
15. Roy, Bikram Jit, Fokhrul Alam MD, Gaziur Rhaman MD, Nripandra Kumar Singha, et al. (2014b) Landing Trends, Species composition and Percentage composition of Sharks and Rays in Chittagong and Cox's Bazar, Bangladesh. Int J Adv Res Biol Sci 1: 81-93.
16. Roy, Bikram Jit, Hasan Ali SM, Nripendra Kumar Singha, Gaziur Rahman Md, et al. (2014c) Sharks and Rays Fisheries of the Bay of Bengal at the landing centers of Chittagong and Cox'sBazar, Bangladesh. International Journal of Agricultural and Soil Science. 2: 48-58.
17. Anderson ED (1985) Analysis of various sources of pelagic shark catches in the Northwest and western central Atlantic Ocean and Gulf of Mexico with comments on catches of other large pelagics. Shark catches from selected fisheries off the US East Coast (NOAA) Technical Report NMFS pp: 1-14.
18. Roy, Bikram Jit, Hasan Ali SM, Nripendra Kumar Singha, Gaziur Rahman Md, et al. (2014c) Sharks and Rays Fisheries of the Bay of Bengal at the landing centers of Chittagong and Cox'sBazar, Bangladesh. International Journal of Agricultural and Soil Science 2: 48-58

24

Seasonal Histological Changes in Gonads of the Catfish (*CLARIAS LAZERA*)

Mahmoud Abdelghagffar Emam[1]* and Badia Abughrien[2]

[1]*Histology and Cytology Department, Faculty of Veterinary Medicine, Benha University, Egypt*
[2]*Histology and Anatomy Department, Faculty of Veterinary Medicine, Tripoli University, Libya*

Abstract

Forty mature catfish of both sexes (n=10 per season; 5 males and 5 females) were used to study the effect of different seasons on the histological and histochemical structure of the gonads. The histological results showed that both testes and ovaries of the catfish were degenerated during winter that was considered as a resting season of the catfish gonadal activity. Both testes and ovaries began to restore their intact and fully mature structure during spring and continue the same during summer where the testes show distended seminiferous lobules with all spermatogenic cells and spermatozoa also, the ovaries showed the different developmental stages including mature follicles therefore, both spring and summer were considered as spawning season of the catfish. During autumn, both testes and ovaries appeared as spent gonads where the testes showed many empty seminiferous lobules and the ovaries showed many atretic follicles therefore, autumn was considered as post-spawning or spent season. The results of Gonado-Somatic Index (GSI) were coincided with the histological structure of the gonads where they show peak value during spring and summer (spawning season) and showed the lowest value during winter (resting season).

Keywords: Histology; Testis; Ovary; Catfish; Seasons

Introduction

Catfish is considered as the cheapest source of high quality animal protein and rich in calcium, phosphate, iodine and vitamins [1]. Generally, fish is not only used for human consumption, but also used as a good source of animal meal [2]. The study of the gonads of the different teleost fish attracts the attention of several investigators in the *African catfish, Clarias gariepinus*; in *Bagrus docmac and Bagrus bayad*; in *Nile tilapia, Oreochromis niloticus*; in *mullet, Mugil cephalus* [3-9]. The purpose of this work was to study the effect of different seasons of the year on the histological and histochemical structure of the gonads of the catfish.

Materials and Methods

Animals

A total of 40 mature catfish of both sexes (n=10 per season; 5 males and 5 females) were collected alive from Toukh fish market in Kalyubia governorate, Egypt during the different seasons of the year (2010-2012). Winter season in Egypt extends from December to February, spring extends from March to May, summer extends from June to August, and autumn extends from September to November. The fish were captured for marketing from River Nile in Al-Qanater, Kalyubia governorate, Egypt. The fish were captured for marketing from River Nile in Al-Qanater, Kalyubia governorate, Egypt.

Preparation of tissue specimens

The catfishes were transported alive to the Histology and Cytology Department, Faculty of Veterinary Medicine, Benha University. Immediately, the fish was weighted, decapitated and the gonads were removed and weighed after opening the fish belly. After weighing, the gonads were immediately fixed in Bouin's solution for 24 hours, dehydrated through ascending grades of ethanol, cleared in xylene, embedded in paraffin wax and sectioned at 5 µm. The sections were stained with Haematoxylin and Eosin, Periodic acid Schiff (PAS), and Crossman's trichrome. Fixative and staining methods were used as outlined by Bancroft et al. [10].

The Gonado-Somatic Index (GSI)

The GSI (according to Dougbag et al.) was used for following up the seasonal variations in the gonads by the formula:

Weight of gonad

GSI = ×100 [11]

Fish body weight

Results

Our results revealed that the gonads of catfish revealed variable structure according to seasons of the year.

Testis

During winter, testis showed very thick tunica albuginea and interstitial connective tissue and showed degenerated spermatogenic cells except spermatogonia (Figure 1). During spring, testicular lobules showed all spermatogenic developmental stages (Figure 2) and continue the same structure during summer where testicular lobules were filled with spermatozoa also, showed thin tunica albuginea and interstitial connective tissue compared to (Figure 3). During autumn, many testicular lobules appeared empty from speermatozoa while, other spermatogenic cells were present and began to degenerate (Figure 4).

Ovary

During winter, ovary showed very thick tunica albuginea and

***Corresponding author:** Mahmoud Abdelghagffar Emam, Histology and Cytology Department, Faculty of Veterinary Medicine, Benha University, Egypt
E-mail: mahlasm@yahoo.com

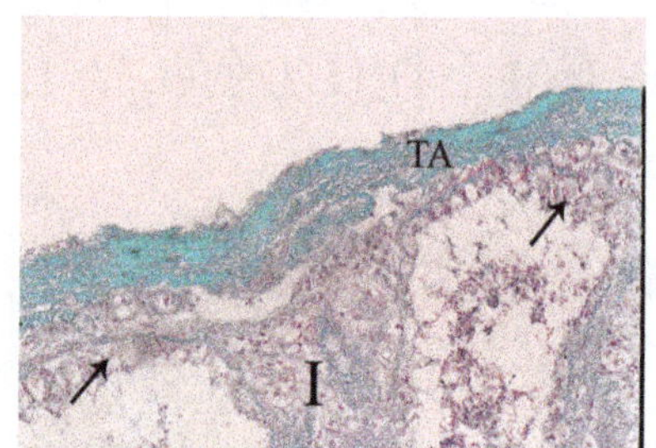

Figure 1: Section of catfish's testis during winter showing, thick tunica albuginea (TA), increased interstitial connective tissue (I), and intact spermatogonia (arrow). Crossman's trichrome (×200).

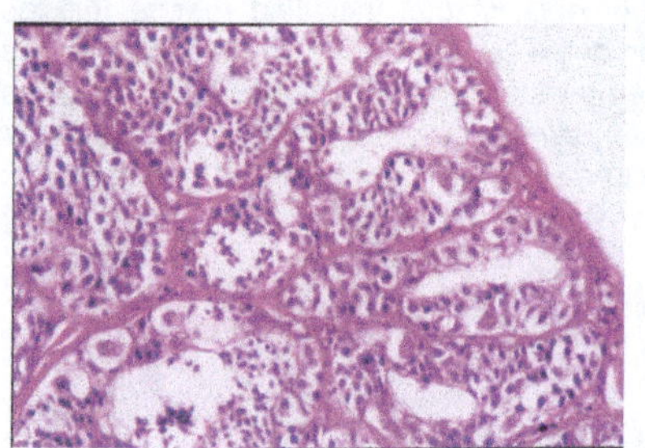

Figure 2: Section of catfish's testis during spring showing testicular lobules filled with all spermatogenic cells. H&E (×200).

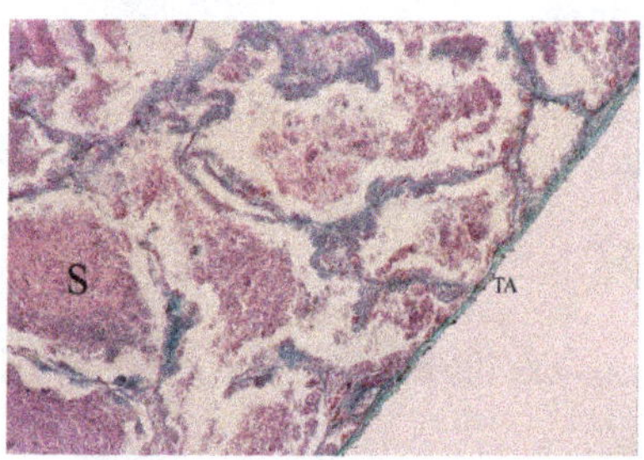

Figure 3: Section of catfish's testis during summer showing, thin tunica albuginea (TA) compared to Figure 1. Testicular lobules are filled with spermatozoa (S). Crossman's trichrome (×100).

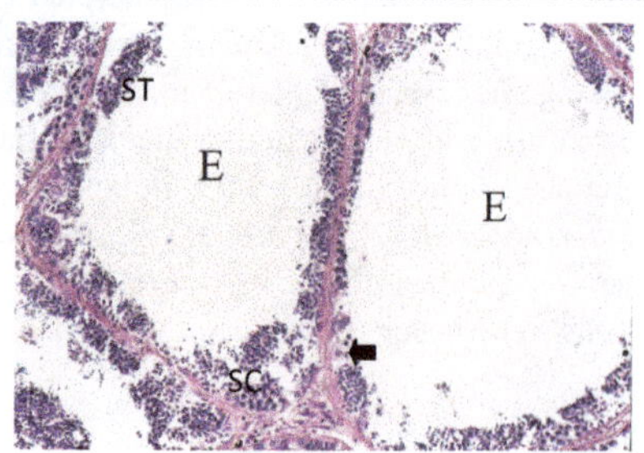

Figure 4: Section of catfish's testis during autumn showing, empty testicular lobules (E). Notice all developmental stages; spermatogonia (arrow), spermatocytes cyst (SC), spermatid cyst (ST). PAS technique (×200).

interstitial connective tissue and showed many degenerated follicles which mainly were of pre-vitellogenic type (Figure 5). During spring, ovary showed all the developmental stages but, the common stages were vitellogenic and post-vitellogenic ones (Figure 6) and continue the same structure during summer where post-vitellogenic (mature) follicles were usually seen, the ovary showed thin tunica albuginea (Figure 7). During autumn, many atretic follicles began to appear and the pre-vitellogenic follicles began to increase in number again (Figure 8).

The results of GSI were coincided with the obtained histological results where the GSI of both sexes reached their lowest values during winter where the gonads, began to increase greatly during spring, reached their highest values during summer and began to decrease again during autumn (Table 1).

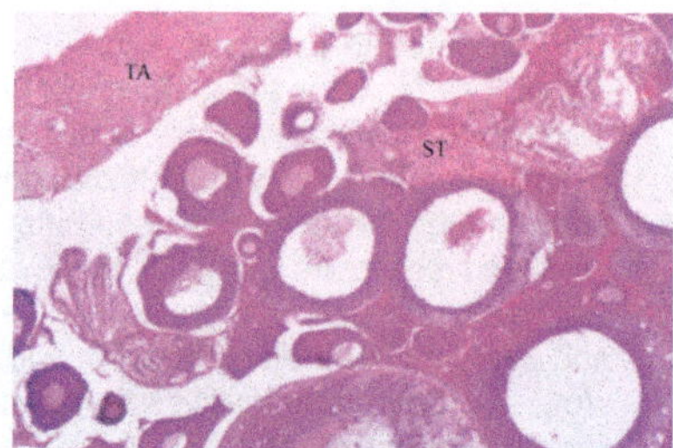

Figure 5: Section of catfish's ovary during winter showing, very thick tunica albuginea (TA) and increased stromal connective tissue (ST). Notice the degenerated follicles. PAS technique (X400).

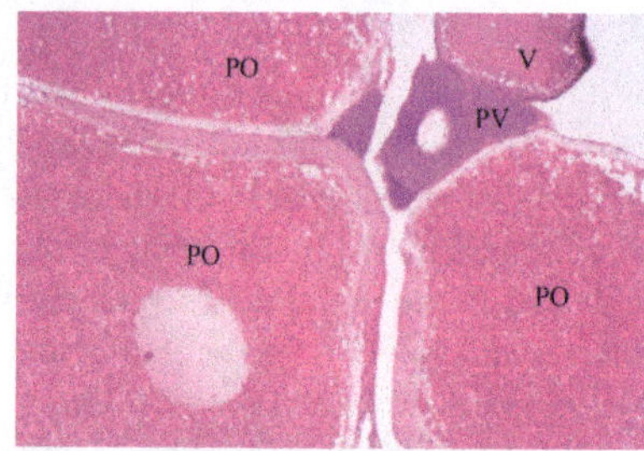

Figure 6: Section of catfish's ovary during spring showing, pre-vitellogenic follicles (PV), vitellogenic follicles (V) and many post-vitellogenic follicles (PO). H&E (X400).

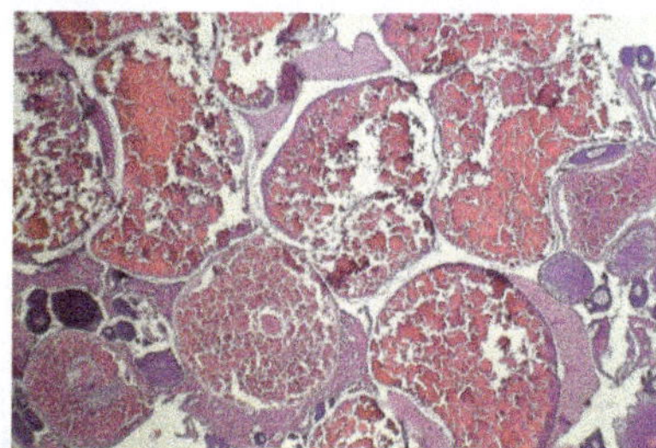

Figure 7: Section of catfish's ovary during summer showing, abundant mature follicles (MF). H&E (X40).

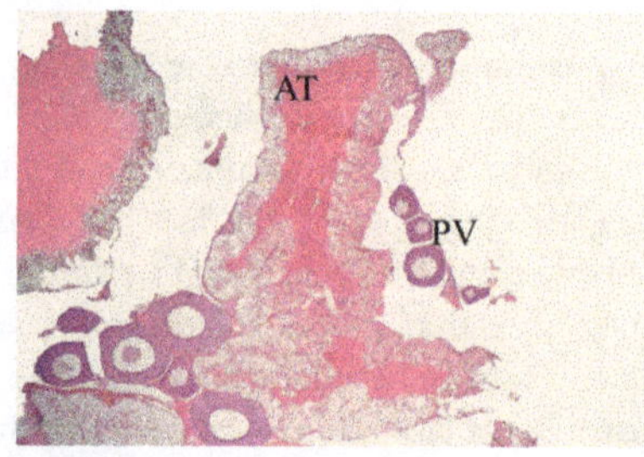

Figure 8: Section of catfish's ovary during autumn showing, predominance of atretic (AT) and pre-vitellogenic follicles (PV). H&E (X200).

	Winter	Spring	Summer	Autumn
Male catfish	1.07	3.6	4.0	2.2
Female catfish	9.3	26.4	30.7	18.8

Table 1: GSI values for male and female catfishes during different seasons of the year.

Discussion

The tunica albuginea of both testes and ovaries had no uniform thickness all over the year but, it reached a maximum thickness during winter and became thin during spring and summer due to pressure exerted on it by the distended testicular lobules or enlarged mature follicles, and began to increase again during autumn.

Our results revealed testicular lobules of variable shapes and sizes according to seasons, where it decreased in size during winter which was cold, rainy and short day light season (resting season) and thereafter increased during spring and reached a maximum size during summer which was hot, dry and long day light season (spawning season) as they were distended by different developmental stages of spermatogenic cells and they showed slight decrease again during autumn which was dry and less hot than spawning season (spent season) but still enlarged and larger as compared to structure during winter. This finding derives support from studies of Yoakim [12] in *S. schall*, Latif and Salem [13] in *L. nebulosus*, Resink et al. [14] in *C. gariepinus* and [6] in *Bagrus* species. Spermatogonia were found throughout the year, but they were abundant during the winter season to replenish the testes after this resting season where the testes showed degeneration and increased interstitial connective tissue. This finding is in conformity to those of Zaki et al. [15] and Resink et al. [14] in *C. gariepinus*, Dziewulska and Domaga [16] in salmonid and Guerriero et al. [17] in *L. cephalus*. In this study spermatocytes were found throughout the year, but they were abundant during the spawning season. Similar observations were also reported by Rosenblum et al. [18] in *I. nebulosus* and Gaber [6] in *Bagrus* species. The spermatids increased during spring to produce spermatozoa and became few during summer as most of them were changed into spermatozoa. The spermatozoa began to appear within lumen of testicular lobules during spring while during summer, the spermatozoa increased and the testicular lobules showed different activity where some lobules were filled with spermatozoa and others were spent or empty as they discharged their content during spawning. Similar findings have also been noticed by Latif and Salem [13] in *L. nebulosus*, Resink et al. [14] in *C. gariepinus* and Guerriero et al. [17] in *L. cephalus*. During autumn, most of the lobules were empty indicating spawned testes, but some testes were filled with spermatozoa.

The pre-vitellogenic follicles were found throughout the year, but were common in autumn as the spent ovaries and in winter as the resting ovary and less abundant during spring and summer (spawning seasons). This result was supported by studies of Ismail [19] in *C. lazera* and Dougbag et al. [11], Salem [20], and Abel El Hafez et al. [21] in other fish. The vitellogenic follicles decreased in autumn as the spawning begun to end and rarely observed during winter as it was resting season but, abundant during spring and summer (spawning season) in order to turn rapidly into mature follicles. Maddoch and Burton [22] reported that the oocytes undergoing vitellogenesis indicate spawning activity. The mature or post-vitellogenic follicles were common and abundant during spring and summer as they were in the the spawning and ready to spawn and ovulate, but rarely observed during the winter (resting) season. The atretic follicles were present throughout the year but they were abundant during autumn (spent ovaries) and winter (resting ovaries) and these atretic follicles indicated the spawned individuals. This finding is similar to those obtained by Yoakim [12] in *S. schall*, Gaber [6] in *Bagrus* and Merson et al. [23] in summer flounder.

Another method of studying the seasonal variations of the gonads was the values of Gonado-somatic index (GSI) for both male and female catfish which was used as an indicator of gonadal development as when the GSI reached a maximum value, this gave a perfect indication to the time of spawning. This was supported by Ismail [19] in *C. lazera*; Dougbag et al. [11] in *T. niloticus*, Khallaf et al. [24] and Gaber [6] in *Bagrus* species.

The results about GSI were coincided with the obtained histological results where the GSI of both sex reached their minimum values during winter season where the gonads were in resting season, began to increase greatly during spring season, reached their maximum values during summer where the gonads were in the spawning season and began to decrease again during autumn season. This indicated a long spawning season of the catfish extending from spring to summer and might extend into autumn depending on the temperature, but its peak was reached during summer season.

Conclusion

The histological results were coincided with GSI as both of them revealed that winter is the resting season of catfish's gonadal activity, spring and summer are the spawning seasons of catfish's gonads, and autumn is post-spawning or spent season of catfish's gonads. These information help in understanding the pattern of catfish reproduction in Egypt that may help in aquaculture of catfish.

References

1. Dadzie S, Wangila BB (1980) Reproductive biology, length weight relationship and relative condition of ponds raised, Tilapia zilli. J Fish Biol 17: 243-253.
2. Anderson DP,Mitchum DL (1974) Atlas of trout histology textbook. Wyoming Game & Fish Dept., Cheyenne, WY 82009, USA.
3. Deleeuw R, Goos HJ, Van oordt PG (1987) The regulation of gonadotropin release by neuro hormones and gonadal steroids in the *African catfish, Clarias gariepinus*. Aquaculture 63: 43-58.
4. Van Oordt PG, J Peute, R Van Den Hurk,WJ Viveen (1987) Annul corelative changes in gonads and pituitary gonadotropes of feral *African catfish, Clarias gariepinus*. Aquacultue 63:27-41.
5. Schulz RW, Renes IB, Zandbergen MA, J Peute, W Dijk, et al. (1997) Pituitary gonadotrophs are strongly activated at the beginning of spermatogenesis in *African catfish, Clarias gariepinus*. J Biol Rep 57: 139-147.
6. Gaber SA (2000) Biological, Histological and Histochemical studies on the Reproductive Organs and Pituitary gland of *Bagrus docmac and Bagrus bayad* in the Nile water, with special reference to the ultrastructure of supporting tissues. Ph.D. Thesis. Zagazig University, Egypt.
7. Mousa AM (1998) Immunocytochemical and Histological studies on the reproductive endocrine glands of the *Nile tilapia, Oreochromis niloticus* (Teleostei, cichlidae). J. Egypt Ger soc zool 27: 109-134.
8. Mousa SA,Mousa MA (1999) Immunocytochemical and histological studies on the hypophyseal-gonadal system in the freshwater *Nile tilapia, Oreochromis niloticus*, during sexual maturation and spawning in different habitats. J Exp Zool 343-354.
9. Mousa SA,Mousa MA (1998) Immunocytochemical studies on the Gonadotropic cells in the Pituitary gland of male *mullet, Mugil cephalus* during the annual reproductive cycle in both natural habitat and capitivity. J Egypt Ger Soc Zool 25: 59-74.
10. Bancroft JD,Cook HC, Stirling RW, Turner DR (1994) Manual of histological Techniques and their diagnostic application. 2nd Ed.,Churchill Livingston, Edinburgh, London, Madrid, Melbourne, New York and Tokyo.
11. Dougbag A, El-Gazzawy E, Kassem A, El-Shewemi S, Abd El-Aziz M et al. (1988) Histological and histochemical studies on the testis of Tilapia niloticus. II. Seasonal variations. Alex J Vet Sci 4: 59-69.

12. Yoakim EG (1971) Seasonal variations in the pituitary gland and gonads of the Nile catfish (Syndontus schall) in relation to its reproductive cycle. Ph.D. Thesis. Fac of Sci., Ain Shams University.

13. Latif AF, Salem SA (1983) Sexual cycle of Lethrinus nebulosus (Forsk.) in the Red Sea. II. Microscopic peculiarities of the testis. Egypt. J Histol 6: 141-158.

14. Resink JW, R Van Den Hurk,Voorthuis PK,Terlou M, Leeuw DE et al. (1987) Quantitative enzyme histochemistry of Steroid and glucuronide Synthesis in testes and seminal vesicle, and its correlation to plasma gonadotropin level in Clarias gariepinus. Aquaculture 63:97-114.

15. Zaki MI, Dowidar MN, Abdala A (1986) Reproductive biology of Clarias gariepinus (Syn, lazera) Burchell (clariidae) in lake Manzalah, Egypt. II structure of the testes. Folia Morphologica 43: 307-313.

16. Dziewulska K, Domaga AJ (2003) Histology of salmonid testes during maturation. Reprod Biol 3:47-61.

17. Guerriero G,Ferro R, Ciarcia G (2005) Correlations between plasma levels of sex steroids and spermatogenesis during the sexual cycle of the chub, Leuciscus cephalus L. (Pisces:Cyprinidae). Zoological Studies 44: 228-233.

18. Rosenblum PM,Pundy J, Callard IP (1987) Gonadal morphology, enzyme histochemistry and plasma streoid levels during the annual reproductive cycle of male and female brown bullhead catfish, Ictalurus nebulosus. J Fish Biol 31: 325-341.

19. Ismail RS (1992) Physiological study on spawning in some fishes (Clarias lazera). Master Thesis. Fac. of Vet. Med., Zagazig University, Egypt.

20. Salem SA (1991) On the sexual cycle of Lethrinus bungus (Ehrenb.) in the Egyptian Red Sea coast. I. Microscopic Peculiarities of the ovary. Egypt, J Histol. 14: 55-62.

21. Abd El-Hafez EA,Mahmoud DM, Ahmed SM,Hassan AH (2007) Histomorphological changes in the ovaries of Oreochromis niloticus during breeding and non breeding seasons. The 31st conference of the Egyptian society of Histol and Cytol P 28.

22. Maddock DM,Burton MP (1998) Gross and histological observation of ovarian development and related condition changes in American plaice. J Fish Biol 53: 928-944.

23. Merson RR,Casey CS,Martinez C, Soffientino B, Chandlee M et al. (2000) Oocyte development in summer flounder: seasonal changes and steroid correlates. J Fish Biol 57: 182-196.

24. Khallaf EA, El-Saadany MM, Authman M (1991) Oogenesis of *Bagrus* bayad (Forsk.). J Egypt Ger Soc Zool 4: 1-4.

Pathological Findings of Experimental *Aeromonas hydrophila* Infection in Golden Mahseer (*Tor putitora*)

Rohit Kumar[1], Veena Pande[2], Lalit Singh[1], Lata Sharma[1], Neha Saxena[1], Dimpal Thakuria[1], Atul K Singh[1] and Prabhati K Sahoo[1]*

[1]ICAR-Directorate of Coldwater Fisheries Research, Bhimtal, Nainital, Uttarakhand, India
[2]Department of Biotechnology, Kumaun University, Bhimtal, Nainital, Uttarakhand, India

Abstract

Introduction: Mahseers, belong to the family *Cyprinidae*, are well renowned for the excellent game as well as food fishes. Various aspects simultaneously considered towards the domestication of this fish species also include the study of fish health issues, preventive and remedial programmes for sustainable aquaculture. Experimental studies can be conducted in the control condition to study the pathological symptoms during bacterial infection. Considering the bacterial pathogens as a major constraint for aquaculture, the present study was carried out to standardize the bacterial concentration of *A. hydrophila* for the experimental challenge of Golden Mahseer for the first time, and various clinical changes were also monitored during the infection.

Materials and methods: A total of 140 live fishes were acclimatized to the laboratory conditions for experimental study. A constant water quality was monitored during the acclimation period and throughout the experiment. The test group was injected with *A. hydrophila* and various physiological, biochemical and tissue level changes were observed during the experiment in comparison to phosphate buffer saline (PBS) injected control group.

Results: LD_{50} value of *A. hydrophila* for Golden Mahseer was successfully standardized. Clinical signs including weakness, slower movement, swimming closer to the surface, fin haemorrhages and red patches at the gut region were observed. Enlargement of spleen followed by tissue necrosis along with signs of haemorrhagic septicaemia was also seen in infected fishes. The necrosis of hepatocytes was observed in the histological section of liver tissue. The bacterial infection increases the superoxide dismutase activity and cortisol level in Golden Mahseer.

Conclusion: The LD_{50} value of *A. hydrophila* for experimental challenge studies on Golden Mahseer is 1.74 × 10^5 cfu per 100 g of body weight. The symptoms of severe bacterial haemorrhage were observed. Necrosis of hepatocytes was observed in infected fishes. The superoxide dismutase activity and cortisol level also increased in infected fishes. *A. hydrophila* was confirmed to be the etiological agent which was re-isolated using spread plate method and confirmed by 16s rRNA sequencing.

Keywords: Bacterial infection; Experimental study; Bacterial haemorrhagic septicaemia; Disease symptoms

Introduction

Mahseers, belong to the family *Cyprinidae*, are well known for the excellent game as well as food fishes all over the world especially the South-Asian countries [1-3]. In India, eight mahseer species are available among which Golden Mahseer, *Tor putitora* is considered the candidate species for hill aquaculture in the mid-Himalayan range [3-5]. The artificial breeding, seed production and scientific management of Golden Mahseer are in high priorities [6,7]. Scientific management including water quality, feed and health management are also equally important for sustainable aquaculture [8,9]. However, the data regarding the health management, mainly diseases is least available for this important fish species [2,10].

Aquaculture is primarily affected by microbial pathogens especially of bacterial origin [11], wherein, *Aeromonas hydrophila* and other *Aeromonas* are responsible for the majority of diseases [12]. *A. hydrophila* is reported being primary bacterial pathogen affecting freshwater fishes like *Clarias gariepinus* [13], *Labeo rohita* [14], *Sparus aurata* [15], *Magalobrama amblycephala* [16]. The severity of pathogenesis or disease outbreak can be moderate to high in fishes [17], but every type of disease results in deterioration of product and economic losses.

Aeromonad septicaemia is characterized by diverse pathological symptoms such as dermal ulceration, fin haemorrhages, fin rots, red sores, haemorrhages and necrosis of the visceral organs, etc. [17-19]. The acute form of the disease may result in fatal sepsis without any symptoms [17] while chronic infections may show the symptoms like hemorrhagic septicaemia with ulceration, inflammation, and dermal lesions [18]. Liver and kidney tissues are the primary targets of bacterial accumulation and sepsis [17,20]. The liver may become pale and show greyish to greenish coloration with foci while kidney may engorge and become friable [17,21]. Despite the knowledge of severity and symptoms of the *Aeromonad septicaemia* in other species, factors such as host-pathogen interaction, temperature requirement of the bacteria, course of pathogenesis, and immunity to pathogenesis may vary for fish species [12,22].

The present study was aimed to examine the effect of *Aeromonas hydrophila* infection on Golden Mahseer. Various macroscopic and microscopic changes in tissues and stress levels were investigated to understand the fish response to bacterial infection. As the liver tissue is

***Corresponding author:** Prabhati K. Sahoo, Principal Scientist, ICAR-Directorate of Coldwater Fisheries Research, Bhimtal (263136), Nainital, Uttarakhand, India
E-mail: psahooin@yahoo.co.in

the primary organ affected by bacterial infection, microscopic changes in hepatic cells were also examined to ascertain the level of cell necrosis due to pathogenesis.

Materials and Methods

Ethical approval

The institutional ethical committee approved the protocols for maintenance, handling during experiments, and sacrificing of the fishes.

Experimental fish

Live fishes (size=33.32 ± 5.96 g; 146 ± 15.17 mm) were collected from Kosi River, Ramnagar (29.40°N 79.12°E) using a cast net. The fishes were initially acclimatized in ponds for three weeks and then in fibre reinforced plastic tanks of 2000 L capacity (experimental tanks) for two weeks before the experiment. Fishes were fed twice, daily with a formulated diet [23] based on 5% of the body weight. Constant aeration and water flow were maintained in experimental tanks. Water temperature and physicochemical parameters were recorded at regular intervals using multi-parameter auto-analyser (Hach®, Colorado, US) and colorimetric methods.

Bacterial strain and determination of LD_{50}

The bacterial strain (originated from carp fishes) was obtained from Fish Pathology Laboratory (ICAR-CIFA, Bhubaneswar) and grown in our laboratory using TSB medium (Himedia, Mumbai, IN) for 24 h at 25°C. Single colony isolated on TSB agar was inoculated in TSB broth. The culture of *A. hydrophila* was grown overnight, and 100 µl of culture was inoculated into the fresh broth. Optical density was measured at regular intervals up to OD_{600}=0.7 and culture was transferred to ice. The broth culture was centrifuged at 5000 × g for 15 min in a cooling (4°C) centrifuge (Eppendorf, Hamburg, DE). The pellets were washed twice with sterile phosphate buffer saline (PBS), pH 7.4 (Himedia, Mumbai, IN) and centrifuged. The final suspension was used to prepare 10 fold dilutions in 20 mL PBS having 10^6 cfu, 10^5 cfu, 10^4 cfu and 10^3 cfu mL^{-1}. The dilutions were spread plated on TSB agar plates in duplicate for each dilution. The colony forming unit (cfu) count at optical density 600 nm was standardized with the same procedure before the experiment in three consecutive trials.

For the pathogenicity challenge test, we followed intra-peritoneal injection method using 1 ml Insulin syringe. After acclimatization in the aquarium for five days with proper aeration and feed, fish samples (33.32 ± 5.96 g) were separated into five different groups (12 fishes in each group). Four groups received 10^6 cfu, 10^5 cfu, 10^4 cfu and 10^3 cfu of bacterial culture per 100 g body weight respectively while the fifth group was injected with 100 µl of PBS. The samples were observed up to 96 h, and dead fishes were removed for routine bacteriological examination. Final LD_{50} values were estimated according to the method of Miller and Tainter [24]. The results were calculated as mean values from two independent experiments.

Pathogenicity experiment

A consecutive experiment was conducted in two different groups of fishes with 35 samples in each using the standardized LD_{50} dose to test the pathogenicity. Test group were injected with *A. hydrophila* intraperitoneally according to their body weight while the control group was injected with 100 µl of PBS. The fishes were observed for any physiological and pathological symptoms at various time intervals. Three samples from test group were anesthetized using tricaine methanesulfonate (200 mg L^{-1}) (Sigma-Aldrich, St. Louis, US) and blood was drawn from a caudal vein in 3.8% sodium citrate containing vials. The blood samples were centrifuged at 7000 × g for 10 min and plasma was collected in the fresh tube. Plasma was stored at -80°C until further use. The plasma samples were used to estimate the levels of superoxide dismutase (SOD) and cortisol using ELISA and colorimetric kits (Cayman, Ann Arbor, MI), respectively. The statistical analyses were performed by the standard curve plot of known concentration of SOD and cortisol. The data was analysed using independent sample t-test (α=0.05). Liver, head kidney and spleen tissues from each sample collected at different time intervals were also plated on TSB agar plates (amp+) to detect the presence of bacteria.

Identification of reisolated *A. hydrophila* was also done by PCR amplification and sequencing of 16s rRNA. The genomic DNA was isolated from the randomly selected clones (6 no) grown on TSB agar plates and 16s rRNA was PCR amplified using the universal primer pairs (16s rRNA F- AGAGTTTGATCCTGGCTCAG; 16s rRNA R-GGTTACCTTGTTACGACTT). The PCR reaction was performed in a total reaction volume of 50 µl containing 1 × PCR Buffer, 10 pM of each primer, 200 µM each dNTP, 50 ng template DNA and 1U Taq DNA Polymerase (Invitrogen, Waltham, USA). The reaction conditions were 94°C for 4 min; 35 Cycles of denaturation at 94°C (30 s), annealing at 58°C (30 s) and extension at 72°C (90 s) followed by a final extension at 72°C for 10 min. The PCR products (Figure 1) were sequenced using Big Dye Terminator v3.1 cycle sequencing kit in ABI3730 Genetic Analyser at Scigenom Laboratory, Kochi. The nucleotide sequences so obtained were matched with NCBI GenBank Blastn toolkit using default settings.

Histopathology

Liver tissues collected at 48 h were processed for histological examination. Tissue samples were fixed in Bouin's Fixative and prepared for histological examination using recommended procedure [25]. Tissues were further embedded in paraffin wax and sectioned at 5 µm in a rotary microtome (Thermo Fisher Sci, Massachusetts, USA). The slides were stained with haematoxylin and eosin and observed under Leica DM2500 (Leica, Wetzlar, DE) to study any pathological changes/difference in tissues after bacterial infection.

Results

Bacterial culture and experimental conditions

Aeromonas hydrophila is successfully grown at 25°C, but no growth was observed below 18°C and above 30°C. Correlation study between OD_{600} and cfu resulted that there are ~10^6 cfu mL^{-1} of *A. hydrophila* at OD_{600}=0.7. All samples were healthy, and their movement was active.

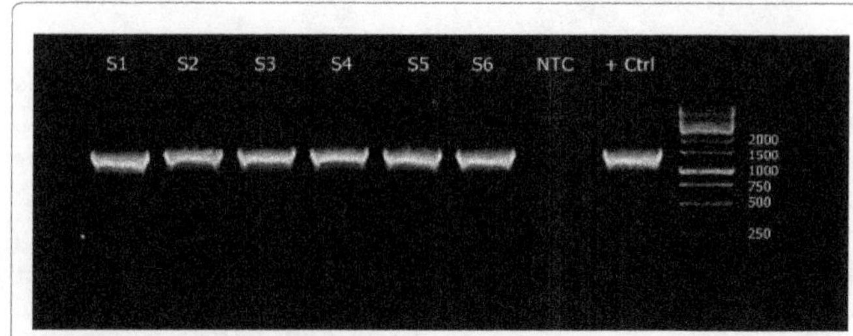

Figure 1: PCR amplicon (16s rRNA) amplified from bacterial samples to identify the etiological agent in the test group (S1 to S6–re-isolated samples from the test group, NTC-non-template control, +ctrl-the stock culture of *A. hydrophila*).

During the experiment, the ambient temperature was 22-24°C, water pH was 7.5-8.0, total hardness 80-100 mg L^{-1}, dissolved oxygen ranged between 5.2-7.4 mg L^{-1} and ammonia was <0.1 mg L^{-1}. The LD_{50} value determined using log-probit calculation method of Miller and Tainter [24] is 1.74×10^5 cfu per 100 g of body weight of the fish (Table 1 and Figure 2).

Identification of *Aeromonas* using 16s rRNA

An amplicon of 1240 bp was successfully amplified from 6 randomly selected bacteria isolated from liver (S1-S2), head kidney (S3-S4) and spleen tissues (S5-S6) (Figure 1). The bacteria were confirmed as *A. hydrophila* based on nucleotide sequence blastn similarity (score-2202; Ident-99%; e-value-0.0) with 16s rRNA sequence of *A. hydrophilia* in NCBI GenBank database.

Macroscopic and microscopic changes on bacterial infection

Clinical signs including weakness, slower movement, swimming closer to the surface, fin haemorrhages and red patches at the gut region were observed. In the morphological examination, the hyperemia of fins was first observed at 6 h post infection (p.i.); initially on pelvic fins followed by prominent haemorrhages at caudal fin at 48 h. Darkening of the skin and red patches were observed on the ventral side of the gut region along with prominent haemorrhagic signs at the anal orifice (Figure 3). The dorsal skin and eyes did not reveal any significant symptoms of infection. Gills also did not show any haemorrhagic symptoms up to 12 h p.i.; however gills became pale after 24 h and haemorrhages appeared at 48 h p.i. The liver was seen to be yellowish brown having haemorrhagic symptoms on the surface after 6 h. The liver was pale, and greyish white foci were observed at 12 and 24 h p.i. The degenerative changes in the liver with focal necrosis of hepatocytes were observed histologically at 48 h p.i. (Figure 4). Dark red colouration and engorged spleen were observed however the spleen was friable at 48 h p.i. Haemorrhagic foci were found in internal organs while intestine was filled with yellow coloured mucoid liquid. Gall bladder was enlarged and filled with emerald-green secretion. The haemorrhage symptoms over internal organs at macroscopic and microscopic levels revealed a visceral haemorrhagic septicaemia (Figures 4A-4C).

Stress and antioxidant activity matrices

The bacteria challenged (test group) fishes showed the stress symptoms like slow swimming and loss of appetite. The symptoms were notable after 12 h p.i. and were consistent up to 96 h p.i. or morbidity. Mean superoxide dismutase activity in control samples was 2.26 U mL^{-1}. The superoxide dismutase (SOD) activity of test samples increased significantly over control group [$t(29)=3.9$, $p<0.05$]. SOD activity increased at 3 h and was maximum at 24 h (9.81 U mL^{-1}); followed by a decrease in activity at 48 h or 96 h p.i. (7.22 and 4.40 U mL^{-1} respectively) (Figure 5a). The plasma cortisol concentration of Golden Mahseer in PBS inoculated fishes was 43.24 ng mL^{-1}. The plasma cortisol level at various time intervals showed a significant increase in test group [$t(29)=2.02$, $p<0.05$]. The level of cortisol was highest at 24 h p.i. (79.56 ng mL^{-1}) and gradually decreases at 48 h and 96 h p.i. (62.98 and 42.71 ng mL^{-1} respectively) (Figure 5b).

Group	Dose	Log Dose	Sample Size	Moribund fishes	Moribund fishes (%)	*Corrected %	Probits
1	1.00E+07	7.11	12	12	100	98	7.05
2	1.00E+06	6.11	12	10	83.33	83	5.95
3	1.00E+05	5.11	12	7	58.33	58	5.2
4	1.00E+04	4.11	12	1	8.33	8	3.59
5	1.00E+03	3.11	12	0	0	2	2.95

*Corrected % Formula for 0 and 100% mortality: For 0 % dead–100(0.25/n); For 100 % dead–100(n-0.25/n).

Table 1: Results of the lethal doses of *Aeromonas hydrophila* for the determination of LD_{50} after intraperitoneal injection in fishes.

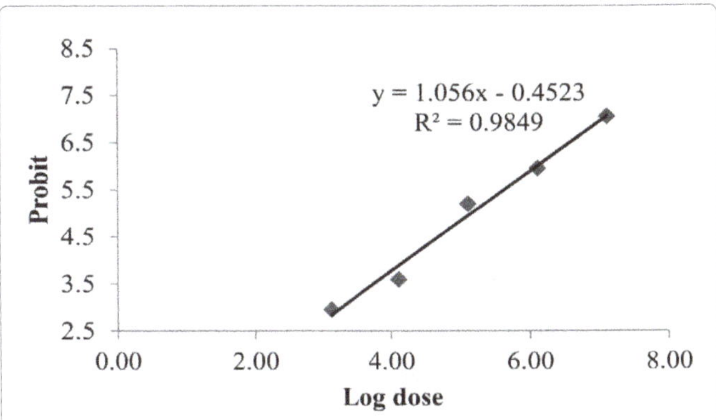

Figure 2: The plot of log doses of bacterial cfu per 100 g of body weight versus probits to calculate the LD_{50} dose of *Aeromonas hydrophila*.

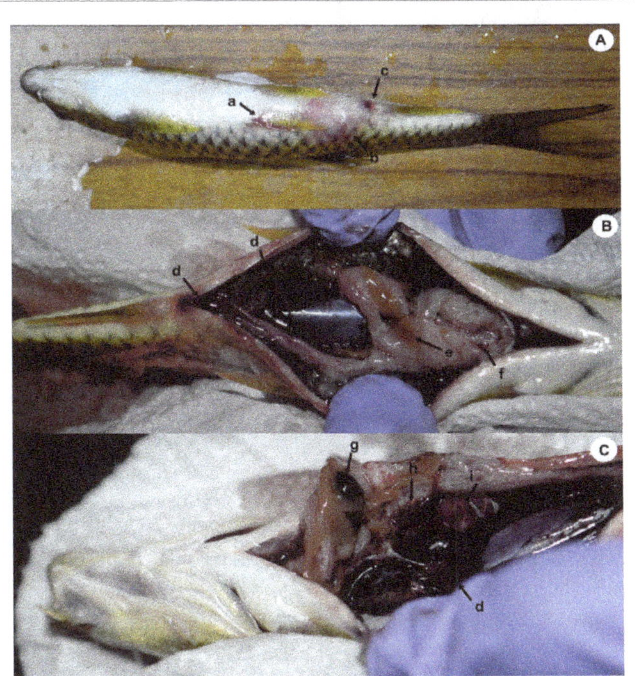

Figure 3: The pathological symptoms of Aeromonas infection in test samples. a–hyperemia at the base of pelvic fins; b–haemorrhages signs at the abdominal muscle and darkening of skin; c–haemorrhages signs at anal orifice; d–internal bleeding and haemorrhages; e–Greyish white foci in liver and colour is paler; f–Reddish colour haemorrhagic foci on intestine; g–enlarged gall bladder filled with emerald-green secretion; h–enlargement of spleen with deep red colouration; i–Haemorrhage symptoms on gut muscles.

Discussion

Aeromonas hydrophila is the primary bacterial pathogen for many fish and aquatic organism, and thus, it is a great concern for the aquaculture industry [12,26]. The fish under the present study is also a valued and potential species for aquaculture in the Himalayan cold water region. Therefore, the present study was aimed to examine the effect of *Aeromonas hydrophila* infection on Golden Mahseer.

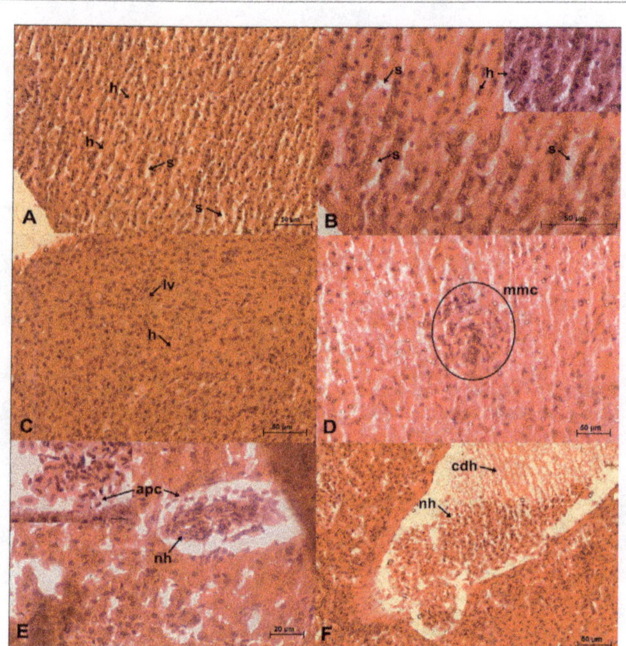

Figure 4: Liver microphotographs of Golden Mahseer showing normal hepatocytes (A&B) and hepatocytes after bacterial infection (C-F). A–Normal hepatocytes (h) and sinusoids (s) H&E × 100; B–Normal hepatocytes (h) and sinusoids (s) at higher resolutions H&E × 400 (Inset H&E × 1000); C–Shrunken hepatocytes and loss of vacuolation (lv) due to glycogen depletion H&E × 100; D–Formation of melanomacrophage centre (mmc) due to bacterial infestation (bi) marked by circle H&E × 200; E–Necrosis of hepatocyte (nh) and inset showing accumulation of phagocytic cells (apc) at the site of necrosis H&E × 400; E–Heavy necrosis of hepatocytes (nh) along with the complete degradation of hepatocyte (cdh) H&E × 100.

Initially, we standardized the suitable temperature of 25°C for the growth of *Aeromonas hydrophila* in the laboratory condition as the ambient temperature in hill environment remains low (-2°C-25°C in different seasons). Many authors observed the growth of this bacterium within the range of 27°C-30°C [12,27] which we could not see in our conditions. The fish-pathogenic *Aeromonas* are also reported to be psychrophilic in nature and was isolated from diseased rainbow trout that lives at very low temperature (5°C-20°C) [27,28].

The level and dose of pathogenicity of the bacteria in several carp species were also found to be variable [26,29-31]. Therefore, the optimum dose of bacterial inoculation and pathogenicity was determined for the first time in this species and how this dose of bacteria interacts with the physiological mechanism was determined. Determination of LD_{50} in study organism before the experimental challenge is advantageous for successful experiment and induction of clinical signs and symptoms [32]. Where higher dose may cause mortality, the suboptimal treatment may not produce the desired symptoms. In the present study, an LD_{50} value of 1.74×10^5 cfu per 100 g of body weight was standardized for *A. hydrophila* in Golden Mahseer. LD_{50} value of *A. hydrophila* was previously determined for varies model and non-model fish species [26,30,33,34]. The standardized LD_{50} value of *A. hydrophila* for *invivo* experiments would serve as a baseline data for future immune response studies.

The intraperitoneal inoculation of *A. hydrophila* produced similar signs and symptoms, reported during the disease progression in other fish species [33-35]. Clinical signs including weakness, slower movement, swimming closer to the surface, fin haemorrhages and red patches at the gut region were observed during 24-48 h p.i. Similar observations were documented in *A. hydrophila* infected catfish [21]. The fishes showed noticeable signs of haemorrhages and abdominal swelling. Large external ulcerative lesions also developed in the injected area and around the anal orifice. Histological lesions were observed in liver (Figure 4) and the spleen, and kidney tissues were found diffused due to acute infection. Sundus and Jamal highlighted similar observations like structural fragmentation and necrosis of liver tissue and friable kidney [36]. *Aeromonas hydrophila* may itself initiate pathological processes and induce inflammation [32]. However, the role of extracellular products released during the infection may also play a significant role in causing degeneration and necrosis [37]. Also, histological symptoms were not performed in other tissues as the liver was the primary site of infections due to *Aeromonas* infection [17]. The infection was also very high in the liver in comparison to other tissues when plated on Agar plates at each time points of sample collection.

Dermal lesion with haemorrhagic foci may be correlated to bacterial haemorrhagic septicaemia mainly caused by *Aeromonas* sp. [21]. The study findings are similar to Cipriano and Laith & Najiah who observed the dermal ulcerative lesions, focal haemorrhages and inflammation in chronic infection of *Aeromonas* sp [18,21]. The histological findings of the study are similar to Laith & Najiah [21] who reported that *A. hydrophila* causes haemorrhagic sepsis, liver necrosis along with the accumulation of phagocytic cells at the site of necrosis. The results also agree with the finding of Afifi that various toxins and extracellular products such as hemolysin, protease and elastase released

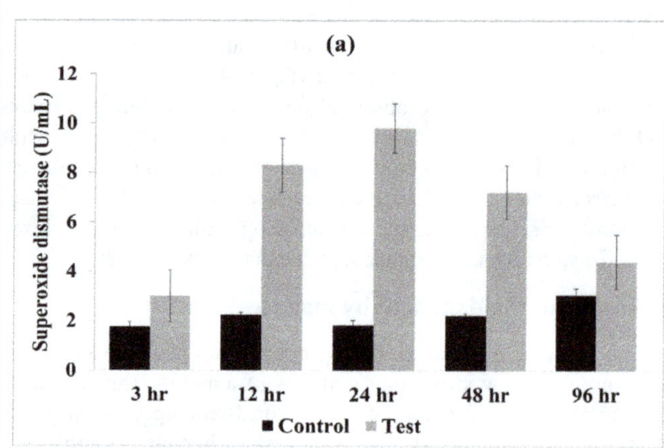

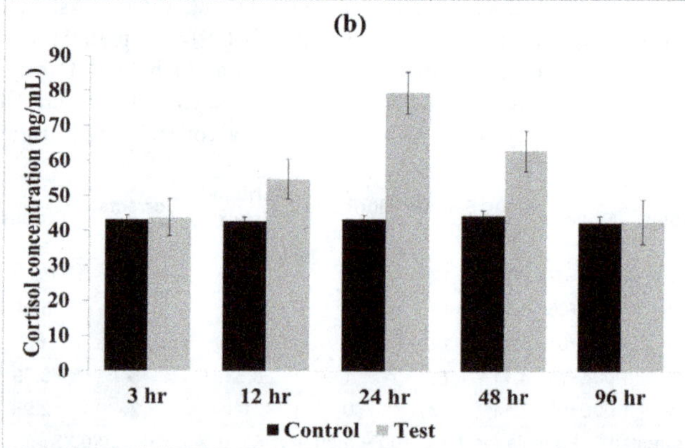

Figure 5: (a) The superoxide dismutase activity in control and test samples, (b) Concentrations of cortisol in plasma of test and control samples.

by *A. hydrophila* may cause severe hepatocyte necrosis [38]. Similarly, necrosis of hepatocytes and the presence of non-hepatocytic cells, probably the phagocytic cells (Figure 4E) in liver histology (Figure 4F) also indicate the systemic bacterial infection in the liver which is also shown previously in various pathological studies [17,20,39].

Pathogenicity of *Aeromonas* was also determined using the stress related SOD and cortisol parameters. Both the SOD and cortisol were elevated at a particular time and scaled down during the longer periods when the fish overcame the pathogenicity. Superoxide dismutase (SOD) acts as cyclic enzymes that catalyze the dismutation of superoxide radicals [40]. The amount of superoxide dismutase present in the cellular and extracellular environment is significant in the prevention of oxidative stress related diseases and can help to predict the defence mechanism of fish species [41,42]. Production of reactive oxygen species (ROS) is also an important mechanism of the vertebrate immune system [43]. They also possess various antioxidant enzymes to counteract the ROS and their adverse effects [41]. In the present study, a significant increase in SOD activity was observed at 24 h p.i. Wang studied the tissue-level activity and gene expression of two superoxide dismutase (MnSOD and Cu/Zn SOD) after *A. hydrophila* infection and found that the activity was highest in liver and kidney at 24 h p.i [44]. However, a subsequent decrease in SOD activity suggests the failure of SOD to remove superoxide radicals. These results are similar to that of Reddy [45]. The failure of SOD leads to accumulation of superoxides that ultimately may result in internal tissue necrosis. The necrotic lesions observed in liver tissue at 48 h p.i. may be the combined effect of infection and ROS species. Although, cortisol inhibit the immune response by suppressing the inflammatory substances; the amount of cortisol may serve as a useful indicator of stress level during infection in fishes [46]. The cortisol level was significantly ($p<0.05$) increased in blood plasma of Golden Mahseer after *A. hydrophila* challenge. The highest level was detected at 24 h p.i. which could be the combined effect of stress as well as bacterial infection. Increase in plasma cortisol level was reported in many piscine systems after bacterial infection [46,47]. Small and Bilodeau also reported the similar findings on cortisol level for channel catfish (*Ictalurus punctatus*) after exposure to *Edwardsiella ictaluri* [48]. High cortisol level was observed in Atlantic cod after experimental infection with *Aeromonas salmonicida* [49]. Increased cortisol level is commonly associated with stressful conditions of the fish principally during disease condition [50]. Endotoxin may also increase the plasma cortisol level that may suppress the immune response of the fish [51,52]. Hence, the increased level of plasma cortisol in Golden Mahseer may be associated with the stressful condition of fish as well as bacterial endotoxins. This is the first report on plasma cortisol level of Golden Mahseer in control and stress conditions. Handling during the bacterial and PBS inoculation as well as sampling may cause stress to the fishes, and its effect may not be neglected. However in the present study the cortisol level increased almost two folds in diseased fishes compare to control which is in agreement with the findings of Olsen and Haukenes and Barton who found the increased plasma cortisol level in diseased fishes [50,53].

Conclusion

In the present study, we determined the LD_{50} concentration (1.74×10^5 cfu per 100 g body weight) of *A. hydrophila* and examined various pathophysiological symptoms during the disease progression. The sign and symptoms of severe bacterial haemorrhage were observed, and *A. hydrophila* was confirmed to be the etiological agent by re-isolation of a bacterium using spread plate method followed by sequencing of 16s rRNA sequencing from infected fishes (test group). Two major stress indicators were also studied in test samples during the progression of infection. Superoxide dismutase activity and cortisol level were significantly high in test samples as compared to control. Severe liver necrosis and tissue damage were observed at 48 h p.i. Experimental conditions standardized in the present study will serve as a baseline data for further studies on Golden mahseer. Also, the knowledge of various symptoms during pathogenesis help to understand the course of pathogenesis and overall immune response of fishes which can be very beneficial to control and prevent the disease outbreaks in farmed fishes.

Acknowledgement

The authors acknowledge to Indian Council of Agricultural Research (ICAR) and Department of Biotechnology (DBT), Government of India, New Delhi, for laboratory facilities and financial support (under project ID BT/PR7162/AAQ/3/614/2012) to carry out this research activity.

References

1. Mohindra V, Khare P, Lal KK, Punia P, Singh RK, et al. (2007) Molecular discrimination of five mahseer species from Indian peninsula using RAPD analysis. Acta Zool Sin 53: 725-732.
2. Shahi N, Mallik SK, Sarma D (2014) Golden mahseer, Tor putitora-a possible candidate species for hill aquaculture. Aquaculture Asia pp: 22-28.
3. Bhatt JP, Pandit MK (2015) Endangered Golden mahseer Tor putitora Hamilton: a review of natural history. Rev Fish Biol Fisheries.
4. FAO (Food and Agriculture Organization of the United Nations) (2014) Food and Agriculture Organization of the United Nations Crop Production and Trade Statistics.
5. Talwar PK, Jhingran AG (1991) Inland Fisheries of India and Adjacent Countries. Oxford and I B H Publication Co. Calcutta pp: 1-1158.
6. Pandey AK, Patiyal RS, Upadhyay JC, Tyagi M, Mahanta PC (1998) Induced spawning of the endangered golden mahseer, Tor putitora, with ovaprim at the state fish farm near Dehradun. Indian J Fish 45: 457-459.
7. Bhatt JP, Nautiyal P, Singh HR (2004) Status (1993-1994) of the endangered fish himalayan mahseer Tor putitora (Hamilton) (Cyprinidae) in the Mountain Reaches of the River Ganga. Asian Fish Sci 17: 341-355.
8. Pillay TVR, Kutty MN (2005) Aquaculture, Principles and Practices. Blackwell Publishing Ltd, Oxford, UK pp: 630.
9. Leung TLF, Bates AE (2013) More rapid and severe disease outbreaks for aquaculture at the tropics: Implications for food security. J Appl Ecol 50: 215-222.
10. Barat A, Kumar R, Goel C, Singh AK, Sahoo PK (2015) De novo assembly and characterization of tissue-specific transcriptome in the endangered golden mahseer, Tor putitora. Meta Gene 7: 28-33.
11. Zorrilla I, Chabrillon M, Arijo S, Diaz-Rosales P, Martinez-Manzanares E, et al. (2003) Bacteria recovered from diseased cultured gilthead sea bream (Sparus aurata L.) in southwestern Spain. Aquaculture 218: 11-20.
12. Janda JM, Abbott SL (2010) The Genus Aeromonas: Taxonomy, Pathogenicity, and Infection, Clin Microbiol Rev 23: 35-73.
13. Angka SL, Lam TJ, Sin YM (1995) Some virulence characteristics of Aeromonas hydrophila in walking catfish (Clarias gariepinus). Aquaculture 130: 103-112.
14. Sahu S, Das BK, Pradhan J, Mohapatra BC, Mishra BK, et al. (2007) Effect of Mangifera indica kernel as a feed additive on immunity and resistance to Aeromonas hydrophila in Labeo rohita fingerlings. Fish Shellfish Immun 23: 109-118.
15. Reyes-Becerril M, Lopez-Medina T, Ascencio-Valle F, Esteban MA (2011) Immune response of gilthead seabream (Sparus aurata) following experimental infection with Aeromonas hydrophila. Fish Shellfish Immun 31: 564-570.
16. Tran NT, Gao ZX, Zhao HH, Yi SK, Chen BX, et al. (2015) Transcriptome analysis and microsatellite discovery in the blunt snout bream (Megalobrama amblycephala) after challenge with Aeromonas hydrophila. Fish Shellfish Immun 45: 72-82.
17. Yardimci B, Aydin Y (2011) Pathological findings of experimental Aeromonas hydrophila infection in Nile tilapia (Oreochromis niloticus). Ankara Univ Vet Fak Derg 58: 47-54.

18. Cipriano RC, Bullock GL, Pyle SW (2001) Aeromonas hydrophila and Motile Aeromonad Septicemias of Fish. Fish Disease Leaflet 68, US Department of the Interior Fish & Wildlife Service, Washington.

19. Austin B, Austin DA (2007) Bacterial Fish Pathogens, Disease of Farmed and Wild Fish. Springer Praxis, Godalming, UK.

20. Sun S, Zhu J, Jiang X, Li B, Ge X (2014) Molecular cloning, tissue distribution and expression analysis of a manganese superoxide dismutase in blunt snout bream Megalobrama amblycephala. Fish Shellfish Immunol 38: 340-347.

21. Laith AR, Najiah M (2013) Aeromonas hydrophila: Antimicrobial Susceptibility and Histopathology of Isolates from Diseased Catfish, Clarias gariepinus (Burchell). J Aquac Res Development 5: 2.

22. Wu CJ, Wu JJ, Yan JJ, Lee HC, Lee NY, et al. (2007) Clinical significance and distribution of putative virulence markers of 116 consecutive clinical Aeromonas isolates in southern Taiwan. J Infect 54: 151-158.

23. Akhtar MS, Pal AK, Sahu NP, Ciji A, Mahanta PC (2013) Thermal tolerance, oxygen consumption and haemato-biochemical variables of Tor putitora juveniles acclimated to five temperatures. Fish Physiol Biochem 3: 1387-1398.

24. Miller LC, Tainter ML (1944) Estimation of LD_{50} and its error by means of log-probit graph paper. Proc Soc Exp Bio Med 57: 261.

25. Bancroft JD, Layton C, Suvarna SK (2013) Bancroft's theory and practice of histological techniques. Churchill Livingstone Elsevier, Edinburgh, UK, pp: 69-156.

26. Sarkar MJA, Rashid MM (2012) Pathogenicity of the bacterial isolate Aeromonas hydrophila to catfishes, carps and perch. J Bangladesh Agric Univ 10: 157-161.

27. Shahi N, Mallik SK, Sahoo M, Das P (2013) Biological characteristics and pathogenicity of a virulent Aeromonas hydrophila associated with ulcerative syndrome in farmed rainbow trout, Oncorhynchus mykiss (Walbaum), in India. ISR J Aquacult-Bamid 65: 926-936.

28. Molony B (2001) Environmental requirements and tolerances of Rainbow trout (Oncorhynchus mykiss) and Brown trout (Salmo trutta) with special reference to Western Australia: A review. Fish Res Rep West Aust 130: 1-28.

29. Ali MF, Rashid M, Rahman MM, Haque MN (2014) Pathogenicity of Aeromonas hydrophila in Silver Carp Hypophthalmichthys molitrix and its Control Trial. IOSR J Agric Vet Sci 7: 2319-2372.

30. Sahoo PK, Pillai BR, Mohanty J, Kumari J, Mohanty S, et al. (2007) In vivo humoral and cellular reactions, and fate of injected bacteria Aeromonas hydrophila in freshwater prawn Macrobrachium rosenbergii. Fish Shellfish Immun 23: 327-340.

31. Sahoo PK, Mahapatra KD, Saha JN, Barat A, Sahoo M, et al. (2008) Family association between immune parameters and resistance to Aeromonas hydrophila infection in the Indian major carp, Labeo rohita. Fish Shellfish Immun 25: 163-169.

32. Rey A, Verjan N, Ferguson HW, Iregui C (2009) Pathogenesis of Aeromonas hydrophila strain KJ99 infection and its extracellular products in two species of fish. Vet Rec 164: 493-499.

33. Rodrıguez I, Novoa B, Figueras A (2008) Immune response of zebrafish (Danio rerio) against a newly isolated bacterial pathogen Aeromonas hydrophila. Fish Shellfish Immun 25: 239-249.

34. Harikrishnan R, Rani MN, Balasundaram C (2003) Hematological and biochemical parameters in common carp, Cyprinus carpio, following herbal treatment for Aeromonas hydrophila infection, Aquaculture 221: 41-50.

35. Pullium JK, Dillehay DL, Webb S (1999) High Mortality in Zebrafish (Danio rerio). Contemp Top Lab Anim Sci 38: 80-83.

36. Sundus AAA, Jamal KHA (2012) Detection and study of the experimental infection of Aeromonas strain in the common carp (Cyprinus carpio L.). The Iraqi J Vet Med 36: 222-230.

37. Khalil AH, Mansour EH (1997) Toxicity of crude extracellular products of Aeromonas hydrophila in tilapia, Tilapia nilótica. Lett Appl Microbiol 25: 269-272.

38. Afifi SH, Al-Thobiati S, Hazaa MS (2000) Bacteriological and histopathological studies on Aeromonas hydrophila infection of Nile Tilapia (Oreochromis niloticus) from fish farms in Saudi Arabia. Assiut vet Med J 84: 195-205.

39. Ray SD, Homechaudhuri S (2014) Morphological and functional characterization of hepatic cells in Indian major carp, Cirrhinus mrigala against Aeromonas hydrophila infection. J Environ Biol 35: 253-258.

40. Fridovich I (1986) Biological effects of the superoxide radical. Arch Biochem Biophys 247: 1-11.

41. Catap ES, Jimenez MRR, Tumbali MPB (2015) Immunostimulatory and anti-oxidative properties of corn silk from Zea mays L. in Nile tilapia, Oreochromis niloticus. Int J Fish Aquac 7: 30-36.

42. Umasuthan N, Bathige SD, Revathy KS, Lee Y, Whang I, et al. (2012) A manganese superoxide dismutase (MnSOD) from Ruditapes philippinarum: comparative structural and expressional-analysis with copper/zinc superoxide dismutase (Cu/ZnSOD) and biochemical analysis of its antioxidant activities. Fish Shellfish Immunol 33: 753-765.

43. Das A, Sahoo PK, Mohanty BR, Jena JK (2011) Pathophysiology of experimental Aeromonas hydrophila infection in Puntius sarana: early changes in blood and aspects of the innate immune-related gene expression in survivors. Vet Immunol Immunopathol 142: 207-218.

44. Wang X, Wang L, Ren Q, Yin S, Liang F, et al. (2016) Two superoxide dismutases (SODs) respond to bacterial challenge identified in the marbled eel Anguilla marmorata. Aquaculture 451: 316-325.

45. Reddy TK, Prasad TNVKV, Reddy SJ (2014) Studies on combined effect of Aeromonas hydrophila and cadmium on lipid peroxidation and antioxidant status in selected tissues of Indian freshwater major carp, Catla catla: role of silver nanoparticles. IOSR J Pharm 4: 1-7.

46. Balm PHM, Lieshout EV, Lokate J, Wendelaar-Bonga SE (1995) Bacterial lipopolysaccharide (LPS) and interleukin 1 (IL-1) exert multiple physiological effects in the tilapia Oreochromis mossambicus (Teleostei). J Comp Physiol B 165: 85-92.

47. Swain P, Nayak SK, Nanda PK, Dash S (2008) Biological effects of bacterial lipopolysaccharide (endotoxin) in fish: A review. Fish Shellfish Immun 25: 191-201.

48. Small BC, Bilodeau AL (2005) Effects of cortisol and stress on channel catfish (Ictalurus punctatus) pathogen susceptibility and lysozyme activity following exposure to Edwardsiella ictaluri. Gen Comp Endocrinol 142: 256-262.

49. Magnadottir B, Gisladottir B, Audunsdottir SS, Bragason BTH, Gudmundsdottir S (2010) Humoral response in early stages of infection of cod (Gadus morhua L.) with atypical furunculosis. Icel Agric Sci 23: 23-35.

50. Olsen YA, Falk K, Reite OB (1992) Cortisol and lactate levels in Atlantic salmon Salmo salar developing infectious anaemia (ISA). Dis aquat Org 14: 99-104.

51. Barton BA, Iwama GK (1991) Physiological changes in fish from stress in aquaculture with emphasis on the response and effects of corticosteroids. Ann Rev Fish Dis 1: 3-26.

52. Barton BA (2002) Stress in Fishes: A diversity of responses with particular reference to changes in circulating corticosteroids. Integr Comp Biol 42: 517-525.

53. Haukenes AH, Barton BA (2004) Characterization of the cortisol response following an acute challenge with lipopolysaccharide in yellow perch and the influence of rearing density. J Fish Biol 64: 851-862.

Livelihood Status of Hilsa (*Tenualosa ilisha*) Fishermen of Greater Noakhali Regions of Bangladesh

Jahangir Sarker Md*, Borhan Uddin AMM, Shamsul Alam Patwary Md, Mehedi Hasan Tanmay, Farhana Rahman and Moshiur Rahman*

Department of Fisheries and Marine Science, Noakhali Science and Technology University, Sonapur, Noakhali-3814, Bangladesh

Abstract

The present study aimed to elucidate the livelihood status of Hilsa (*Tenualosa ilisha*) fishermen at Lakshmipur and Noahkali Districts in Bangladesh during August, 2014 to January, 2015 through questionnaire survey method. During Hilsa fishing, Pangas (*Pangasius pangasius*), Koral (*Lates calcarifer*) and Poa (*Johnius coitor*) were also found to capture as by catch using *Chandi jal* (set gill net). 85% fishermen were observed to use mechanized (5-40 HP; Horse Power) boats in the study areas. Hilsa fishing was noticed mostly during October-November while such activities was almost absent during February-May (off period) which made fishermen to start migrating temporarily to the nearest urban areas for their livelihood. Although the Government of Bangladesh formulates an act to have fishing license yet 20% of the fishermen were found having valid fishing license in the study areas. Average daily net income of the fishermen during fishing period was 600 BDT whereas their real field daily income through fishing supposed to be 1695 BDT if they are supported by nets, boats, fuels etc., which means almost 64% of their daily income is taken by the aratdars. Therefore, subsidy as a means of nets, boats, fuels, engines etc., might be recommended for the better livelihood of the fishermen in greater Noakhali region.

Keywords: Hilsa; *Tenualosa ilisha*; Livelihood; Lakshmipur; Noakhali; Bangladesh

Introduction

Fisheries sector plays an immensely important role on the socio-economic development of Bangladesh from time immemorial as it is the part of the cultural heritage of the country. Fisheries sector contributes about 3.00% of the total export earning, 4.37% to GDP and 23.37% to agricultural sector [1]. Annual fish production was 34,10,254 MT in 2012-2013 fiscal year and the production of Hilsa fish is 3.51 MT which contributes 11% to the total fish production and 1% to the national GDP of the country.

Fish is one of the principal dietary ingredients to the people of Bangladesh providing about 60% of their animal protein intake. Hilsa contributes a major part to this. There is a proverb in Bengali is "*Macher raja Ilish*" meaning "Hilsa is the king of fish." It is the national fish of Bangladesh because of its importance. The Hilsa fishery of the country is characterized by the usual common property conditions where the available resources are exploited by a large number of fishermen.

The Indian River shad (*Tenualosa ilisha*, Hamilton, 1822), locally known as 'Ilish' constitutes the largest single species fishery of Bangladesh [2]. There are three Hilsa species found in the Bay of Bengal, *Tenualosa ilisha, Hilsa kelee,* and *Hilsa toli.* The majority of Hilsa fish captured belongs to *Tenualosa ilisha.* Though the Hilsa is generally regarded as an anadromous fish, there is evidence that it is in fact a diadromous fish, which means it migrates both ways between ocean and river. This anadromous fish is mainly found from the Meghna River estuary region through Noakhali, Lakshmipur and Chandpur district to the upper Padma River.

Lakshmipur is one of the southern coastal districts and Noakhali is situated in the central coastal zone of Bangladesh along the northeastern coast of Bay of Bengal. Both of the districts has the presence of large number of ponds, canals, floodplains and also the vicinity of the area to Meghna River estuary ensures the significance of the district in total culture and capture fish of the country. These two districts contribute greatly to the national annual production of Hilsa providing a large number of captured Hilsa to the country. That's why Ramgoti upazila of Lakshmipur district and Subarnachar upazila of Noakhali district are selected as two of the ideal Hilsa fish capture areas for the present research study.

Most of the Hilsa fishers live below the poverty line, and most work in teams as labourers/fishers. The wealthier fishers own the boats and nets. During fishing season, the fishers are dependent on fishing for their livelihood and do not have any alternative sources of income to support their families. Being an isolated community, fishermen are deprived of many amenities of life mostly in off season. Considering the above fact, the present study was carried out to assess the livelihood status of the Hilsa fishermen and the impacts of off season on the socio-economic condition of Hilsa fishermen in Lakshmipur and Noakhali region of Bangladesh.

Materials and Methods

Investigation was carried out in Ramgoti upazila of Lakshmipur district and Subarnachar Upazila of Noakhali district for six months from August, 2014 to January, 2015.

The study was based on the collection of primary and secondary data. Total 400 fishermen were involved in Hilsa fishing (250 in Laksmipur

***Corresponding author:** Jahangir Sarker Md, Department of Fisheries and Marine Science, Noakhali Science and Technology University, Sonapur, Noakhali-3814, Bangladesh, E-mail: swaponj@yahoo.com

Moshiur Rahman, B.Sc. Department of Fisheries and Marine Science, E-mail: moshi7403@gmail.com

district and 150 in Noakhali district). Therefore, primary data (10% of the total population) for this study were collected from randomly selected 40 (forty) Hilsa fishermen (25 fishermen from Ramgoti upazila of Lakshmipur district and 15 fishermen from Subarnachar upazila of Noakhali district) comprising of younger, middle aged and elderly experienced fishers. Pre-evaluated and post evaluated questionnaire were used for the data collection from fishermen in accordance with the objective set for the study.

Primary data were collected through personal interview and Focus Group Discussion (FGD) and secondary data were collected through Crosscheck Interviews (CI) with key informants, from project reports and documents. Necessary relevant information on the socio-economic condition of fishermen was collected from regional offices.

After collection of data, these were edited and coded. All the collected data were summarized and scrutinized carefully and recorded. All the collected information were accumulated and analyzed by MS-Excel and then presented in textual, tabular and graphical forms to understand the present livelihood status of the fishermen in the studied area.

Results and Discussion

Human capital

Age distribution: Knowledge on the age structure of fishermen is very important in estimating the productive potential of human resources. Different categories of age groups are 16-25 years, 26-35 years, 36-45 years and 46-55 years were considered to examine the age distribution. It was appeared that out of 40 fishermen age group of 26-35 years was the highest (57%) and 46-55 years was the lowest (5%) (Table 1). Most of the fishermen from 26-35 years were married and majority of them are head of their family. Besides all of them are full of energy and have more opportunity to earn more. Age group ranged from 46-55 years was the lowest that are considered the oldest fishermen and they are gradually losing their energy after these years. Khatun reported that most of the fish farmers belonged to the age groups of 36 to 50 years (46%) in Charbata union Noakhali district [3]. Minar found that most of the fishermen belonged to the age groups of 31 to 40 years (56.00%) in the Kirtonkhola River nearby to the Barisal town which was different to the present findings [4].

Family size: The family sizes of the fishermen were divided into three classes as small, medium and large. From the study it was found that, about 50% fishermen family were medium sized (5-6 members), about 35% fishermen family were small sized (2-4 members) and 15% family were large sized (>6 members) (Table 1). Ali found that most of the fish farmer (45%) belonged in the 4 to 5 member's family in Mymensingh district which is similar to the present findings [5]. The family size has considerable influence on the income and expenditure of the family.

Educational status: Education has significant impact on the society. In the present study seven categories were used to determine the level of education. Out of 40 interviewed fishermen, 35% had primary (up to class 5), small portion (5%) of them can read only, 10% Illiterate, 10% can only write their name, 10% are literate to SSC level and the most important thing is that secondary education (Up to Class 8) completed fishermen was 30% out of 40 fisherman interviewed (Table 1). Pravakar found that about 10% had no education, 16% had primary (Up to class 5) level, 48% had secondary level, 16% had higher secondary level and 10% had bachelor level of education in Shahrasti Upazila under Chandpur district which is slightly similar to the present findings [6]. According to the fishermen sampled to the present study it is found that fishermen want to change the trends about illiteracy. They want to be educated so that they would not like to be cheated by the others due to lack of their education. Hilsa fishermen were consistently being cheated by the aratdars and loan providers as they were forced to sell their captured Hilsa to them. They revealed that majority of the fishermen were illiterate as most of the fishermen were involved in the fishing activity which was the only way to economically support their family in their early stage of life and also the lack of awareness about education. Another important factor was observed that there is a few numbers of educational institutions in the areas of fishing community. Most of the fishermen families are unable to bear the educational expenses of their school going children during banning period or lean season due to low or zero income.

Religious status: From the present study, it was found that Muslims constituted 85% of the fishermen community with 15% Hindus (Table 1). There was no Buddhists or Christians which was similar to Minar [4]. The dominance of Muslims in the fishing community indicates that Muslims are gradually coming to fishing profession by breaking the previous superstitions of the society. Pravakar found that about 75% and 25% of the pond fish farmers were Muslims and Hindus respectively in Shahrasti Upazila under Chandpur district [6].

Physical capital

Housing condition: The nature of house indicates the social status of the people. Ramgoti Upazila of Lakshmipur and Subarnachar Upazila of Noakhali were not developed as like as the main town of Lakshmipur and Noakhali district respectively. From the survey, it was found that 65% households of the fishermen were tin shed, 25% households were semi-pacca and 10% households were cemented building (Table 1).

Title	Types	Percentage (%)
Age Distribution (Years)	16-25	20
	26-35	57
	36-45	18
	46-55	5
Family Size	Small (2-4 members)	35
	Medium (5-6 members)	50
	Large (>6 members)	15
Educational Status	Illiterate	10
	Can write their name only	10
	Can read only	5
	Primary (Up to Class 5)	35
	Secondary (Up to Class 8)	30
	Secondary School Certificate (SSC)	10
	Higher Secondary School Certificate (HSC)	0
Religious Status	Muslim	85
	Hindus	15
Housing Conditions	Tin Shed	65
	Semi Pacca	25
	Cemented	10
Sanitation	Yes	35
	No	65
Subsidy	Get VGF	25
	Not Get VGF	75

Table 1: Information of fishermen in Lakshmipur and Noakhali districts in Bangladesh.

Khatun observed in their study that most of the house of pond fish farmers (78%) was made of tin-shed, 12% houses were katcha (straw components), 8% half cemented building and 2% cemented building of Charbata, Noakhali [3]. This is because of low income of the fishermen throughout the country.

Sanitary facilities: Fishermen are the poorest group of people in the country. It was observed that sanitary conditions of the fishermen were very poor. Only 35% of fishermen families were found who use sanitized toilets (Table 1). So they are in unhygienic condition which revealed that the sanitary conditions of the fishermen were not satisfactory like fisherman of the Kirtonkhola River nearby to the Barisal town where Minar found that 74% of toilets were kacha while 10% were semi-pacca and 16% of the fishermen had no sanitary facilities [4]. Poor sanitation system reflects poor socio-economic condition and lower income.

Livelihood strategies of Hilsa fishermen: Hilsa fishermen are dependent on fishing as a source of income and nutrition. They are engaged in fishing from generation to generation. The fishermen in the study areas were reported to go for fishing during day, night and even for 24 hours together. In day period fishermen usually prefer for fishing between 5 am to 11 am. In night, they prefer fishing between 7 pm to 4 am. From the present study, both professional and seasonal fishermen were found. Present analysis showed that professional fishermen constituted 65% and seasonal fishermen 35% to the total sample population (Table 2). Paul found that about 70% fishermen in Birulia and 64% fishermen in Boroibari were full-time fishermen respectively on the other hand about 26% fishermen in Birulia and 34% fishermen in Boroibari were part-time fishermen respectively [7].

According to the present study, about 100% of fishermen used *Chandi jal* (set gill net) with mesh size of 5 inches for Hilsa fishing, but width varies from 5.5-9.15 m which depends on the depth of seabed and distance from the sea shore [8]. Sometimes fishermen illegally use *Current jal* (drift gill nets). Sazzad studied on Hilsa fishery in the river Meghna in Chandpur district with sixty sampled boats for three gears motorized *gulti jal, Chandi jal* and *Current jal* [8]. For catching Hilsa, about 57% of fishermen spent 6 hours per day, whereas 30% and 13% of fishermen were involved in fishing for 5 hours and 4 hours respectively (Figure 1) in the present study. Bhaumik and Saha found that about 24.0% of the fishers engaged in fishing operation for 241-260 days, 39.9% spend 12 hours per day and 29.0% of them caught 131-150 kg fish/month [9].

Fishermen normally capture mature Hilsa in the fishing season (August-January). From the survey, it was found that about (55%) of fishermen capture Hilsa that weight (0.750-1.0) kg whereas a few (8%) were found about >0.5 kg of Hilsa (Table 3). According to the Hilsa fishers, the size of Hilsa fish are gradually increasing than the previous years, as Government of Bangladesh has prohibited all types of Hilsa fishing during banning period (11 days, 13 to 23 October, 2014). The fishers also mentioned that Pangas (*Pangasius pangasius*), Koral (*Lates calcarifer*) and Poa (*Johnius coitor*) comprise 10% of the total composition of fish capture while 90% of that is Hilsa.

Sazzad studied on Hilsa fishery in the river Meghna and found that Hilsa fishing was profitable [8]. Considerable differences in price were noticed between fishermen and consumers and this happened due to the involvement of a number of middlemen in the marketing chain of Hilsa. An important observation in the study area was the dispute between the fishermen and the aratdars regarding price of the fish. Generally boat, net and loan are provided by the aratdar with a number of terms and conditions, among them two most important is, first they have to be paid 10% of the total catch of Hilsa by the fishermen, another one is, rest of the catch have to be sold to those aratdars or the loan providers. These systems reduce the selling price of Hilsa caught by the fishermen as well as reduce the income of them. The marketing chain from fishermen to consumers passes through local aratdar or broker to local market where retailer sells the fish to consumers which is similar to Pravakar in Shahrasti Upazila under Chandpur district [6].

Types of fishermen	No. of fishermen (n=40)	Total fishermen (%)
Professional fishermen	26	65
Seasonal fishermen	14	35
Total	40	100

Table 2: Types of fishermen in the study areas.

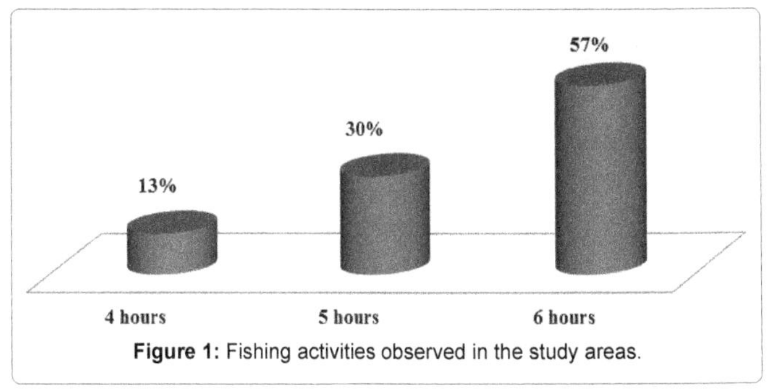

Figure 1: Fishing activities observed in the study areas.

Sizes of Hilsa fish	No of fishermen	Total fishermen (%)
>0.5 kg	3	8
0.5-0.750 kg	8	20
0.750 kg-1.0 kg	22	55
<1.0 kg	7	17

Table 3: Sizes of Hilsa fish during harvesting in the study areas.

Types of boat: Boat was the main fishing gear for Hilsa fishermen. Two types of boats were found in the study area used in Hilsa fishing; mechanized and non-mechanized boat. It is denoted that 85% of the fishermen use mechanized boat while the rest (15%) used non-mechanized boat (Table 4). These boats have difference in storage as well as in carrying capacity. Mechanized boats have higher storage capacity and carrying capacity than non-mechanized boats. Mechanized boats went to the deep sea areas for 3/5/7 (most of them were gone for 3 days) days according to their storage and carrying capacity while the non-mechanized boats move from dawn to dusk nearby sea shore area. It was found that 75% mechanized boat has storage capacity and 25% of non-mechanized boat has storage capacity (Table 4). They use only ice for two types of storage facilities; on-boat and on-shore. Ice concentration is not same for on-boat and on-shore. Generally they use ice at a concentration of 350-400 kg/MT of Hilsa in on-boat storage. But in on-shore, the ice concentration for storage is 1:1 (i.e., 1 kg ice for 1 kg Hilsa fish).

Engine capacity (Ranges of HP): The capacity of the mechanized boats used in the study area ranged from 10 to 40 HP (Horse Power). The selected fishermen were grouped into three categories based on their use of HP of the engine. The categories are 10 to 20 HP; 20 to 30 HP and 30 to 40 HP. From the study area it is found that 20 to 30 HP had the highest number (45%) of fishermen whereas 30 to 40 HP had the lowest number (10%) (Table 4). It is also noticed that 30 to 40 HP of engine were only used for deep sea fishing when they were gone for 3/5/7 days. Among them, some of the fishermen have some special preference in using the engine according to the made by which country or not. Mainly Japanese and Chinese engines are their

Title	Types	Percentage (%)
Boat types	Mechanized	85
	Non mechanized	15
Storage Facilities	Yes	75
	No	25
Engine Capacity (HP)	10-20	35
	20-30	45
	30-40	10
	>40	10
Engine Preferences	Japanese	30
	Chinese	70
Carrying capacities (Metric Ton)	Less than 0.5	30
	0.5-1.0	25
	1.0-1.5	15
	1.5-2.0	20
	Above 2.0	10
Having License	Yes	20
	No	80

Table 4: Information about the boat used by the fishermen.

special preference. It is found that Chinese engines are used in highest percentage (70%) while Japanese engines are used in lowest percentage (30%) (Table 4).

Carrying capacities: Different sizes of boats have different capacities. Boats are categorized into five classes such as less than 0.5 MT; 0.5-1.0 MT; 1.0-1.5 MT; 1.5-2.0 MT and above 2.0 MT. It is found that in the study area less than 0.5 MT has the highest percentage (30%) while above 2.0 MT has the lowest percentage (10%) (Table 4). For the small trip, generally 6 am to 5 or 6 pm, fishermen need lower storage capacity boat. But for long trip, generally 3 to 7 days, they need higher storage capacity boat as they go to the deep sea for Hilsa fissing.

License availability: Although license is the key element for Hilsa fishing in both riverside and deep sea areas in the developed countries, but there is no essential obligation of license for Hilsa fishing in Bangladesh. As a result large amounts of Hilsa fish including jatka (having 23 cm sizes of Hilsa) are captured by large number of fishermen. Though the Government of Bangladesh took some necessary steps to control the jatka fishing but their thought is that license is not essential for Hilsa fishermen. In the study area it is found that 80% fishermen has no license while the rest 20% has license for Hilsa fishing (Table 4). According to UFO of Cox's Bazar, fishermen must have license for Hilsa fishing in the Cox's Bazar sea shore area or BFDC area.

Fishermen activity observed in the study area: Hilsa fishermen of Bangladesh do not get involved in Hilsa fishing all the year round. The activities of the fishermen are shown in Table 5. Less Hilsa fishing activity was observed during December to January due to the gradual decrease of Hilsa capture rate. All the brood generally migrates seaward after the release of their egg. Almost no fishing activity was observed during February to March due to absence of Hilsa in sea shore or riverside area. In this time, the released eggs turn into hatchling to juvenile (jatka). It is the restricted period for all fishermen announced by the Government of Bangladesh. No fishing activity was observed during April to July because of the protection for juvenile Hilsa (jatka). It is the time when all jatka get time to grow up to mature Hilsa. There was another no fishing activity was observed during banning season (13 to 23 October, 2014 for 11 days). In this period all brood Hilsa move toward river to release their eggs in freshwater. All types of Hilsa fishing are strictly prohibited and this is announced to fishermen's village by respective Upazila Fisheries Officer and also broadcast to the mass media. During August to September, some of matured Hilsa move toward the sea shore or riverside area, so all Hilsa fishermen start their fishing activity. October to November is the high fishing activity period because this is the peak season for Hilsa fishing. All mature Hilsa start to move freshwater river to release their eggs and fishermen capture most of the Hilsa during this time.

Cost benefit analysis of a fisherman

Income of fishermen: Level of income of an individual family determines socio-economic status in a society. Annual income of a fisherman comes from main occupation as well as secondary occupation. There are various sources of income such as fishing, agriculture, service, day labourer, business, cattle rising, poultry and selling its product, rickshaw pulling etc.

It was also found that, the real field income of a fisherman is BDT. 1695.00 per day. Thus a fisherman's deserved annual income is Tk. 440,700.00. But a fisherman only obtain as his wage is BDT. 600.00 from aratdar which means his obtained annual income is BDT. 156,000.00. So he is deceived at BDT. 1095.00 per day and Tk. 284,700.00 per year from his aratdar (Table 6). Therefore a fisherman's income is Tk. 156,000.00 per year and his cost is Tk. 78,840.00 per year (Table 7), thus his net annual income (Obtained Annual income-Total cost) is BDT. 77,160.00. A fisherman could manage his family maintenance at

Month	Activity
December-January	Less Hilsa fishing activity
February-March	Almost no Hilsa fishing activity.
April-July	No Hilsa fishing activity
August-September	Start fishing activity
October-November	High fishing activity (It is a peak season)
13 October, 2014 to 23 October, 2014 (28 Ashwin to 10 Kartik, 11 days)	No fishing activity (The banning period for Hilsa fishing)

Table 5: Fishermen activities observed in the study areas.

Gross income (BDT)					
Real field income				Obtained income	
Per day average catch	Amount (BDT)	Total income (BDT)	Annual income (BDT)	Per day (BDT)	Annual income (BDT)
Hilsa (3.5 kg@BDT 400.00)	1400.00	**1695.00**	440,700.00	**600.00**	156,000.00
Koral (500 gm@BDT 300.00/kg)	150.00				
Pangas (500 mg@BDT 150.00/kg)	75.00				
Poa (200 gm@BDT 350.00/kg)	70.00				

Table 6: Calculated income of a fisherman (group comprising 10 members during fishing; each item is divided by 10) observed in the study area excluding banning period (260 working days).

Cost (BDT)		
Item of cost	**Cost (BDT)**	**Total cost (BDT)**
Fuel and oil (2 liter@72 BDT/ day/man)	37,440.00	**78,840.00**
Food (80 BDT/day/man)	20,800.00	
Ice (60 BDT/day/man)	15,600.00	
Subscription to the Pirates	5,000.00	

Table 7: Calculated operational cost (BDT) of a fishermen (group comprising 10 members during fishing; each item is divided by 10) observed in the study area excluding banning period (260 working days).

Tk. 297.00 per day of his total fishing year (260 working days) by this earning. This is mostly the lowest income for a fisherman to maintain his family like foods, clothes, medicine or disease treatment, education, homestead maintenance, etc., according to the present life style for a family.

The selected fishermen were grouped into four categories based on the level of their net annual income. The 1st category described the fishermen having net annual income less than Tk. 50,000. The 2nd, 3rd and 4th categories had income levels of Tk. 50,001-Tk. 75,000; Tk. 75,001-Tk. 100,000; and above Tk. 100,000 respectively (Table 8). It denoted that the annual income less than Tk. 50,000 categories had the highest number (45%) of fishermen while above Tk. 100,000 categories had the lowest number (10%). The net annual income of 25% fishermen had Tk. 50,001-Tk. 75,000 and 20% fishermen had Tk. 75,001-Tk. 100,000. This indicates that the fisherman who cannot manage his family having his net annual income less than Tk. 50,000 are the most in number.

Impacts of off-season on the fishermen

Fishers do not have opportunities for alternative income generating activities by the government during off-season and they suffer much during those periods. They find themselves their own alternative income generating sources. Present studies have identified several alternative livelihood strategies already adopted by the Hilsa fishermen. During off-season, 15% of fishermen had no off-season income, some of them (25%) were engaged in rickshaw pulling, some of them (20%) were engaged in construction work and most of the fishermen (40%) were engaged in day labor (Table 9) which was more or less similar with the findings of Ali. It is also mentioned that boat and net making categorized fishermen were very poor in percentage but they were involved during off-season [5].

In reduction of poverty and to improve livelihood in this area a very limited effort was found from the side of the Government organizations (GOs). In the present days, the Government had distributed some VGF (Vulnerable Group Feeding) card to the poor fishermen in time of off-season. From the present study, it was found that 25% of fishermen got subsidy (VGF card) from government and 75% of fishermen had no VGF card (Table 1) though they were actual fishermen. Through VGF card they got rice 40 kg rice per month for four month only but they claim that they don't get full benefit of VGF card provided by the Government. They only get 30-35 kg rice per month and the remaining 5-10 kg rice did not distributed at all.

Net Annual Income/Year		
Categories (BDT)	Number of fishermen (n=40)	% of total fishermen
Less than 50,000	18	45
50,001-75,000	10	25
75,001-100,000	8	20
Above 100,000	4	10
Total	**40**	**100**

Table 8: Fishing income/year (BDT) of the fishermen in the study area.

Types of work	No of fishermen (n=40)	% of fishermen	Wage per day (BDT)	Income per year (BDT)
Day labor	16	40	350.00	36,750.00
Rickshaw puller	10	25	350.00	36,750.00
Construction worker	8	20	400.00	42,000.00
No work	6	15	0.00	0.00

Table 9: Fishermen involvement in other activities during off-season (considering 105 days).

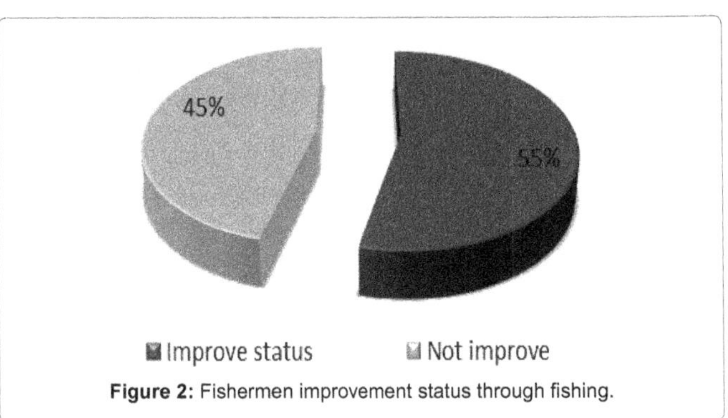

Figure 2: Fishermen improvement status through fishing.

Livelihood constraints: Day by day increasing price and unavailability of fishing materials like boat, net, ice were main fishing constraints for fishermen whereas most of the fishermen do not get boat during their prime need and a small number of fishermen do not get net, ice, etc.

During off-season, the poor fishermen household suffered food storage and try to consume less expensive foods items, they mostly depended on vegetables and their fish consumption reduce to 1-2 day/week from 4-5 days/week. Fishermen also suffered from various problems such as, inadequate credit facility, lack of marketing facilities, lack of knowledge of fishing, lack of appropriate gears etc., which were similar with the findings of Alam and Hossain [10,11].

Livelihood outcomes: Livelihood outcomes can be thought of as the inverse of poverty. Contributing to the eradication of poverty and food insecurity depends on equitable access to resources, access of disadvantaged groups to sufficient, safe and nutritionally adequate food [12]. In spite of poor resources livelihood outcomes of fish farming are positive and most of them increased their income, food security and basic needs. The survey found that 55% of fishermen had improved their socio-economic condition through fishing. They had better food, clothes, housing conditions and children education. But 45% farmers had not yet been improved their status (Figure 2). Similar results also reported by Halder [13].

Livelihood outcome factors are food security, nutrition, health, income, education, housing facilities, environment, safety etc. The fishermen community goes under food insecurity for 4-5 months in every year. Reason for the food insecurity was off-season of fish catches. Food crisis become severe in the months of April to July. Only a few fishermen those who had agricultural land had food security for the whole year. Educational status of the fishermen in the study area was not good and most of the people were illiterate. But the primary education percentage of the children of fishermen was increasing gradually.

Conclusion

The study was focussed on the livelihood of the Hilsa fishermen of Ramgoti upazila under Lakshmipur district and Subarnachar upazila under Noakhali district, Bangladesh.

Fishers were found to be mostly poor and neglected in the society and are exploited by the rich people/Mohajan/Aratdar in different ways. Many fishers do not have fishing equipment (boat and net) and as such they undertake fishing in Mohajan's boat as labourers or on catch share basis (paying Mohajan's at 10% according to their catch). About 100% of fishermen used *Chandi jal* (set gill net) of which 85% fishers were used mechanized boat. Among of all fishermen 20 to 30 HP of engine of mechanized boat had the highest number (45%). There is no essential obligation of using license for Hilsa fishing. October to November is the peak season for fishing activity. According to the present study, a fisherman is deceived at Tk. 1095.00 per day by his aratdar. It indicates that this is too hard for a fisherman to maintain his family like food, cloth, medicine or disease treatment, education, homestead maintenance, etc., according to the present life style for a family

The socio-economic condition of the fishermen in the adjacent area was not satisfactory. The fishermen were deprived of many amenities. Thus surviving is the most important thing for fishermen; they find themselves their own alternative income generating sources during off-season.

As fishermen play an important role in catching Hilsa fish under severe stressful conditions, so Government should take some important steps by providing some extra providence (VGF card, providing soft loan, fishing gears and nets, etc.,) during off-season to improve their socio-economic conditions. Government of Bangladesh should ensure that license holders are getting more incentives in all cases like boat making, net weaving, fuel, diesel and engine purchasing and providing VGF than others. Government should implement an act to settle a fixed wage per day for Hilsa fishers and should ensure punishment to those aratdars who are violating the act. Government should check that superscription to the pirates by frequent patrolling by coast guard.

References

1. Do F (2014) Jatka Conservation Week, Department of Fisheries, Ministry of Fisheries and Livestock, Dhaka, Bangladesh p: 130.
2. FAO (2004) Hilsa Investigation in Bangladesh, Marine Fishery Resources Management in the Bay of Bengal, Reports-BOBP/REP 37, Colombo, Srilanka.
3. Khatun S, Adhikary RK, Rahman M, Sikder MNAM, Hossain MB (2013) Socioeconomic Status of Pond Fish Farmers of Charbata, Noakhali, Bangladesh. Intern Journ of Life Sci Biotech and Pharm Resea 1: 356-365.
4. Minar MH, Rahman AFMA, Anisuzzaman M (2012) Livelihood status of the fisherman of the Kirtonkhola River nearby to the Barisal town. Journ of Agroforest and Environ 6: 115-118.
5. Ali H, Azad MAK, Anisuzzaman M, Chowdhury MMR, Hoque M, et al. (2009) Livelihood status of the fish farmers in some selected areas of Tarakanda upazila of Mymensingh district. Journ of Agroforest and Environ 3: 85-89.
6. Pravakar P, Sarker BS, Rahman M, Hossain MB (2013) Present Status of Fish Farming and Livelihood of Fish Farmers in ShahrastiUpazila of Chandpur District, Bangladesh. American Eurasi Journ of Agricul and Environ Sci 13: 391-397.
7. Paul B, Faruque H, Ahsan DA (2013) Livelihood Status of the Fishermen of the Turag River, Bangladesh. Middle-East Journ of Sci Rese 18: 578-583.
8. Sazzad A (1993) An economic study of Hilsa fishing in the river Meghna in chandpur district, MS Thesis, Agricultural Economics, BAU, Mymensingh p: 83.
9. Bhaumik U, Saha SK (1994) Perspectives on socio-economic status of the fishermen engaged in fishing in the estuaries of Sunderbans. Journ of Ecol 12: 181-185.
10. Alam MS (2004) Gender role and gender participation in beel fishery in some selected area in Gazipur sadar under Gazipur district, MS Thesis, Department of Fisheries Management, BAU, Mymensingh p: 54.
11. Hossain MM (2007) Utilization pattern of Mokash beel for livelihood of the local fishermen of Kaliachoir Upazila under Gazipur district, MS Thesis, Department of Aquaculture, BAU, Mymensingh p: 80.
12. Scoones I (1998) Sustainable Rural Livelihoods: A framework for analysis. IDS working paper No: 72, Brighton, IDS, UK p: 20.
13. Halder P, Ali H, Gupta N, Aziz MSB, Monir MS (2011) Livelihood status of fresh fish, dry fish and vegetable retails at Rajoir Upazila of Madaripur district, Bangladesh. Bangladesh Research Publications Journal 5: 262-270.

27

Snails and Fish as Pollution Biomarkers in Lake Manzala and Laboratory A: Lake Manzala Snails

Hanaa MM El-Khayat[1], Hoda Abdel-Hamid[1], Hanan S Gaber[2], Kadria MA Mahmoud[1] and Hassan E Flefel[1]

[1]*Department of Environmental Researches and Medical Malacology, Theodor Bilharz Research Institute, Imbaba, PO Box-30, Giza, Egypt*
[2]*National Institute of Oceanography and Fisheries, Cairo, Egypt*

Abstract

Physiological, hematological and biochemical parameters have been used as biomarkers for water quality in snail samples collected from Lake Manzala.

The results showed significant increase in AST, ALT, and ALP in *Planorbis* and *Physa* snail samples collected from Dakahlya site in Lake Manzala. Most of snails are collected from of Port-Said and Dakahlya sites showed significant increase in urea. On the other hand, alteration in creatinine values in samples from different lake sites was recorded. Significant increase of total protein level and total bilirubin was obtained in all samples. Most of snail samples showed significant decrease in hemocytes count. The oxidative enzymes (CAT, GGT and GST) recorded alteration in their activity.

Regarding Histopathological observations, in the foot region of *Biomphalaria* snails collected from Port Said and Dakahlya governorates are the most affected. The head foot showed splitting in the longitudinal and oblique muscle fibers and increased empty spaces within muscle. Shrinkage, focal areas of necrosis, large fat vacuoles and enlargement were observed in the salivary gland. Snail's ganglia showed enlargement of neurosecretory neurons, degeneration with large vacuoles and fibrosis. Hepatopancreas became much more distorted with necrosis, atrophy, degeneration and fat vaculation especially in Port Said and Damietta samples. Also, hepatopancreatic acini filled with different developmental stages of *S. mansonai* cercariae were observed in *Biomphalaria* snails collected from Port Said. Severe degenerative changes were observed in most of gonad's cells including ova and sperms especially in snails collected from Damietta. Also, *Biomphalaria* snails collected from Lake Mazala showed accumulation of heavy metals in the head foot tissues. In conclusion, the severe alteration and degeneration recorded in the physiological and hematological parameters and also histopathological observations are clear evidence for the pollution of the water from which these snails were collected.

Keywords: Lake manzala; Aspartate aminotransferasel; Alanine aminotransferase; Alkaline phosphatase; Total protein; Bilirubin; Hemocytes; Oxidative enzymes and histopathology

Abbrevations: AST: Aspartate Aminotransferase; ALT: Alanine Aminotransferase; ALP: Alkaline Phosphatase; ALB: Albumin; CAT: Catalase; GST: Glutathione-S-Transferase; GGT: Gamma Glutamyltransferase; TBRI: Theodor Bilhaz Institute; HE: Hematoxylin and Eosin Stain; A/G ratio: Albumin/Globuline ratio

Introduction

Lake Manzala is considered one of the most important lakes in Egypt. It is exposed to high levels of pollutants from industrial, domestic and agricultural resources [1-3]. Ali reported that Lake Manzala receives about 4000 million cubic meters of untreated industrial, domestic and agricultural waste water annually [4].

The use of physiological and biochemical parameters as indicators of water quality has been developed to detect sublethal impacts of pollutants. Prominent among these biomarkers are physiological variables, such as plasma levels of metabolites [5], haematological data [6,7], levels of hormones [8-11] and biochemical variables such as detoxifying enzyme activities [12,13]. Interesting reports concerning the mechanisms of metal uptake, accumulation, transport, and elimination of metals in molluscs are usually focused on chemical, biochemical, molecular, and physiological aspects [14-21]. El-Khayat assessed genetic variation and genetic pattern of *Lymnaea* snails collected from irrigation canals in four different Governorates using ISSR markers, with the characterization of environmental parameters of the collecting *Lymnaea* sites. The authors showed high polymorphism by using for the first time the ISSR PCR technique for studying genetic variations of *L. natalensis* snails in Egypt and concluded that *L. natalensis* snails can survive associated with other snails, plants, and insects and can be tolerate the heavy metals in water [22].

Similarly, histopathological changes have been widely used as biomarkers in the health evaluation of animal organisms. The discharge of toxic elements into the rivers, estuaries and coastal waters poses serious pollution and consequently affects the fish, flora and fauna as snail.

Moreover, freshwater molluscs play an important role in aquatic ecosystems, providing food for many fish species and vertebrates [23].

This work aims to record the alterations of the Physiological, hematological and histopathological parameters in snails collected from Lake Manzala as a bio-indicator for water pollution.

***Corresponding author:** Hanaa MM El-Khayat, Department of Environmental Researches and Medical Malacology, Theodor Bilharz Research Institute, Imbaba, PO Box-30, Giza, Egypt, E-mail: hanaamahmoud@hotmail.com

Materials and Methods

Snail samples were collected from 8 sites in Lake Manzalafrom 3 governorates; Port-Said (Kobry El-Lansh, Kaar El-Bahr and El-Khankak), Dakahlya (Gammalya, Matarya and Nasayma) and Damietta (Ananyya and Sayala). The snails collected were kept in water from their habitat and examined for natural infection. The negative (uninfected) *Biomphalaria* snails and other collected species (*Physa* and *Planorbis*) were contributed in the physiological studies. On the other hand, both negative and positive *Biomphalaria* samples were examined histologically.

Biochemical studies

Determination of liver and kidney functions: The assessment of aspartate aminotransferase (AST), alanine aminotransferase (ALT), alkaline phosphatase (ALP), urea, creatinine, total and direct bilirubin, albumin (ALB) and total protein were examined in snail tissue extracts. They were assayed biochemically using biosystem autoanalyzer, Backmann at Theodor Bilhaz Institute (TBRI) hospital laboratories. Snail tissues were dissected out, homogenized in bi-distilled water (1:1 w/v) using motor homogenizer and centrifuged at 5000 rpm for 20 min at 4°C and the supernatants were taken and kept at -20°C till used as described by [24].

Creatinine was determined according to [25]. In this method, creatinine reacts with picrate to form a coloured complex and the rate of formation of the complex is measured photometrically at 492 nm.

Urea was determined by using the coupled urease/glutamate dehydrogenase (GLDH) enzyme system according to [26].

Determination of antioxidant enzymes: The antioxidant enzymes catalase (CAT), Glutathione-S-Transferase (GST) and Gamma Glutamyltransferase (GGT) were assayed in snail tissue extract using spectrophotometer. Snail's tissues were dissected out. Each snail tissue from each treatment was homogenized in bi-distilled water (10:1 w/v) using motor homogenizer. Homogenates were centrifuged at 5000 rpm for 20 min at 4°C and the supernatants were taken and kept at -20°C till used.

Determination of snail hemolymph components: Snail hemolymph was collected in accordance to the technique of [27]. The hemolymph was obtained via small hole made in the shell into which capillary tube was inserted then it was drawn into tube by capillary suction. The hemocytes of the samples hemolymph were determined by haemocytometer. For total and differential counting, monolayer of hemocytes were stained with Giemsa stain for 20 minutes, according to the methods of [28] and counted by light microscopy.

Histopathological examinations: Snail specimens collected from Lake Manzala were dissected, removed from their shells gently and fixed in 10% buffered neutral formalin solution. Five-micron thick paraffin sections were prepared, stained by hematoxylin and eosin (HE) and then examined microscopically and photographed for histopathology observations [29].

Statistical analysis: Data are expressed as means ± SD. The results were computed statistically (SPSS software package, version 20) using the T-test analysis. Values of $p<0.05$ were considered statistically significant.

Results

Biochemical parameters

The present results showed significant increase in AST, ALT, and ALP in *Planorbis* and *Physa*, respectively in most samples collected from Nasayma site in Lake Manzala. Most of snails collected from Port-Said and Dakahlya sites showed significant increase in urea. Results of creatinine in samples from different lake sites showed alteration, ranged between non-significant decrease and increase (Table 1). Significant increase of total protein level was obtained in all field samples while total bilirubin showed the highest levels in *physa* and *planorbis* samples collected from Nasayma, Dakahlya and in *Biomphalaria* samples collected from Matarya, Dakahlya. Also, results showed higher levels of indirect bilirubin than direct. Most snail samples showed approximately normal A/G ratio (Table 2).

Antioxidant enzymes

Significant alterations in catalase (CAT) level were noticed in all snail samples collected from Lake Manzala as compared with lab bread controls (except in *Planorbis* collected from Kobry El-Lansh and *Biomphalaria* from Gammalya and *Biomphalaria* from Nasayma). The recorded alterations in the snail samples was increased by 18 to185%, or decreased by -13 to -90% (Table 3).

Glutathione-S-transferase (GST) alteration was demonstrated in all samples includes decrease in activity ranging from -21% to -83% ($P<0.001$) and increase in activity ranging from 13% to 119%.

The same result was noticed in Gamma-glutamyl transpeptidase (GGT) in snail samples as compared with lab bread controls, some samples showed decrease change activity ranging from -1% to -35% and other samples showed increase change activity ranging from 6% to 666%, (Table 3).

Determination of hemolymph components

The majority of snail samples showed significant decrease in total and differential cell count as compared with lab bread controls (Table 4). The higher percent of decrease in the total cell count (-72%) was recorded in *Biomphalaria* collected from Nasayma, Dakahlya. Hemoglobin concentration showed alteration; increased to 2.6 g in *Planorbis* collected from Kobry El-Lansh and decreased to 0.8 g in *Physa* collected from Annanya, Damietta.

The histopathological observations

A knowledge of the normal histology and structure of snails is guided by [30].

Head foot: The normal foot region has an outer cuticular layer as a protective layer of the foot. Inner to this lining there is a tall columnar epithelium with basal nuclei in its cell. Amongst the columnar epithelium there are modified sacs like cells in the form of unicellular glands which open through the cuticular layer exterior to the foot surface. These unicellular glands are involved in mucous secretion. Embedded in between there are transversely muscle fibers, called as longitudinal muscle fibers. Major part of the foot muscles are made up of thickly arranged oblique muscle fibers.

Histopathological observations in foot region of *Biomphalaria* snail samples showed necrotic change (shrinkage) in the mucous secreting unicellular glands (Figure 1b) and hyaline substances are shown in samples collected from Port Said (Figure 1c) and splitting fiber tissues in Dakahlya and Damietta snails (Figure 1d,1e). Also, results showed oblique splitting muscle fibers, increased empty spaces and atrophy within muscles of snail head in Dakahlya samples (Figure 1f,1g,1h).

Salivary gland: The normal salivary gland of *B. alexandrina* snail composed of two lobs found in the buccal mass as shown in (Figure 2a).

Parameter			AST (Unites/ml)		ALT (Unites/ml)		ALP (IU/L)		Glucose		Creatinine (mg/dl)		urea (mg/dl)	
Treatments			Mean ± SD	Change %	Mean ± SD	Change %	Mean ± SD	Change %	Mean ± SD	Change %	Mean ± SD	Change %	Mean ± SD	Change %
Control lab			21.9 ± 4		41.5 ± 10		38.4 ± 8		45.0 ± 8		0.55 ± 0.22		9.4 ± 1	
Bort-Said	Kaar El-Bahr	*Biomphalaria*	26.2 ± 2	20	64.3 ± 24	55	42.5 ± 4	11	76.1 ± 27	69	0.35 ± 0.0	-36	13.8 ± 1	47
		Planorbis	30.2 ± 4	38	50.6 ± 7	22	56.8 ± 11	48	72.4 ± 11	61	0 ± 0.0	-100	12.4 ± 0	32
	El-Khankak	*Biomphalaria*	11 ± 0**	-50	57.9 ± 0	40	106 ± 0.0	176	85.7 ± 0.0	90	1.46 ± 0.0	165	30 ± 0.0*	219
	Kobry El-Lansh	*Biomphalaria*	37.4 ± 5*	71	74.8 ± 12	80	94.0 ± 29	145	99.2 ± 6*	120	0 ± 0.0	-100	27.3 ± 1.2**	190
		Planorbis	40.5 ± 6*	85	65.7 ± 11	58	101.4 ± 11*	164	96.8 ± 13*	115	1.53 ± 1.25	178	25.9 ± 1.1**	176
Dakahlya	Gammalya	*Biomphalaria*	31.1 ± 0**	42	65 ± 15	57	65.5 ± 10	71	91.0 ± 23	102	0.40 ± 0.04	-27	22.9 ± 13**	144
	Nasayma	*Biomphalaria*	29.6 ± 3*	35	44.1 ± 10	6	52.0 ± 4	35	80.1 ± 19	78	1.31 ± 0.94	138	15.7 ± 3	67
		Planorbis	56.4 ± 2**	158	85.8 ± 9*	107	105.6 ± 10*	175	131.1 ± 6**	191	0.38 ± 0.0	-31	19.3 ± 2*	105
		Physa	53.8 ± 3**	146	66.4 ± 10	60	162.3 ± 4.7**	323	119.3 ± 9*	165	1.41 ± 0.0	156	20.4 ± 2*	117
	Matarya	*Biomphalaria*	32.1 ± 2*	47	44 ± 17	6	66.3 ± 3.3*	73	85.3 ± 3.1*	90	0.12 ± 0.0	-78	22.8 ± 8	143
Damietta	Annanya	*Physa*	34.2 ± 5	56	53.6 ± 7	29	112.7 ± 4**	193	66.4 ± 12	48	0.08 ± 0.0	-85	14.5 ± 3	54
	Sayala	*Planorbis*	30.6 ± 4	40	51.2 ± 8	23	57.9 ± 11	51	67.4 ± 16	50	0 ± 0.0	-100	12.3 ± 1	31

*, ** & *** significant compared to control value at $p<0.05$, $p<0.01$ & $p<0.001$, respectively.

Table 1: Aspartate amino transferase (AST), alanine amino transferase (ALT) alkaline phosphatase (ALP), glucose, creatinine and urea in tissue extract of snails collected from Lake Manzala.

Parameter			Total protein (g/dl)		Albumin (g/dl)		Globulin (g/dl)		A/G Ratio ◊	Total Bilirubin (Umol/l)		Direct Bilirubin (mg/dl)		In-Direct Bilirubin (U/mg)	
Treatments			Mean ± SD	Change %	Mean ± SD	Change %	Mean ± SD	Change %		Mean ± SD	Change %	Mean ± SD	Change %	Mean ± SD	Change %
Control lab			6.21 ± 0.1		3.27 ± 0.02		2.94 ± 0.11		1.13	2.8 ± 0.1		0.2 ± 0.04		2.5 ± 0.1	
Bort-Said	Kaar El-Bahr	*Biomphalaria*	10.35 ± 0.4**	67	6.84 ± 0.30**	109	3.54 ± 0.11*	20	1.93	4.4 ± 0.0*	1.6	0.6 ± 0.0	200	3.8 ± 0.0*	52
		Planorbis	11.97 ± 0.6**	93	5.52 ± 0.57*	69	6.45 ± 0.04***	119	0.85	4.8 ± 0.8	2	0.1 ± 0.0	-50	4.7 ± 0.8	88
	El-Khankak	*Biomphalaria*	12.42 ± 0.04**	100	7.53 ± 0.26**	130	4.89 ± 0.30*	66	1.63						
	Kobry El-Lansh	*Biomphalaria*	12.39 ± 0.8**	100	6.42 ± 0.08***	96	3.87 ± 0.74	32	1.22	6.3 ± 0.3**	3.5	0.3 ± 0.0	50	6.0 ± 0.3**	140
		Planorbis	10.86 ± 0.6**	75	6.24 ± 0.51*	91	4.62 ± 0.12**	57	1.34	6.5 ± 1.2*	3.7	1.1 ± 0.1**	450	5.4 ± 0.4**	116
Dakahlya	Gammalya	*Biomphalaria*	11.49 ± 0.3**	85	6.72 ± 0.40**	106	4.77 ± 0.12**	62	1.44	5.3 ± 0.1**	2.5	0.1 ± 0.4	-50	5.1 ± 0.2**	104
	Nasayma	*Biomphalaria*	23.04 ± 0.5***	271	13.95 ± 0.30***	327	9.12 ± 0.22***	210	1.53	4.5 ± 0.0*	1.7	2.5 ± 0.0**	1150	2.0 ± 0.0	-20
		Planorbis	24.27 ± 0.7***	291	12.66 ± 0.04***	287	11.61 ± 0.63**	295	1.12	10.4 ± 0.0**	7.6	0.4 ± 0.0	100	10.0 ± 0.0*	300
		Physa	8.94 ± 0.2**	44	6.24 ± 0.04***	91	2.7 ± 0.15	-8	2.35	14.7 ± 1.8*	11.9	1.8 ± 1.3	800	12.9 ± 0.6**	416
	Matarya	*Biomphalaria*	11.67 ± 0.59**	88	7.95 ± 0.35**	143	3.72 ± 0.25	27	2.16	5.0 ± 0.1**	2.2	0.4 ± 1.5	100	4.4 ± 0.2**	76
Damietta	Annanya	*Physa*	9.9 ± 0.6*	59	5.1 ± 0.41*	56	4.8 ± 0.16**	63	1.05	5.7 ± 0.9*	2.9	2.1 ± 0.4*	950	3.6 ± 1.3	44
	Sayala	*Planorbis*								4.8 ± 0.8	2	0.3 ± 0.2	50	4.5 ± 0.7	80

*, ** & *** significant compared to control value at $p<0.05$, $p<0.01$ & $p<0.001$, respectively.
◊ A/G = Ratio of albumin / globulin concentration

Table 2: Total protein, Albumin, globulin, A/G ratio, total Bilirubin, direct and indirect in tissue extract of snails collected from Lake Manzala.

Parameter			CAT (Unites/g)		GST (Unites/g)		GGT (Unites/g)	
Treatments			Mean ± SD	Change %	Mean ± SD	Change %	Mean ± SD	Change %
Field collected snails								
Control lab			9.06 ± 0.18		3.07 ± 0.5		999 ± 24	
Port-Said	Kaar El-Bahr	*Biomphalaria*	3.175 ± 0.37**	-65	1.52 ± 0.18	-50	653 ± 311	-35
		Planorbis	15.15 ± 1.9*	67	5.60 ± 0.13*	82	160 ± 19***	-84
	Kobry El-Lansh	*Biomphalaria*	3.96 ± 0.67**	-56	2.11 ± 0.26	-31	1241 ± 172	24
		Planorbis	7.59 ± 0.00	-16	4.66 ± 0.0	52	1227 ± 0*	23
Dakahlya	Gammalya	*Biomphalaria*	7.92 ± 1.59	-13	2.44 ± 0.33	-21	909 ± 147	-9
		Planorbis	16.7 ± 0.00*	84	5.31 ± 0.0	73	984 ± 0	-1
	Nasayma	*Biomphalaria*	10.69 ± 1.31	18	3.93 ± 0.49	28	716 ± 118	-28
		Planorbis	1.36 ± 0.25***	-85	0.83 ± 0.19*	-73	1152 ± 237	15
		Physa	0.89 ± 0.00**	-90	0.53 ± 0.0	-83	857 ± 0	-14
	Matarya	*Biomphalaria*	7.19 ± 0.05**	-21	3.63 ± 0.72	18	1274 ± 46*	28
		Physa	18.15 ± 0.00**	100	6.63 ± 0.0	116	915 ± 0	-8
		Planorbis	15.52 ± 0.00*	71	5.80 ± 0.0	89	1056 ± 0	6
Damietta	Annanya	*Physa*	7.195 ± 0.54*	-21	4.04 ± 0.93	32	7651 ± 651**	666
		Planorbis	25.785 ± 1.83**	185	6.71 ± 0.41*	119	3901 ± 731*	290
	Sayala	*Planorbis*	17.31 ± 0.00**	91	3.46 ± 2.7	13	1123 ± 266	12

*, ** & *** significant compared to control value at p< 0.05, p<0.01 & p<0.001, respectively.

Table 3: Catalase (CAT), glutathione-S-transferase (GST) and Gamma-glutamyl transpeptidase (GGT) in tissue extract of snails collected from Lake Manzala.

Parameters			Total cell count		Hyalinocytes			Round small hemocytes			Granuolocytes			Hemoglobin (g/100ml)	
Examined			Mean ± SD	Change %	%	Mean ± SD	Change %	%	Mean ± SD	Change %	%	Mean ± SD	Change %		
Control			3.2 ± 1.0		57	1.85 ± 0.5		27	0.8 ± 0.28		16	0.55 ± 0.2		1.8	
Port Said	Kaar El-Bahr	*Biomphalaria*	2.25 ± 0.7*	-30	61	1.4 ± 0.7	-24	24	0.5 ± 0.28	-38	15	0.35 ± 0.4	-36	2.3	28
		Planorbis	1.2 ± 0.3***	-63	57	0.7 ± 0.0***	-62	24	0.25 ± 0.21***	-69	16	0.25 ± 0.3*	-55	2.6	44
	Kobry El-Lansh	*Biomphalaria*	2.35 ± 0.6	-27	60	1.35 ± 0.4*	-27	24	0.55 ± 0.21	-31	16	0.45 ± 0.5	-18	1.8	0
		Planorbis	0.95 ± 0.1***	-70	45	0.4 ± 0.0***	-78	41	0.35 ± 0.07***	-56	14	0.20 ± 0.0***	-64	2.3	28
Dakahlya	Matarya	*Biomphalaria*	2.6 ± 0.1	-19	71	1.85 ± 0.2	0	18	0.45 ± 0.21*	-44	11	0.30 ± 0.3	-45	2.0	11
	Gammalya	*Biomphalaria*	1.4 ± 0.0***	-56	77	1.1 ± 0.0***	-41	16	0.2 ± 0.00***	-75	7	0.10 ± 0.0***	-82	1.7	-6
	Nasayma	*Biomphalaria*	0.9 ± 0.4***	-72											
Damietta	*Sayala*	*Planorbis*	1.95 ± 0.0**	-39	70	1.4 ± 0.0*	-24	15	0.3 ± 0.0***	-63	15	0.3 ± 0.0**	-45		
	Annanya	*Physa*	2.65 ± 0.5	-17	70	1.85 ± 0.6	0	23	0.55 ± 0.21	-31	7	0.25 ± 0.1**	-55	0.8	-56
		Planorbis	1.63 ± 0.04***	-49											

*, ** & *** significant compared to control value at p<0.05, p<0.01 & p<0.001, respectively.

Table 4: Hematologic parameters of snails collected from Lake Manzala.

The histopathological effects of polluted water showed shrinkage and atrophy in the salivary gland of snails collected from Damietta (Figure 2b), focal areas of necrosis (Figure 2c,2d), large fat vacuoles (Figure 2e) and enlargement of the salivary gland (Figure 2f) in snails collected from Port said.

Central ganglia: The central nervous system ganglia are in the form of compact mass of ring surrounding the esophagus of the snail. (Figure 3) showed that all ganglia exhibit presence of enlarged neurosecretory neurons (Figure 3a). Fibrosis (Figure 3b,3c) and degeneration with large vacuoles (Figure 3d) were observed in snail samples collected from Damietta and Dakahlya (Figure 3).

Hepatopancreas: The normal histological structure of *Biomphalaria* hepatopancreas includes glandular tubules interspersed with connective tissues. The entire gland is enclosed within a thin walled sac called as tunica propria. The hepatopancreatic epithelium is rested on thin basement membrane; at least 3-4 types of cells can be recognized in the hepatopancreatic epithelium of the snail, digestive, calcium and excretory cells (Figure 4a). The histopathological changes showed cellular necrosis followed by loss of secretory activity of the epithelial cells in Port Said samples (Figure 3b). Also, atrophy, degeneration and fat vaculation were noticed in Port Said and Damietta samples (Figure 4c,4d). Dilated lumen and more than two hepatopancreatic tubules connected together with one larger lumen in Dakahlya samples (Figure 4e,4f).

Male organs (Prostate gland): The normal histological structures of the male organs of *B alexandrina* composed mainly of sperm duct and the prostate tubules (Figure 5a).

The histopathological observations of Port Said samples showed severe dilated sperm duct and prostate tubules, dilated lumen of prostate tubules which filled with hyaline and degeneration wall with necrotic change (Figure 5b,5c). While Dakahlya samples showed enlarged sperm duct, degenerated prostate tubules and clogged sperms.

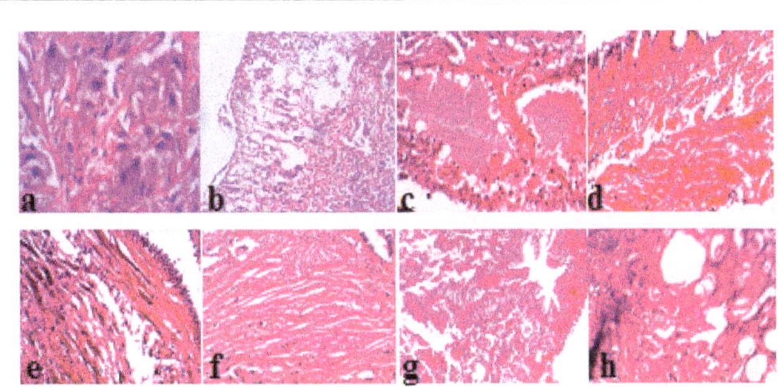

Figure 1: The normal histological structures of the head foot of snail *Biomphalria alexandrina* (a) (X400); shrinkage in the mucous secreting unicellular glands (b) (X100); hyaline degeneration (c) (X400); oblique muscle fiber got splitting and focal areas of necrosis (d &e) (X100); atrophy (f) (X100); empty spaces or vacuoles within muscle (g &h) (X400).

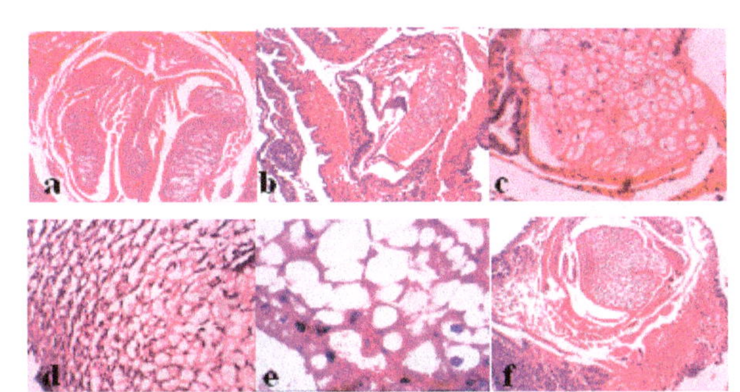

Figure 2: The normal salivary gland of *Biomphalria alexandrina* snail (a) (X40); shrinkage in and degeneration of one lobe (b) (X100); focal areas of necrosis 100x (c &d) (X400); large fat vacuoles (e) (X400); and enlargement of the salivary gland (f) (X400).

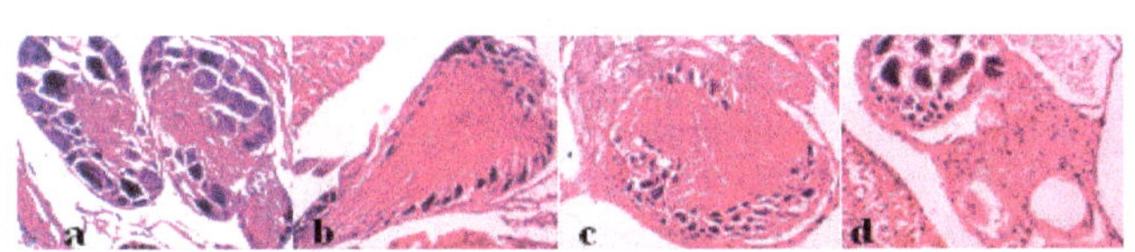

Figure 3: Enlargement of neurosecretory neurons (a) (X400); fibrosis (b, c) (X400); degeneration with large vacuoles (d) (X400).

The prostate gland in Damietta samples showed severe degeneration and atrophy (Figure 5b,5c).

The hermaphrodite gland: Histology of normal hermaphrodite gland of the adult *B. alexandrina* snails as that of any other pulmonate snail consists of number of vesicles known as acini separated from each other by thin vascular connective tissue (Figure 6a). Each acinus is enveloped in a sheath of squamous epithelium. In each acinus both male and female reproductive gametes are produced where mature ova are located at the periphery of the acini and bundles of male sperms are arranged in the center. Various stages of sperm and ovum development (simultaneous) are evident.

Histopathological alteration in Port Said samples included, acini lost their normal architechture and their separating connective tissues are almost degenerated (Figure 5b). The acinar epithilum showed necrotic changes in the form of decreasing cytoplasm of oocytes and partial destruction (Figure 6c). Atrophy and reduction in the number of sperms was also observed (Figure 6d). Degenerative changes were observed in most of the ova, where some of them have faint staining nuclei and others lost their nucleous (Figure 6e,6f). Some acini appear more or less evacuated and large fat vacuoles can be seen in Dakahlya samples (Figure 6g). Damietta showed the most degenerated features

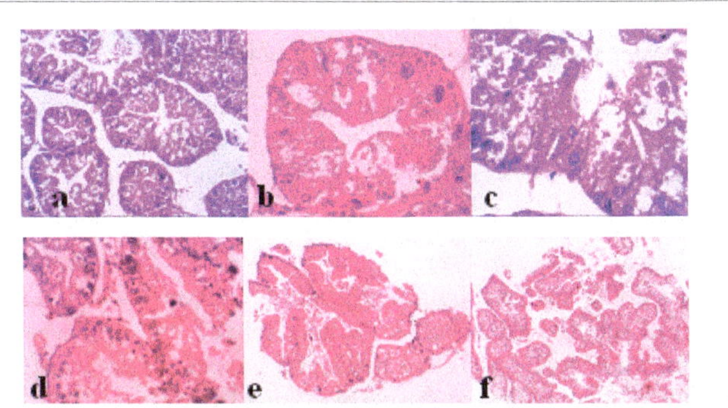

Figure 4: The normal histological structures of hepatopancrease of snail *Biomphalria alexandrina* (a) (X400); vacuolar degeneration of tubules cells (b)(X400); Atrophy, degeneration and fat vaculation (c & d) (x400); severe necrotic change of cells of tubules (b)(X400); Dilated lumen and more than two hepatopancreatic tubules connected together with one larger lumen (e, f)(X100).

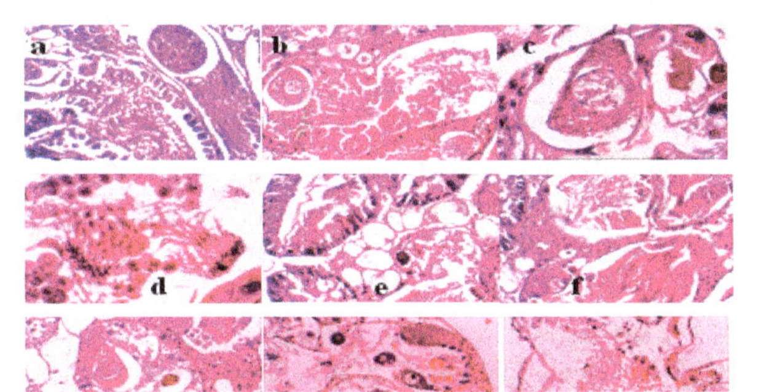

Figure 5: The normal histological structures of the male organs of *Biomphalaria alexandrina* (a) (X100). Severe dilated sperm duct and prostate tubule, dilated lumen of prostate tubules and filled with hyaline degeneration wall of prostate tubules with necrotic change (b & c, X100 & X400), enlarged sperm duct (d, X100), Clogged sperms (e, X100) and degenerated prostate gland (f, X100).

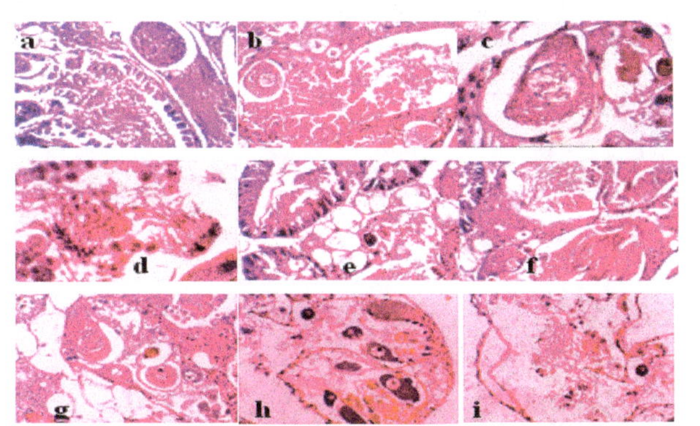

Figure 6: The normal histological structures of hermaphrodite of snail *Biomphalria alexabndrina* (a)(X400); vacuolation and atrophy of different stages of sperm (b)(X400); atretic and absorption of oocytes (c)(X400); atrophy and necrotic change of sperm stages(d)(X400); large fat vacuoles (e)(X400);severe necrotic change(f)(X100); severe fat vacuoles and degenerated hermaphrodite (g)(X100) and degenerated hermaphrodide , atretic oocytes, Atrophy of sperms (h & i, X100).

in the hermaphrodite gland as atretic oocytes and sperms and atrophy of most gland components (Figure 6h,6i).

The infected *Biomphalaria* samples collected from lake manzala

Some of *Biomphalaria* samples collected from Dakahlya and Port Said showed the presence of parasite sporocycts.

The oblique muscle fiber got damaged and mother sporocysts take place within foot muscles, thereby causing splitting, necrosis and increased empty spaces within muscle fibers (Figure 7a-7c).

The digestive gland was destructed while daughter sporocycts which contain many developing cercariae were noticed. The histopathological changes of digestive gland of *B. alexandrina* induced exudation in the lumen of tubules, expansion of hemolymphatic spaces between the tubules, loosing of connective tissue and increase of vaculation and necrotic changes in the digestive cells (Figure 7d,7e,7f).

Accumulation of heavy metals in snail tissues

(Figure 8) showed the accumulation of heavy metals in head foot tissues of *Biomphalaria* snails collected from Port Said, Damietta and Dakahlya samples.

Discussion

Under conditions of pollution mollusks are susceptible to the pathogenic effects of toxicants, which in turn may result in detrimental changes to their immunological and physiological processes [31].

The present results showed significant increase in AST and ALT, and ALP in *Planorbis* and *Physa* snail samples, collected from Nasayma site in Lake Manzala. Moreover, the results showed alterations in CAT, GST and GGT activity in snail samples collected from Lake Manzala.

AST and ALT are vital enzymes in the metabolism and generation of energy from amino acids [32]. Therefore, the elevated transaminases may indicate the high energy demand of the snail under stressful conditions of intoxication. Also, the increase in ALT, AST and ALP enzymes were correlated with alteration in phospholipid metabolism [33] which indicated mainly to hepatocellular disorder [34]. Under physiological stress conditions in animals, the catalytic activity of the urea pathway enzymes is also accelerated [35].

These results are in agreement with [36] who recorded a significant increase of transaminases activity and catalase in the garden snail specimens (*Helix pomatia* L) which were collected from polluted area

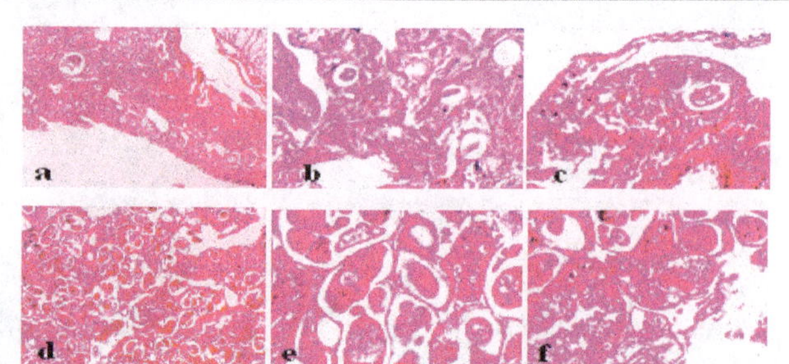

Figure 7: Head foot muscle of *Biomphalaria* snail containing mother sporocyct causing splitting, necrosis and increased empty spaces within muscle fibers (a, b, c X100); hepatopancrease acini filled with different stages of *S. mansonai* cercariae causing degeneration, loosing of connective tissue and increase of vaculation and necrotic changes in the digestive cells (d, X100 & e, f X400).

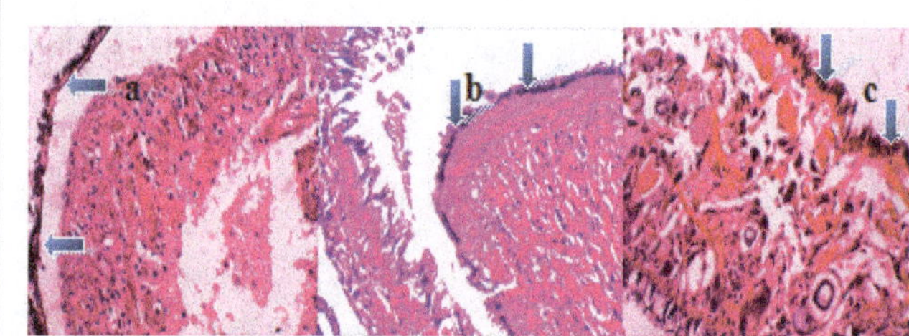

Figure 8: Port Said samples showed mantel layer more dark in color (arrows), separated from head foot (edema) and large number of pigment cells scattered in head foot with necrotic change in the middle of connective tissues (a, X400). Damietta samples showed darkened of mantel (arrows) closed to connective tissues of head foot (b, X400). Dakahlya samples showed atrophy of connective tissues with edema and darkened its outer layer (c, X400).

compared to control. Also, [37] indicated that there are significant elevations in the levels of acid phosphatase and alkaline phosphatase, after using of Profenophos against *B. alexandrina*, which can be explained by the destruction of internal snail cells. Mohamed revealed an elevation in the activities of AST, AlT and AkP enzymes in snails' tissues post treatment with LC_{10} and LC_{25} of Basudin, Selecron and Bayluscide in comparison with control groups [38]. Some other authors recorded increase of the activity of these enzymes, while others recorded decrease in intoxicated animals [36,39,40]. Abdel-Daim reported increased serum AST, ALT, ALP, cholesterol, urea, uric acid, creatinine and tissue MDA after application of deltamethrin subacute intoxication (1.46 µg/L for 28 days) against *Oreochromis niloticus* fish [41]. At the same time they found that tissue levels of GSH, GSH-Px, SOD and CAT were reduced. On the other hand, [42] recorded suppression of the antioxidant enzyme activity and alterations of serum biochemical parameters in freshwater fish Nile tilapia, *Oreochromis niloticus*.

Significant increase of total protein level was recorded also in all samples collected from Lake Manzala. This increase may be attributed to the changes in hepatic protein synthesis [43,44] due to the stress in the polluted habitat. These results go in the same direction as those of [45] who recorded an increase in the total protein concentration in *Helix* snails dependent in the presence of metal dust. Also, [46] highlighted a significant increase in the total protein rate under the effect of a chemical stress at different biological models. Mello observed significant changes in protein metabolism in response to exposure to different concentrations of *E. splendens* var. *hislopii* latex, with significant increases in snails exposed to 0.8 and 1.0 mg/l of the latex, indicating latex toxicity [47]. The same was observed by [48] using other plants and higher concentrations.

Snails collected from most Port-Said and Dakahlya sites showed significant increase in urea. Urea is only synthesized in liver from excess amino acids and excreted by kidney and major illness may increase urea levels [49]. The variation in the nitrogen degradation products showed that the increase of urea content occurred when the uric acid level declined. In accordance with this, the exposure of *Biomphalaria glabrata* to *Euphorbia splendens* var. *hislopii* latex caused the urea content increased which reflects a disturbance in the snail's regulation of their metabolism due to intoxication caused by the latex exposure [50].

Snail samples collected from most sites of Lake Manzala showed significant decrease in total hemolymph cell count, hyalinocytes, round small hemocytes and granuolocytes. The decrease in hemolymph cells may be considered as a haemolysis response to the multiple pollution elements in Lake Manzala. This was mentioned by [51] that haemocytosis represents a response to external stress or certain stimuli and may originate from a variety of biotic or abiotic sources [52]. These results were in agreement with [53] who found that exposure to dyestuff and chemical effluent could result in decreases in RBC count and Hb content which are symptoms of anemia.

The histopathological changes produced by pollutants in organs and tissues can occur before they produce irreversible effects on the biota. So, histological methods can be used in conjunction with other parameters and/or ecotoxicological bioindicators as an early warning system for the survival of the species, as well as for environmental protection.

Histopathological observations in head foot region of *Biomphalaria* snails showed shrinkage in the mucous secreting unicellular glands and hyaline substances in samples collected from Port Said sites, splitting fiber tissues, increased empty spaces and atrophy within muscles of snail head in Dakahlya and Damietta samples. The salivary gland of snails collected from Damietta showed shrinkage and atrophy while there were focal areas of necrosis, large fat vacuoles and enlargement of the salivary gland in snails collected from Port Said. All snails ganglia showed modified and enlargement of neurons, degeneration with large vacuoles and fibrosis in samples collected from Damietta and Dakahlya. The histopathological changes of hepatopancreas included cellular necrosis followed by loss of the epithelial cells were shown in Port Said samples. Also, atrophy, degeneration and fat vaculation were noticed in Port Said and Damietta samples. Dilated lumen and more than two hepatopancreatic tubules connected together with one larger lumen in Dakahlya samples. The prostate gland in snails of Port Said samples showed severe dilated sperm duct and prostate tubules, dilated lumen of prostate tubules which filled with hyaline and degeneration wall with necrotic change. While Dakahlya samples showed enlarged sperm duct, degenerated prostate tubules and clogged sperms. The prostate gland in Damietta samples showed severe degeneration and atrophy. Regarding the hermaphrodite gland in Dakahlya samples, decreasing cytoplasm of oocytes, partial destruction, lost nucleus, large fat vacuoles, atrophy and reduction in the number of sperms were observed. Damietta samples showed the most degeneration features in the hermaphrodite gland as atretic oocytes and sperms and atrophy of most gland components.

All these histopathological damages in snail organs may be due to the pollution of Lake Manzala water by heavy metals which recorded by [22]. Stress responses in invertebrates can occur following acute or chronic exposures to contaminated environments and as such, the overall health status of individuals within those environments, both in terms of histopathological lesions and the presence of infecting organisms, may ultimately reflect the general health status of these sites [54].

The digestive glands of molluscs have been known as target organs for contaminant effects because; this organ plays a major role in contaminant uptake, intracellular food digestion and metabolism of inorganic and organic chemicals in the organisms [55-57]. However, particulate metal uptake is mainly achieved via the digestive tract by endocytosis; further metals are transferred first to lysosomes and then to residual bodies, especially in the digestive cells of the digestive gland [58]. It could also be possible that in the damage in the snail's hepatopancreas including the alteration of liver and kidney enzymes is according to functionality analog with vertebrate's liver that accumulate mostly heavy metals compared to other organs, and which damage it also [59,60].

In agreement of these results, the exposure of the snails *Archachatina marginata* to sublethal concentrations of the metals resulted in a prevalence of hepatocellular foci of cellular alterations (FCA) in the hepatopancreas of snails. Basophilic adenoma and ovotesticular fibrillar inclusions were also observed in the ovotestes of snails exposed to the test metals [61]. Jonnalagadda have been reported histopathological alterations such as degeneration and the gathering of amebocytes in areas between the tubules in the digestive gland of snail *Bellamya dissimilis* exposed to endosulfan [62]. The histopathological examinations of *Lymnaea luteola* exposed to Paraquat (Gramoxone) revealed the following changes: amebocytes infiltrations, the lumen of digestive gland tubule was shrunken; degeneration of cells, secretory cells became irregular, necrosis of cells and atrophy in the connective tissue of digestive gland [63]. Moreover, it is worthy to mention that in the freshwater snails nervous system has been proved to be sensitive to many toxic materials and cytotoxicants that may induce injurious consequences [64-66].

Some of *Biomphalaria* samples collected from Dakahlya and Port Said showed the presence of parasite sporocycts. The most histopathological deleterious effects have been noticed within the tissues caused in the foot and hepatopancreas due to the invasion of larval trematode parasites to the host snail *B. alexandrina*. The oblique muscle fiber got damaged may be due to penetration of miracidia at the time of infection in the nature. Since earlier stages of larval development i.e. sporocyst and mother sporocyst, takes place within foot muscles, thereby causing increased empty spaces within muscle fibers after their entry in to the viscera of the snail. The digestive tubules epithelium got damaged to the extent of loss of normal tubular structure may be due to metabolic and other excretory materials in the form of granules found scattered in the connective tissue. The destruction of the digestive gland was even more severe may be due to the developing of daughter sporocycts which contains many of the developing cercariae.

Similar observations were recorded by [61] in the snail *Archachatina marginata* that the digestive gland tubule becomes compressed thereby resulting reduced tubular lumen of the gland as observed by that more cercaria and rediae were found in between the hepatic tubules and tunica propria causing extension of the space between tubules.

The histological observations of *Biomphalaria* snails collected from Lake Mazala showed accumulation of heavy metals in the head foot tissues. This was proved in the study of [22] who recorded that the metals concentrations were higher in snail tissues and water samples from Lake Manzala. The collected water samples from Damietta sites showed the highest significant Cu & Cd concentration while Port-Said samples showed the highest Pb concentration and Dakahlia showed the highest Zn concentration.

In conclusion, the severe alterations and degeneration recorded in the physiological and hematological parameters and also histopathological observations are clear evidence for the pollution of the water from which these snail samples were collected. This conclusion is confirmed by [67] who recorded highly significant concentrations of Cu, Cd, Pb and Zn in water samples from different Lake Manzala sites. Also, these metals were highly concentrated in snail and fish tissues and the higher metal bioaccumulation was determined in snails collected from sites showed higher water metals concentrations.

Acknowledgement

This study is a joint project (Biomarkers as indicators of environmental pollution: Experimental approach and case studies), kindly funded by the Academy of Scientific Research and Technology through the Bilateral Agreement between Academy of Scientific Research and Technology of the Arab Republic of Egypt and Bulgarian Academy of Sciences (2012-2014).

References

1. Badawy MI, Wahaab RA (1997) Environmental impact of some chemical pollutants on Lake Manzala. Int. J. Environ. Hlth. Res. 7: 161-170.

2. Abdel-Baky TE, Hagras AE, Hassan SH, Zyadah MA (1998) Heavy metals concentration in some organs of *Oreochromis aureus* stein in Lake Manzala. E Egypt. J. Egypt. Ger. Soc. Zool. 25: 237-256.

3. Ibrahim A, Bahnasawy M, Mansy S, El-Fayomy R (1999) Distribution of heavy metals in the Damietta Nile Estuary ecosystem. Egypt. J. Aquat. Biol. Fish. 3: 369-397.

4. Ali MHH (2008) Assessment of some water quality characteristics and determination of some heavy metals in Lake Manzala, Egypt. Egypt. J. Aquat. Biol. Fish. 2: 133-154.

5. DiGiulio RT, Benson WH, Sanders BM, VanVeld PA (1995) Biochemical Mechanisms: Metabolism, Adaptation and Toxicity. Fundamentals of Aquatic Toxicology: Effects, Environmental Fate and Risk Assessment. pp: 523-562.

6. Lohner TW, Reash RJ, Willet VE, Rose LA (2001) Assessment of tolerant sunfish populations (*Lepomis* sp.) inhabiting selenium-laden coal ash effluents. Hematological and population level assessment. Ecotoxicol. Environ. Saf. 50: 203-216.

7. Cazenave J, Wunderlin DA, Hued AC, De Los Angeles-Bistoni M (2005) Haematological parameters in a neotropical fish, *Corydoras paleatus* (Jenyns, 1842) (Pisces, Callichthyidae) captured from pristine and polluted water. Hydrobiologia. 537: 25-33.

8. Hontela A, Daniel C, Ricard AC (1996) Effects of acute and subacute exposures to cadmium on the interrenal and thyroid function in rainbow trout, *Oncorhynchus mykiss*. Aquat. Toxicol. 35: 171-182.

9. Barton BA, Rahn AB, Feist G, Bolling H, Schreck CB (1998) Physiological stress response of the freshwater chondrostean paddlefish (*Polyodon spathula*) to acute physical disturbances. Comp. Biochem. Physiol. 120: 355-363.

10. Hontela A (1998) Interrenal dysfunction in fish from contaminated sites: *In vivo* and *invitro* assessment. Environ.Toxicol.Chem. 17: 44-48.

11. Benguira S, Hontela A (2000) Adrenocorticotrophin and cyclic adenosine 3′, 5′-monophosphate-stimulated cortisol secretion in interrenal tissue of rainbow trout exposed in vitro to DDT compounds. Environ. Toxicol. Chem. 19: 842-847.

12. Paris-Palacios S, Biagianti-Risbourg S, Vernet G (2000) Biochemical and (ultra) structural hepatic perturbation of *Brachydanio rerio* (Teleostei, Cyprinidae) exposed to two sublethal concentrations of copper sulphate. Aquat. Toxicol. 50: 109-124.

13. Teles M, Pacheco M, Santos MA (2003) *Anguilla anguilla* L. liver ethoxyresorufin O-deethylation, glutathione S-transferase, erythrocytic nuclear abnormalities and endocrine responses to naphthalene and beta-naphthoflavone. Ecotoxicol. Environ. Saf. 55: 98-107.

14. Viarengo A (1989) Heavy metals in marine invertebrates: mechanisms of regulation and toxicity at the cellular level. Rev Aquat Sci. 1: 295-317.

15. Rainbow PS, Dallinger R (1993) Metal uptake, regulation and excretion in freshwater invertebrates. Ecotoxicology of metals in invertebrates. pp: 119-131.

16. Roesijadi G, Robinson WE (1993) Metal regulation in aquatic animals: mechanisms of uptake, accumulation and release. Molecular biological and biochemical approach to aquatic toxicology. pp: 387-420.

17. Dallinger R (1995) Mechanisms of metal incorporation into cells. Cell biology in environmental toxicology. Bilbao, Spain: University of the Basque Country Press. pp: 135-154.

18. Dallinger R (1995) Metabolism and toxicity of metals: metallothioneins and metal elimination. Cell biology in environmental toxicology. Bilbao, Spain: University of the Basque Country Press. pp: 171-190.

19. Taylor MG (1995) Mechanisms of metal immobilization and transport in cells. Cell Biology in environmental toxicology. Bilbao, Spain: University of the Basque Country Press. pp: 155-170.

20. Brown MT, Depledge MH (1998) Determinants of trace metal concentrations in marine organisms. Metabolism of trace metals in aquatic organisms. New York. pp: 185-217.

21. Langston WJ, Bebianno MJ, Burt GR (1998) Metal handling strategies in molluscs. Metabolism of trace metals in aquatic organisms. New York. pp: 219-284.

22. El-Khayat HMM, Mahmoud KMA, Abdel-Hamid H, Abu El Einin HM (2015a) Applications of ISSR rDNA in detecting genetic variations in *Lymnaea natalensis* snails with focusing on the characterization of their collecting sites in certain Egyptian Governorates. African Journal of Biotechnology. 14: 1354-1363.

23. Maltchik L, Lanés LEK, Stenert C, Medeiros ESF (2010) Species-area relationship and environmental predictors of fish communities in coastal freshwater wetlands of southern Brazil. Environ. Biol. Fish. pp: 88: 25-35.

24. Reitman S, Frankel S (1957) A colorimetric method for the determination of serum glutamic oxalacetic and glutamic pyruvic transaminases. Am. J. Clin. Pathol. 28: 56-63.

25. Henry RJ, Cannon DC, Winkleman W (1974) Clinical Chemistry: Principles and Techniques. Harper and Row Publishers, New York.

26. Tietz NW (1995) Clinical Guide to Laboratory Tests. WB Saunders Co, Philadelphia, USA. pp: 622-626.

27. Michelson EH (1966) Specificity of hemolymph antigens in taxonomic discrimination of medically important snails. J. Parasitol. 52: 466-472.

28. Abdul Salam JM, Michelson EH (1983) *Schistosoma mansoni*: Immunofluorescent detection of its antigen reacting with *Biomphalaria glabrata* amoebocytes. Exp. Parasitol. 55: 132-137.

29. Bancroft Jd, Stevens A (1996) Theory and Practice of Histological Techniques. Edinburgh: Churchill Livingstone. pp: 766.

30. Emile MA (1980) Snail-Transmitted Parasitic Diseases. Boca Raton: CRC Press.

31. Morley NJ, Lewis JW, Hoole D (2006) Pollutant-induced effects on immunological and physiological interactions in aquatic host-trematode systems: implications for parasite transmission. J. Helminthol. 80: 137-49.

32. Tunholi V, Lustrino D, Tunholi-Alves V, Mello-Silva CC, Maldonado A, et al. (2011) Biochemical profile of *Biomphalaria glabrata* (Mollusca: Gastropoda) after infection by *Echinostoma paraensei* (Trematoda: Echinostomatidae) Parasitol Res, 109: 885-891.

33. Varley H, Gowenlock AH, Bell M (1980) Enzymes. "Practical Clinical Biochemistry". William Heinemann Medical Books, LTD London. 22: 685-770.

34. El-Khayat HMM, Abu Zikri N (2004) Biochemical situation in *Biomphalaria alexandrina* infected with *Schistosoma mansoni* during twelve weeks post infection. J. Egypt. Ger. Soc. Zool. 43: 57-75.

35. Becker W (1980) Metabolic interrelationships of parasitic trematodes and molluscs; especially *Schistosoma mansoni* in *Biomphalaria glabrata*. Z. Parasitenkd. 63: 101-111.

36. Bislimi K, Behluli A, Halili J, Mazreku I, Halili F (2013) Impact of Pollution from Kosova'S Power Plant in Obiliq on Some Biochemical Parameters of the Local Population of Garden Snail (*Helix Pomatia L.*) Resources and Environment. 3: 15-19.

37. Mohamed R (2011) Impact profenophos (pesticide) on infectivity of *Biomphalaria alexandrina* snails with *schistosoma mansoni* miracidia and on their physiological parameters. *Open J Ecol.* 1: 41-47.

38. Mohamed AM, El-Emam MA, Osman GY, Abdel-Hamid H, Ali REM (2012) Biological and biochemical responses of infected *Biomphalaria alexandrina* snails with *Schistosoma mansoni* post exposure to the pesticides Basudin and Selecron and the phytoalkaloid Colchicine. J. Evol. Biol. Res. 4: 24-32.

39. Naplekova NN, Bulavko GI (1983) Enzyme Activity of Soils Polluted by Lead Compounds. Soviet Soil Sci. 15: 33-38.

40. Perez-Mateos M, Gonzales-Carcedo S (1987) Effect of cadmium and lead on Soil Enzyme Activity. Rev. Ecol. Biol. Soil. 1: 11-18.

41. Abdel-Daim MM, Abdelkhalek NKM, Hassan AM (2015) Antagonistic activity of dietary allicin against deltamethrin-induced oxidative damage in freshwater Nile tilapia. Oreochromis niloticus. Ecotoxicol. Environ. Safety. 111: 146-152.

42. Abdelkhalek NKM, Ghazy EW and Abdel-Daim MM (2015) Pharmacodynamic

interaction of *Spirulina platensis* and deltamethrin in freshwater fish Nile tilapia, Oreochromis niloticus: impact on lipid peroxidation and oxidative stress. Environ Sci Pollut Res. Int. 22: 3023-3031.

43. Saad AM, Hussein MF, Bushara HO, Dargie JD, Taylor MG (1984) Erythrokinetics and albumin metabolism in primary experimental *Schistosoma bovis* infections in *Zebu calves*. J. Comp. Pathol. 94: 249-262.

44. Mahmoud MR, El-Abhar HS, Saleh S (2002) The effect of Nigeila sativa oil against the liver damage induced by *Schistosoma mansoni* infection in mice. J. Enthnopharmacol. 79: 1-11.

45. Grara N, Atailia A, Boucenna M ,Khaldi F, Berrebbah H, et al. (2012) Effects of Heavy Metals on the Snails *Helix aspersa* Bioindicators of the Environment Pollution for Human Health. Int. Conf. Appl. Life Sci. Turkey.

46. Masaya M, Yoshinobu H, Ai Y, Maki K, Yasuo O (2002) Determination of cellular levels of nonproteinthiols in phytoplankton and their correlation with susceptibility to mercury. J. Phycol. 38: 983.

47. Mello-Silva CC, Pinheiro J, Vasconcellos MC, Rodrigues MLA (2006) Physiological changes in *Biomphalaria glabrata* Say, 1818 (Pulmonata: Planorbidae) due to the concentration of the latex of *Euphorbia splendens* var. *hislopii* (Euphorbiaceae). Mem. Inst. Oswaldo Cruz. 101: 03-08.

48. Alcanfor JDX (2001) Ação de extratos de plantas do cerrado sobre *Biomphalaria glabrata* (Say; 1818) hospedeiro intermediário de *Schistosoma mansoni* (Sambom; 1907). Goiânia/Goiás. Master Science dissertation. Instituto de Patologia Tropical e Saúde Pública da Universidade Federal de Goiás. P. 84.

49. Bisop MH, Dubenn-Engelkiry JL, Fody MD (1996) Non protein nitrogen. "Clinical Chemistry, Principles, Procedures, Correlations". Publisher, 227 East Washington Square, Philadelphia, PA 19106. Chapter 16: 341-356.

50. Mello-Silva CC, de Vasconcellos MC, Bezerra JCB, Rodrigues MLA and Pinheiro J (2011) The influence of exposure to *Euphorbia splendens* var. *hislopii* latex on the concentrations of total proteins and nitrogen products in *Biomphalaria glabrata* infected with *Schistosoma mansoni*. Acta Tropica, 117: 101-104.

51. Helal IB, ELMehlawy MH, Rizk ET, EL-Khodary GM (2003) Effect of *Euphorbia peplus* plant extract and the antihelmenthic prazequantel on the defence system of *Biomphalaria alexandria* snail. Egypt. J. Aqaat. Biol. & Fish. 7: 501-505.

52. Wolmarans CT, Yssel E (1988) *Biomphalaha glahrata:* Influence of selected abiotic factors on leukocytosis J. mvertebr. PathoL. 57: 10-14.

53. Koprucu SS, Koprucu K, Urail MS (2006) Acute toxicology of synthetic pyrethroid deltamethrin to fingerling European catfish (Silirus glanis L.). Bulletin of Environmental Contamination and Toxicology. 76: 59-65.

54. Stentiford GD, Feist SW (2005) A histopathological survey of shore crab (Carcinus maenas) and brown shrimp (*Crangon crangon*) from six estuaries in the United Kingdom. J Invert Pathol. 88: 136-46.

55. Rainbow PS, Phillips DJH (1993) Cosmopolitan biomonitors of trace metals. Marine pollution bulletin. 26: 593-603.

56. Marigomez I, Soto M, Cajaraville MP, Angulo E, Giamberini L (2002) Cellular and sub cellular distribution of metal in mollusks. Microscopy research and technique. 56: 358-392.

57. Usheva LN, Vaschenko MA, Durkina VB (2006) Histopathology of the digestive gland of the bivalve mollusk Crenomytilus grayanus (Dunker, 1853) from southwestern Peter the Great Bay, Sea of Japan. Russ J Mar Biol. 32: 166-172.

58. Marigómez I, Lekube X, Cajaraville MP, Domouhtsidou G, Dimitriadis V (2005) Comparison of cytochemical procedures to estimate lysosomal biomarkers in mussel digestive cells. Aquat Toxicol. 75: 86-95.

59. Frazier JM (1979) Bioacumulation of cadmium in marine organism. Environ. Helath Perspect. 28: 75.

60. Benedeti L, Balongnani L, Balongnani FA, Marini M, Otaviani E (1982) Effect of pollution on some freshwater species I. Histochemical and biochemical features of lead Viviparus viviparous (Mollusca, Gastropoda). Basic and appl. Histochem. 26: 79.

61. Otitoloju AA, Ajikobi DO, Egonmwan RI (2009) Histopathology and Bioaccumulation of Heavy Metals (Cu & Pb) in the Giant land snail, *Archachatina marginata* (Swainson). Open Environ Poll Toxicol J. 1: 79-88.

62. Jonnalagadda PR, Rao BP (1996) Histopathological changes induced by specific pesticides on some tissues of the fresh water snail, *Bellamya dissimilis*. Bulletin of Environmental Contamination and Toxicology. 57: 648-654.

63. Kanapala VK, Arasada SP (2013) Histopathological Effect of Paraquat (Gramoxone) on the Digestive Gland of Freshwater Snail *Lymnaea luteola* (Lamarck: 1799) (Mollusca: Gastropoda). Int J Sci Res Environ Sci. 1: 224-230.

64. Hernadi L, Vehovszky A (1992) Ultrastructural biochemical and electrophysiological changes induced by 5, 6-dihydroxytryptamine in the CNS of the snail *Helix pomatia* L. Brain Res. 578: 221-234.

65. Boer HH, Moorer-van CM, Muller LJ, Kiburg B, Vermorken JB, et al. (1995) Ultrastructural neuropathological effect of taxol on neurons of the fershwater snail Lymnaea stangnalis. J. Neuro-Oncel. 17: 49-57.

66. Wiemann M, Wittkowaski W, Altrup U, Speckmann EJ (1995) Alterations of neuronal fibers after epileptic activity induced by pentylenetetrazole: fine structure investigated by calcium cytochemistry and neurobiotin labeling (buccal ganglia, Helix pomatia). Cell Tissue Res. 289: 43-53.

67. El-Khayat HMM, Mahmoud KMA, Gaber HS, Abdel-Hamid H, Abu Taleb HMA (2015b) Studies on the effect of pollution on Lake Manzala ecosystem in port-said, damietta and Dakahlya governorates, Egypt. J. Egypt. Soc. Parasitol. (JESP). 45: 155-168.

Population Dynamic and Stock Assesment of White Seabream *Diplodus sargus* (Linnaeus, 1758) in the Coast of North Siani

Ahmed M Al-Beak[1]*, Ghoneim, SI[2], El-Dakar AY[3] and Salem M[2]

[1]*General Authority for Fish Resources Development (GAFRD), Egypt*
[2]*Suez Canal University, Egypt*
[3]*Suez University, Egypt*

Abstract

In the present study fisheries, population dynamic and stock assessment of *Diplodus sargus* in the coast of North Siani (Eastern Mediterranean, Egypt) studied. Length weight relationship, catch length structure, length scale relationship, total length by the end of each year of life, growth in weight, Von Bertalanffy parameters, the values of (total, natural and fishing mortalities), survival rates, Approximate maximum length with the highest biomass of *D. sargus* and approximate maximum age t_{max}. Also Cohort analysis (VPA, age based) which represent the estimated values of the population numbers, survivors, natural and fishing mortalities for each year of life of *D. sargus* were studied.

Keywords: Eastern mediterranean; Age and growth; *Diplodus sargus*

Introduction

The white seabream *Diplodus sargus* [1] is a commercial species found throughout the eastern temperate Atlantic and Mediterranean Seas [2,3] where it occurs in coastal rocky reef areas and Posidonia beds.

Due to its economic importance this species made the subject of study of various scientists in different countries [4-23].

White seabream was a good valuable commercial fish in Egypt, representing nearly 757 tons yearly about 1.1% by value of total catches for the Egyptian Mediterranean yield from year 2001 to year 2012 [24].

The aim of this study is to establish biological key characteristics and population parameters, where it is necessary for management and fish stock assessment in the Eastern Mediterranean and to compare these with data from other Mediterranean regions.

Materials and Methods

All of 991 fishes specimens of *D. sargus* (TL=11-38 cm) where collected during the period from September 2010-April 2012, in El-Arish Marin Seaport from the catches by El-Dabba (Trammel net) gear about 95%, and by the Long line gear about 5%.

Several scales (5-6) were removed from the area below the pectoral fin, making sure that they were not regeneration scales, washed and stored dry in individually labeled envelopes.

Total length (TL) was measured to the nearest mm. And Total weight (TW) recorded to the nearest gram.

Total length-Scale radius relationship computed according to [25] total length-total weight relationship was computed according to [26]. Estimate the growth parameters of the [27] by fitting the [28,29] while "to" was estimated by inverse [27] for t_0 from L_∞ and K, and the asymptotic weight "W_∞" was estimated by converting "L_∞" to the corresponding weight using the obtained formula for length weight relationship.

Length with the highest biomass in an unfished population (L_{opt}), estimated according to [30] from the parameters of the [27] growth function and natural mortality.

Estimate of life span (t_{max}) according to [31], where it is the approximate maximum age that fish of a given population would reach.

Instantaneous total mortality coefficient "Z" estimated by means of the following methods [32-35]. The Powell-Wetherall plot based [36] discussed in [37] and Linearized catch curve based on age composition data [38]. Instantaneous natural mortality coefficient "M" estimated by means of the following methods [39-46]. The fishing mortality coefficient "F" estimated by subtracting the natural mortality coefficient from the total mortality coefficient.

The exploitation rate "E" estimated by the formula suggested by [47]. Estimation of survival rates "S" as a number of fish alive after a specified time interval, divided by the initial number, usually on a yearly basis was done according to [38] equation.

Virtual Population Analysis (VPA) has become one of the most commonly used age-and time-dependent fish population models in fisheries science to analyze the historical data for estimation of population parameters of *D. sargus* in the coast of North Sinai [48,49].

Results

Besides, fishery management plans rely on accurate age determinations; if age estimations are not validated, errors in age determination could result in inaccurate mortality estimates, underestimation of strong year classes and longevity [50].

Total length-scale radius relationship

Microscopic examination of scales growth rings showed a linear

***Corresponding author:** Ahmed M Al-Beak, General Authority for Fish Resources Development (GAFRD), Egypt, E-mail: albeak2020@yahoo.com

regression between Length and scale radius of *D. sargus* represented by a straight line (Figure 1), the following formula representing this relationship:

$$L = 5.305\ S - 2.528 r = 0.983$$

where, "L" is the total length (cm) and "S" is the total scale radius (micrometre division).

Length-weight relationship

The obtained equation found to be representing the relation between lengths and weights of *D. sargus* were:

$W=0.011L^{3.165}$ with $r=0.976$

This relation can be explained graphically as in (Figure 2).

Theoretical growth in length and weight

Theoretical growth in length and weight of *D. sargus* in the coast of north Sinai by solve [27] growth equation for length and weight and fitting the [28,29] plot, were found as follows and constant as in (Table 1).

For length $Lt = 40.71(1 - e^{-0.2497(t+0.2794)})$

For weight $Wt = 1368.1(1 - e^{-0.2497(t+0.2794)})^{3.165}$

Estimation of L_{opt} and t_{max}

Approximate maximum length with the highest biomass of *D. sargus*, caught from North Sinai coast was 26.63 cm, and approximate maximum age t_{max} was 11.73 years.

Population structure

Demographic structure

Length composition: Total length frequency composition of *D. sargus* distributed with 28 size groups from size range 11-11.9 cm to size range 38-38.9 cm, the size group 16-16.9 cm (about 16.25% of the total frequencies), is the domination in the size groups and the size groups 11-11.9 cm and 38-38.9 (about 0.10% of the total frequencies) were the least size group frequencies (Figure 3).

Age composition: Age composition of *D. sargus* with the percentage of fishes of each five age groups we found the age group (I) is dominant in the catch about 48.34%, where Age group (V) is the least represented group in the catch about 4.14% (Figure 4).

Instantaneous mortality and survival rats: Instantaneous total mortality rate "Z" of *D. sargus* found 0.7066 year^{-1}, even as the instantaneous natural mortality "M" and instantaneous fishing mortality "F" was 0.3961 year^{-1} and 0.3105 year^{-1} respectively. Survival rate from age composition data using [38] equation found 0.4857.

Exploitation rate: Exploitation rate "E" of *D. sargus* in North Sinai coast found 0.4394 where it less than the optimum fishing mortality in an exploited stock suggested by [47], approximately 0.5.

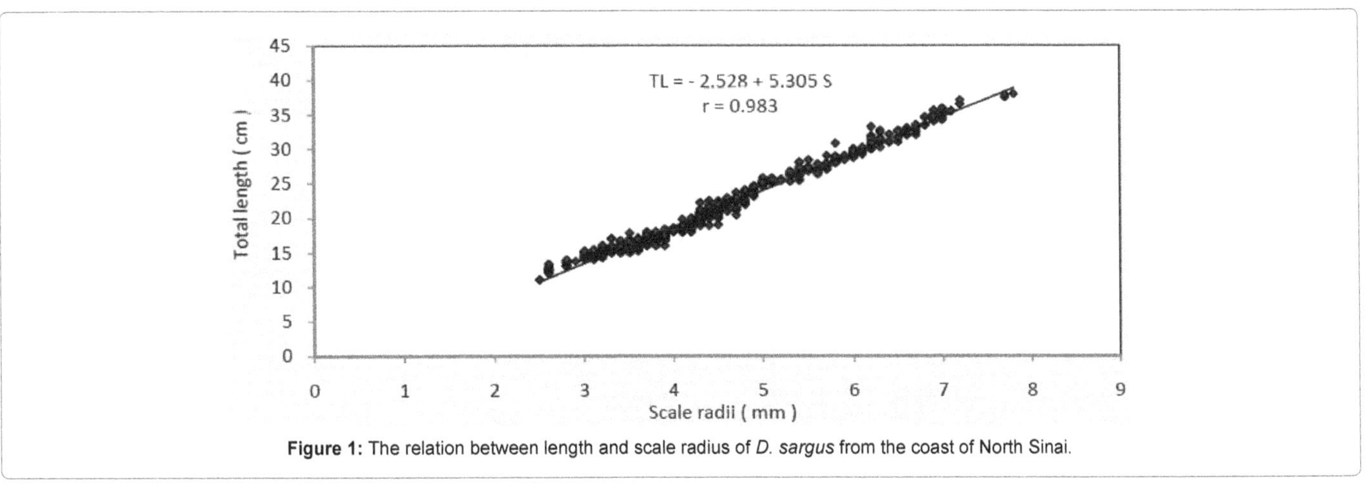

Figure 1: The relation between length and scale radius of *D. sargus* from the coast of North Sinai.

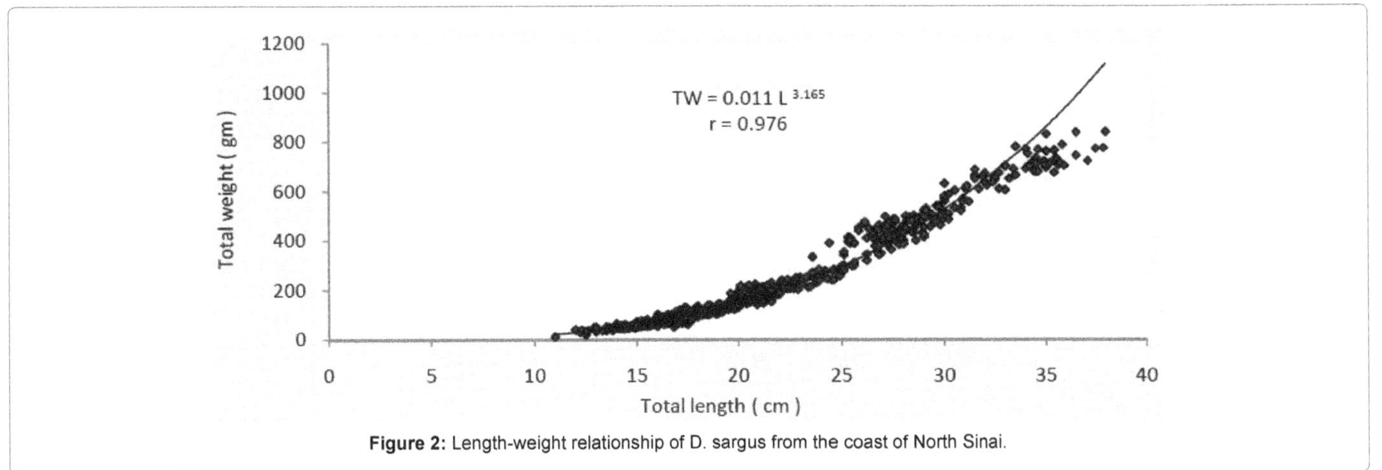

Figure 2: Length-weight relationship of D. sargus from the coast of North Sinai.

Constants	Ford (1933)–Walford (1946)
L_∞	40.71 cm.
K	0.2497 year^{-1}
t_o	-0.2794 year^{-1}
W_∞	1368.1 gm.

Table 1: Constants of Von Bertalanffy's growth equation of *D. sargus* from the coast of North Sinai.

Virtual population analysis

Estimation of age and time-population model done by using [48] cohort analysis as in Figure 5 and we perceive that this model defined cumulative instantaneous rate of fishing mortality "F" of the fish in the population, where it increased to the maximum value at age group 2 (0.4295 year^{-1}) then it decreased, age group 1 and 5 have minimum value. Population number and survivors had decreased from age group 1 to 5 by the natural losses of biomass of cohort and fishing losses, while catches number increased in smaller ages (1 and 2) and decreased in older ages (4 and 5) where that means there was fishing pressure of smallest fishes.

Discussion

Biological management of fisheries resources is generally aimed at preventing overfishing and optimizing yield. Age and growth parameters are the most important study to our understanding of the species biology was enable to control of fishing.

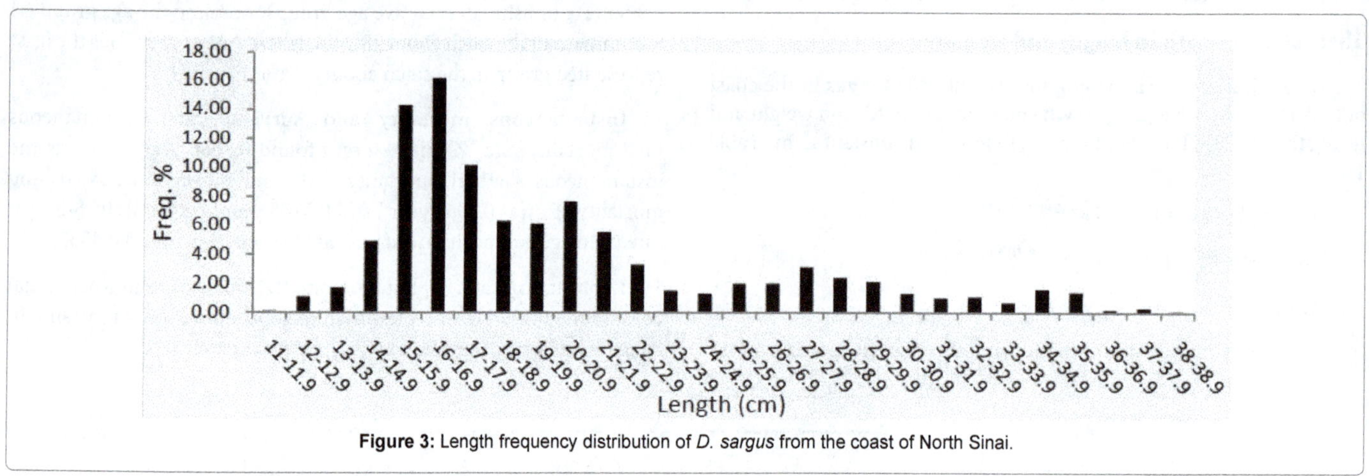

Figure 3: Length frequency distribution of *D. sargus* from the coast of North Sinai.

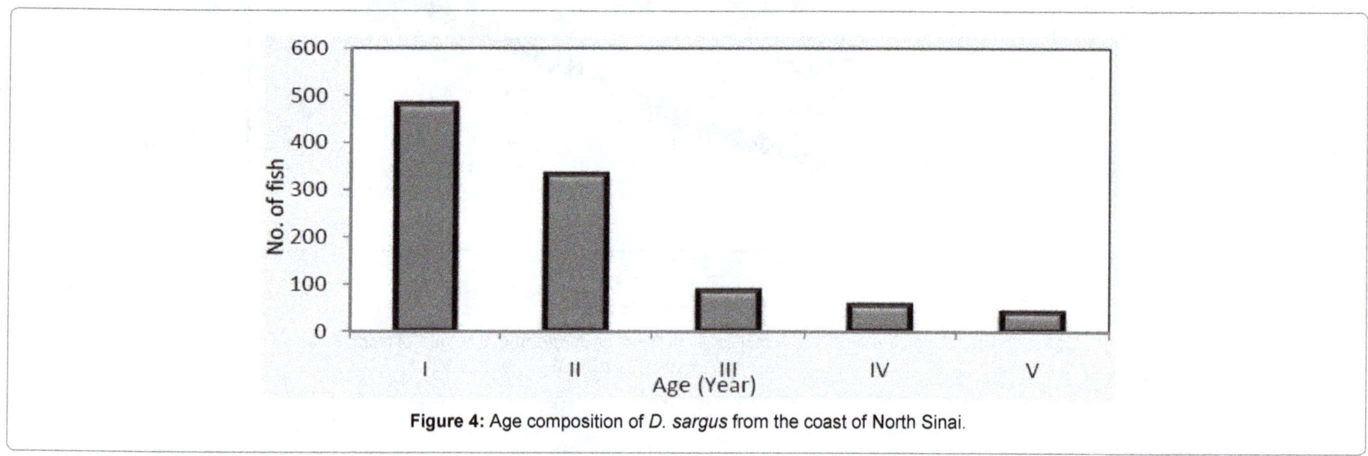

Figure 4: Age composition of *D. sargus* from the coast of North Sinai.

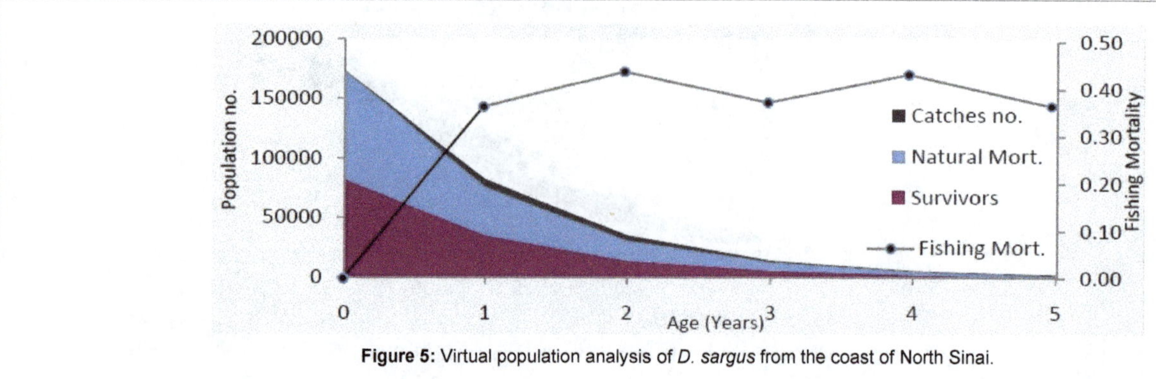

Figure 5: Virtual population analysis of *D. sargus* from the coast of North Sinai.

In the present work, data for body length and scale radius show a linear on their scatter diagram. For length-weight relationship of *D. sargus* in the coast of North Sinai fishery we found "b" parameter (b=3.165), this was agreement with [51] in the Egyptian Mediterranean water b=3.144, [52] in the Gulf of Lion b=3.123, [53] in the Azores b=3.181 for both Males and Females, for Males was b=3.032 and for Females was b=3.054. Less than [54] in the South-East coast of South Africa b=3.242 and more than [55] in the Egyptian Mediterranean water b=2.859, [21] in the Gulf of Tunis b=3.129 for Males, for Females b=2.994 and for all individuals b=3.051, [23] in Abu Qir bay b=2.942 and [56] in the Eastern cost of Algeria b=2.987.

Growth parameters of the [27-29] by fitting the plot are arranged (Table 2). From this table we can perceive a varied diverse between authors and we can deduce that there are difference between growths in different locations; it may be return to the water surface temperature and food abundance.

The approximate maximum age (t_{max}) of *D. sargus* in the coast of North Sinai are less than [57] in South Africa t_{max}=21 years and t_{max}=14 years. Similar with [17] t_{max}=12 years, t_{max}=13.4 years and [23] in Abu Qir bay t_{max}=11.45 years, these means that *D. sargus* have the same maximum age in the nearest geographic locations.

Length frequency distributions provide snapshots of the combination of fish species present and the sizes of individuals at particular locations and times. The smallest fish length in the catch of *D. sargus* in present work was 11 cm TL, while the biggest length was 38 cm TL, this result indicates that the *D. sargus* in the coast of North Sinai was fishing in small lengths at the first year of life.

In addition, if the length composition of our sample reflects the commercial fishery catches, we must point out that more than 70% of the fish caught were smaller than length at first maturity. Therefore, in order to improve the stock management of *D. sargus* and the conservation of this species in the coast of North Sinai, an increase in the minimum legal length authorized for capture is strongly recommended.

Author and date	Method	L_∞	K	t_o	W_∞	Location
[52]	Scales	46.7	0.12	-1.63	2089	N/W Medit.
		0			7	
[60]	Otoliths	48.4	0.18	-0.06	-	N/E Atlantic
		8			-	
[9]	Otoliths	41.7	0.25	-0.08	-	N/W Medit.
		0			-	
[57]	Otoliths	30.9	0.25	-1.05	-	South Africa
		4			-	
[17]	Otoliths	47.3	0.14	-1.97	-	Canary Islands
		0			-	
[22]	Otoliths	40.9	0.18	-1.28	-	South Portugal
	Scales	39.5	0.15	-1.89	-	
[61]	Otoliths	41.2	0.18	-0.86	524	South
[55]	Scales	32.7	0.13	-1.84	-	Egypt
[23]	Scales	31.3	0.26	-0.73	-	Abu Qir Bay
		8			-	
[56]	All	36.3	0.15	-0.49	-	Eastern Algeria
	Males	35.1	0.16	-0.43	-	
	Females	35.4	0.16	-0.6	-	
Present study	**Scales**	**40.7**	**0.25**	**-0.28**	**1368**	**E. Medit.**

Table 2: Von Bertalanffy's growth parameters (L∞, K, t_o and W∞) for *D. sargus* for various authors and in different locations.

The most dominant age in the catch of *D. sargus* is age group I where it contributes about 48.34%, this result was similar with [52,56] that indicate not only there was a fishing pressure on this fish but also that occur in different locations.

Before estimating the fishing and natural mortalities separately, it is convenient to estimate the total mortality. Instantaneous total mortality coefficient "Z" of *D. sargus* in the coast of North Sinai fishery was 0.7066 year^{-1}. [53] estimated the total mortality for *D. vulgaris* in the South coast of Portugal, which was 0.625 year^{-1}, [23] found total mortality was 1.092 year^{-1} for *D. sargus* and 1.049 year^{-1} for *D. vulgaris.*

Estimating natural mortality "M" is one of the most difficult and critical elements of a stock assessment [58]. Natural mortality "M" of *D. sargus* in present study was 0.3961 year^{-1}, [23] estimated natural mortality which was 0.606 year^{-1} for *D. sargus* and 0.600 year^{-1} for *D. vulgaris.* The same species may have different natural mortality rates in different areas depending on the density of predators and competitors, whose abundance is influenced by fishing activities [59].

Estimates of fish mortality rates are often included in mathematical yield models to predict yield levels obtained under various exploitation scenarios. Fishing mortality "F" in present study was 0.3105 year^{-1}.

Exploitation rate is the fraction of an age class that caught during the life span of a population exposed to fishing pressure, the exploitation rate was 0.4394, [22] found exploitation rate was 0.445.

Virtual Population Analysis (VPA) cohort analysis was first developed as age based methods. It is commonly used for studying the dynamics of harvested fish populations [49] the feature of VPA that is most important for practical use is that, given a high fishing pressure, estimates of population size obtained tend to converge rapidly toward their true value, and hence usually provide, given a reasonable estimate of M, reliable estimates of recruitment [48]. Present study could be considered as a base for future studies that help to predict the future catch in North Sinai coast and demonstrate that the *D. sargus* died by natural mortality more than those which die by fishing mortality. It could also be seen that, the increase in fishing mortality as the fish increases in age was accompanied by a decrease in the population numbers of the species understudy. On the other hand, the natural mortality decreases as the fish gets older. These results are in agreement with [23] in Abu Qir bay, Alexandria, Egypt.

From previous results we can conclude that the white seabream *D. sargus* in the coast of North Sinai facing more stress which fishing effort is more based on age groups I and II with small lengths from 14 to 17 cm. Also, it must to catch these fishes at ages more than 2 till 3 years to give it chance to grow to economical preferred size and to reduce overfishing of its first year of life.

References

1. Linnaeus C (1758) Systema naturae per regna tria naturae, secundum classes, ordinus, genera, species, cum characteribus, differentiis, synonymis, locis. Tomus I. Editio decima, reformata. Impensis Direct. Laurentii Salvii, Holmiae. P. 824.

2. Fisher W, Schneider M, Bauchot ML (1987) Fiches FAO d'identification des espèces pour les besoins de la pêche. Méditerranée et mer Noire. Zone de pêche. Végétaux et Invertébrés (FAO species identification cards for needs of fishing. Mediterranean and Black sea. Fishing area. Plants and Invertebrates). Rome, FAO. pp. 1-760.

3. Lenfant P, Planes S (1996) Genetic differentiation of white seabream within the

Lion's Gulf and the Ligurian Sea (Mediterranean Sea). Journal of Fish Biology. 49: 613-621.

4. Girardin M (1978) Les Sparidae (Pisces, Teleostei) du Golfe du Lion-Ecologie et Biogeographie. Universit'e des Sciences et Techniques du Languedoc, Laboratoire Ichthyologie et de Parasitologie G'en'erale, Montpellier, Diplome D'Estudes Approfundies D' ' Ecologie G'en'erale et Apliqu'ee-Option Ecologie Aquatique. P. 146.
5. Wassef EA (1985) Comparative biological studies of four *Diplodus* species (Pisces: Sparidae). *Cybium* 9: 203-215.
6. Rosecchi E (1987) L'Alimentation de *Diplodus annularis*, *Diplodus sargus*, *Diplodus vulgaris* et *Sparus aurata* (Pisces, Sparidae) dans le golfe de Lion et les lagunes littorales. Rev. Trav. Inst. Pe^chesmarit. 49: 125-141.
7. Abou-Seedo F, Wright JM, Clayton DA (1990) Aspect of the biology of Diplodus sargus kotschyi (Sparidae) from Kuwait bay. Cybium. 14: 217-223.
8. Harmelin JG, Leboulleux V (1995) Microhabitat requirements for settlement of juvenile sparid fishes on Mediterranean rocky shores. Hydrobiologia. Pp. 300-320.
9. Gordoa A, Moli B (1997) Age and growth of the *sparids D. vulgaris, D. sargus and D. annularis* in adult populations and the differences in their juvenile growth patterns in the North Western Mediterranean Sea. *Fish. Res.* 33: 123-129.
10. Sala E, Ballesteros E (1997) Partitioning of space and food resources by three fish of the genus Diplodus (Sparidae) in a Mediterranean rocky infralittoral ecosystem. *Mar. Ecol. Prog. Ser.* 152: 273-283.
11. Macpherson E, Biagi F, Francour P, Garcia RA, Harmelin J, et al. (1997) Mortality of juvenile fishes of the genus Diplodus in protected and unprotected areas in the Western Mediterranean Sea. *Mar. Ecol. Prog. Ser.* 160: 135-147.
12. Macpherson E (1998) Ontogenetic shifts in habitat use and aggregation in juvenile sparid fishes. J. Exp. Mar. Biol. Ecol. 220: 127-150.
13. Planes SE, Macpherson, Biagi F, Garcia RA, Harmelin J, et al. (1999) Spatio temporal variability in growth of juvenile sparid fishes from Mediterranean littoral zone. J. Mar. Biol. Assoc. UK. 79: 137-149.
14. Gonçalves JMS (2000) Biologica Pesqueirae Dinamica Populacionalde *Diplodus vulgaris* (Geoffr) e Spondylio soma cantharus (L) (Pisces, Sparidae) na costa Sudoeste de Portugal. Universdado do Algarve, UCTRA, Faro, Ph.D. Thesis. P. 369.
15. Vigliola, L, Harmelin-Vivien ML (2001) Post-settlement ontogeny in three Mediterranean reef fish species of the genus Diplodus. *Bull. Mar. Sci.* 68: 271-286.
16. Mariani S (2001) Cleaning behaviour in Diplodus spp: chance or choice? A hint for future investigations. J. Mar. Biol. Assoc. UK. 81: 715-716.
17. Pajuelo JG, Lorenzo JM (2002) Growth and age estimation of *Diplodus sargus cadenati* (Sparidae) off the Canary Islands. *Fish. Res.* 59: 93-100.
18. Lanfant J (2003) Demographic and genetic structure of white seabream populations (*Diplodus sargus*, Linnaeas, 1758) inside and outside a Mediterranean reserve. *C.R.* Biologies 326: 751-760.
19. Morato TP, Afonso P Lourinho, RDM Nash, RS Santos (2003) Reproductive biology and recruitment of the sea bream in the Azores. *J. Fish. Biol.* 63: 59-72.
20. Pajuelo JG, Lorenzo JM (2004) Basic characteristics of the population dynamics and state of exploitation of Moroccan white seabream, *Diplodus sargus cadenati* (Sparidae) off Canarian Archipelago. *J. Appl. Ichthyol.* 20: 15-21.
21. Mouine NP, Francour M Ktari, NCH-Marzouk (2007) The reproductive biology of *Diplodus sargus sargus* in the Gulf of Tunis (central Mediterranean). *Scientia Marina*. 71: 461-469.
22. Abecasis D, Bentes L, Coelho R, Correia C, Lino PG, et al. (2008) Ageing Seabream: A comparative study between scales and otoliths. Fisheries Research. 89: 37-48.
23. Mahmoud HH, Osman AM, Ezzat AA, AM Saleh (2010) Fisheries biology and management of *Diplodus sargus sargus* (Linnaeus, 1758) in Abu Qir Bay, Egypt. *Egy. J. Aquat. Res.* 36: 123-131.
24. GAFRD (2012) Fish statistics year book, General Authority for Fish Resources Development. Egypt. Pp: 106.
25. Whitney RR, Carlender KD (1956) In temperature of body Scale regression for competing body length of fish. J. wild. Management. 20: 21-27.
26. Le Cren ED (1951) The length-weight relationship and seasonal cycle in gonadal weight and condition in perch (Perca fluviatilis). J. Animal Ecol. 20: 201-219.
27. Von Bertalanffy L (1938) A quantitative theory of organic growth. (Inquiries on growth laws II). Hum. Biol. 10: 181-213.
28. Ford E (1933) An account of the herring investigation conducted at Ply Mouth. J. Marin. Biol. Ass. UK. Pp: 305-384.
29. Walford LA (1946) A new graphic method of describing the growth of animals. Biol. Bull. Mar. Biol. 90: 141-147.
30. Beverton RJH (1992) Patterns of reproductive strategy parameters in some marine teleost fishes. J. Fish Biol. 41: 137-160.
31. Taylor CC (1958) Cod growth and temperature. J. Cons. CIEM. 23: 366-370.
32. Heinke F (1913) Investigations on the plaice. General report 1. Plaice fishery and protective regulations. Rapp Pv Réun. Cons. Int. Explor. Mer. 17: 1-153.
33. Jackson CHN (1938) The analysis of animal population. J. Anim. Ecol. 8: 238-264.
34. Chapman DG, Robinson DS (1960) The analysis of a catch curve. Biometrics. 16: 354-368.
35. Beverton RJH, Holt SJH (1956) A review of methods for estimating mortality rates in exploited fish populations, with special reference to sources of bias in catch sampling. Rapp Pv Réun. CIEM. 140: 67-83.
36. Powell DG (1979) Estimation of mortality and growth parameters from the length frequency of a catch. Rapp Pv Réun. CIEM. 175: 167-169.
37. Wetherall JA, Plovina JJ, Ralston S (1987) Estimating growth and mortality in steady-seate fish stocks from length-frequency data. ICLARM. Conf. Proc. 13: 53-74.
38. Ricker WE (1975) Computation and interpretation of biological statistics of fish populations. Bull. Fish. Res. Broad of Canada. 191: 2-6.
39. Ursin E (1967) A mathematical model of some aspects of fish growth, respiration and mortality. J. Fish. Res. Bd. Can. 24: 2355-2453.
40. Alverson DL, MJ Carney (1975) A graphic review of the growth and decay of population cohorts. J. Cons. int. Explor. Mer. 36: 133-143.
41. Pauly D (1980) On the interrelationships between natural mortality, growth parameters, and mean environmental temperature in 175 fish stocks. J. Cons. Int. Explor. Mer. 39: 175-192.
42. Hoenig JM (1983) Empirical Use of Longevity Data to Estimate Mortality Rates. Fishery Bulletin. 82: 898-903.
43. Chen S, Watanabe S (1989) Age Dependence of Natural Mortality Coefficient in Fish Population Dynamics. Nippon Suisan Gakkaishi. 55: 205-208.
44. Jensen AL (1996) Beverton and Holt life history invariants result from optimal trade-off of reproduction and survival. Canadian Journal of Fisheries and Aquatic Sciences. 53: 820-822.
45. Lorenzen K (1996) The relationship between body weight and natural mortality in juvenile and adult fish: a comparison of natural ecosystems and aquaculture. Journal of Fish Biology. 49: 627-647.
46. Hewitt DA, Hoenig JM (2005) Comparison of two approaches for estimating natural mortality based on longevity. Fishery Bulletin. 103: 433-437.
47. Gulland JA (1971) The fish resources of the Oceans. Fishing News Books Ltd. England. Pp: 255.
48. Pope JG (1972) An investigation of accuracy of virtual population analysis using cohort analysis. Res. Bull. ICNAF. 9: 65-74.
49. Xiao Y, YG Wang (2007) A revisit to Pope's cohort analysis. Fish. Res. 86: 153-158.
50. Beamish RJ, GA Macfrlane (1983) The forgotten requirement for age validation in fisheries biology. Trans. Am. Fish. Soc. 112: 735-743.
51. El Maghraby AM, Botros GA (1981) Maturation, spawning and fecundity of two sparid fish *Diplodus sargus* L and *Diplodus vulgaris*, *Geoffer* in the Egyptian Mediterranean waters. *Bull. Nat. Inst. of Oceanogr. Fish.* ARE 8: 51-67.
52. Man-Wai R, Quignard JP (1982) The seabream *Diplodus sargus* (Linne 1758) in Gulf of Lion: growth of the seabream and characteristics of landings from the

commercial fishing grounds of Sete and Grau-du-Roi. Rev. Trav. Inst. Peches Marit. Nates. 46: 173-194.

53. Morato TP, Afonso P, Lourinho JP, Barreiros RS, Santos, et al. (2001) Weight length relationship for 21 coastal fish species of Azores, North Eastern Atlantic. Fish. Res. 50: 297-302.

54. Mann BQ (1992) Aspect of the biology of tow inshore sparid fishes *(Diplodus sargus capensis* and *Diplodus cervinus hottentotus*) off the South-East coast of South Africa. M.Sc. Thesis. Rhodes University.

55. Lahlah M (2004) Ecological studies on two fish species inhabiting coastal Seaweed meadous in Alexandria waters. Ph.D. Thesis. Alex. Univ. Fac. of Scince.

56. Benchalel W, MH Kara (2012) Age, growth and reproduction of the white seabream *Diplodus sargus sargus* (Linneaus, 1758) off the eastern coast of Algeria. *J. Appl. Ichthyol.* 29: 640-70.

57. Mann BQ, CD Buxton (1997) Age and growth of Diplodus sargus capensis and D.cevinus hottentotus (Sparidae) on the Tsitsikamma coast, S. Africa. Cybium 21: 135-147.

58. Harmelin-Vivien ML, Hewitt DA, Lambert DM, Hoenig JM, Lipcius RN (2007) Direct and indirect estimates of natural mortality for Chesapeake Bay blue crab. Transactions of the American Fisheries Society. 136: 1030-1040.

59. Sparre P, Venema SC (1998) Introduction to tropical fish stock assessment. Part I, FAO Fish. Tech. pap. 1: 306.

60. Pastor CM, VML Cuadros (1996) Edad, crecimiento y reproducci´on de Diplodus sargus Linnaeus (1758) (Sparidae) en aguas asturianas (norte de Espana). Bollt. Instituto Espanol Oceanografia. 12: 65-76.

61. Erzini K, Bentes L, Coelho R, Correia C, Lino P, et al. (2001) Fisheries biology and assessment of demersal species (Sparidae) from the South of Portugal. UE. DG XIV-98/082. Final report. pp: 1-263.

Macrobenthic Community Structure - An Approach to Assess Coastal Water Pollution in Bangladesh

Jahangir Sarker Md[1]*, Shamsul Alam Patwary Md[1], Borhan Uddin AMM[1], Monjurul Hasan Md[1], Mehedi Hasan Tanmay[1], Indrani Kanungo[1] and Mohammed Rashed Parvej[2]

[1]*Department of Fisheries and Marine Science, Noakhali Science and Technology University, Noakhali, Bangladesh*
[2]*Bangladesh Fisheries Research Institute, Mymensingh, Bangladesh*

Abstract

A research on the assemblages of benthic macro faunal community in the coastal areas of Bangladesh was conducted during February-March, 2015 following the standard methods to assess the status of environmental pollution. The abundance (r=0.846) and species richness (r=0.864) of the macrobenthic communities were significantly influenced by the water salinity of the sampling sites ($p \le 0.05$). Both the study areas namely the Bakkhali River Estuary and the Meghna River Estuary showing the highest (3909 ± 540 ind./m^2) and lowest (2236 ± 689 ind./m^2) density of benthic macrofaunal abundance respectively might be considered as moderately polluted areas according to the results obtained from Shannon-Wiener index of species diversity (2.69 ± 0.13 and 2.00 ± 0.11 respectively) and Margalef's species richness (2.21 ± 0.43 and 1.36 ± 0.11 respectively). Therefore, it is plausible that the macrobenthic community explained in the present study might be a key future outline to assess the status of coastal water pollution of those concerned areas of Bangladesh.

Keywords: Macrobenthos; Bakkhali river estuary; Meghna river estuary; Shannon-wiener index; Margalef's species richness

Introduction

Benthos is the organism that inhabit in bottom of an aquatic body. Benthic communities are usually dominated by different species of polychaete, oligochaete worms, gastropods, bivalvia and various minor insect larvae. Benthic organisms such as macro, meio and micro fauna and flora play an important role in food chains in an aquatic ecosystem [1]. Macrobenthic organisms may be influenced positively or negatively by physico-chemical parameters of the environment depending on their sources [2]. According to environmental conditions benthic communities vary considerably [3]. The amount of nutrients released from the sediment by benthic communities may vary [4]. Various physical and chemical conditions of the water body such as depth, current of the water, organic contents of the sediments, contaminations of bed sediments environment, toxicity of sediments influence the abundance and distribution of macrobenthos [5]. Macrobenthos are the most commonly used organisms for bio-monitoring in lotic habitat worldwide [6]. It is evident that macrobenthos play an important role in improving and preserving water quality through mineralization and recycling of organic matters [7,8]. The physical and chemical status of the riverine ecosystem becomes recognizable through the elasticity of the community structure of the benthic organisms [9,10]. That's why benthic macro-invertebrates make ideal subject for biological assessment of water quality [11].

Bangladesh is blessed with an extensive coastline of about 710 Km [12]. The southeastern and southwestern coast of this country is mostly covered by a complex estuarine ecosystem with strong interactions of biotic and abiotic factors. The main estuarine systems of the country are Brahmaputra-Megna (Gangetic delta), Karnaphuly, Matamuhuri, Bakkhali and Naf rivers, which are comprised of mangroves, salt marshes, sea grass, seaweeds, fisheries, coastal birds, animals, coral reefs, deltas, salt beds, minerals and sand dunes. The estuarine environment, which serves as feeding, breeding and nursery grounds for a variety of animals, varies according to the volume of discharge of the river and tidal range. These diverse living resources in the estuarine environment play an important role which is economically significant in many ways. Although coastal and estuarine resources contribute a vital role in terms of both the ecosystem and the economy, study of the estuarine coastal environment in Bangladesh is still lacking [13].

Khan [14] conducted an investigation on the abundance and distribution of macrobenthic organisms in the Mouri River, Khulna to determine the level of river pollution. They identified twenty (20) different species in their study area where polychaeta dominated all over the river. Abu Hena [15] conducted a primary research work on the composition of macrobenthos in the Bakkhali Channel System, Cox's Bazar to investigate the relationship between soil parameters and the macrobenthos composition in their study area. But there is almost no information on the long term study of benthos particularly in the coastal waters of Bangladesh. Therefore the present study was designed to explore diversity of benthos in the South-Western coastal waters of Bangladesh with the following objectives.

Objectives

- To know the diversity of benthic macro fauna in the Meghna River Estuary (at Chairman Ghat, Noakhali) and the Bakkhali River Estuary (Cox's Bazar).
- To compare the benthic macro faunal abundance in between two selected estuaries.
- To assess the environmental conditions of the Meghna River Estuary and the Bakkhali River Estuary.

***Corresponding author:** Jahangir Sarker Md, Department of Fisheries and Marine Science, Noakhali Science and Technology University, Noakhali, Bangladesh
E-mail: swaponj@yahoo.com

Materials and Methods

Sediment samples were collected from the Meghna River Estuary (at Chairmanghat, Noakhali) and the Bakkhali River Estuary (Cox's Bazar) during February-March, 2015 (Figure 1). 6 sampling stations (3 from each estuary with triplicate fashion) were selected to carry out the present study. Among the two study sites the Bakkhali river estuary, Cox's Bazar, is situated in the southern region of Bangladesh. The approximate geographical location of this estuary is between 20085'40" to 21046'92" N latitude and 91096'60" to 92034'37"E longitude (Figure 1). The estuary is directly influenced by semi-diurnal tides and climatology impacted by monsoon winds where it's bottom consists mostly of muddy and sandy particles [16]. 3 sampling stations from the Bakhkhali River Estuary namely S-IB, S-2B, S-3B and another 3 stations from the Meghna River Estuary namely S-4M, S-5M and S-6M were selected. Besides on the coast of Bangladesh the Meghna River Estuary is a coastal plain estuary. The bathymetry, tides and outflow from the Meghna River are the important driving forces of that estuary [17]. The approximate geographical location of this estuary is between (22035'14.7"N and 91001'31.8"E to 22035'24.6"N and 91001'47.4"E) (Figure 1).

For macrobenthic fauna, samples were collected using a small boat during February to March, 2015. Sediment samples were collected using an Ekman dredge having a mouth opening of 0.02 m^2. Collected sediment samples were sieved through 500 μm mesh screen to retain macrobenthos. The sieved organisms were preserved immediately with 10% formalin solution in the plastic container with other residues. Preserved samples were then brought back to laboratory for further analysis. In the laboratory, small amount of "Rose Bengal" was added to increase visibility of organisms. Identification of macrobanthic fauna were done using simple microscope up to possible taxonomic level [18-20] and their counting were made as total individual per m^2 (ind./m^2). During sampling, in situ water quality parameters were measured at each sampling site. The water salinity (ppt), temperature (°C), pH, and DO (Dissolved Oxygen, mg/L) alkalinity (ppm) were measured using refractometer (NewS-100, TANAKA, Japan), thermometer (centigrade scale), pH meter (HANNA Instruments), DO meter (HANNA Instruments) and Hach hardness and alkalinity kit respectively.

The total number of macro invertebrates was counted in a sample and then number of macro-invertebrates per square meter occurrence was computed using the following formula Welch [21],

$$N = \frac{O}{a.s} * 10000$$

Where

N=Number of macro-invertebrates 1 sq. m. of profoundal bottom

O=No. of macro-invertebrate (actually counted) per sampled area,

a=Transverse area of Ekman dredge in sq. cm, and

s=Number of sample taken at one sampling site.

Species diversity index (H)

Species richness index (d); and evenness index were calculated according to following equations

The data harvested from monthly samples were blended to provide the value of Shannon-Wiener Index (Species diversity, H) according to [22],

$$H = -\sum_{i}^{s} =_1 PiLnPi$$

Where

S=Total number of species in a sample,

Pi=ni/N=Proportion of individuals of the total sample belonging to the ith species.

N=Total number of individual of all the species,

ni=Number of individuals belonging to the ith species.

The Margalef's index

Species Richness (D) is simple ratio between total species (S) and

Figure 1: Map showing the location of two study sites namely the Meghna River Estuary, Noakhali and Bakkhali River Estuary, Cox's Bazar of Bangladesh.

total numbers of individual (N) [23]. It can be used to compare one community with another. The index is

$$D = \frac{S-1}{\ln N}$$

Where

D=Margalef's index

S=Number of species in sample

ln=log normal

N=Total number of individuals in sample

Simpson index (D)

The Simpson Index value also ranges between 0 and 1, and the greater the value, the greater the sample diversity [24]

$$\text{Simpson Index } D = \frac{1}{\sum_{i=1}^{s} Pi2}$$

In the Simpson index, P is the proportion (n/N) of individuals of one particular species found (n) divided by the total number of individuals found (N), Σ is still the sum of the calculations, and s is the number of species.

Species Evenness: According to Cox [25,26] the species evenness is

$$E = \frac{H}{H\max}$$

Where

The value of E is between 0.

Hmax=ln(S) and S=S=Total number of species in a sample

H=Shannon Diversity value

Equitability

Shannon diversity divided by the logarithm of number of taxa [27] was measured by using following formula:

$$J = \frac{Hs}{Logs}$$

Where

J=Equability index

Hs=Shannon and Weiner Index

S=Number of species in a population

The dominance index (D)

The dominance index [27] was measured to determine whether or not particular fisheries species dominate in a particular aquatic system and can be useful index of resource monopolization by a superior competitor, particularly in communities that have been invaded by exotic species. This index was determined by using following formula:

$$D = \sum_{i=1} \left(\frac{ni}{n}\right) 2$$

Where

ni=number of individuals of species i

n=total number of individuals

Menhinick's richness index

The ratio of the number of taxa to the square root of sample size [28].

$$I_{Menhinick} = S / \sqrt{N}$$

Where

S=Number of species in sample

N=Total number of individuals in sample

Brillouin index

It is measured by using following formula [29]

$$I_{Brillouin} = \frac{\ln(N!) - \sum \ln(n_i!)}{N}$$

Where

N! is N factorial, i.e., N × (N-1) × (N-2) × (N-3) × ... × 3 × 2 × 1

Fisher's alpha

A diversity index, defined implicitly by the following formula [27].

$$S = a \times \ln(! + n / a)$$

Where

S=number of taxa,

n=number of individuals and

a=Fisher's alpha.

Berger-Parker dominance

According to Harper the Berger-Parker dominance is simply the number of individuals in the dominant taxon relative to n [27].

Paleontological Statistics (PAST) version 3.15, a software package for paleontological data analysis written by Ryan [30] was used to run the analysis. PAST has grown into a comprehensive statistical package that is used not only by paleontologists, but in many fields of life science, earth science, and even engineering and economics.

Results and Discussion

The abundance of macrobenthos was studied during February to March, 2015 in the Bakkhali river estuary situated in Cox's Bazar district and the Meghna river estuary situated at Chairman Ghat in Noakhali district. Among the observed water quality parameters (Table 1) temperature (°C) is the important one because it has a major influence on biological activity and growth and the higher the water quality the greater the biological activity (Washington State Department of Ecology, 1991). The mean temperature (°C) observed in the Bakkhali River estuary and Meghna River estuary were 28.33 ± 1.53 and 27.33 ± 1.53 respectively. Due to runoff of huge freshwater from other upper rivers of Bangladesh through the Meghna River to the Bay of Bengal, the average salinity difference of this estuary is lower than the Bakkhali River estuary (Figure 2). The average salinity of the Meghna river estuary was 5.67 ± 0.58 ppt which was lower than the average salinity of the Bakkhali River estuary (22.00 ± 2.65 ppt) during the study period (Table 1). There was no significant difference in pH values observed between two study sites. The average pH measured in the study sites were 6.98 ± 0.45 in the Bakkhali River estuary and 7.87 ± 0.81 in the Meghna River estuary (Table 1 and Figure 3). The values of dissolved oxygen observed in the present study influenced the abundance of the macrobenthic community both in the Meghna River and the Bakkhali River estuary (Figure 4). This result is supported by Islam [31], who reported that Species richness of macrobenthic

Table 1: *In situ* water quality parameters measured from the Bakkhali River Estuary and the Meghna River Estuary.

Sites	Stations	Temperature (°C)	Salinity (ppt)	Dissolved Oxygen (ppm)	pH	Alkalinity (ppm)
Bakkhali River Estuary	*S-1B*	27	19	6.8	7.5	135
	S-2B	28	24	13.7	6.75	120
	S-3B	30	23	7.1	6.7	114
	Mean ± Sd	28.33 ± 1.53	22 ± 2.65	9.20 ± 3.9	6.98 ± 0.45	123 ± 10.82
Meghna River Estuary	*S-4M*	26	6	12.57	7.15	174
	S-5M	27	5	12.89	8.75	168
	S-6M	29	6	9.15	7.72	180
	Mean ± Sd	27.33 ± 1.53	5.67 ± 0.58	11.54 ± 2.07	7.87 ± 0.81	174 ± 6.00

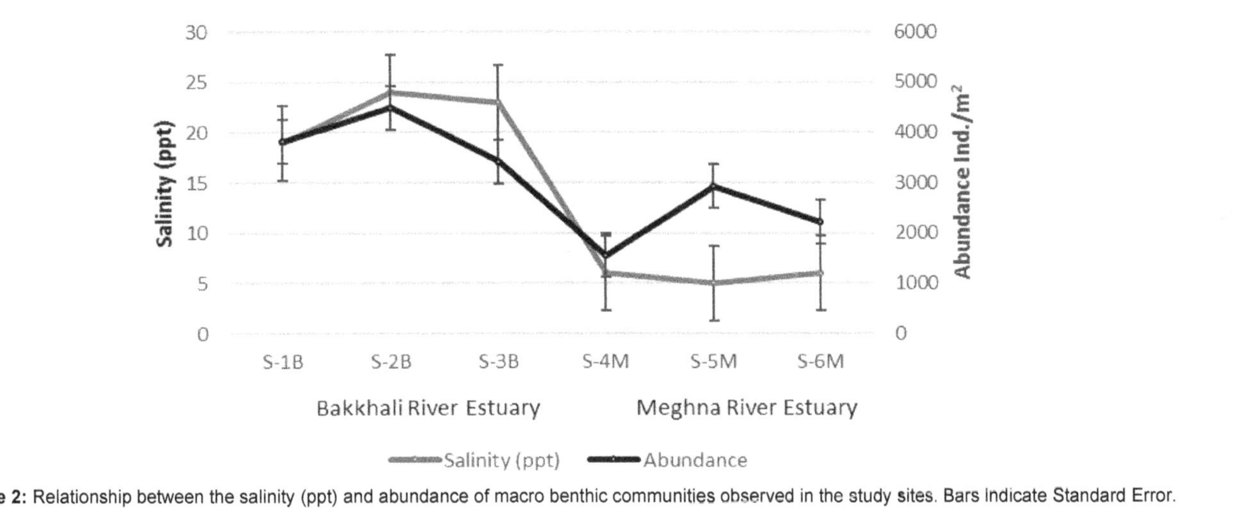

Figure 2: Relationship between the salinity (ppt) and abundance of macro benthic communities observed in the study sites. Bars indicate Standard Error.

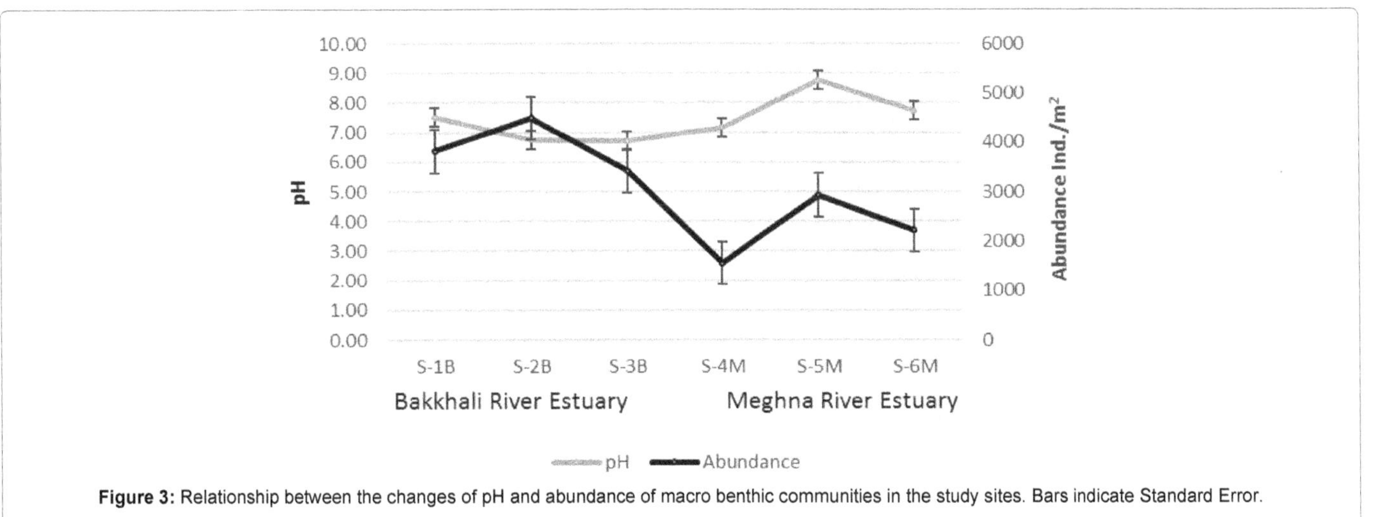

Figure 3: Relationship between the changes of pH and abundance of macro benthic communities in the study sites. Bars indicate Standard Error.

community was positively ($p < 0.005$) influenced by dissolved oxygen and percentage of silt while it was negatively ($p < 0.005$) influenced by percentage of sand and particle density. The abundance of benthic community was significantly ($p < 0.05$) influenced by water salinity (Table 1). Hossain and Marshall [32] also identified that species richness increased onwards, though abundance (density) showed no distinct directional trend. Diversity indices were generally positively correlated with salinity and pH ($p < 0.05$) and negatively with clay and organic matter. Hossain and Marshall [32] suggested that species distribution and community structuring is more strongly influenced by sediment particle characteristics than by the chemical properties of the water (pH and salinity).

5 major groups of macrobenthos (Polychaeta, Oligochaeta, Arthropods, Gastropods and Bivalvia) identified in Bakhlali river estuary (3909 ± 540) was higher than the Meghna river estuary (2236 ± 689) where the existence of bivalvia and gastropoda were found absent in the Meghna river estuary during the study period (Table 2). Polychaete and bivalvia showed the highest (49.42 %) and lowest (5.54 %) density respectively (Table 2) among 28 families (Table 3) identified from 5 major microbenthic groups. The average benthic

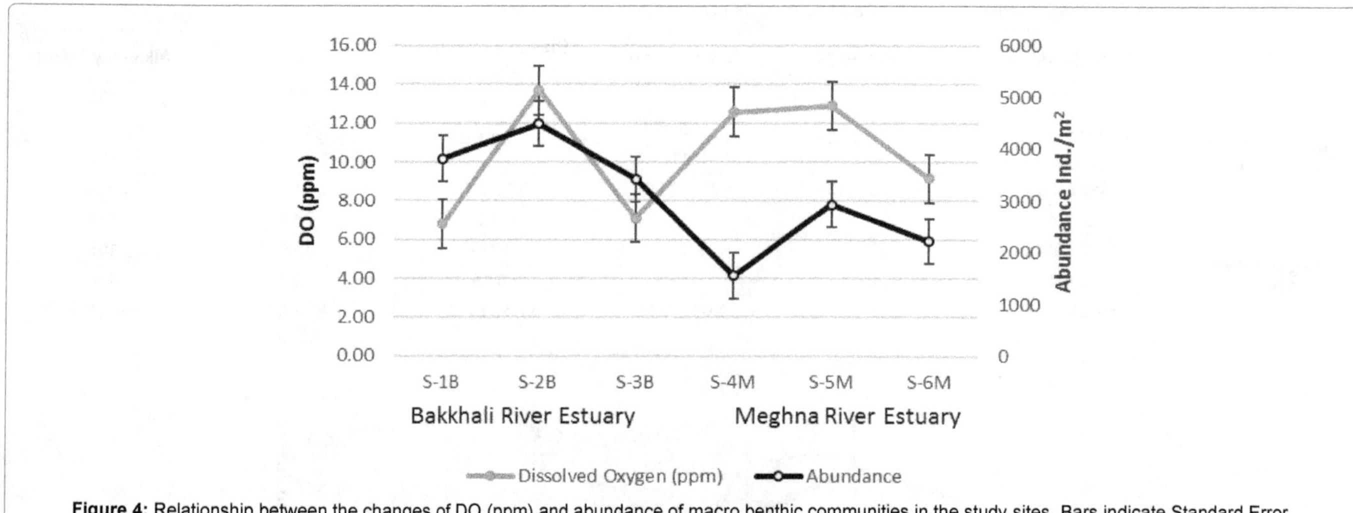

Figure 4: Relationship between the changes of DO (ppm) and abundance of macro benthic communities in the study sites. Bars indicate Standard Error.

Table 2: Abundance of Benthic groups (individuals/m^2) found in the study sites.

Benthos Groups	Bakkhali River Estuary			Meghna River Estuary			Mean ± SD	Total	Percentage (%)
	S-1B	*S-2B*	*S-3B*	*S-4M*	*S-5M*	*S-6M*			
Polychaete	1555	2089	1733	711	1733	1289	1518.33 ± 432.73	9110	49.42
Oligochaete	221	266	177	267	311	222	244 ± 42.83	1464	7.94
Arthropods	266	445	311	311	533	489	392.50 ± 100.91	2355	12.77
Bivalvia	311	400	311	0	0	0	170.33 ± 172.90	1022	5.54
Gastropoda	1066	933	711	0	0	0	451.67 ± 463.38	2710	14.70
Unidentified	400	355	177	266	355	221	295.67 ± 80.07	1774	9.62
Total	3819	4488	3420	1555	2932	2221	3072.50 ± 1070.42	18435	100
Mean ± SD	3909 ± 540			2236 ± 689					

organisms found in the sampling stations of the Bakkhali River Estuary and the Meghna River Estuary were (3909 ± 540 Ind./m^2 and 2236 ± 689 Ind./m^2 respectively) was similar to the work done by Ibrahim who identified that the benthic communities were more dominant during pre-monsoon season (25,836.8 ind./m^2) as compared to post-monsoon season (21,573.1 ind./m^2) in the coral areas of Karah Island, Terengganu, Malaysia. Besides khan identified that the population density varied from 96 to 9410 ind./m^2 in the Mouri River of Bangladesh. Amongst 28, the 10 most abundant microbenthic families recorded from the two study sites were Lumbrinereidae-12.30%, Cerithidae-10.12%, Nerediidae-8.43%, Goniadidae-7.47%, Naididae-7.47, Capitellidae-6.26%, Neptydae-5.07%, Ocypodidae- 4.83, Isaeidae-4.82% and Sternaspidae-2.89% (Table 3). Lumbrinereidae was found dominant both in the Meghna River Estuary and the Bakkhali River Estuary where Cerithidae was dominant in the Bakhkhali river estuary (Table 3). The present findings were quite similar to the findings of Hossain [33] who investigated the polychaetes faunal biodiversity of the Meghna River estuarine bed. Similar results from other study [33] on faunal composition, seasonal abundance of polychaete (ind./m^2), species richness and species biodiversity (Swandwip, Hatiya, Bhola, Barisal and Chandpur) of Bangladesh revealed that Polychaetes was the most dominant among the macrobenthic groups constituted 56.72% of the total macrobenthos. However, results from the present study on microbenthic species composition was a little bit higher than that of the results postulated from Abu Hena on Bakhlali river estuary (Polychaeta, 9.966-30.31%; Oligochaeta, 3.68-30.31%; Crustacea, 0.02-58.40%; Bivalvia, 1.40-82.09% and Gastropoda, 0.08-4.25% and similar to the results of Asadujjaman [14,15,18,34].

Among the recorded 28 macrobenthic families from the study sites, the maximum number of families was found in S-1B and S-2B of the Bakkhali River Estuary where 21 families were common in both stations (Table 3). Abundance of polychaetes were found to be the highest (9110 ind./m^2) among all macro-benthic communities (Table 2). Maximum value (2089 ind./m^2) macrobenthos was found at sampling station S-2B of the Bakkhali River Estuary and minimum (711 ind./m^2) at station S-4M of the Meghna River Estuary (Table 2). A total 15 families were identified under taxonomic group of Polychaeta (Table 3). Olygochaetes were common at all the stations and occupied fourth position as regards to the abundance of total macro-benthos (Table 2). The maximum value (311 ind./m^2) was recorded at station S-5M of the Meghna River Estuary whereas the minimum (177 ind./m^2) value was recorded at station S-2B of the Bakkhali River Estuary (Table 2). A total 2 families were identified under the group of Oligochaeta (Table 3).

Arthropods constituted 12.77% of total macro-benthos (Table 2) and ranked 3rd. The maximum value (533 ind./m^2) was found at station S-5M of the Meghna River Estuary and minimum (266 Ind./m^2) at station S-1B of the Bakkhali River Estuary (Table 2). A total of 4 families were identified under the taxonomic group of Arthropoda (Table 3).

Gastropods constituted 14.70% of total macro-benthos (Table 2) however absent in the Meghna river estuary (Table 2). Gastropods had its highest density (1066 ind./m^2) at station S-1B and lowest (711 Ind./m^2) at station S-3B of the Bakkhali River Estuary (Table 2). A total of 3 families were recorded during the study period under this group (Table 3).

Table 3: Abundance of benthos families (individuals/m^2) observed in the present study.

Family	Bakkhali River Estuary			Meghna River Estuary			Total	Mean	Standard Deviation	Percentage (%)
	S-1B	*S-2B*	*S-3B*	*S-4M*	*S-5M*	*S-6M*				
Capitellidae	356	267	311	44	133	44	1155	192.50	137.09	6.26
Goniadidae	222	400	400	0	311	44	1377	229.50	174.31	7.47
Lumbrinereidae	89	0	178	*356	*756	*889	2268	378.00	366.30	12.30
Nereidae	311	400	177	222	311	133	1554	259.00	99.26	8.43
Onupidae	44	0	0	0	0	0	44	7.33	17.96	0.24
Spionidae	89	0	0	0	0	0	89	14.83	36.33	0.48
Sternaspidae	311	89	133	0	0	0	533	88.83	122.46	2.89
Syllidae	44	178	0	0	0	0	222	37.00	71.28	1.20
Magelonidae	44	89	267	0	0	89	489	81.50	99.21	2.65
Neptydae	44	313	267	44	178	89	935	155.83	115.79	5.07
Paraonidae	0	0	0	0	44	0	44	7.33	17.96	0.24
Maldanidae	0	44	0	0	0	0	44	7.33	17.96	0.24
Sabellidae	0	89	0	44	0	0	133	22.17	37.17	0.72
Glyceridae	0	44	0	0	0	0	44	7.33	17.96	0.24
Cossuridae	0	133	0	0	0	0	133	22.17	54.30	0.72
Orbiniidae	0	44	0	0	0	0	44	7.33	17.96	0.24
Naididae	178	222	177	267	311	222	1377	229.50	52.10	7.47
Tubificidae	44	44	0	0	0	0	88	14.67	22.72	0.48
Mysidae	44	44	44	0	0	44	176	29.33	22.72	0.95
Isaeidae	222	400	267	0	0	0	889	148.17	172.54	4.82
Ampeliscidae	0	0	0	133	177	89	399	66.50	77.98	2.16
Ocypodidae	0	0	0	178	356	356	890	148.33	175.01	4.83
Veneridae	133	222	89	0	0	0	444	74.00	91.69	2.41
Trapezidae	89	44	0	0	0	0	133	22.17	37.17	0.72
Tellinidae	89	133	222	0	0	0	444	74.00	91.69	2.41
Trachidae	222	222	0	0	0	0	444	74.00	114.64	2.41
Cerithidae	*533	*711	*622	0	0	0	1866	311.00	345.30	10.12
Littorinidae	311	0	89	0	0	0	400	66.67	124.88	2.17
Unidentified	400	356	177	267	356	222	1778	296.33	87.74	9.64
Total	3819	4488	3420	1555	2933	2221	18436	3072.67	1070.42	100.00

*Indicates the highest number of the stations.

Maximum value (400 ind./m^2) of bivalvia was found at station S-2B and minimum (311 Ind./m^2) at S-1B and S-3B (Ind./m^2). Bivalvia ranked 7th and contributed 1.15% of total Macro-benthos (Table 2). Total 3 families of benthos were identified under this taxonomic group (Table 3). The percentages of Polycheate were higher than the other benthic groups at all the stations (Table 3). The stations (S-4M, S-5M, S-6M) of the Meghna river estuary showed higher percentages of Olygochaete than the stations of Bakkhali river estuary.

Some water quality parameters were strongly correlated with the abundance of the benthic macrofaunal communities. The abundance of the benthic macrofaunal communities were significantly positively correlated with salinity ($r=.846$; $p \leq 0.05$) and negatively correlated with alkalinity ($r=-.842$, $p \leq 0.05$). On the other hand, the abundance of the benthic macrofaunal communities were significantly positively correlated with hardness ($r=0.857$, $p \leq 0.05$) and TDS ($r=0.887$; $p \leq 0.05$). The abundance of benthic communities were negatively significant with DO ($r=-.106$, $p \leq 0.05$) (Table 4).

Macrobenthic species composition in Bakkhali river estuary showed the maximum numbers of Polycheate (46%) followed by gastropods-23%, atrhropods-9%, bivalvia-8% and oligochaete-6% where the Meghna river estuary (Figure 5) showed the maximum number of Polychaete (56%) followed by arthropods-20%, oligochaete-12% (Figure 5). The density of macrobenthos found in 6 stations of the two study sites were tabulated in Table 5 with percentage and ranked according to the abundance of the macrobenthos in each station. Polycheate ranked number 1 in all stations while other benthic groups were fluctuated within the stations (Table 5).

A biodiversity index seeks to characterize the diversity of a sample or community by a single number. The concept of the "species diversity" involves two components: the number of species or richness and the distribution of individuals among species. However, Shannon–Wiener diversity index considers the richness and proportion of each species while Evenness and Dominance indices represent the relative number of individuals in the sample and the fraction of common species respectively. Quality of an aquatic ecosystem is dependent on the physico-chemical qualities of waters and it is reflected on biological diversity. Different diversity indices were recorded in Table 6. About 28 families were identified from sampling stations during the study period. Among the 28 families, the highest number of families were identified from Station S-2B (22 families) followed by the S-1B (21 families), S-3B (15 families), S-6M (11 families), S-5M (10 families) and S-4M (09 families) (Table 6). The Dominance-D value of the sampling stations was found 0.06878. The highest Dominance-D value was found in station S-6M (0.2155) followed by the S-4M (0.1546), S-5M (0.1392), S-3B (0.0913), S-2B (0.07605) and S-1B (0.07552) (Table 5). The Simpson_D value of the sampling stations was identified as 0.9312. The highest and lowest values (Table 6) of Simpson_D were

Table 4: Relationship between water quality parameters and abundance of macrobenthic communities.

	Temperature	Salinity	DO	pH	Alkalinity	Abundance
Temperature		0.445	-0.491	-0.357	-0.443	0.289
Salinity			-0.318	-0.720	-0.980**	0.846**
DO				0.190	0.298	-0.106
pH					0.645*	-0.287
Alkalinity				.		-0.842**
Abundance						

**=highly correlated; *=moderately correlated ($p \leq 0.05$)

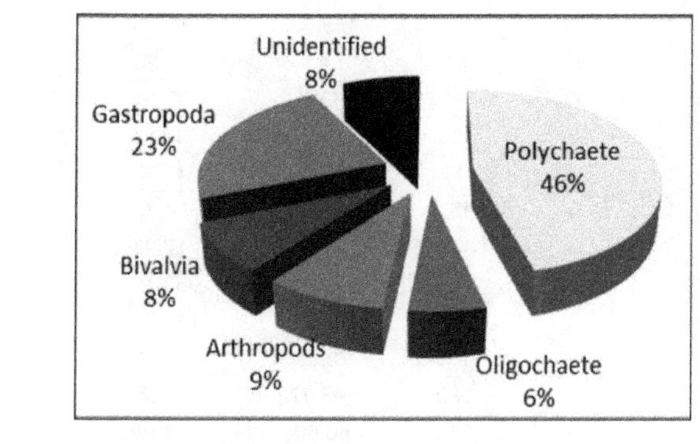

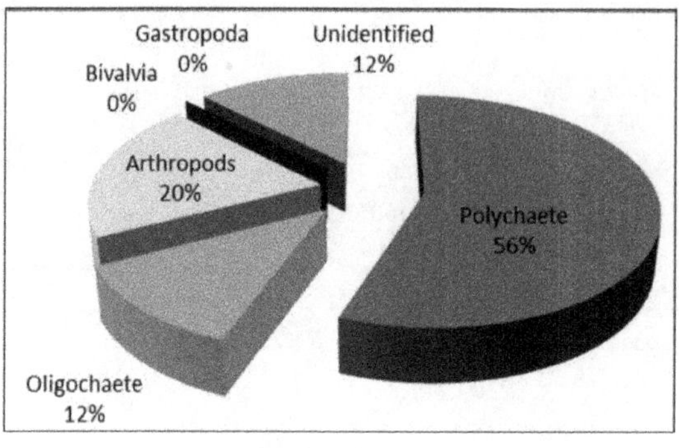

Figure 5: Percentage of macro benthic groups in the Bakkhali River Estuary (A) and the Meghna River Estuary (B).

identified in the stations S-1B (0.9245) and S-6M (0.7845) respectively. Another important diversity index is Shannon_H diversity index. The value of Shannon_H diversity index was recorded 2.89 while the value of Equitability J was found highest in the station of S-3B (0.94) and the lowest (Table 6) was in the station of S-6M (0.80). On the other hand another diversity index Fisher Alpha showed highest value in the station of S-2B (3.01) and the lowest was in the station of S-4M (1.27) (Table 6). After analyzing the diversity indices of the sampling stations of Bakkhali River Estuary and Meghna River Estuary, the significant differences were found between the two study sites which are shown in Figure 6. Different diversity indices showed significant differences between the two study sites. The diversity values of Shannon H (Bakkhali River Estuary-2.85, Meghna River Estuary-2.11), Evenness (Bakkhali River Estuary-0.67, Meghna River Estuary-0.63), Simpson Index (Bakkhali River Estuary-0.93, Meghna River Estuary-0.84), Mechinick (Bakkhali River Estuary-0.24, Meghna River Estuary-0.16), Margalef (Bakkhali River Estuary-2.67, Meghna River Estuary-1.36) and Fisher alpha (Bakkhali River Estuary-3.16, Meghna River Estuary-1.55) were found higher in the Bakkhali River Estuary and lower in the Meghna River Estuary (Figure 6). Only the Dominance-D value and Berger-Parker value of the Meghna River Estuary were found higher than the Bakkhali River Estuary (Figure 6).

Shannon Weiner diversity Index is a commonly used diversity index that takes into account both abundance and evenness of species present in the community. In the present study after analyzing the whole samples (18) from six sampling stations of two sampling sites, overall H value was found 2.89 (Table 6). The highest Shannon diversity index (2.781) was found at S-2B in the study period and lowest was found at S-6M (1.909) (Table 6). In biological communities, Shannon-Wiener diversity index varies from 0 to 5. According to this index, values less than 1 characterize heavily polluted condition, and values in the range of 1 to 2 are the characteristics of moderate polluted condition while the value above 3 signifies stable environmental conditions [35,36]. Higher value of Shannon_H indicated that the sampling stations have high number of individuals. Significant difference was found in the mean Shannon diversity index among the stations of the study sites (Table 6). This finding is similar to the findings of Bhandarkar, who investigated on the potential of benthic macro-invertebrates community assemblages in predicting the water quality status. Bhandarkar [37] identified that (Shannon-Weiner index value ranges from 1.2 to 2.9 in three ecosystems), all the selected sampling sites fall under moderate pollution. The Shannon equitability index values showed a greater equitability in the apportionment of individuals among the species in all the sites.

According to Margalef [38], the higher diversity values reflect the suitability of habitat for the organism and have been reported to be correlated with longer food chain and complex food web of the ecosystems and also more stable community. Margalef index has no limit value and it shows a variation depending upon the number of species. In the present study the values of Margalef diversity index were between 1.089 and 2.497 at station S-4M and S-2B respectively. Menhinick index, like Margalef's index, attempts to estimate species richness but at the same time it is independent on the sample size. In the present investigation, it ranged from 0.185 to 0.34 (Table 6). The low diversity associated with station S-5M, as described by the Shannon, Margalef and Menhinick indices, may be attributed to lesser number of species and environmental degradation due to anthropogenic pressures, besides other biotic factors [39]. The mean Margalef's value of the Bakkhali River Estuary (2.21) is higher than the mean value of the Meghna River Estuary (1.17). Table 5 showed how the values differ among the stations of the study sites. Again, Hossain [33] identified that the Shannon species diversity index of polychaetes varied from

Table 5: Density of macrobenthos found in the Bakkhali River Estuary and the Meghna River Estuaryo.

Study Sites	Stations	Benthos Group	Individuals/m²	Percentage (%)	Rank of Abundance
Bakkhli River Estuary	S-1B	Polychaete	1556	41.21	1
		Oligochaete	222	5.88	6
		Arthropods	266	7.04	6
		Bivalvia	265	7.02	6
		Gastropoda	1067	28.26	2
		Unidentified	400	10.59	5
		Total	3776	100	
	S-2B	Polychaete	2089	46.55	1
		Oligochaete	266	5.93	6
		Arthropods	445	9.92	6
		Bivalvia	400	8.91	6
		Gastropoda	933	20.79	3
		Unidentified	355	7.91	6
		Total	4488	100	
	S-3B	Polychaete	1733	50.67	1
		Oligochaete	177	5.18	6
		Arthropods	311	9.09	6
		Bivalvia	311	9.09	6
		Gastropoda	711	20.79	3
		Unidentified	177	5.18	7
		Total	3420	100	
Meghna River Estuary	S-4M	Polychaete	711	45.72	1
		Oligochaete	267	17.17	4
		Arthropods	311	20.00	4
		Bivalvia	0	0.00	0
		Gastropoda	0	0.00	0
		Unidentified	266	17.11	4
		Total	1555	100	
	S-5M	Polychaete	1733	59.11	1
		Oligochaete	311	10.61	5
		Arthropods	533	18.18	4
		Bivalvia	0	0.00	0
		Gastropoda	0	0.00	0
		Unidentified	355	12.11	5
		Total	2932	100	
	S-6M	Polychaete	1289	58.04	1
		Oligochaete	222	10.00	6
		Arthropods	489	22.02	3
		Bivalvia	0	0.00	0
		Gastropoda	0	0.00	0
		Unidentified	221	9.95	6
		Total	2221	100	

* 1-5% = rank 7; 06-10% = rank 6; 11-15% = rank 5; 16-20% = rank 4; 21-25% = rank 3; 26-40% = rank 2; >41%=rank 1.

0-1.36. It was the highest at Swandwip during post monsoon and lowest at Bhola during monsoon in Bangladesh. Abu Hena [15] also identified that the Shannon diversity index ranged from 0.65-1.04 among the sampling stations at the Bakkhali River Estuary, Cox's Bazar.

Species evenness refers to how close in numbers of each species in an environment are. Mathematically it is defined as a diversity index, a measure of biodiversity which quantifies how equal the community numerically. The higher value shows lower variation in number of species. Usually it has been also defined as the ratio of observed diversity to maximum diversity, the latter being said to occur when the species in a collection are equally abundant [40]. Evenness index value for collected 18 samples was 0.62, where the highest (0.85) and the lowest (0.61) values of Evenness recorded from S-3B and S-6M, respectively (Table 6). No significant difference was found in mean value of evenness value among the stations and as well as within the two study sites (Table 6).

In the present study diversity of benthic organisms were in the sequence of Polychaete (49.42%) > Gastropods (14.70%) > Arthropods (12.77%) > Oligochaete (7.94%) > Bivalvia (5.54%). On the other hand there was an inverse relationship between these two indices in the three sampling stations of the Bakkhali River Estuary. This findings is similar to the investigation of Bu-Olayan and Thomas who observed that the diversity of benthic organisms were in the sequence of Annelida > Mollusca > Crustacea > others group in Kuwait Bay of the Arabian

Table 6: Diversity Indices Observed in the Present Study.

Stations	S-1B	S-2B	S-3B	Mean ± SD	S-4M	S-5M	S-6M	Mean ± SD	All Stations
Taxa_S	21	*22	15	19.33±3.79	9	10	11	10 ± 1.0	28
Individuals	3819	*4488	3420	3909 ± 540	1555	2933	2221	2236 ± 689	18436
Dominance D	0.08	0.08	0.09	0.08 ±0.01	0.15	0.14	*0.22	0.17± 0.04	0.07
Simpson D	*0.93	0.92	0.91	0.92 ± 0.01	0.85	0.86	0.78	0.83 ± 0.04	0.93
Shannon_H	2.76	*2.78	2.54	2.69 ±0.13	1.98	2.12	1.91	2.00 ± 0.11	2.89
Evenness	0.75	0.73	*0.85	0.78 ± 0.06	0.81	0.83	0.61	0.75 ± 0.12	0.62
Brillouin	2.74	*2.77	2.53	2.68 ± 0.13	1.97	2.11	1.89	1.99 ± 0.11	2.88
Menhinick	*0.34	0.33	0.26	0.31 ± 0.04	0.23	0.18	0.23	0.22 ± 0.03	0.21
Margalef	2.43	*2.50	1.72	2.21 ± 0.43	1.09	1.13	1.30	1.17 ± 0.11	2.85
Equitability_J	0.91	0.90	*0.94	0.91 ± 0.02	0.90	0.92	0.80	0.87 ± 0.07	0.86
Fisher_alpha	2.93	*3.01	2.02	2.65 ± 0.55	1.27	1.29	1.51	1.36 ± 0.13	3.37
Berger-Parker	0.14	0.16	0.18	0.16 ± 0.02	0.23	0.26	*0.40	0.30 ± 0.09	0.12

*Indicates the highest value among the stations.

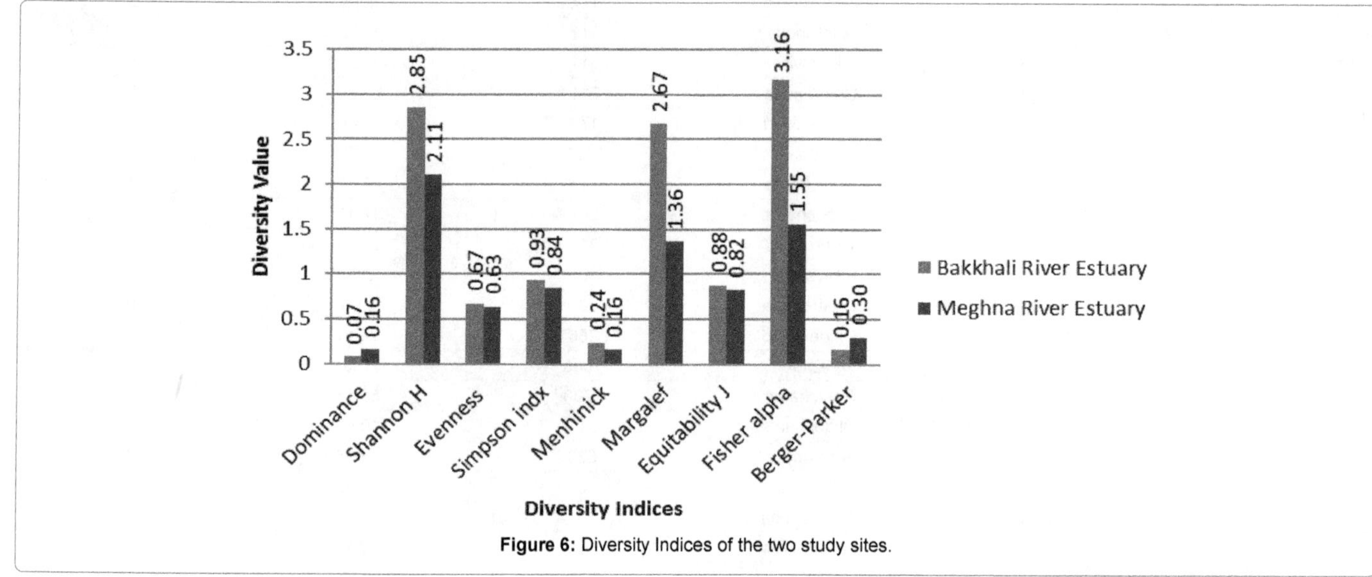

Figure 6: Diversity Indices of the two study sites.

Gulf. Evenness index (-) was found to be increased with increasing Bu-Olayan H and Thomas [41] identified that low diversity indices correspond to the increase in trace metal level in benthic species collected from four sites, wherein high abundance of certain benthic species and high trace metal levels due to manmade perturbations were observed altering the diversity indices and those indices would validate benthic organisms as an indicator to trace metal pollution in Kuwait marine ecosystem, however primary or secondary data regarding trace metal concentrations is absent in the present study area [41].

Conclusion

The coastline along the South-Eastern part of Bangladesh has high fisheries resources and the fisheries production of the estuarine areas of that coast is higher than other coastal areas of the country. Environmental pollution is believed to be the major constraints of fisheries production of an area. Although similar research is untouched to assess the environmental pollution of a water body, macrobenthic structure of those water bodies is used as indicator. Therefore, due to the lack of detailed study on the macrobenthic structure of those areas, present study was designed to assess the pollution status of the Bakkhali River estuary and the Meghna River estuary. A total 28 families under 05 major groups/taxa of macrobenthic communities were identified and the dominant group both in number of families (16 families) and individuals (49.42%) was the Polychaeta followed by Gastropoda (14.70%), Arthropods (12.77%), Oligochaete (7.94%) and the lowest was Bivalvia (5.54%). The abundance (r=.846) and species richness (r=.864) of the macrobenthic communities were significantly influenced by the water salinity of the sampling sites ($p \leq 0.05$). Both the study areas namely the Bakkhali River Estuary and the Meghna River Estuary showed the highest (3909 ± 540 ind./m^2) and lowest (2236 ± 689 ind./m^2) density of microbenthic communities respectively. These two study sites might be considered as moderately polluted areas according to the results obtained from Shannon-Wiener index of species diversity (2.69 ± 0.13 and 2.00 ± 0.11 respectively) and Margalef's species richness (2.21 ± 0.43 and 1.36 ± 0.11 respectively). Therefore, it can be concluded that the macrobenthic community explained in the present study might be a key future outline to assess the status of coastal water pollution of those concerned areas of Bangladesh.

Acknowledgment

The authors realized thanks to the laboratory staffs of the Department of Fisheries and Marine Science, Noakhali Science and Technology University (NSTU) for their help in sample collections.

References

1. Sinelgrove PVR (1998) Biodiversity Conservation 7: 1123-1132.
2. Aura CM, Raburu PO, Herrmann J (2011) Macro invertibrate's community structure in Rivers Kipkaren and Sosiani, River Nzoia Basin, Kenya. J Eco Nat Environment 3: 39-46.
3. McLusky DS (1989) The estuarine ecosystem. Chapman and Hall, London, pp: 133.
4. Newrkla P, Gunatilaka A (1982) Benthic community metabolism of three Austrian pre-alpine lakes of different tropic conditions and its oxygen dependency. Hydrobiologia 92: 531-536.
5. Pearson TH (1970) The benthic ecology of Loch Linnhe and Loch Eil, a Sea-Loch system on the west coast of Scotland. The physical environment and distribution of the macrobenthic fauna. J Exp Mar Biol Ecol 5: 1-34.
6. Bonada N, Prat N, Resh VH, Statzner B (2006) Development in aquatic insect Bio-monitoring: a comparative analysis of recent approaches. Annual Review of Entomology 51: 495-523.
7. Bilgrami KS, Munshi D (1985) Ecology of river Ganges: Impact on human activities and conservation of aquatic biodata (Patna to Farakka). Allied Press, Bhagalpur.
8. Venkateswarlu V (1986) Ecological studies on the rivers of Andhra Pradesh with special reference to water quality and Pollution. Proc Indian Sci Acad 96: 495-508.
9. Wilhm RL, Dorris TC (1968) The biological parameters for water quality criteria. Bio Science 18: 477-492.
10. Cairns JR, Dickson KL (1971) A simple method for the biological assessment of the effects of waste discharges on aquatic bottom dwelling organisms. J Wat Pollut Control 43: 755-772.
11. Hynes HBN (1970) The ecology of stream insects. Annual Review of Entomology 15: 25-42.
12. Pramanik MAH (1988) Methodologies and techniques of studying coastal systems: Case Studies II, Space and Remote Sensing Organization (SPARSO), Bangladesh pp: 122-138.
13. Abu Hena MK, Ashraful MAK (2009) Coastal and estuarine resources of Bangladesh: Management and conservation issues. Maejo International Journal of Science and Technology 2: 313-42.
14. Khan AN, Kamal D, Mahmud MM, Rahman MA, Hossain MA (2007) Diversity, Distribution and Abundance of Benthos in Mouri River, Khulna, Bangladesh. Int J Sustain Crop Prod 2: 19-23.
15. Abu Hena MK, Kohinoor SMS, Siddique MAM, Ismail J, Idris MH, et al. (2012) Composition of Macrobenthos in the Bakkhali Channel System, Cox's Bazar with Notes on the Soil Parameter. Pakistan Journal of Biological Sciences 15: 641-646.
16. Belaluzzaman AM (1995) Ecology of the Intertidal Macrobenthic Fauna in Cox's Bazar Coastal Area, MSc Thesis, Institute of Marine Sciences, University of Chittagong, Bangladesh pp: 199.
17. Jakobsen F, Azam MH, Kabir M, Mahboob-Ul (2002) Residual Flow in the Meghna River Estuary on the Coastline of Bangladesh, Estuarine, Coastal and Shelf Science 4: 587-597.
18. Ibrahim S, Hussin WMRW, Kassiml Z, Joni ZM, Zakaria MZ, et al. (2005) Seasonal Abundance of Benthic Communities in Coral Areas of Karah Island, Terengganu, Malaysia, Turkish Journal of Fisheries and Aquatic Sciences 6: 129-136.
19. Alam MS (1993) Ecology of the Intertidal Macrobenthos of Halishahar coast, Chittagong, Bangladesh, Ph. D. Thesis, Department of zoology, University of Chittagong, Bangladesh pp: 243.
20. Al-Yamani, Faiza Y, Skryabin, Valeriy, Boltachova, et al. (2012) Illustrated Atlas on the Zoo benthos of Kuwait, Kuwait Institute for Scientific Research.
21. Welch PS (1948) Limnology. Mc graw Hill book Company, New York.
22. Wilhm JL, Dorris TC (1966) Species diversity of benthic marco-invertebrates in a stream receiving domestic and oil refinery effluents. Am. Midl. Nat 76: 427-449.
23. Margalef R (1968) Perspectives in Ecological Theory. University of Chicago Press, Chicago, IL p: 111.
24. Simpson EH (1949) Measurement of diversity. Nature pp: 688.
25. Cox WG (1996) Laboratory Manual of General Ecology. Ed Wm C Brown Publsishers.
26. Stiling PD (1996) Ecology theories aladapplications. Ed Prentice Hall, New Jersey.
27. Harper DAT (1999) Numerical Palaeobiology. John Wiley & Sons.
28. Magurran AE (2004) Measuring biological diversity. Blackwell.
29. Maurer BA, McGill BJ (2011) Measurement of species diversity. Biological diversity: frontiers in measurement and assessment. Oxford University Press, Oxford, New York pp: 55-64.
30. Ryan PD, Harper DAT, Whalley JS (1995) PALSTAT, Statistics for palaeontologists. Chapman & Hall, Kluwer Academic Publishers.
31. Islam M, Shafiqul, Sikdar M, Nurul Azim, Al-Imran M, et al. (2013) Intertidal Macrobenthic Fauna of the Karnafuli Estuary: Relations with Environmental Variables, World Applied Sciences Journal 21: 1366-1373.
32. Hossain M, Belal, Marshall, David J (2014) Benthic infaunal community structuring in an acidified tropical estuarine system, Aquatic Biosystems.
33. Hossain MB (2009) Macrozoobenthos of the meghna river estuarine bed with special reference to polychaete faunal biodiversity, International journal of sustainable agricultural technology, Science publication, Ghurpukur Research institute (GPRI), Bangladesh 3: 11-16.
34. Asadujjaman M, Hossain M, Belal, Shamsuddin M, Amin A, et al. (2012) The effect of industrial waste of Memphis and Shelby country on primary planktonic producers, Bioscience 20. 905-912.
35. Mason CF (1988) Biology of Fresh Water Pollution. Longman scientific and technical.
36. Stub R, Appling JW, Hatstetter AM, Hass IJ (1970) The effect of industrial waste of Memphis and Shelby country on primary planktonic producers, Bioscience 20: 905-912.
37. Bhandarkar SV, Bhandarkar WR (2013) A study on species diversity of benthic macro invertebrates in freshwater lotic ecosystems in Gadchiroli district Maharashtra, International journal of Life Sciences 1: 22-31.
38. Margalef R (1956) Information Y diversidad especifi caenlas communidades de organisms. Invest Pesg 3: 99-106.
39. Ravera O (2001) A comparison between diversity, similarity and biotic indices applied to the macro invertebrate community of a small stream: The Ravella River (Como Province, Northern Italy). Aquatic Ecol 35: 97-107.
40. Margalef DR (1958) Information theory in ecology. Gen Syst 3: 36-71.
41. Bu-Olayan AH, Thomas BV (2005) Validating species diversity of benthic organisms to trace metal pollution in Kuwait Bay, off the Arabian Gulf, Applied Ecology and Environmental Research 3: 93-100.

Reproduction of *Mugil cephalus* (Percoidei: Mugilidae) off the Central Mexican Pacific Coast

Elaine Espino-Barr[1]*, Manuel Gallardo-Cabello[2], Marcos Puente-Gómez[1] and Arturo Garcia-Boa[1]

[1]*Instituto Nacional de Pesca, Playa Ventanas s/n, Manzanillo, Colima, México*
[2]*Instituto de Ciencias del Mar y Limnología, Universidad Nacional Autónoma de México, México*

Abstract

Reproduction of *Mugil cephalus* of the Pacific coast of Mexico was studied. Fish were captured with gill nets and cast nets; they are a common low priced product for local consumption. The study of the reproduction period and ages of first maturity helps manage the fishery. Fish were obtained from local commercial fishery from August to December 2007, January to March 2008 and November 2012 to October 2013. Size and weight, sex and gonad maturity were registered. The male:female ratio was 0.88:1. Mature organisms occurred all year round. Sexual maturation (L_{50}) of males and females was observed at a mean size of 34.0 cm in males (4.64 years of age) and 35.0 cm in females (4.98 years of age). First maturity length (L_{25}) was both 30.0 cm in males and females corresponding to 3.4 years of age in both cases. The allometric relationship with the hepatosomatic index was $LW=4.00 \cdot 10^{-3} \cdot TL^{2.771}$ (r^2=0.849). Condition factor indexes of Clark and Safran EW showed a maximum increment during June, August and December; Fulton and Safran TW in July and September to November. The gonadosomatic index showed its highest values from November to January. The hepatosomatic index reached its maximum values in June, July and August. The gastric repletion index reached its highest values in June, February and October. The mean oocytes diameter was 0.38 mm (range 0.22 to 0.52 mm, standard deviation=0.13). Fecundity ranged from 1'422,076 to 1'747,736 oocytes in females between ages 3 and 12 years old, and mean relative fecundity was 2,830 oocytes$\cdot g^{-1}$ (1,500 to 2,900 oocytes$\cdot g^{-1}$). This study is the base line for the fishery management of *M. cephalus* in Central Mexican Pacific, where the main regulations need information on the first maturity size and reproductive season.

Keywords: Fecundity; Maturity period; Fish reproduction; Gonadosomatic index; Hepatosomatic index; Gastric repletion index; Condition factor

Introduction

The striped mullet *Mugil cephalus* Linnaeus 1758 (Figure 1) has a worldwide distribution between 42°N and 42°S [1]. Through the biogeographic areas and provinces of the American continent, the only exceptions are the cold temperatures of the Northeast Pacific in the Pacific Province and the Magellanic Province in South America [2]. In the Western Atlantic it distributes from Nova Scotia to Argentina including the Gulf of Mexico [3]. In the Eastern Pacific it distributes from California to Chile, including the Gulf of California and Galapagos Islands [4].

New research made by Whitfield et al. establishes that *M. cephalus* is a cosmopolite species that flourishes in a high variety of habitats and that can be considered as a eurytopic species complex, and could be used as a biological marker in the health levels of different ecosystems where it inhabits [5].

This species is important for the meat consumption and the "roe" (female mature gonads), which reaches a higher price than the meat: the roe is $300.00 Mexican pesos per kilogram ($18.00 US dollars) and the meat $30.00 Mexican pesos ($1.80 US dollars) per kilogram.

M. cephalus ranks in 22nd place in Mexican fisheries, with a capture of 12 280 [6]. This species has been studied in many parts of the world where well established fisheries are located. In the case of Mexico, analysis has also been carried out on this fishery and biological aspects [7-20]. However, most of the studies in Mexico have been carried out in the Tamiahua lagoon, Tamaulipas, on the Atlantic Ocean, or in Mazatlán, Sinaloa and Nayarit on the northern Mexican Pacific Ocean.

In the coast of Jalisco and Colima, *M. cephalus* does not reach a high catch volume as in these other places, but it is part of a multispecific fishery and important to know the health status of their populations. Traditionally *Mugil curema* is fished in a higher amount in the coasts of Jalisco and Colima, than *M. cephalus*. In 2014, *M. curema* was fished up to 626 tons (79% of the total Mugilidae species), and 167 tons of *M. cephalus* (21% of the total Mugilidae species) [6]. Therefore *M. curema* has been analyzed and some studies of its population dynamics were done [21-26].

The objectives of the present study were to analyze monthly frequency of the gonadic maturity stages and massive spawning

Figure 1: Stripped mullet *Mugil cephalus*.

***Corresponding author:** Elaine Espino-Barr, Instituto Nacional de Pesca, Playa Ventanas s/n, Manzanillo, Colima, México, E-mail: elespino@gmail.com period; monthly values of the gonadosomatic and hepatosomatic index; monthly values of the gastric repletion index; monthly values of the condition factor of Fulton, Clark and Safran; values of total and relative fecundity; and to compare our results to those reported by other authors [27-29].

These studies will give a solid background for closed seasons and gill net mesh sizes, based on the minimum reproductive size. These fishing measures will allow the species to reproduce at least once, protecting the fishery from overexploitation.

Materials and Methods

From August to December 2007, January to March 2008 and November 2012 to October 2013, *M. cephalus* samples were monthly collected in the Cuyutlan Lagoon, Colima, Mexico (103°57'-104°19' W and 18°57'-19°50' N) and in Cruz de Loreto Lagoon, Jalisco, Mexico (105°27'-105°33' W and 19°58'-20°05' N) (Figure 2). The fishing gears were gill-nets of 2.0, 2.5, 3, 3.5 and 4 inches mesh size (5.08, 6.35, 7.62, 8.89 and 10.16 cm). Total length was measured to the nearest millimeter (TL, cm) from the snout tip to the caudal fin extreme in 262 organisms (fishermen deliver this species intact); the total (TW, g) and eviscerated (EW, g) weight of 784 specimens (weighed to the nearest 0.1 g) were measured.

The function $W=a.L^b$ was used to obtain weight-length relationship and sex was recorded macroscopically for each specimen. Sexual maturation was determined *in visu* on fresh organisms taken to the laboratory the same day they were caught. Sokolov and Wong, Holden and Raitt, Aboussouan and Lahaye and Espino-Barr suggest a scale to determine sexual maturity and describe the stages as follows [30-33]:

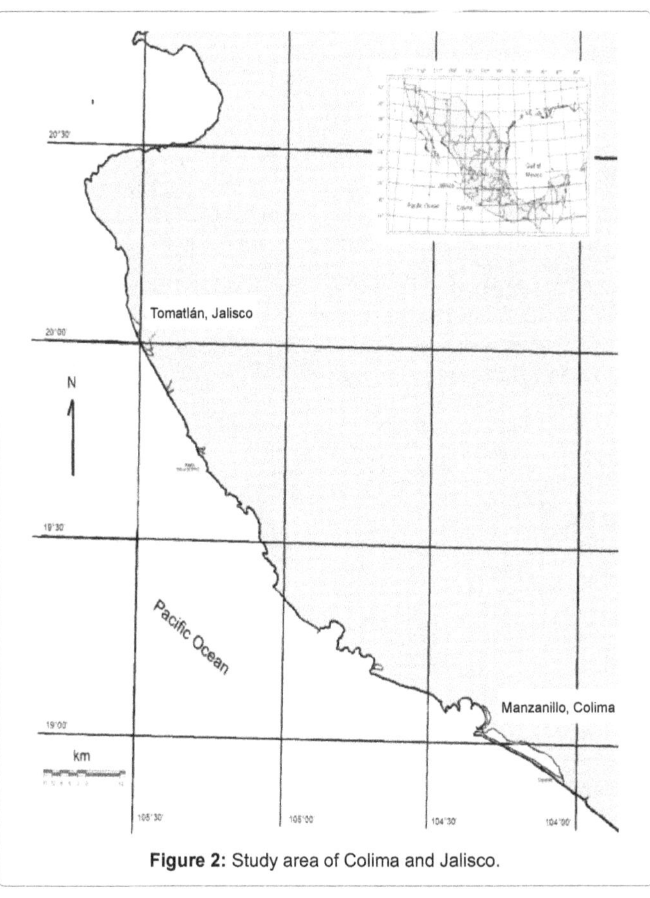

Figure 2: Study area of Colima and Jalisco.

- Phase I: Sexually immature organisms, in which sex cannot be distinguished, the gonads are very fine filaments.
- Phase II: Organisms have not yet matured sexually, the testis start to develop and are light colored and ovaries are pale pink, and oocytes cannot be observed.
- Phase III: Sexual maturity begins; sexual glands can be perfectly identifiable. Oocytes are beginning to form and are opaque, the color of the ovaries start to turn dark pink, testis also show darker and opaque color.
- Phase IV: Mature, sexual glands are well developed, ovaries are rose-orange, oocytes are big and transparent, and testes are whitish.
- Phase V: This stage corresponds to the spawning, both, ovules and sperm that are expelled if the visceral cavity is pressed, gonads show and intense blood supply and both ovaries and testes show brighter colors.
- Phase VI: Corresponds to the post-spawning, sexual products have been expelled, both ovaries and testes are empty, gonads coloring tend towards a dark pink.

Sparre and Venema suggest that the first spawn length is calculated as the 50% of the accumulative frequency (L_{50}) of phases IV and V of the sexual maturity scale mentioned above, considering that the lowest spawning length (L_{25}) is also registered, to compare with data reported in other studies [34,35]. This kind of analysis are carried out for both males and females and both (L_{25}) as (L_{50}) are useful because deliver information on the stages close to the reproduction. Gaertner and Laloe and Sparre and Venema represent this function by the equation [34,36]:

$H_p=1/[1+e^{a+b \cdot TL}]$,

where: H_p=the percentage of mature organisms (males or females) and *a* and *b* are constants.

Transforming this equation logarithmically, we obtain:

$\ln 1(1/H_p-1)=a-b \cdot TL$, and the length at which 50% of the population is sexually mature (L_{50}) corresponds to: $L_{50}=a/b$.

To include L_{50}, the original equation is modified:

$Y=1/[1 + a\,(1-TL/L_{50})]$

The minimum TL of first spawning (L_{25}) was also recorded to be compared with other authors' findings.

The formula mentioned by Rodríguez-Gutiérrez to calculate the gonadosomatic index (GSI) for males and females of *M. cephalus* [35], considers the gonad weight (GW) in relation to the fish total weight,

i.e., $GSI=100 \cdot GW/TW$ (TW=total weight).

To measure physical fitness of fish, we obtained the condition factor

$K=(EW \cdot TL^{-3}) \cdot 100$,

$K=(TW \cdot TL^{-3}) \cdot 100$ and

$a=TW \cdot TL^{-b}$ and

$a=EW \cdot TL^{-b}$ [27-29].

The hepatosomatic index (HSI), expressed as the percentage of liver weight (LW) with respect to the total weight was calculated as:

HSI=100·LW/TW [35].

The stomach repletion index is the relation between the stomach weight and the body weight, calculated individually and averaged monthly.

The gravimetric method was used to calculate total fecundity (F) and relative fecundity (Fr) using the weight of 45 females in phase V of gonadic maturity. To estimate F, two subsamples of 0.1 g were obtained of each individual and put in modified Gilson fluid for preservation [37]. All oocytes were counted with the help of a stereoscopic microscope and measured with an ocular micrometer.

The following formula was used to determine fecundity:

$F=n \cdot G_i/g_i$,

where n=number of oocytes in the subsample; Gi=weight of the gonad (g) and gi=weight of the subsample (g) [31].

The relationship between fecundity and total length and weight was calculated with the formula

$F=a \cdot x^b$,

where x=individual weight or length, a=intercept or initial number of oocytes, b=slope or oocyte number changing rate.

The data obtained by Espino-Barr et al. in the study of otoliths, were used to obtain –for each age- the relations between TL, TW, LW, testis weight (TeW), ovary weight (GW), and fecundity [19].

Results

M. cephalus sex cannot be differentiated by their body morphology, so organisms have to be opened and eviscerated; 262 individuals were sexed. The use of gill nets of different size (2.0-4.0 μn) in commercial fishery allowed catching individuals of various age groups.

Once organisms begin their gonadal maturity, sex identification is performed quite easily. Ovaries are elongated and during the season close to spawning, oocytes are easily observed, with a bright yellow pink coloration.

The relationship from each age group to the values of the total length (TL, cm), total weight (TW, g), eviscerated weight (EW, g), liver weight (LW, g), gonad weight (TeW, g and GW, g), and fecundity (number of oocytes) are shown in Table 1.

Testes are cream colored long triangular tape type and smaller than ovaries.

Oocyte diameters were 0.38 mm (from 0.22 to 0.52 mm ± 0.12 mm standard deviation: SD). The f ecundity values oscillated from 1'422,076 to 1'747,736 oocytes in females from 3 to 12 years of age, length from 28.5 cm to 48.8 cm, and total weight 239 g to 1,165 g (Table 1). Relative fecundity showed average values of

Fr=2,830 oocytes·g^{-1} (ranging from 1,500 to 2,900 oocytes·g^{-1}).

The sample size was of 262 organisms of *M. cephalus,* of which 123 were female (46.95%), 109 male (41.60%), and 30 undetermined (11.45%). The male: female proportion was 0.88: 1.0.

According to the values obtained of the monthly frequencies for the gonadal maturity scale, it was observed that immature male organisms corresponding to Phase II prevailed from March to August in 100% with high values also in September and October. Immature females were present with more than 50% in February, May, August and December (Figures 3a and 3b). Phase III: Maturing, females were not present and males showed up during January and November, while phase IV, mature, was observed in July, September and October for females and in January for males. Phase V or spawning stage was observed in females in November, January and February. Males were in phase V in September and from November to January. Phase VI, post-spawning, females were observed during all year round, with high values in March, April, June, July, September and October, while males were present in February, October, November and December (Figures 3a and 3b).

Length at first maturity was L_{25}=30.0 cm in females (Figure 4a) and in males (Figure 4b), corresponding to 3.4 years of age. First reproduction length was L_{50}=35.0 cm in females (Figure 4a) and L_{50}=34.0 cm in males (Figure 4b), which correspond to 4.98 and 4.64 years of age, respectively.

The gonadosomatic index (GSI) showed the highest values in November and January for total length and total and eviscerated weight (Figures 5a and 5b), followed by a very light increase of GSI values during September and October only observed for total length and total weight. GSI values were very low the rest of the year.

We obtained the following allometric relationships of the hepatosomatic index (HSI) $LW=4.00 \cdot 10^{-3} \cdot TL^{2.771}$ (r^2=0.849). The index b shows that, in terms of length, the liver weight is lower than a cubic proportion, which results in a negative allometric growth of the fish

Age	TL (cm)	TW (g)	EW (g)	LW (g)	TeW (g)	GW (g)	F (eggs)
0	15.62	40.000	33.000	0.767	-	-	-
1	20.43	89.000	74.000	1.604	-	-	-
2	24.71	156.000	130.000	2.706	-	-	-
3	28.53	239.000	198.000	4.018	1.440	2.379	1,422,076
4	31.94	333.000	277.000	5.481	2.966	4.216	1,485,136
5	34.98	436.000	362.000	7.037	5.308	6.684	1,537,943
6	37.69	544.000	451.000	8.640	8.557	9.756	1,582,684
7	40.11	653.000	542.000	10.253	12.743	13.374	1,620,990
8	42.26	762.000	633.000	11.836	17.799	17.426	1,653,846
9	44.18	869.000	722.000	13.374	23.653	21.827	1,682,328
10	45.90	973.000	808.000	14.855	30.202	26.489	1,707,202
11	47.42	1,072.000	890.000	16.248	37.203	31.245	1,728,711
12	48.79	1,165.000	967.000	17.571	44.642	36.097	1,747,736

Table 1: Length (TL, cm), total weight (TW, g), eviscerated weight (EW, g), liver (LW, g), testis weight (TeW, g), ovary weight (GW, g) and fecundity (number of oocytes) for each age group (years).

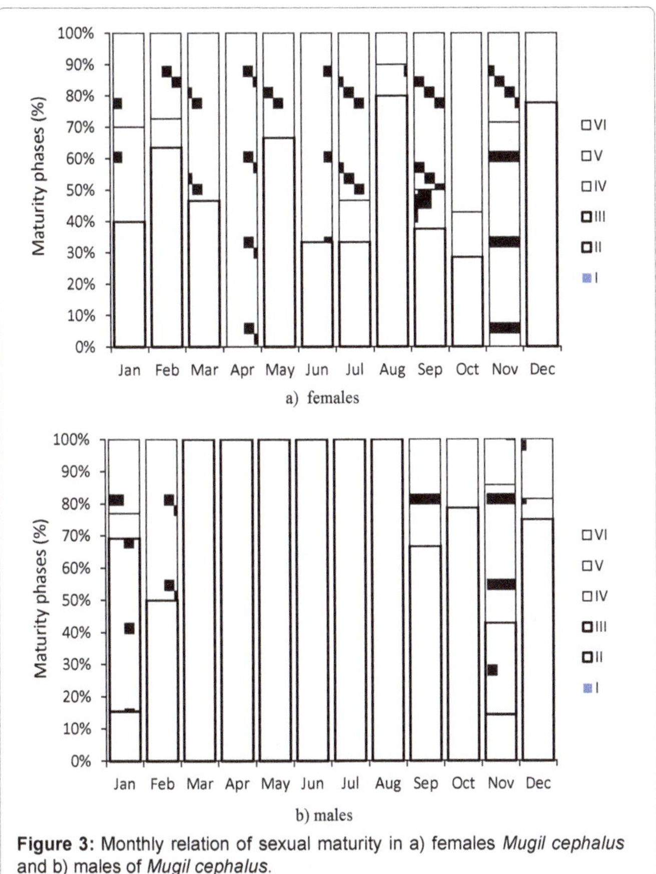

Figure 3: Monthly relation of sexual maturity in a) females *Mugil cephalus* and b) males of *Mugil cephalus*.

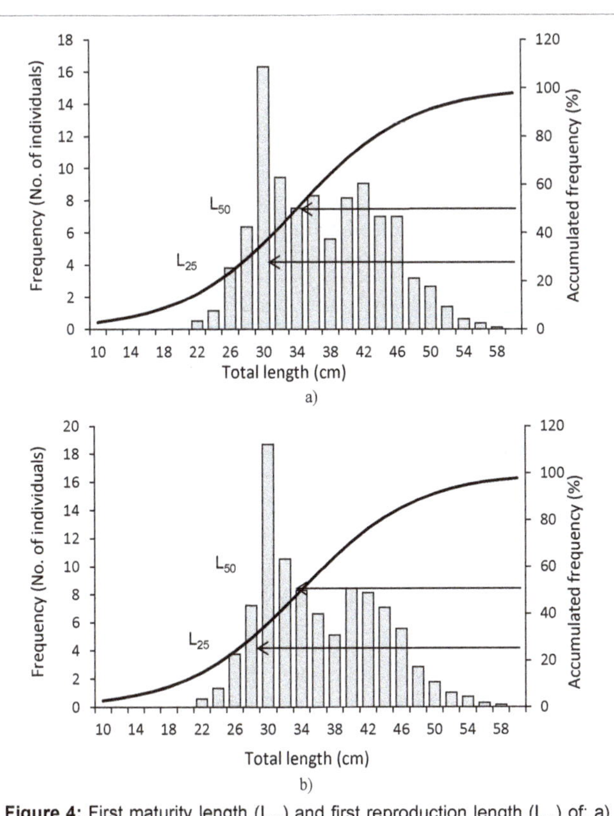

Figure 4: First maturity length (L_{25}) and first reproduction length (L_{50}) of: a) females and b) males of *Mugil cephalus*.

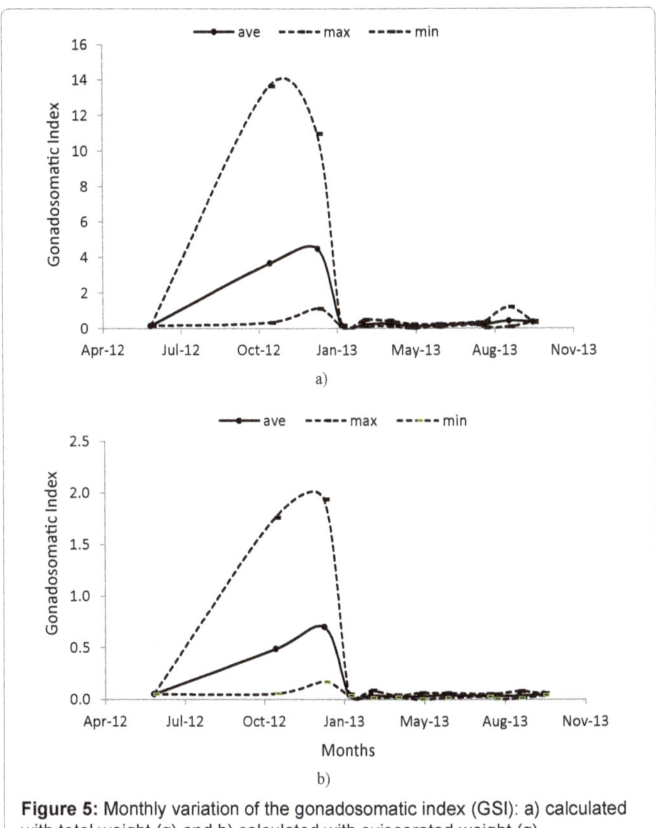

Figure 5: Monthly variation of the gonadosomatic index (GSI): a) calculated with total weight (g) and b) calculated with eviscerated weight (g).

and a decrease of its fatty reserves as it ages. HSI variations are shown in Figures 6a and 6b; maximum values are observed in May, June and July and lower values in February and March.

The differences in the stomach repletion index (Figures 7a and 7b) showed higher values during June, February and October; lower values are observed in December, January and September.

Figure 8 shows data of the condition factor; the highest values are obtained in January, August and December for Clark index and Safran EW. For Fulton index and Safran TW the highest values correspond to July, September, October and November.

Discussion

The highest length growth rates of *M. cephalus* calculated by Espino-Barr et al. are in groups zero and three years of age, a second period corresponds between ages 4 and 7 years, and a third period between ages 8 and 12 years, which show the lowest length growth rate [19]. As length growth rate starts to decrease, total weight, gonad weight and fatty reserve index start to rise. In this way, two main seasons were registered in the life cycle of *M. cephalus*: first, from ages zero and three when most of the energy obtained through food is used to increase its length (reducing depredation and interspecific competence), and second from ages four to twelve, when this energy is oriented to form the sexual products and fatty reserves (Figure 9) [33,38,39].

Sexual proportion was 0.88:1 male:female, values slightly higher of 1:1.1 male:female were found for *M. cephalus* in Tamiahua lagoon in the Atlantic sea [40].

During all year round *M. cephalus* specimens were observed in post spawning phase (Figure 3), which indicates that during every months

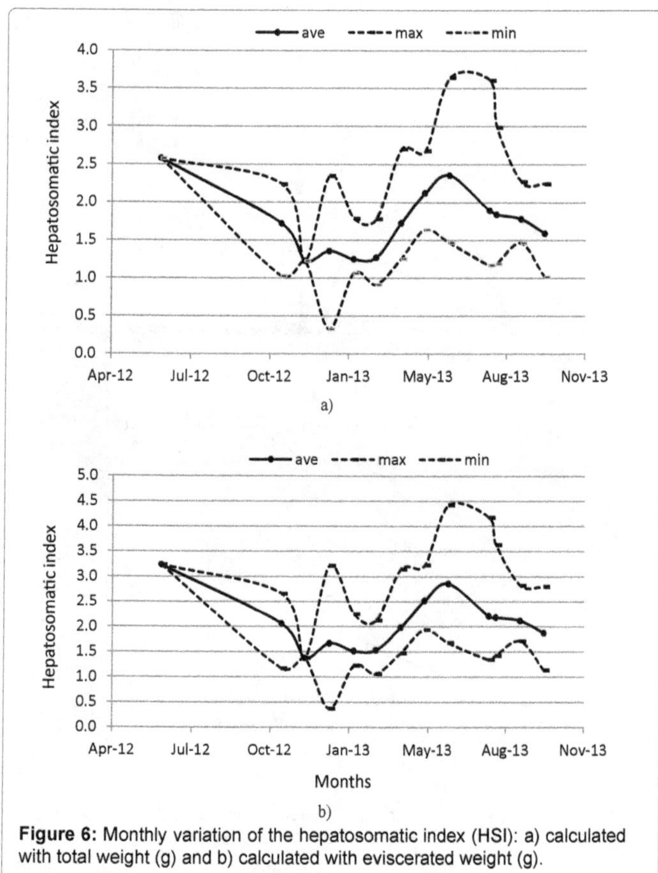

Figure 6: Monthly variation of the hepatosomatic index (HSI): a) calculated with total weight (g) and b) calculated with eviscerated weight (g).

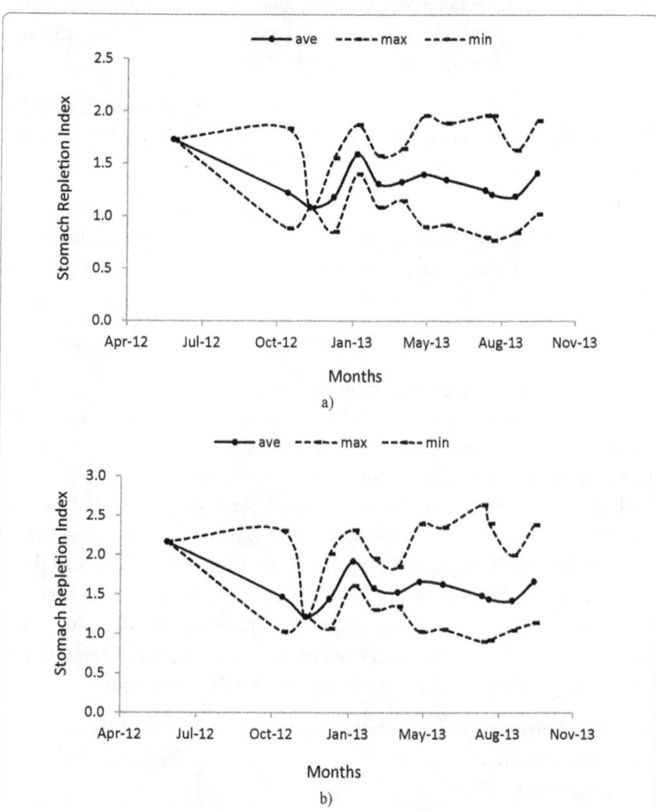

Figure 7: Monthly variation of the stomach repletion index (GRI): a) calculated with total weight (g) and b) calculated with eviscerated weight (g).

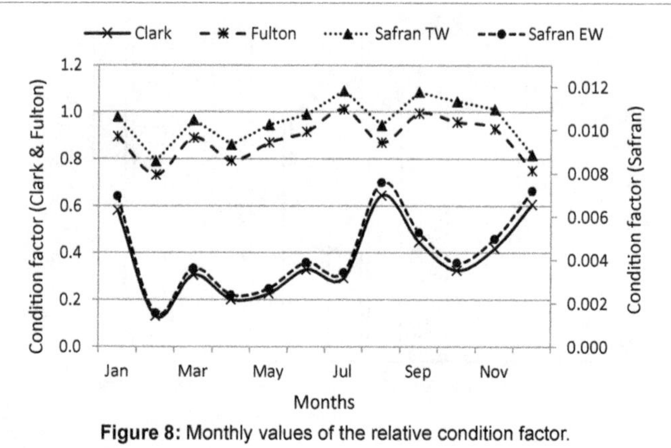

Figure 8: Monthly values of the relative condition factor.

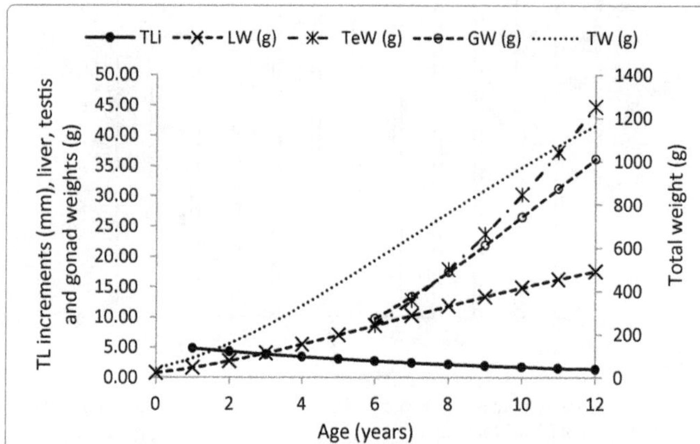

Figure 9: Relationship between age and total length increment (TLi, cm), total weight (TW, g), liver weight (LW, g), testes weight (TeW, g), gonad weight (GW, g) of *Mugil cephalus*.

mature organisms are present and that the reproduction carries out during every month of the year. However, most of the organisms in spawning phase occur during November, December and January. In the case of *M. cephalus* of Tamiahua, Veracruz, spawning occurs during autumn and winter, coinciding with the northern winds [40].

The highest first reproduction size of *M. cephalus* (L_{50}) was 42.52 cm in males and females which corresponds to an age of three years, in the Gulf of México (Table 2), followed by Briones-Avila who obtained values of 38.00 cm in males and females of specimens from Nayarit and Sinaloa in the Mexican Pacific [7,41,42]. Render et al. found sizes of 36.95 cm in organisms of Louisiana in USA [43]. Hubbs reported lengths of 33.00 cm in males and 35.00 cm in females with two years of age in Florida, USA [44]. Organisms of smaller lengths have been reported by Arnold [45,46], Jacot et al. in the coasts of Florida, and by Oren in Texas and Florida, USA [47]. In the present study males reached first reproduction length at 34.0 cm (4.64 years old) and females at 35.0 cm of total length and 4.98 years of age

The gonadosomatic index (GSI) reached the highest values in November and January in this study (Figures 5a and 5b). Similar spawning seasons to the present study (November to February), were found in Tamiahua lagoon Veracruz, Mexico [40], in North Carolina and to lower Florida [48], in Hawaii [49], in southwestern Taiwan [50], west coast of Taiwan [51], northeastern coasts of Taiwan [52], south west of Bay of Bengal [53]; in all these areas the massive spawning period of *M. cephalus* is from the end of autumn and during winter.

	Males		Females		Author
Area	TL (cm)	Age (yr)	TL (cm)	Age (yr)	
Florida (E), USA	30.36	2	30.36	2	Jacot [45]
Florida, USA	33.00	2	356.00	2	Hubbs [44]
Gulf of Mexico	24.00	-	25.80	-	Arnold and Thompson [46]
Gulf of Mexico	42.52	3	42.52	3	Márquez-Millán [7]
Texas, USA	20.00-35.50		25.00-35.00		Oren [47]
Florida (W), USA	23.00-29.00	3	24.00-31.00	3	Oren [47]
Florida (E), USA	23.60	2	25.00	3	Oren [47]
Louisiana, USA	36.95		36.95		Render [43]
Gulf of Mexico	28.00	3	29.90	3	Ibáñez-Aguirre and Gallardo-Cabello [40]
Nayarit, Mexico	38.00	-	38.00	-	Briones-Ávila [41]
Sinaloa, Mexico	38.00	-	38.00	-	Briones-Ávila [42]
Mexican Central Pacific	30.00-34.00	3.4-4.64	30.00-35.00	3.4-4.98	This study

Table 2: First maturity (L_{25}) and reproduction (L_{50}) length of *Mugil cephalus* in different places.

Country	Locality	Period	J	F	M	A	M	J	J	A	S	O	N	D	Author
Australia	east coast	March-July	-	-	x	x	x	x	x	-	-	-	-	-	Kailola [58]
Australia	west coast	March-September	-	-	x	x	x	x	x	x	x	-	-	-	Kailola [58]
South Africa	Natal estuaries	May-August	-	-	-	-	x	x	x	x	-	-	-	-	Marais [54]
Russia	Primorye	May-September	-	-	-	-	x	x	x	x	x	-	-	-	Novikov [55]
France	France	July-November	-	-	-	-	-	-	x	x	x	x	x	-	Keith and Allardi [56]
Spain	Minorca (Balearic Archipelago)	August-November	-	-	-	-	-	-	-	x	x	x	x	-	Cardona [57]
India	southwest Bay of Bengal	October-December	-	-	-	-	-	-	-	-	-	x	x	x	Jeyaseelan [53]
Taiwan	northeastern coasts	December-January	x	-	-	-	-	-	-	-	-	-	-	x	Hsu [52]
Taiwan	West coast	december	-	-	-	-	-	-	-	-	-	-	-	x	Shung [51]
Taiwan	southwestern Taiwan	October-February	x	x	-	-	-	-	-	-	-	x	x	x	Chang [57]
Hawaii	Hawaii	December-February	x	x	-	-	-	-	-	-	-	-	-	x	Honebrink [49]
USA	North Carolina to lower Florida	October-February	x	x	-	-	-	-	-	-	-	x	x	x	Scotton [48]
USA	Delaware Bay	June-August			-	-	-	x	x	x	-	-	-	-	Scotton [48]
Mexico	Tamiahua Lagoon, Veracruz	December-February	x	x	-	-	-	-	-	-	-	-	x	x	Ibáñez-Aguirre and Gallardo-Cabello [40]
Mexico	Central Pacific (Jalisco and Colima)	-	x	-	-	-	-	-	-	-	-	-	x	x	Present study

Table 3: Spawning seasons of *Mugil cephalus* in different places.

However in other parts of the world, spawning of *M. cephalus* occurs mainly during summer, as is the case of Delaware Bay [48], Natal estuaries of South Africa [54] and Primorye, Russia [55]. In other areas the maximum reproduction of *M. cephalus* takes place during summer and early winter, as occurs in France [56] and the Balear Island of Menorca, Spain [57]. Also spawning of *M. cephalus* can occur during spring and summer as is the case of the east and west coast of Australia [58] (Table 3).

The hepatosomatic index obtained in this study was

$b=2.771$ ($r^2=0.849$),

which indicates a negative allometric growth; since fish decrease their fatty reserves as they grow older. In the case of *M. cephalus* in the Tamiahua lagoon, Veracruz, the relationship obtained for the hepatosomatic index showed an isometric growth b=3.0, which shows that the increments of the liver weight are directly proportional to the cubic length [40].

Monthly values of hepatosomatic index (Figures 6a and 6b) showed that the liver accelerates its activity of reserving fatty acids during the periods before spawning; therefore, their weight increases considerably. The highest activity of fatty acid reserves is in June, July and August and starts to decrease in February and March after the spawning period of November, December and January. A similar phenomenon was observed in Tamiahua, where the fatty acids accumulation occurred in previous spawning periods [40]. However the largest accumulation of fatty acids was in the abdominal cavity instead of the liver. Ibáñez and Gallardo-Cabello reported this same observation in *M. cephalus* in Tamiahua, where the accumulation of glycogen is so intense in the periods previous spawning, that the stored reserve of hepatic acids is insufficient, which shows a sub estimated hepatosomatic index of the fatty acids reserve in the animal's body [40].

Variations in the stomach repletion index (Figures 7a and 7b) showed higher values during June, February and October, which are the months previous to spawning and higher accumulation of fatty acids reserve in the liver. Once the spawning has occurred, in the months of November, December and January, values of the gastric repletion values decline significantly.

Figure 8 shows the values of the condition factor; the highest values are obtained in January, August and December for the Clark index and Safran with eviscerated weight values (EW). For the Fulton index and Safran with total weight (TW) the highest values correspond to July, September, October and November. In all cases the highest were in the months previous to spawning or at its beginning. Similar values were obtained in Tamiahua where the highest value of the condition factor increments before the spawning period and decreases at its end; in May is when the higher recovery of the condition factor occurs [40].

Table 4 shows the fecundity values of *M. cephalus* in different countries; the highest values correspond to the Black Sea, where it can reach values over 7 million of oocytes [47]. Also Berg reports in the same area 7 million of oocytes in organisms of 52.00 cm of total length and 13 years of age [47]. In the Hawaiian Islands, Keith et al. found a maximum fecundity of 7 million of eggs; in Mauritania [59], Brulhet found a maximum of 6 million, and Popescu of 5 million in organisms from the Danuvian delta [47]. Grant and Spain reported 4 million 800 thousand in Australia [60]. Values of fecundity of 3 million 790 thousand eggs were reported in SW Korea by Yang and Kim, and of 3 million in Taiwan by Tung and Hsu [52,61,62]. Solís found a maximum value of fecundity for *M. cephalus* in Tamaulipas, Mexico in the Atlantic Ocean of 2 million 919 thousand oocytes in females of 48.00 to 56.00 cm of total length [63]. In Australia, Thomson and Kesteven reported 2'781,000 and Tosh 2 million and a half in females of *M. cephalus* [47,64,65]. In Mauritania and Senegal, Landret found a value of 2'322,400 oocytes [47]. In our study values of 1'582,684 to 1'747,736 oocytes in females of the coast of the Central Mexican Pacific (Jalisco and Colima) were found in organisms of 37.70 to 48.80 cm of total length and 6 to 12 years of age. The lowest values were reported by Ibáñez and Gallardo-Cabello of 898,512 oocytes in females of 38.00 cm total length and 6 years of age, from the Tamiahua lagoon, Veracruz, Mexico, although these authors mention to have found fecundity values of 1'483,056 oocytes in females older than 6 years of age [40]. At last, Shehadeh et al. found the lowest value reported in the Table 4, of 795,000 oocytes in *M. cephalus* females from Hawaii [47].

This great variability of the values found for the fecundity of *M. cephalus* can be because of the difference in length and age of the studied organisms, as there is a positive relation between fecundity and bigger and older aged females, even in the same area.

Relative fecundity values obtained in this study were 50,272 oocytes per gram, ranged from 9,425 to 150,011, higher values than those obtained by Ibáñez and Gallardo-Cabello for *M. cephalus* in the Tamiahua lagoon, Veracruz of 1,680 oocytes per gram and a range of 680 to 4,776 oocytes [40].

It was observed that in the same study area as *M. cephalus*, *M. curema* showed a fecundity of 9,612 to 238,795 in females of 0 to 5 years of age and lengths of 10.54 cm to 27.79 cm, and an average relative fecundity of 1,120 oocytes per gram (850 to 1,176 oocytes per gram), far below the values of *M. cephalus.*

Conclusions

- Sex ratio was 0.88:1 males: females.
- Average length of sexual maturity (L_{50}) was 34.0 cm in both males and females with 4.64 years (males) and 4.98 years (females); average length of first maturity (L_{25}) for both was 30.0 cm corresponding to 3.4 years of age.
- The gonadosomatic index was at its maximum values in November, December and January. A second very small period occurs during September and October. Mature organisms occur throughout the year.
- The allometric relationship between the liver weight and the fish length is negative (b=2.771). Monthly values of the hepatosomatic index are higher in June, July and August
- The gastric repletion index shows its highest values in June, February and October.
- The condition factor reaches its highest values in June, August and December with Clark and Safran EW indexes and July,

Author	Area	Fecundity (oocytes)	Organism size
Thomson [65]	Australia	1'275,000-2'781,000	-
Kesteven [64]	Australia	1'275,000-2'781,000	-
Grant and Spain [60]	Australia	1'600,000-4'800,000	-
Tosh [47]	Australia	2'000,000-2'500,000	-
Jacob and Krishnamurthi [47]	India	1'320,000	50 cm (TL)
Tung [62]	Taiwan	700,000-3'000,000	-
Hsu [52]	Taiwan	700,000-3'000,000	
Yand and Kim [47]	SW Korea	3'790,000	78.7 cm, 5 years old
Keithy Allardi [56]	Francia	500-2,800/gram	-
Popescu [47]	Black Sea: Danubian delta	5'065,800-5'085,440	-
Nikolskii [47]	Black Sea	3'089,000-7'206,000	-
Berg [47]	Black Sea	5'000,000-7'000,000	52 cm, 13 years old
Brulhet [47]	Mauritania	4'000,000-6'000,000	-
Landret [47]	Mauritania and Senegal	2'322,400	50 cm FL
Shehadeh [48]	Hawaii	340,000-795,000	induced spawning
Nash [47]	Hawaii	1'000,000 (effectively released)	-
Keith [47]	Hawaii	5'000,000-7'000,000 oocytes	-
Solís [63]	Tamaulipas, Mexico	1'341,000-2'919,000 oocytes (6,510 oocytes/g)	48 and 56 cm (TL)
Ibáñez-Aguirre and Gallardo-Cabello [40]	Veracruz, Mexico	405,767-898,512 (1,680 oocytes/g, 680-4,776)	-
This study	Central Mexican Pacific	1'422,076-1'747,736 oocytes (2,830 oocytes·g^{-1}, 1,500 - 2,900 oocytes·g^{-1})	28.5 cm to 48.8 cm (TL)

Table 4: Fecundity values by different authors and countries.

September, October and November with Fulton and Safran TW indexes.

- Total fecundity was 1'422,076 to 1'747,736 oocytes for females of 3 to 12 years old.
- Relative fecundity was average: 2,830 oocytes·g^{-1} from 1,500 to 2,900 oocytes per female from 3 to 12 years old.
- Average oocytes diameter was 0.38 mm (from 0.22 to 0.52 mm).

Recommendations

The studies of the reproduction of the *Mugil cephalus* must be continued and published, because they are an important base line to compare if significant variations occur in the average size of sexual maturity (L_{50}) and at first maturity (L_{25}), which may indicate overexploitation of this resource.

Development of models of maximum sustainable yield and simulation capture should be taken into account to reach a rational management of this fishery, by capturing mature organisms that have already reproduced at least once and that will provide through recruitment new organisms to the population, preventing over-exploitation of the resource. Also, ban seasons should be imposed and the law obeyed as in the mesh size of the fishing gear, to assure that only adults will be fished.

Acknowledgement

We want to express our gratefulness to fishermen, and also EG Cabral-Solís, D Brambila-López and A Pérez-Muñoz that provided samples to complete the information of this study.

References

1. De Silva SS (1980) Biology of grey mullet: a short review. Aquaculture 19: 21-36.
2. Barletta M, Valenca D (2016) Biogeography and distribution of Mugilidae in the Americas. Crosetti & Blaber (ed.) pp: 42-63.
3. Harrison IJ (2002) Mugilidae. In: Carpenter K (ed.) FAO species identification Guide for fisheries purposes. The living marine resources of the Western Central Atlantic. Bony fishes Part 1 (Acipenseridae to Grammatidae). Rome FAO pp: 1071-1085.
4. Harrison IJ (1995) Mugilidae. In: Fischer W, Krupp F, Schneider W, Sommer C, Carpenter KE (eds.) FAO guide for species identification for purposes of Fisheries. Pacifico Centro Oriental, FAO, Roma pp: 1293-1298.
5. Whitfield AK, Panfili J, Durand JD (2012) A global review of the cosmopolitan flathead mullet Mugil cephalus Linnaeus 1758 (Teleostei: Mugilidae), with emphasis on the biology, genetics, ecology and fisheries aspects of this apparent species complex. Reviews in Fish Biology and Fisheries 22: 641-681.
6. SAGARPA (2015) Statistical Yearbook fishing 2014. National Commission of Aquaculture and Fisheries Secretariat of Agriculture, Livestock, Rural Development, Fisheries and Food.
7. Márquez-Millán R (1974) Observations on mortality totally growth in length of mullet (Mugil cephalus) in the lagoon Tamiahua, sea, México. INP, Scientific series 2: 1-17.
8. García S (1980) Contribution to the study of the fishery for mullet (Mugil cephalus L) in the lagoon Tamiahua, View. INP of the Ministry of Fisheries p: 28.
9. Díaz-Pardo E, Hernández-Vázquez S (1980) Growth, reproduction and feeding habits of the smooth Mugil cephalus in Laguna de San Andrés, Tamps. Proceedings of the National School of Biological Sciences 23: 109-127.
10. Romero MAS, Castro JL (1983) Aspects of the biology of mullet (Mugil cephalus Linnaeus) in the Dead Sea, Chiapas, Mexico. An Esc Nac Cienc Biol, México 23: 95-112.
11. Pérez-García M, Ibáñez AL (1992) Morphometry of fish Mugil cephalus and M. curema (mugiliformes: Mugilidae) in Veracruz, Mexico Rev Biol Trop 40: 335-339.
12. Ibáñez AL, Leonart J (1996) Relative growth and comparative morphometrics of Mugil cephalus L. and M. curema V. in the Gulf of Mexico. Scientia Marina 60: 361-368.
13. Ibáñez AL, Gallardo-Cabello M (1996a) Total and natural mortality of Mugil cephalus and M. curema (Pisces: Mugilidae), in Tamiahua Lagoon, Veracruz. I. Selectivity. Hidrobiológica 6: 9-16.
14. Ibáñez AL, Gallardo-Cabello M (1996b) Age determination of the grey mullet Mugil cephalus L. and the white mullet M. curema. V. (Pisces: Mugilidae) in Tamiahua lagoon, Veracruz. Marine Science 22: 329-345.
15. Sánchez-Rueda O, Mar IG, Aguirre ALI, García AM (1997) Sediment in the stomach contents of Mugil cephalus and M. curema (mugiliformes: Mugilidae) in the lagoon Tamiahua, Mexico. Rev Biol Trop 45: 1163-1166.
16. Briones-Ávila E (1992) Diagnosis of fishery mullet (Mugil cephalus) in Sinaloa. INP CRIP-Mazatlán. Newsletter 25: 44-51.
17. Briones-Ávila E (1998) Fishery biology Mugil cephalus in Agua Brava Lagoon, Nayarit, during 1992-1994. Master's Thesis. Autonomous University of Sinaloa, México.
18. Gallardo-Cabello M, Espino-Barr E, Cabral-Solís EG, Puente-Gómez M, Garcia-Boa A (2012) Study of the otoliths of Mugil cephalus (Pisces: Mugilidae) in Mexican Central Pacific. J of Fish and Aqua Sci 7: 346-363.
19. Espino-Barr E, Gallardo-Cabello M, Garcia-Boa A, Puente-Gómez M (2015a) Growth analysis of Mugil cephalus (Percoidei: Mugilidae) in Mexican Central Pacific. Glob J Fish Aquac 3: 238-246.
20. Ibáñez AL (2016) Age and growth of Mugilidae. In: Crosetti D and Blaber S (eds.) Biology, Ecology and Culture of Grey Mulletv (Mugilidae). Taylor & Francis Group, Boca Raton, Fla, USA p: 539.
21. Espino-Barr E, Cabral-Solís EG, Gallardo-Cabello M, Ibáñez AL (2005) Age determination of Mugil curema Valenciennes, 1836 (Pisces: Mugilidae) in the Cuyutlan Lagoon, Colima, Mexico. Intl J Zool Res 1: 21-25.
22. Gallardo-Cabello M, Solís EGC, Espino-Barr E, Ibáñez AL (2005) Growth analysis of white mullet Mugil curema (Valenciennes, 1836) (Pisces: Mugilidae) in the Cuyutlán Lagoon, Colima, México. Hidrobiológica 15: 321-325.
23. Ibáñez AL, Cabral-Solís EG, Gallardo-Cabello M, Espino-Barr E (2006) Comparative morphometrics of two populations of Mugil curema (Pisces: Mugilidae) on the Atlantic and Mexican Pacific coasts. Sci Mar 70: 139-145.
24. Cabral-Solís EG, Espino-Barr E, Gallardo-Cabello M, Ibáñez AL (2007) Fishing impact on Mugil curema stock of multi-species gill net fishery in a tropical lagoon, Colima, Mexico. J Fish Aqua Sci 2: 235-242.
25. Cabral-Solís EG, Gallardo-Cabello M, Espino-Barr E, Ibáñez AL (2010) Reproduction of Mugil curema (Pisces: Mugilidae) from the Cuyutlán lagoon, in the Pacific coast of México. Agricultural Research Advances 14: 19-32.
26. Espino-Barr E, Gallardo-Cabello M, Cabral-Solís EG, Puente-Gómez M, Garcia-Boa A (2013) Otoliths analysis of Mugil curema (Pisces: Mugilidae) in Cuyutlan Lagoon, Mexico. Agricultural Research Advances 17: 35-64.
27. Fulton T (1902) Rates of growth of sea-fishes. Sci Invest Fish Div Scot Rept 21: 3720.
28. Clark F (1928) The weight-length relationship of the Californian sardine (Sardina coerulea) at San Pedro. Fish Bull USA 12: 22-44.
29. Safran P (1992) Theoretical analysis of the weight-length relationship in fish juveniles. Mar Biol 112: 545-551.
30. Sokolov VA, Wong M (1972) Research into pelagic fish in the Gulf of California (sardine, thread herring and anchovy) in 1970. Series information 1: 1-35.
31. Holden MJ, Raitt DFS (1975) Manual Fisheries Science. Part 2: Methods for research and application resources. ONU/FAO. Tec Doc fisheries 115: 1-207.
32. Aboussouan A, Lahaye J (1979) The potential of ichthyologiwues populations. Fertility and echthyoplancton. Cybium 6: 29-46.
33. Espino-Barr E, Vega AG, Hernández HS, Vega HG (2008) Manual fisheries biology. Autonomous University of Nayarit p: 168.
34. Sparre P, Venema SC (1995) Introduction to the assessment of tropical fish stocks. FAO Doc. Tec. Fishing, Roma p: 420.

35. Rodríguez-Gutiérrez M (1992) Techniques for quantitative assessment of gonadal maturity in fish p: 79.

36. Gaertne D, Laloe F (1986) Etudebiometrique the tiller a'premier maturity of sexualle Geryonmaritae, Maning and Holthuis, 1981 Senegal. Oceanologica Act 9: 479-487.

37. Simpson AC (1951) The fecundity of the plaice. Fishery Investigations, London 18: 1-27.

38. Espino-Barr E, Gallardo-Cabello M, Cabral-Solís EG, Puente-Gómez M, Garcia-Boa A (2015b) Reproduction of Gerres cinereus (Percoidei: Gerreidae) off the Mexican Pacific coast. Marine and Coastal Sciences Magazine 7: 83-98.

39. Gallardo-Cabello M, Espino-Barr E, Garcia-Boa A, Puente-Gómez M, Cabral-Solís EG (2015) Reproduction of Diapterus brevirostris (Percoidei: Gerreidae) in the Mexican Pacific coast. Glob J Fish Aquac 3: 221-229.

40. Ibáñez AL, Gallardo-Cabello M (2004) Reproduction of Mugil cephalus and M. curema (Pisces: Mugilidae) from a coastal lagoon in the Gulf of Mexico. Bull of Mari Sci 75: 37-49.

41. Briones-Avila E (1990) The period of gonadal maturity in mullet (Mugil cephalus L.) in Sinaloa and Nayarit. Beef. VIII Congreso Nacional Oceanografía. Mazatlán, Sinaloa. México.

42. Briones-Avila E (1994) The regulation of the fishery smooth (Mugil cephalus and white mullet) in Sinaloa and Nayarit. Res. IV National Congress of Ichthyology. Morelia, Michoacán.

43. Render JH, Thompson BA, Allen RL (1995) Reproductive development of striped mullet in Louisiana estuarine waters with notes on the applicability of reproductive assessment methods for isochronal species. Trans Am Fish Soc 124: 26-36.

44. Hubbs CL (1921) Remarks on the life history and scale characters of American mullets. Trans Am Microsc Soc 40: 26-27.

45. Jacot AP (1920) Age, growth and scales characters of the mullet Mugil cephalus and Mugil curema. Trans. Am Microsc Soc 39: 199-230.

46. Arnold EL, Thompson JR (1958) Offshore spawning of the striped mullet, Mugil cephalus, in the Gulf of Mexico. Copeia 158: 130-132.

47. Oren OH (Ed) (1981) Aquaculture of grey mullets. International Biol Prog: 26, Cambridge Univ Press, Cambridge p: 507.

48. Scotton LN, Smith RE, Smith NS, Price KS, de Sylva DP (1973) Pictorial guide to fish larvae of Delaware Bay: with information and bibliographies useful for the study of fish larvae. Delaware Bay Report Series. College of Marine Studies, University of Delaware p: 205.

49. Honebrink R (1990) Fishing in Hawaii: a student manual. Education Program, Division of Aquatic Resources, Honolulu, Hawaii p: 79.

50. Chang CW, Tzeng WN, Lee YC (2000) Recruitment and hatching dates of grey mullet (Mugil cephalus L.) juveniles in the Tanshui Estuary of Northwest Taiwan. Zoological Studies 39: 99-106.

51. Shung SH (1977) Studies on the catch and fishery biology of Mugil cephalus in 1975. Bulletin of Taiwan Fisheries Research Institute 28: 123-133.

52. Hsu CC, Han YS, Tzeng WN (2007) Evidence of flathead mullet Mugil cephalus L. spawning in waters northeast of Taiwan. Zool Stud 46: 717-725.

53. Jeyaseelan MJP (1998) Manual of fish eggs and larvae from Asian mangrove waters. United Nations Educational, Scientific and Cultural Organization. Paris p: 193.

54. Marais JFK (1976) The nutritional ecology of mullets in the Swartkops estuary. PhD thesis, University of Port Elizabeth, Port Elizabeth.

55. Novikov NP, Sokolovsky AS, Sokolovskaya TG, Yakovlev YM (2002) The fishes of Primorye. Vladivostok, Far Eastern State Tech. Fish. Univ p: 552.

56. Keith P, Allardi J (2001) Atlas freshwater fish France. National Museum of Natural History, Paris. Natural heritage 47: 1-387.

57. Cardona L (2000) Effects of salinity on the habitat selection and growth performance of Mediterranean flathead grey mullet Mugil cephalus (Osteichthyes, Mugilidae). Estuarine, Coastal and Shelf Science 50: 727-737.

58. Kailola PJ, Williams MJ, Stewart PC, Reichelt RE, McNee A, et al. (1993) Australian fisheries resources. Bureau of Resource Sciences, Canberra, Australia p: 422.

59. Keith P, Vigneux E, Bosc P (1999) Atlas freshwater fish France. National Museum of Natural History, Paris. Natural heritage 39: 136.

60. Grant CJ, Spain AV (1975) Reproduction, growth and size allometry of Mugil cephalus Linnaeus (Pisces: Mugilidae) from north Queensland inshore waters. Aust J Zool 23: 181-201.

61. Yang WT, Kim UB (1962) A preliminary report on the artificial culture of grey mullet in Korea. In Proc Indo-Pacif Fish Counc 9: 62-70.

62. Tung IH (1948) On the egg development and larval stages of the grey mullet, Mugil cephalus Linnaeus. Rep Inst Fish Biol Minist Econ Aff Nat Taiwan Univ 3: 187-215.

63. Solís RJM (1966) Fertility mullet (Mugil cephalus Linnaeus). Inst Nal Inv Biol Pesq Sría. Ind y Com Trabajos de Divulgación 11: 1-6.

64. Kesteven G (1942) Studies in the biology of Australian mullet. Account of the fishery and preliminary statement of the biology of Mugil dobula Günther. Aust Counc Sci Industr Res Bull 157: 1-147.

65. Thomson JM (1963) Synopsis of biological data on the grey mullet (Mugil cephalus L.). C.S.I.R.O. Div Fish Oceanogr Sydney, Australia.

31

Morphological and Molecular Characterization of *Diplozoon kashmirensis; D. aegyptensis* and *D. guptai* Collected from Fishes of Kashmir Valley-India

Fayaz Ahmad[1], Khalid M Fazili[2], Tanveer A Sofi[1*], Bashir A Sheikh[1], Ajaz A Waza[2], Rabiya Rashid[2] and Tantry Tariq Gani[3]

[1]*Department of Zoology, University of Kashmir, Srinagar, India*
[2]*Department of Biotechnology, University of Kashmir, Srinagar, India*
[3]*Sheri Kashmir Institute of Medical Science-Soura, Srinagar, India*

Abstract

The study reports the results of molecular characterization of the Internal Transcribed Spacer (ITS) of ribosomal DNA of 3 *Monogenean* species using polymerase chain reaction (PCR), nucleotide sequencing and construction of phylogenetic trees from different fish hosts of Kashmir. The present study shows that the size of the amplified product is 873bp long for *D. kashmirensis,* 1120bp long in *D. aegyptensis* and 687bp long in *D. guptai* revealing that there are intraspecific differences in their base pair lengths. Guanine and Cytocine (G+C) content of three *Diplozoon* species was found nearly constant for three species i.e., 47% (*D. kashmirensis*); 47% (*D. aegyptensis*) and 48% (*D. guptai*), this GC richness contributes to physical attributes of RNA structures, as there is correlation between GC content and optimal growth temperature. An important observation during the present study has been noticed that *Schizothorax niger* is infected by all the three species of *Diplozoidae; D. kashmirensis; D. aegyptensis* and *D. guptai,* but when all six fishes were collected simultaneously, parasitism by all the parasite species was never observed. Phylogenetic trees Maximum Parsimony (MP), Maximum Likelihood (ML) and Neighbor Joining (NJ) showed that *D. kashmirensis* and *D. aegyptensis* share a common host *Carassius carassius* and *S. niger.*

Keywords: *Diplozoon*; Ribosomal DNA; Schizothorax; Kashmir; Phylogenetic trees.

Introduction

Monogeneans belonging to the *Diplozoidae* are common parasites on the gills of cyprinid fish. The life cycle is direct, including free-swimming oncomiracidia, larval stage (diporpa) and adult. Two larvae (diporpae) permanently fuse into a pair to form the sexually maturated adult. In the adult, the vitellaria and almost all the internal organs are situated in the anterior part of the body. The female and male reproductive organs and terminal part of the gut are situated in the posterior part. The attachment apparatus of adults consist of four pairs of clamps and a pair of small central hooks situated on the ventral side of the opisthaptor. Due to the complicated determination of several groups of *monogenean* parasites, molecular markers based on species-specific variability in the ribosomal DNA region (rDNA) their cytogenetics have been designed and shown to be useful for precise species identification [1-5]. The interspecific nucleic acid variability of Internal Transcribed Spacers of rDNA (ITS) has also been used to distinguish *diplozoid* parasites [6-11].

From the available data, it has been concluded that morphological and metrical differences in the clamp size, pharynx size, prohaptoral length, opisthohaptoral length, sucker distance, testis, ovary and egg size were the major criteria for species determination. Species deterimination of trematodes is difficult and demands great skill and experience. As the structures of taxonomic importance (central hooks, clamps etc.,) grow gradually and the measurements of sclerotized structures are variable, species determination of trematodes in different developmental stages is not always clear. There are still some unclear descriptions of trematode species that differ only by host species, and some studies that did not employ recommended criteria [12,13]. Molecular biology techniques have been used as objective methods to distinguish between parasite species. The rDNA genes, particularly the 28S gene, have been found generally useful in molecular taxonomy and phylogeny of parasites [14-16]. However, there are no published molecular studies of trematode genomes from the Kashmir valley. The present study reports the results of molecular analysis of the Internal Transcribed Spacer (ITS) of ribosomal DNA of 3 *Monogenean* species namely *Diplozoon kashmirensis* Kaw, *Diplozoon aegyptensis* Fischthal et Kuntz, *Diplozoon guptai* [17] using Polymerase Chain Reaction (PCR), nucleotide sequencing and construction of phylogenetic from different fish hosts of the Kashmir valley.

Materials and Methods

Parasite material

Parasite specimens of *Diplozoon* spp. were collected from the *Carassius carassius; Cyprinus carpio communis; C. c. specularis; Schizothorax niger; S. esocinus; S. curvifrons* and *S. plagiostomus* of Kashmir and were used for DNA extraction. Samples were immediately fixed in 70% alcohol after collecting from the gills, gill cover, mouth cavity, eyes & fins of host fish. These samples were remained in alcohol until the present study.

DNA isolation

Parasite specimens of three *Diplozoon* species were collected from fish hosts of *Carassius carassius*; *Cyprinus carpio communis*;

***Corresponding author:** Tanveer A. Sofi, Department of Zoology, University of Kashmir, Srinagar-190006, India, E-mail: stanveer96@gmail.com

Schizothorax curvifrons; *Schizothorax esocinus*; *Schizothorax niger* and *Schizothorax plagiostomus* from Wular lake, Anchar lake, Dal lake, Manasbal lake, River Jhelum and River Sindh of Kashmir valley preserved in 100% ethanol for genomic DNA extraction and stored at-20°C for good quality of DNA. For DNA extraction ethanol was removed from parasites as per the protocol given by [18] and as such, these specimens were air dried to remove ethanol. The resultant DNA was examined on 1.5% agrose-TAE gels, stained with ethidium bromide (EtBr) and visualized under UV light.

Results

Morphological characterization

The present specimens having rectangular opisthaptor with four pairs of clamps and two individuals in form of a cross belongs to genus *Diplozoon* Nordmann, 1832. When compared *D. kashmirensis* with *D. aegyptensis*, they showed similarity in comparative size of clamps, size of eggs, form of oral suckers but differs in the proportion of body length to its breadth, shape of ovary, shape and position of testis. *D. kashmirensis* resembles *D. guptai* in egg size, absences of sticky glands in the anterior part, position of testis with respect to ovary, extent of vitellaria and in the arrangement of intestine in hind portion of the worm but shows strong variations as regards total body length, size ratios, clamp size, the shape & size of testis and also prepharynx size. The variations of the three species can be regarded as intraspecific variations due to geographical isolation and are not sufficient for the creation of new species and thus the present specimens are described as *D. kashmirensis* Kaw, 1950; *D. aegyptensis* Fischthal et Kuntz, 1963 and *D. guptai* [17-24] (Table 1).

PCR amplification

The PCR amplified products of ITS regions of rDNA were successfully obtained using the primers (Table 2). PCR amplification was carried out to amplify ITS region of *Diplozoon* species (Table 3). The size of the amplified product was found to be 873bp long for *D. kashmirensis*; 1120bp long in *D. aegyptensis* and 687bp long in case of *D. guptai* (Figure 1). In BLAST search of these sequences, they showed similarity with other *Diplozoon* spp. (Table 3). In bioinformatics analysis, the results tallied with those of the earlier study; hence, the same are not repeated here in. Based on morphological studies, these species were identified as belonging to three *Diplozoon* species. The present results of the molecular analysis corroborate the species identification of these forms. Therefore, it can be assumed that the present species recovered from the different fish hosts of water bodies

Species Particulars													
Particulars	**Total Body Length**	**Forebody**	**Hind body**	**Length ratio between fore & hind body**	**Clamp size**	**Testis**	**Egg size**	**Anterior suckers**	**Prepharynx**	**Pharynx**	**Host**	**Site**	**Locality**
***D. kashmirensis* Kaw, 1950**	2.3-4.32	1.4-2.64 x 0.71-1.51	0.9 x 1.72 x 0.5-0.69	1:0.646	0.15 x 0.075 0.166 x 0.076 0.154 x 0.075 0.140 x 0.074	0.16 x 0.27-0.29	0.27 x 0.29 x 0.07-0.09	0.063-0.074 x 0.045-0.063	0.065	0.065-0.075	Schizothorax sp.	Gills	Dal Lake, Kashmir
***D. aegyptensis Fischthal* et Kuntz, 1963**	4.529 (3.62-5.77)	2.665 x 0.558 (1.879-3.452 x 0.299-0.836)	1.128 x 0.178 (0.867-1.871 x 0.130-0.245)	-----	0.070 x 0.097 0.065-0.079 x 0.092-0.102	0.136 x 0.08 (0.103-0.19 x 0.063-0.093)	0.292 x 0.107 (0.524-0.313 x 0.081-0.132)	0.038 x 0.07 (0.029-0.046 x 0.065-0.079)	0.027 (0.020-0.034)	0.062 x 0.044 (0.051-0.075 x 0.040-0.050)	Labeo forskalii	Gills	Giza Fish Market, Giza Fish Market, Egypt
***D. aegyptensis* Fayaz et Chishti 1993**	4.2 (3.95-4.25)	-----	-----	-----	0.114 (0.10-0.124) x 0.045 (0.04-0.048)	0.155 (0.14-0.17) x 0.105 (0.10-0.11)	0.25 (0.22-0.28) x 0.082 (0.076-0.088)	0.048 (0.032-0.064)	0.054 (0.041-0.068)	0.064 (0.056-0.072) x 0.047 (0.044-0.05)	Schizothorax niger	Gills	Dal lake, Kashmir
***Diplozoon guptai* Fayaz et Chishti 1999**	1.873 (1.28-2.55)	1.66 x 0.604 (0.755-0.144 x 0.44-0.65)	0.603 x 0.395 (0.46-0.76 x 0.289-0.48)	1:0.523 (1:0.4-0.639)	0.102 x 0.045 0.106 x 0.032 0.093 x 0.038 0.081 x 0.035	0.813 x 0.0786 (0.076-0.12 x 0.052-0.128)	0.245-0.07 (0.228-0.268 x 0.06-0.08)	0.056 x 0.05 (0.04-0.08 x 0.032-0.072)	0.0356 (0.03-0.042)	0.057 x 0.025 (0.044-0.069 x 0.02-0.028)	Schizothorax niger; S. esocinus; Labeo sp. and Carasius carassius	Gills	Dal and Anchar Lake, Kashmir

Table 1: Comparative Morphological characteristics of Diplozoon species (measurements in mm).

Species	**Primer Designed**	**GenBank Accession Number**	**Author and Year**
***Diplozoon kashmirensis* Kaw, 1950**	Forward	AF 369758 to AF 369761	Sicard et al., 2001
	Cer5.8S 2249:5′GCTCACGTGACGATGAAGAG3′		
***Diplozoon aegyptensis* Fischthal et Kuntz, 1963**	Reverse		
***Diplozoon guptai* Fayaz and Chishti, 1999**	Cer28S 3116 :5′TTCGCTATCGGACTCGTGCC3′		

Table 2: Primers used for Trematodes.

Monogenea	**Initial Denaturation**	**Denaturation for 30 cycles**	**Annealing**	**Extension**	**Final extension**
Diplozoon kashmirensis; D. aegyptensis* and *D. guptai	95°C for 10 minutes	30 cycles at 95°C for 30 seconds	55°C for 30 seconds	72°C for 75 seconds	72°C for 10 minutes

Table 3: PCR assay of Monogeneans which were carried out in a thermocycler (Eppendorf Mastercycler Personal) under different conditions.

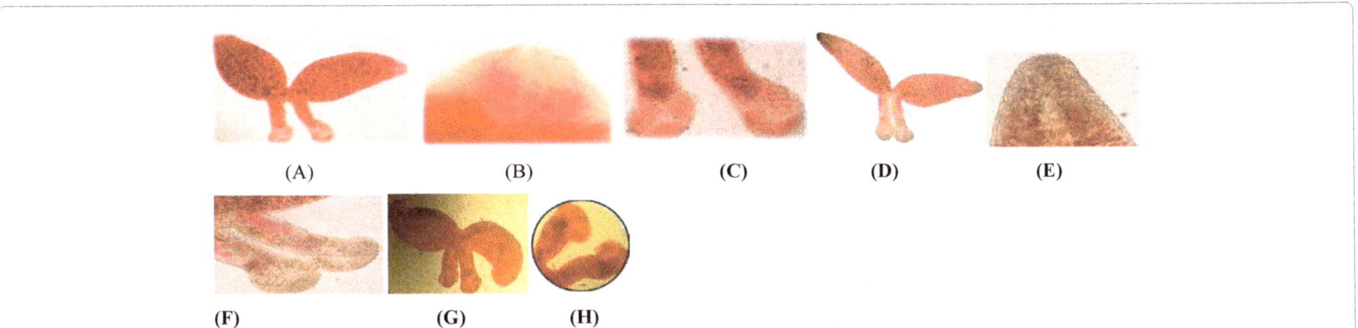

Figure 1: (A). Whole specimen of *D. kashmirensis*, **(B).** Anteroir end showing prohaptor of *D. kashmirensis*, **(C).** Posteror end showing posthaphtors of *D. kashmirensis*, **(D).** Whole specimen of *D. aegyptensis*, **(E).** Forebody showing suckers & pharynx of *D. aegyptensis*, **(F).** Hindbody showing clamps & eggs of *D. aegyptensis*, **(G).** Whole specimen of *D. guptai*, **(H).** Posterior body showing posthaphtors of *D. guptai*.

S no	Monogenean Species	Host	GenBank Accession No.	Family	Base pair length	Authors	Country	Year
1.	***D. kashmirensis*** **Kaw, 1950***	Carassius Carassius, Cyprinus carpio cummunis, Schizothorax niger, S. esocinus, S. curvifrons	AF973616	Diplozoidea	873 bp	Present study	India	2015
2.	***D. aegyptensis*** **Fischthal et Kuntz, 1963***	Carassius Carassius, Schizothorax niger;	AF973617	Diplozoidea	1120 bp	Present study	India	2015
3.	***D. guptai*** **Fayaz and Chishti, 1999***	Schizothorax niger	AF973618	Diplozoidea	687 bp	Present study	India	2015
4.	***D. bliccae*** **(Glaser, 1965)**	Blicca bjoerkna	AF369761	Diplozoidea	988 bp	Sicard et al.	France	2001
5.	***D. paradoxum*** **Nordmann, 1832**	Abramis brama	AF369759 and AJ563372	Diplozoidea	769 bp	Matejusova	Czech Republic	2004
6.	**D. homoion Bychowsky & Nagibina, 1959**	Rutilus rutilus, Scardinius erythrophthalmus	AF369760	Diplozoidea	996 bp	Sicard et al.	France	2001

Table 4: Monogenean trematode species used for molecular comparison of ITS rDNA sequences along with their hosts, country and GenBank accession numbers for corresponding sequences (*Query sequence).

of Kashmir valley is *D. kashmirensis* Kaw, *Diplozoon aegyptensis* Fischthal et Kuntz, and *Diplozoon guptai* [24].

[Reagents for PCR: Taq DNA polymerase 3 U/μl, dNTP mixture 100 mM, primers 20 pmols, 10 × TaqDNA Polymerase buffer (Genei), PCR water (Sterile milli-Q)].

Sequences deposited in GenBank

GenBank: AF973616; *Diplozoon kashmirensis*, complete sequence.

GenBank: AF973617; *Diplozoon aegyptensis*, complete sequence.

GenBank: AF973618; *Diplozoon guptai*, complete sequence.

The three *monogenean* species of Trematodes viz., *Diplozoon kashmirensis* Kaw, 1950; *Diplozoon aegyptensis* Fischthal et Kuntz, 1963 and *Diplozoon guptai* [17] which were recovered during the present study are used for molecular study for the first time.

Nucleotide sequences

PCR products were visualized and documented, and the sizes of the sequences were estimated. The sequence obtained from three different *Diplozoon* species were submitted to GenBank and their accession number acquired were AF973616; AF973617 and AF973618 (Table 4). Sequences were compared with other sequences of *monogenean* species from GenBank. When the BLAST search was performed, the query sequence showed maximum similarity with 28S rDNA sequence of *Diplozoon* spp. The nucleotide sequences obtained and shown in (Figures 2-5) are as raw sequences (Table 5).

Pairwise alignment

Pairwise alignments of *Diplozoon* species were made by using different softwares such as Gene Runner, DNA Dynamo, Chromas Pro. *D. kashmirensis* showed maximum similarity with those of *D. bliccae* where as *D. aegyptensis* showed maximum similarity with *D. paradoxum* and in case of *D. guptai* that showed maximum similarity to *D. homoin* (Tables 6-8).

Construction of phylogenetic tree

Phylogenetic trees were obtained by comparing the 28S rDNA sequences of the query parasite and other available sequences for related *monogenean* parasites. The E value was found to be zero up to the 100th sequence of BLAST search and the query coverage 95% and above. The species of *D. kashmirensis* and *D. aegyptensis* appeared to be the most closely related species, with well-supported clade by Neighbour joining and MP trees (Figures 6-8).

Above Table shows that *Diplozoon kashmirensis* having GenBank accession number AF973616 mostly resembles with *Diplozoon bliccae* with an accession number AF369761.1. Out of 867 base pairs of *Diplozoon kashmirensis*, 807bp match with that of *Diplozoon bliccae* i.e., 93.08% similarity with 15 gaps (1.73%).

From the Table 6 it is clear that *Diplozoon aegyptensis* having GenBank accession number AF973617 shows 94.13% similarity with *Diplozoon paradoxum* with an accession number AF369759.1. Out of 988 base pairs of *Diplozoon aegyptensis*, 930 bp match with that of *Diplozoon paradoxum* with 11 gaps (1.11%).

The present observation shows that *Diplozoon guptai* having GenBank accession number AF973618 shows 86.16% similarity with that of *Diplozoon homoion* having GeneBank accession mnumber AF369760.1 (Table 7). 585 bp of *Diplozoon guptai* matches with *Diplozoon homoion* with 15 gaps, out of total 679 base pairs.

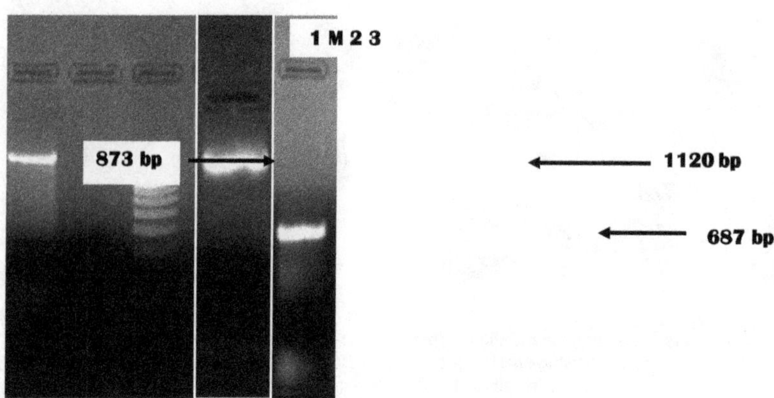

Figure 2: Polymerase Chain Reaction (PCR) products of Trematodes (Monogenea) M=marker; bp=base pairs (100 bp ladder), 1=*Diplozoon kashmirensis* Kaw, 1950, 2=Diplozoon aegyptensis Fischthal et Kuntz, 1963 and 3=*Diplozoon guptai* [24].

1 TGCTTACTGA CTTGAGCATC GATTTCTTGA ACGTGAATTG CGGCATTACC CTCTAATGAT
61 GCCACGCCTA GCCGAGTATC GGCATTAAAT CTAGCACGAC GCTTATTTGG TCCTGGCTTA
121 GAAAGTTGTC AGCCGTCGTG TTGTACTTGG CAACGTGTTG TTCTGTTGTC AAGTCGGCGG
181 TATTATTGAC GCTTGCCAAA TGTAATGGAG AGTTTGTATA TGCGAAATAT CTTCCGGTAG
241 CCTGTTGGTG TTGGCTACGC TGCCCCGTGT ATTTTTTATT TGCATTTTTG TGCATACCGA
301 TGGGGTGGTT AGCTTCTCGT CAGCAGTGCG TCCTTGCCGG TGGTGTCGTG GAATGGGAAT
361 TTCAATAAGC ATTTCTGAAT CCTAATTGTG AAATTGTCAT TTTATGTGCT GTTCTCTTGA
421 GCCGCATGGC CCACTTGTTG TGCGATGACC AGTGACGCTT TGAATGCGAG TGCATGCATG
481 CCAGGTCTCA GCCTATTTGT GATCGCGACA GTGCTTTGCT TGTGTTCTGC GTTTAATTTT
541 TGTCACTGTT TCCCGCGAAT GAGCGAGTCT GGCCCGAGAC GAGAGCATGT GCCCATGTCG
601 TGCTGTGCAG ACATTACTAC TCCATTCTTC GCTAAGTGTG TATCGGTGTC ACCCGTATTT
661 TACTGTACTT CTGTGGTGTA TGCACCTGAC CAAGGATTAG GCGTGATCAC CCGCTGAGCT
721 TAAGCATATC AATGGGCGGA GGAAAAGAAA CTAACCACTA TTCCCTTAGT AACGTCGAGT
781 GAACACCGAT TAGCAAAGCA CCGAAGCTGC GGTCTTTTGG CCGTTCGGCA ATCCGGTGTT
841 TAGGTTATCA TACTCAGGCG ATGTACTGTG GTC

Figure 3: Raw nucleotide sequences of *Diplozoon kashmirensis* Kaw, 1950.

1 AACTGCAAAC TGCCTTGAGC AAATTAGTTG TGAAAGTAAA TTACGGCAGG AGGCTCCCCC
61 TGATAACACG CCTAGCCCCG TGTCGGCATT AAATCGATCA CGACGCTTAA TTGGTTGTGG
121 CTTAGTTTGT TGTCAGCCGT CGTGTTGTAC TTGGCAACGT GTTGTTCAGT TGTCAAGTAG
181 ACGGTATTAT TGACGCTTGC CAAATGTAAT GGAGAGTTAG NDATGCGAAA TATCCGCTGG
241 TAGCCTGTTG GTGTTGGCAA CGCTGGCCCG TGTATGGTTT ACTTGTTTTT TTGTGCATAC
301 TCATGGGGGC GGTTAACTTC GCGTCATCAG AGCGTGTTTG CCGGAAGTGT ATTGCAGTGG
361 CGTGGGAATT TCAATGAGCA TTTGTGAATG GTAATTGTTA AATTGCCATT TTATGTGCTG
421 TTCTCTTGAG CCTTTTGGCC CACGGGTTGT GCGGTGACCA GTGTTGCTTT GAATGCGTGC
481 GCATGCATGC CAGGTCGCAG CCTATTGTGA TCGCGACAGT GCTTTGCTTG TGTTCTGCGT
541 TTAATTTTTG TCACTCCCGC ACTGGTCGCT AAGTGCATGT CCCGAGATGA GATTGTGTGC
601 CCATGTCATG CTGGGCTGAC ATTACTACTC CACTGGTCGC TAAGTGCATG TCGGTGTCAT
661 CAGTATTCTA CTGTACTGCT GTGTTGTGTG TGCACCTGAC CTCGGATTAG GCGTGATTAC
721 CCGCTGAACT TAAGCATATC AATAAGCGGA GGATTAGAAA CTAACCAGGA TTCCCTTAGT
781 AACGGCGAGT GAACAGGGAT TAGCCCAGTT CCGAAGCTGC GGTCTTTTGG CCGTTCGGCA
841 ATGTTATGTT TAGGTTGGCA TACTCAGGCG ATGTACTGTG CTAAGTCCAT TCATGAATAT
901 GGCTAGCTAT CTGTTCCAGA GAGGGTGAAA GGCCCGTGAG CATAGTACGT TGTTCTGTCT
961 TAGCCAACCG TTGAGTCGGG GGTTTACTTG AGGCAGCCCA AAAAGTAGAC GGTATTATTG
1021 ACGCTTGCCA AATGTAATGG AGTTAGTGTG ACCCGAGATG AGATTGGTTG GCATACGCAG
1081 GCGATGTACT GTGCTAAGTC CAGGTGTTTG CATTATTAGT

Figure 4: Raw nucleotide sequences of *Diplozoon aegyptensis* Fischthal et Kuntz, 1963.

1 TGCTGCAAAC TGCCTTGAAA ATCTTCTTCT TGAACGCGAA TCGCGGTATT AGGTACTGCC
61 TGATGCCACG CCTAGCCGAG TGTTGGCATT ATATCTATCA CGACGCTTAA TTGGTCGTGG
121 CTTAGGCGGT TGTCCTCCGT CGTGTTTTAC TTTGCAACGT GTTGCTCAGT TGTACTGTCG
181 ACGGTATTAT TGACGCTTGC CAAATGTAAT GGAGAGTGTG TATATGCGAA ATTTCTGCCG
241 GTAGCCTGTT GGCTGCGGCG ACGCTGCCCC GTGGCCGGTT TACTTGCATT TTTGTATCTA
301 CCGATTGGGG CGGTTAGCTT GTATTCATCA GCCCGTGTTT GCCGGTGGTG ACTCGTGGTG
361 GCGTGGGAAT TTCAATAAGC ATTACTGAAT GGTAATTAAT AAATTGCCAT TATATATGCT
421 GTGCGCTTGA GCCTTTTGGC CCACGGGTTG TATTGTGACC AGTGTTGCTT TGAATGCGCT
481 CGCAAGCATG CCAGGTCTCA GCCTATGGTG ATCGAGACAG TTCTTTGCTT GTGTTATGCG
541 TTTAGGTGTT GTCACCTCTA CTTGCATATG TGCTAGTGTG TACGCGGAAT GAGCTTTTGT
601 GCCCATGTCA TGCTGTGCTG ACGCTACTTC TCCACTGGTC CAGAAGTGCA TGTCGGGGTC
661 ACCATAACTT TGCTGTATTG TGGGTGC

Figure 5: Raw nucleotide sequences of *Diplozoon guptai* [24].

	Diplozoon kashmirensis	*Diplozoon aegyptensis*	*Diplozoon guptai*
Length	873 bp	1120 bp	687 bp
A	177	237	123
C	191	224	148
G	226	312	188
T	279	345	228
G+C	47%	47%	48%
Total No. of Amino Acids	280	353	219
Molecular Weight	30827 Da	38825 Da	24323 Da

Table 5: Summary of base pairs and amino acids of *Diplozoon kashmirensis* Kaw, 1950, *Diplozoon aegyptensis* Kuntz, 1963 and *Diplozoon guptai* Fayaz et Chishti, 1999.

<table>
<tr><td>D. kashmirensis

D. bliccae</td><td>6

6</td><td>ACTGCCTTGAGCATCGACTTCTTGAACGTAAATTGCGGCATTAGGCTCTGCTGATGCCAC
||| ||||||||| ||||||||| |||||||||| |||| ||||||||
GCTGACTTGAGCATCGATTTCTTGAACGTGAATTGCGGCATTACCCTCTAATGATGCCAC</td><td>65

65</td></tr>
<tr><td>D. kashmirensis

D. bliccae</td><td>66

66</td><td>GCCTAGCCGAGTGTCGGCATTAAATCTATCACGACGCTTAATTGGTCGTGGCTTAGTTTG
|||||||||| |||||||||||| ||||||||| |||||||||||| |
GCCTAGCCGAGTATCGGCATTAAATCTAGCACGACGCTTATTTGGTCCTGGCTTAGAAAG</td><td>125

125</td></tr>
<tr><td>D. kashmirensis

D. bliccae</td><td>126

126</td><td>TTGTCAGCCGTCGTGTTGTACT---CAACGTGTTGTTCAGTTGTCAAGTCGACGGTATTA ||||||||||||||||||||||||||||||||
|||||||||||||||||
TTGTCAGCCGTCGTGTTGTACTTGGCAACGTGTTGTTCTTTTGTCAAGTCGGCGGTATTA</td><td>185

185</td></tr>
<tr><td>D. kashmirensis

D. bliccae</td><td>186

186</td><td>TTGACGCTTGCCAAATGTAATGGAGAGTTTGTATATGC--AATATCTGCCGGTAGCCTGT ||||||||||||||||||||||||||||||||||||||||||||
|||||||
TTGACGCTTGCCAAATGTAATGGAGAGTTTGTATATGCGAAATATCTTCCGGTAGCCTGT</td><td>245

245</td></tr>
<tr><td>D. kashmirensis

D. bliccae</td><td>246

246</td><td>TGGTGTTGGCTACGCTGCCCCGTGTATGGTTTATTTGCATTTTTGTGCATACCGATGGGG ||||||||||||||||||||||||
|||||||||||||||||||||||||||
TGGTGTTGGCTACGCTGCCCCGTGTATTTTTTATTTGCATTTTTGTGCATACCGATGGGG</td><td>305

305</td></tr>
<tr><td>D. kashmirensis

D. bliccae</td><td>306

306</td><td>TGGTTAGCTTCTCGTCATCAGTGCGTGTTTGCCGGTGG-GTCGTGGCGTGGGAATTTCAA |||||||||||||| ||||||| |||||||||||||||
|||||||||||
TGGTTAGCTTCTCGTCAGCAGTGCGTCCTTGCCGGTGGTGTCGTGGAATGGGAATTTCAA</td><td>365

365</td></tr>
<tr><td>D. kashmirensis

D. bliccae</td><td>366

366</td><td>TAAGCATTTCTGAATGGTAATTGTGAAATTGTCAT---ATGTGCTGTTCTCTTGAGCCTT ||||||||||||| ||||||||||||||||||||||||||||||||||
||
TAAGCATTTCTGAATCCTAATTGTGAAATTGTCATTTTATGTGCTGTTCTCTTGAGCCGC</td><td>425

425</td></tr>
<tr><td>D. kashmirensis

D. bliccae</td><td>426

426</td><td>TTGGCCCACGGGTTGTGCGGTGACCAGTGTTGCTTTGAATGCGAGCGCATGCATGCCAGG ||||||| ||||||| |||||||| ||||||||||||
||||||||||||
ACGGCCCACTTATTGTGCGATGACCAGTGACGCTTTGAATGCGAGTGCATGCATGCCAGG</td><td>485

485</td></tr>
<tr><td>D. kashmirensis

D. bliccae</td><td>486

486</td><td>TCGCAGCCTATTTGTGATCGCGAC-GTGCTTTGCTTGTGTTCTGCGTTTAATTTTTGTCA ||||||||||||||||||||||||||||||||||||||||||||
|||||||
TCTCAGCCTATTTGTGATCGCGACAGTGCTTTGCTTGTGTTCTGCGTTTAATTTTTGTCA</td><td>545

545</td></tr>
<tr><td>D. kashmirensis

D. bliccae</td><td>546

546</td><td>CTGTTTCTTGCGAATGAGCGAGTCTGGCCCGAGACGAGATTATGTGCCCATGTCGTGCTG |||||| ||||||||||||||||||||||||||
||||||||||||||||
CTGTTTCCCGCGAATGAGCGAGTCTGGCCCGAGACGAGAGCATGTGCCCATGTCGTGCTG</td><td>605

605</td></tr>
<tr><td>D. kashmirensis

D. bliccae</td><td>606

606</td><td>TGCAGACATTACTACTCCATTGGTCGCTAAGTGCATATCGGTGTC--CCGTATTCTACTG |||||||||||||||||| |||||||| ||||||||||||||||
|||||
TGCAGACATTACTACTCCATTCTTCGCTAAGTGTGTATCGGTGTCACCCGTATTTTACTG</td><td>665

665</td></tr>
<tr><td>D. kashmirensis

D. bliccae</td><td>666

666</td><td>TACTGCTGTGGTGTGTGCACCTGACCTCGGATTAGGCGTGATTACCCGCTGAACTTAAGC |||| |||||||| ||||||||| ||||||||||||
|||||||| ||||||
TACTTCTGTGGTGTATGCACCTGACCAAGGATTAGGCGTGATCACCCGCTGAGCTTAAGC</td><td>725

725</td></tr>
</table>

<table>
<tr><td>D. kashmirensis

D. bliccae</td><td>726

726</td><td>ATATCAATAAGCGGAGGAAAAGAAACTAACCAGGATTCCCTT-GTAACGGCGAGTGAACA ||||||| ||||||||||||||||||| ||||||||||||| |||||||||
ATATCAATGGGCGGAGGAAAAGAAACTAACCACTATTCCCTTAGTAACGTCGAGTGAACA</td><td>785

785</td></tr>
<tr><td>D. kashmirensis

D. bliccae</td><td>786

786</td><td>GGGATTAGCCCAGCACCGAAGCTGCGGTC--TTGGCCGTTCGGCAATGTGGTGTTTAGGT |||||| |||||||||||||||||||||||||||||||| ||||||||||
CCGATTAGCAAAGCACCGAAGCTGCGGTCTTTTGGCCGTTCGGCAATCCGGTGTTTAGGT</td><td>845

845</td></tr>
<tr><td>D. kashmirensis

D. bliccae</td><td>846

846</td><td>TGGCATACTCAGGCGATGTACTGTGTAG
| ||||||||||||||||||||
TATCATACTCAGGCGATGTACTGTGCCC</td><td>873

873</td></tr>
</table>

Table 6: Pairwise alignments of the 28S rDNA ITS consequences of Diplozzon kashmirensis and Diplozoon bliccae, numbering refers to ITS sequences.

<table>
<tr><td>D. aegyptensis

D. paradoxum</td><td>1

4</td><td>TGCAAACTGCCTTGAGCATCGACTTCTTGAACGTAAATTGCGGCATTAGGCTCTG-CTGA
||||||||||||||| ||||||| ||||||||| ||||||||||||| ||||
TGCAAACTGCCTTGAGCCTCGACTTCCCGAACGTAAATTACGGCATTAGGCTCTGCCTGA</td><td>59

63</td></tr>
<tr><td>D. aegyptensis

D. paradoxum</td><td>60

64</td><td>TGCCACGCCTAGCCGAGTGTCGGCATTAAATCTATCACGACGCTTAATTGGTCGTGGCTT
|||| |||||||||||||||||||||||||||||||| ||||||||||||
TGCCCGACCTAGCCGAGTGTCGGCATTAAATCTATCACGACATAATATTGGTCGTGGCTT</td><td>119

123</td></tr>
<tr><td>D. aegyptensis

D. paradoxum</td><td>120

124</td><td>AGTTTGTTGTCAGCCGTCGTGTTGTACTTGGCAACGTGTTGTTCAGTTGTCAAGTCGACG
|||||||| ||||||||||||||||| |||||||||||||||||||||||| ||||
AGTTTGTTAAAAGCCGTCGTGTTGTACTTAACAACGTGTTGTTCAGTTGTCAAGTAGACG</td><td>179

183</td></tr>
<tr><td>D. aegyptensis

D. paradoxum</td><td>180

184</td><td>GTATTATTGACGCTTGCCAAATGTAATGGAGAGTTTGTATATGCGAAATATCTGCCGGTA
||||||||||||||||||||||||||||||||||| | |||||||||||| || ||||
GTATTATTGACGCTTGCCAAATGTAATGGAGAGTTAG-NDATGCGAAATATCCGCTGGTA</td><td>239

242</td></tr>
<tr><td>D. aegyptensis

D. paradoxum</td><td>240

243</td><td>GCCTGTTGGTGTTGGCTACGCTGCCCCGTGTATGGTTTATTTGCATTTTTGTGCATACCG
||||||||||||||| |||||| ||||||||||||||| ||||||||||||||||||||
GCCTGTTGGTGTTGGCAACGCTGTCCCGTGTATGGTTTACTTGCATTTTTGTGCATACCG</td><td>299

302</td></tr>
<tr><td>D. aegyptensis

D. paradoxum</td><td>300

303</td><td>AT-GGGGTGGTTAGCTTCTCGTCATCAGTGCGTGTTTGCCGGTGGTGT----C-GTGGCG
|| |||| |||||||||| |||||||||| ||||||||||||||||||||| | ||||||
ATGGGGGCGGTTAGCTTCGCGTCATCAGAGCGTGTTTGCCGGTGGTGTATTGCAGTGGCG</td><td>353

362</td></tr>
<tr><td>D. aegyptensis

D. paradoxum</td><td>354

363</td><td>TGGGAATTTCAATAAGCATTTCTGAATGGTAATTGTGAAATTGTCATTTTATGTGCTGTT
|||||||||||||| ||||||| |||||||||||||| |||||| |||||||||||||||||
TGGGAATTTCAATGAGCATTTGTGAATGGTAATTGTTAAATTGCCATTTTATGTGCTGTT</td><td>413

422</td></tr>
<tr><td>D. aegyptensis

D. paradoxum</td><td>414

423</td><td>CTCTTGAGCCTTTTGGCCCACGGGTTGTGCGGTGACCAGTGTTGCTTTGAATGCGAGCGC
||||||||||||||||| |||||||||||||||||||||||||||||||||||||||| ||||
CTCTTGAGCCTTTTGGCTTTCGGGTTGTGCGGTGACCAGTGTTGCTTTGAATGCGTGCGC</td><td>473

482</td></tr>
<tr><td>D. aegyptensis

D. paradoxum</td><td>474

483</td><td>ATGCATGCCAGGTCGCAGCCTATTTGTGATCGCGACAGTGCTTTGCTTGTGTTCTGCGTT
||||||||||||||||||||| |||||||||||||||||||||||||||||||||||||||
ATGCATGCCAGGTCGCAGCCTA-TTGTGATCGCGACAGTGCTTTGCTTGTGTTCTGCGTT</td><td>533

541</td></tr>
<tr><td>D. aegyptensis

D. paradoxum</td><td>534

542</td><td>TAATTTTTGTCACTGTTTCTTGCGAATGAGCGAGTCTGGCCCGAGACGAGATTATGTGCC
|||||||||||||||| ||||| ||| ||||| || |||||| ||||| ||||||
TAATTTTTGTCACTGCCGCTTGCGTATGTGCGAGTGTGACCCGAGATGAGATTGTGTGCC</td><td>593

601</td></tr>
<tr><td>D. aegyptensis

D. paradoxum</td><td>594

602</td><td>CATGTCGTGCTGTGCAGACATTACTACTCCATTGGTCGCTAAGTGCATATCGGTGTCACC
|||||| ||||||| ||||||||||||||| ||||||||||||||| |||||||| |
CATGTCATGCTGTGCTGACATTACTACTCCACTGGTCGCTAAGTGCATGTCGGTGTCATC</td><td>653

661</td></tr>
<tr><td>D. aegyptensis

D. paradoxum</td><td>654

662</td><td>CGTATTCTACTGTACTGCTGTG--GTGTGTGCACCTGACCTCGGATTAGGCGTGATTACC
||||||||||||||||||||| |||||||||||||||||||||||||||||||||||||
AGTATTCTACTGTACTGCTGTGTTGTGTGTGCACCTGACCTCGGATTAGGCGTGATTACC</td><td>711

721</td></tr>
<tr><td>D. aegyptensis

D. paradoxum</td><td>712

722</td><td>CGCTGAACTTAAGCATATCAATAAGCGGAGGAAAAGAAACTAACCAGGATTCCCTTAGTA
||||||||||||||||||||||||||||||||||||||||||||||||||||||||||||
CGCTGAACTTAAGCATATCAATAAGCGGAGGAAAAGAAACTAACCAGGATTCCCTTAGTA</td><td>771

781</td></tr>
<tr><td>D. aegyptensis

D. paradoxum</td><td>772

782</td><td>ACGGCGAGTGAACAGGGATTAGCCCAGCACCGAAGCTGCGGTCTTTTGGCCGTTCGGCAA
|||||||| ||||||||||||||||||||||||||||||||||||||||||||||||
ACGGCGAG----CAGGGATTAGCCCAGCACCGAAGCTGCGGTCTTTTGGCCGTTCGGCAA</td><td>831

841</td></tr>
<tr><td>D. aegyptensis

D. paradoxum</td><td>832

842</td><td>TGTGGTGTTTAGGTTGGCATACTCAGGCGATGTACTGTGCTAAGTCCATTCATGAATATG
||||||||||||||||||||||||||||||||||||||||||||||||||||||||||||
TGTGGTGTTTAGGTTGGCATACTCAGGCGATGTACTGTGCTAAGTCCATTCATGAATATG</td><td>891

901</td></tr>
<tr><td>D. aegyptensis

D. paradoxum</td><td>892

902</td><td>GCTAGCTATCTGGCCCAGAGAGGGTGAAAGGCCCGTGAGCATAGTGCGTCGTTCTGTCTT
|||||||||||||||||||||||||||||||||||||||||||||| ||| ||||||||||||
GCTAGCTATCTGGCCCAGAGAGGGTGAAAGGCCCGTGAGCATAGTACGTTGTTCTGTCTT</td><td>951

961</td></tr>
<tr><td>D. aegyptensis

D. paradoxum</td><td>952

962</td><td>AGTCAACCGTTGAGTCGGGTTGTTTAGGAATGCAGCC
|| ||||||||||||||||||||||| |||||||
AGCCAACCGTTGAGTCGGGTTGTTTAGTAATGCAGCA</td><td>988

998</td></tr>
</table>

Table 7: Pairwise alignment of the 28S rDNA ITS consequences of *Diplozoon aegyptensis* and Diplozoon paradoxum, numbering refers to ITS sequences.

<table>
<tr><td>D. guptai</td><td>9</td><td>ACTGCCTTGAGCATCGACTTCT--AACGTAAATCGCGGTATTAGGCTCTGCCTGATGCCA |||| |||||||||| ||||||||| ||| |||| |||| |||| |||||||</td><td>68</td></tr>
<tr><td>D. homoion</td><td>6</td><td>ACTGACTTGAGCATCGATTTCTTGAACGTGAATTGCGGCATTACCCTCT-AATGATGCCA</td><td>64</td></tr>
<tr><td>D. guptai</td><td>69</td><td>CGCCTAGCCGAGTGTTGGCATTATATCTATCACGACGCTTAATTGGTCGTGGCTTAGTTT |||||||||| | |||||| |||| ||||||||| ||||| |||||||</td><td>128</td></tr>
<tr><td>D. homoion</td><td>65</td><td>CGCCTAGCCGAGTATCGGCATTAAATCTAGCACGACGCTTATTTGGTCCTGGCTTAGAAA</td><td>124</td></tr>
<tr><td>D. guptai</td><td>129</td><td>GTTGTCAGCCGTCGTGTTTTACTTTGCAACGTGTTGCTCAGTTGTAAAGTCGACGGTATT |||||||||||||| |||| ||||||||| || |||| ||||| ||||||</td><td>188</td></tr>
<tr><td>D. homoion</td><td>125</td><td>GTTGTCAGCCGTCGTGTTGTACTTGGCAACGTGTTGTTCTGTTGTCAAGTCGGCGGTATT</td><td>184</td></tr>
<tr><td>D. guptai</td><td>189</td><td>ATTGACGCTTGCCAAATGTAATGGAGAGTGTGTATATGCGAAATTTCTGCCGG-AGCCTG ||||||||||||||||||||||| ||||||||||| ||| ||||||||||</td><td>248</td></tr>
<tr><td>D. homoion</td><td>185</td><td>ATTGACGCTTGCCAAATGTAATGGAGAGTTTGTATATGCGAAATATCTTCCGGTAGCCTG</td><td>244</td></tr>
<tr><td>D. guptai</td><td>249</td><td>TTGGCGTTGGCGACGCTGCCCCGTGTATGGTTTACTTGCATTTTTGTGCATACCGATTGG |||| ||||| ||||||||||||| |||| ||||||||||||||||| |||</td><td>308</td></tr>
<tr><td>D. homoion</td><td>245</td><td>TTGGTGTTGGCTACGCTGCCCCGTGTATTTTTTATTTGCATTTTTGTGCATACCGA-TGG</td><td>303</td></tr>
<tr><td>D. guptai</td><td>309</td><td>GGCGGTTAGCTTGTCGTCATCAGTGCGTGTTTGCCGGTGGTGATTTGTGGTGGCGTGGGA
|| ||||||| ||||| ||||||| |||||||| ||| |||| |||||</td><td>368</td></tr>
<tr><td>D. homoion</td><td>304</td><td>GGTGGTTAGCTTCTCGTCAGCAGTGCGTCCTTGCCGGTGG-----TGTCGTGGAATGGGA</td><td>358</td></tr>
<tr><td>D. guptai</td><td>369</td><td>ATTTCAATAAGCATTACTGAATGGTAATTAATAAATTGCCATTATATATGCTGTTCTCTT ||||||||||| ||||| |||| ||||| |||| ||| ||||||||||</td><td>428</td></tr>
<tr><td>D. homoion</td><td>359</td><td>ATTTCAATAAGCATTTCTGAATCCTAATTGTGAAATTGTCATTTTATGTGCTGTTCTCTT</td><td>418</td></tr>
<tr><td>D. guptai</td><td>429</td><td>GAGCCTTTTGGCCCACGGGTTGTGCGGTGACCAGTGTTGCTTTGAATGCGTGCGCATGCA ||||| ||||||| ||||||| |||||||| |||||||||| | ||||||</td><td>488</td></tr>
<tr><td>D. homoion</td><td>419</td><td>GAGCCGCATGGCCCACTTGTTGTGCGATGACCAGTGACGCTTTGAATGCGAGTGCATGCA</td><td>478</td></tr>
<tr><td>D. guptai</td><td>489</td><td>TGCCAGGTCGCAGCCTA-TTGTGATCGCGACAGTGCTTTGCTTGTGTTCTGCGTTTATTT ||||||| |||||| |||||||||||||||||||||||||||| ||||| ||</td><td>547</td></tr>
<tr><td>D. homoion</td><td>479</td><td>TGCCAGGTCTCAGCCTATTTGTGATCGCGACAGTGCTTTGCTTGTGTTCTGCGT--AATT</td><td>538</td></tr>
<tr><td>D. guptai</td><td>548</td><td>GTTGTCACTGCTACTTGCATATGTGCGAGTGTGTACCCGGAATGAGATTTTGTGCCCATG
||||||| | | || ||| ||||| || |||| | |||| |||||||||</td><td>607</td></tr>
<tr><td>D. homoion</td><td>539</td><td>TTTGTCACTGTTTCCCGCGAATGAGCGAGTCTGG-CCCGAGACGAGAGCATGTGCCCATG</td><td>597</td></tr>
<tr><td>D. guptai</td><td>608</td><td>TCATGCTGTGCTGACATTACTTCTCCACTGGTCGATAAGTGCATGTCGGTGTCACCAGTA
|| ||||||| |||||||| |||| | ||| ||||| | ||||||||| |||</td><td>667</td></tr>
<tr><td>D. homoion</td><td>598</td><td>TCGTGCTGTGCAGACATTACTACTCCATTCTTCGCTAAGTGTGTATCGG-GTCACCCGTA</td><td>657</td></tr>
<tr><td>D. guptai</td><td>668</td><td>CTTTGCTGTA-TT--GTG-T
||| ||||| || ||| |</td><td>687</td></tr>
<tr><td>D. homoion</td><td>658</td><td>TTTTACTGTACTTCTGTGGT</td><td>677</td></tr>
</table>

Table 8: Pairwise alignments of the 28s rDNA ITS consequences of Diplozzon guptai and Diplozoon homoion, numbering refers to ITS sequences.

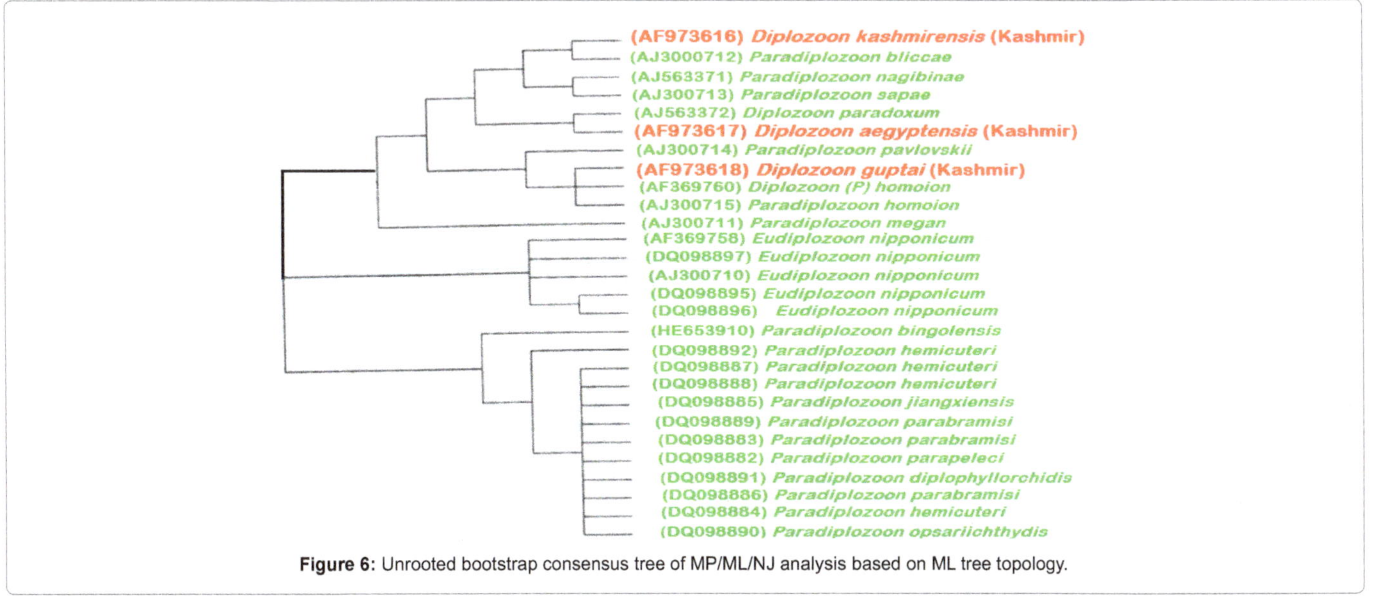

Figure 6: Unrooted bootstrap consensus tree of MP/ML/NJ analysis based on ML tree topology.

Discussion

The rDNA second Internal Transcribed Spacer (ITS2) was amplified using primers Cer5.8S2249 and Cer28S3116 [7] for 3 species of *diplozoids*. Analysis of the ITS2 region following sequencing clearly allowed us discrimination at the species level and produced the same results as species identification made by using morphological structures. During the present study it was observed that the alignment of nucleotide sequences with those of other *Diplozoon* species of *D. bliccae; D. paradoxum* and *D. homoion* [2,6,7], clearly revealed the boundaries of the 5.8S and 28S rDNA genes, as the sequences in these species closely resembles to those of *D. kashmirensis*, *D. aegyptensis* and

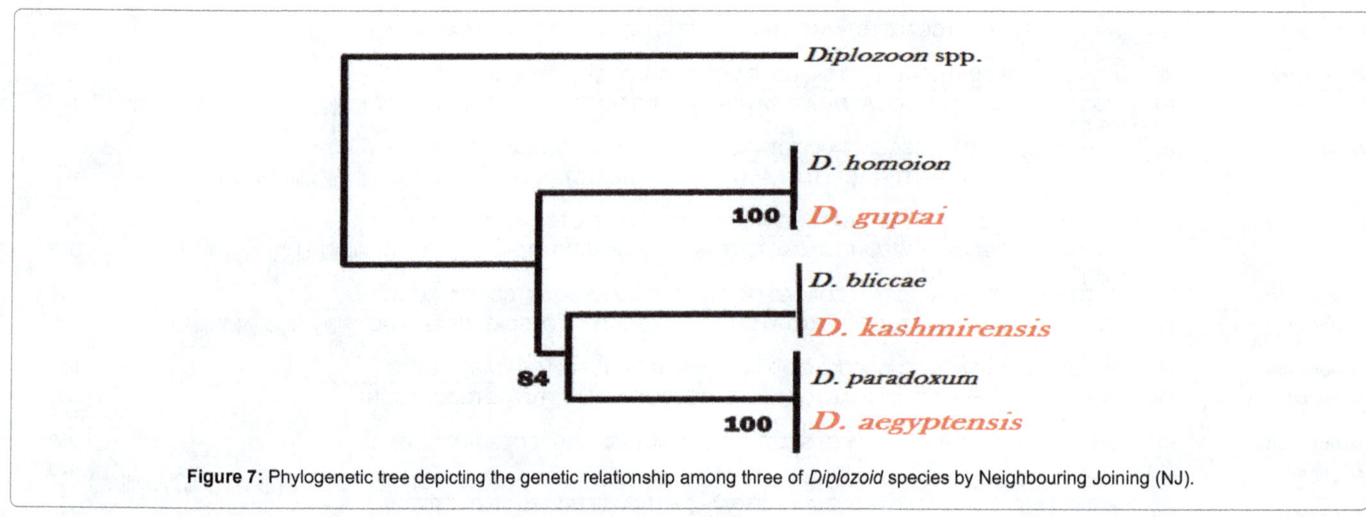

Figure 7: Phylogenetic tree depicting the genetic relationship among three of *Diplozoid* species by Neighbouring Joining (NJ).

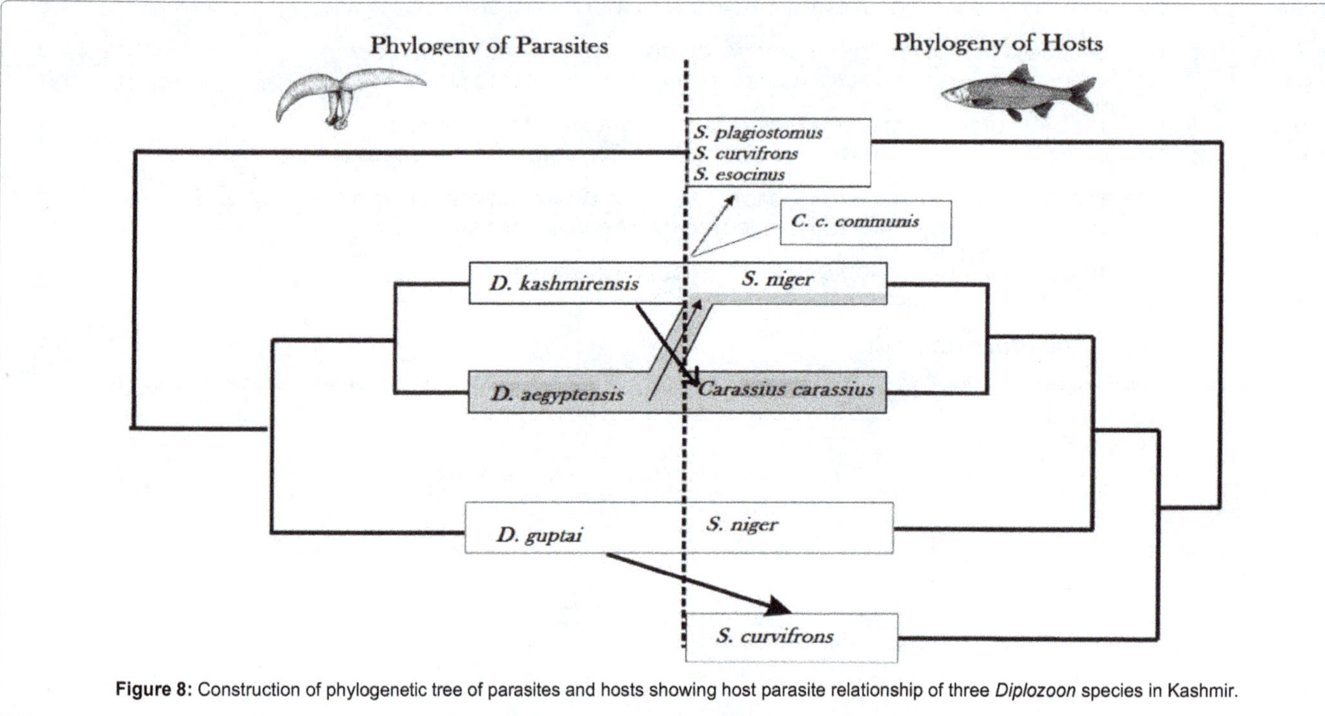

Figure 8: Construction of phylogenetic tree of parasites and hosts showing host parasite relationship of three *Diplozoon* species in Kashmir.

D. guptai. As noted in comparison of ITS2 sequences of *Monogenean* species, the first part of the ITS2 is also highly conserved, with only 6 variable sites in the first 65 nucleotides of the *diplozoid* sequences.

Species discrimination of *diplozoids* based on the shape of clamp sclerites and the length of the central hook can be difficult because of similarities in the shape of certain sclerites and overlapping ranges of central hook measurements. The PCR product of 3 species of *diplozoids: D. kashmirensis*; *D. aegyptensis* and *D. guptai* were clearly discriminated on the basis of nucleotide sequences which were different in their length of base pairs. The length of the PCR product could be useful to distinguish *Diplozoids* from the genus Eu *Diplozoon* and Para *Diplozoon* from other *Diplozoids* [2,6]. Length differences in the ITS2 have also been recorded in the genus *Gyrodactylus* [2,6] but are not generally as large as those found in the ITS1 region of *Lamellodiscus* and *Gyrodactylus* [2,6,19,20]. During the present study there are length difference of PCR products of three *Diplozoon* species i.e., *D. kashmirensis* contains 873bp; *D. aegyptensis* contains 1120bp and *D. guptai* contains 687bp of 28S rDNA genes, so on the basis of length of base pairs the three *diplozoid* species can be discriminated. ITS region have been found to be useful species markers for *monogenean* parasites [1,2-6] so, this method was performed to distinguish the *diplozoid* species. During the present study, the intraspecific variations within *diplozoid* species were studied and differences were detected in the ITS regions, but [2,6] studied that ITS region lacks intraspecific variation in groups of Monogenea which is due to the same species recovered from different hosts.

Diplozoids are generally considered parasites of Cyprinid species but the host specificity differs and relates to geographical origin. In Eurasia, *diplozoid* occurrence is restricted to host fishes from the Cyprinidae and Perciformes families [2,6,8,10,21,22]. However, in Africa they also parasitize members of the Characidae [20,22]. All *diplozoid* species described in the present study are also host specific.

ML, MP and NJ trees showed that *D. kashmirensis; D. aegyptensis* and *D. guptai* are closely related species, and this mirrors the close relationship of their hosts, thus all of these species are found in cyprinids from the same genus Schizothorax. These species have been described morphologically based on clamp shape, total body length, sucker, and pharynx length [17,24]. The present observations on molecular characterization demonstrate sufficient genetic variations between parasites from different hosts to confirm the validity of these species and that they appear to be host specific, as are many *monogenean* parasites. It may be speculated that the similarity of these species is a result of a relatively recent divergence of one from the other following a host-switching event. An important observation during the present study has been noticed that *Schizothorax niger* is infected by all the three species of *Diplozoidae*: *D. kashmirensis; D. aegyptensis* and *D. guptai*, but on all six fishes collected, simultaneous parasitism by all the parasite species was never observed. Two types of factors can be involved in the constitution of such a host-parasite system.

(a) Competition hypothesis: the installation of a first *Diplozoon* species prevents any other species from settling on the same gill. (b) Since natural hybridization has been reported between the two fishes, the introgression of genes from *Carassius carassius* into the genome of *S. niger* allows a host capture of the latter by *D. aegyptensis* and *D. guptai* but excludes the infestation by its natural parasite *D. kashmirensis*.

Conclusion

The present study has confirmed the existence of 3 species of *diplozoids* from 6 species of *cyprinid* fishes from the water bodies of Kashmir valley. All the species were clearly distinguished by differences in nucleic acid sequences within the second ribosomal DNA Internal Transcribed Spacer region (ITS2). Analysis of additional specimens from different cyprinid hosts by molecular methods may be helpful to clarify the systematics of this fascinating family *Diplozoidae*.

Acknowledgment

The authors extend their thanks to the authorities of the Department of Zoology and Biotechnology, University of Kashmir for providing laboratory facilities. TAS is also highly thankful to Prof. Fayaz Ahmad for giving valuable suggestions while compiling this paper.

References

1. Cunningham CO (1997) Species Variation within the Internal Transcribed Spacer (ITS) Region of *Gyrodactylus* (Monogenea: Gyrodactylidae) Ribosomal RNA Genes. The Journal of Parasitology 83: 215-219.
2. Matejusova I, Gelnar M, McBeath AJA, Collins CM, Cunningham CO (2001a) Molecular markers for gyrodactylids (Gyrodactylidae: Monogenea) from five fish families (Teleostei). Int J Parasitol 31: 738-745.
3. Huyse T, Volckaert FAM (2002) Identification of host-associated species complex using molecular and morphometric analyses, with the description of *Gyrodactylus rugiensoides* n. sp. (Gyrodactylidae, Monogenea). Int J Parasitol 32: 907-919.
4. Zietara MS, Huyse T, Lumme J, Volckaert FAM (2002) Deep divergence among subgenera of *Gyrodactylus* inferred from rDNA ITS region. Parasitology 124: 39-52.
5. Simkova A, Matejusova I, Cuningham CO (2006) Molecular phylogeny of the Dactylogyridae sensu. Kritsky & Boeger (Monogenea) using the D1-D3 domains of large ribosomal subunit rDNA. Parasitology 133: 43-53.
6. Matejusova I, Koubkova B, D'Amelio S, Cunningham CO (2001b) Genetic characterization of six species of diplozoids (Monogenea; Diplozoidae). Parasitology 123: 465-474.
7. Sicard M, Desmarais E, Lambert A (2001) Molecular characterisation of Diplozoidae populations on five Cyprinidae species: consequences for host specificity. C R Acad Sci Paris, Sciences de la vie/Life Sciences 324: 709-717.
8. Matejusova I, Koubkova B, Gelnar M, Cunningham CO (2002) *Paradiplozoon homoion* Bychowsky & Nagibina, versus *P. gracile* Reichenbach-Klinke, (Monogenea): two species or phenotypic plasticity? Syst Parasitol 53: 39-47.
9. Sicard M, Desmarais E, Vigneux F, Shimazu T, Lambert A (2003) Molecular phylogeny of the Diplozoidae (Monogenea, Polyopisthocotylea) parasitizing 12 species of Cyprinadea (Teleostei): new data about speciation. Combes C, Jourdane J. Taxonomy, ecology and evolution of metazoan parasites. Universitaires de Perpignan, Perpignan pp. 199-211.
10. Matejusova I, Koubkova B, Cunningham CO (2004) Identification of European diplozoids (Monogenea, Diplozoinae) by restriction digestion of ribosomal RNA internal transcribed spacer. J Parasitol 90: 817-822.
11. Gao Q, Chen MX, Yao WJ, Gao Y, Song Y, et al. (2007). Phylogeny of diplozoids in five genera of the subfamily Diplozoinae *Palombi*, as inferred from ITS-2 rDNA sequences. Parasitology 134: 695-703.
12. Jiang NCH, Wu BH, Wang SX (1985) Four new species of parasitic *Diplozoon* from freshwater fishes of the subfamily Gobioninae. Acta Zootaxonomica Sinica 10: 239-245.
13. Kritscher E (1991) *Diplozoon bileki* nov. spec. (Plathelhminthes: Monogenea: Diplozoidae), ein neues Doppeltier von den Kiemen von *Barbus plebejus euboicus* Stephanidis, (Pisces: Cyprinidae), gesammelt auf der Insel Euba (Grienchenland). Annalen des Naturhistorischen Museum in Wien, Serie B Botanik und Zoologie 92: 251-255.
14. Blair D, Barker SC (1993) Affinities of the Gyliauchenidae: Utility of the 18S rRNA gene for phylogenetic inference in the digenea (Platyhelminthes). International journal for Parasitology 23: 527-532.
15. Cunningham CO, McGillivray DM, Mackenzie K (1995) Phylogenetic analysis of *Gyrodactylus salaries* Malmberg, based on the small subunit 18S ribosomal RNA gene. Molecular and Biochemical Parasitology 71: 139-142.
16. Zhu X, Gasser RB, Chilton NB (1998) Differences in the 5.8S rDNA sequences among ascarid nematodes. International Journal for Parasitology 28: 617-622.
17. Fayaz A, Chishti MZ (1999) Fish Trematode Parasites of Kashmir. Genus *Diplozoon* Nordmann, (Monogenea, Polyophisthocotylea) Orient Sci 4: 79-91.
18. Sambrook J, Russell DW (2001) *Molecular cloning: a laboratory manual*. Cold Spring Harbor Laboratory Press, Cold Spring Harbor, New York.
19. Cable J, Harris PD, Tinsley RC, Lazarus CM (1999) Phylogenetic analysis of *Gryrodactylus* spp. (Platyhelminthes: Monogenea) using ribosomal DNA sequences. Canadian Journal of Zoology 77: 1439-1449.
20. Desdevises Y, Jovelin R, Jousson O, Morand S (2000) Comparision of ribosomal DNA sequences of *Lamellodiscus* spp. (Monogenea, Diplectanidae) parasiting *Pagellus* (Sparidae, Telostei) in the North Mediterranean Sea: species divergence and coevolutionary interactions. International Journal for Parasitology 30: 741-746.
21. Khotenovsky IA (1985) *Fauna of the USSR. Monogenea*. Nauka, Leningrad.
22. Yildirim YB, Zeren A, Genc E, Erol C, Konas E (2010) Parasitological investigation on commercially important fish and crustacean species collected from the TIGEM (Dortyol Turkey) ponds. J Anim Vet Adv 9: 1597-1602.
23. Lambert A, Le Brun N (1988) Hypothese sur l'origine biogeographique de *Diplozoon* (Monogenea, Polyopisthocotylea). Ann Parasit Hum Comp 63: 99-102.
24. Bakshi SA (1999) Fish parasites of some lakes of Kashmir with an analysis of seasonality of incidence and their maturation with regard to different ecological factors. Doctoral Thesis, University of Kashmir, India.

32

Spatial and Temporal Variability of Phytoplankton Assemblages and Physico-Chemical Characterization in Three Similar Dams

Narges Rostamian[1], Ebrahim Masoudi[2], Mohammad Hasan Gerami [3]*, Sirvan Azizpour[1] and Sana Ullah[4]

[1]*Department of Fisheries, University of Environment, Karaj, Iran*

[2]*Department of Fisheries, Gorgan University of Agricultural and Natural Resources, Gorgan, Iran*

[3]*Young Researchers and Elite Club, Shiraz Branch, Islamic Azad University, Shiraz, Iran*

[4]*Fisheries and Aquaculture Lab, Department of Animal Sciences, Quaid-i-Azam University Islamabad, Pakistan*

Abstract

The present preliminary study was undertaken from April to September 2013 in order to assess the limnological factors and phytoplankton communities in three dams, having depth of 6 m and area of 10 ± 2 Hectares, at Node Khanduz (Dam 1), Seyed Abad (Dam 2) and Marzban (Dam 3) in Azad Shahr, Gorgan, Iran. During the study period, a total number of 8 families and 28 genera were identified from all sampling sites. Of these 28 genera, 6 genera were belonging to family Bacillariophyceae, 5 genera were from Cyanophyceae, 2 genera from Charophycea, 2 genera from Chrysophyceae, 2 genera from Euglenophyceae, 8 genera from Chlorophyceae, 2 genera from Dinophyceae and 1 genus was belonging to Xanthophyceae. It was concluded that all the dams were having very well balanced phytoplankton communities yet changes in individuals' composition and numbers were significantly varying among the three studied dams. Further studies focusing on other factors such as presence of heavy metals in the dams and of lengthy periods are recommended.

Keywords: Phytoplanktons; Food chain; Bio-indicator; Assemblage; Composition

Introduction

Phytoplanktons are the primary producers forming the first trophic level in the food chain. It is the basic available food in water, for all consumers such as zooplankton and fish [1]. The life cycle of phytoplankton varies from a few hours to a few days; therefore they are very sensitive to environmental changes [2]. Diversity of planktonic organisms is quite high in fertile standing water bodies. Several phytoplankton species are also employed as bio-indicator for water bodies' specifications such as pollutant or contaminant [3-6].

Dams have been constructed all across the globe that provide water and fulfill other necessities of men [7]. These might be constructed for multiple purposes such as for transport, domestic or agricultural use, defense, ritual or industrial use, social aggrandizement, swimming, fish farming or the creation of the picturesque [8-10]. Dams are thought to have profound effects on the composition and abundance of both terrestrial and aquatic organisms such as phytoplankton assemblages are effected by low water exchange ratio, prevailing environmental conditions and dam operations [11,12]. Although several studies have been carried out on phytoplankton communities in dams, closed lakes or ponds around the world [1,7,13,14] but data on comparison of phytoplankton communities in similar habitat are scarce. It is assumed that same habitat should have similar phytoplankton communities [15]. Therefore this study was designed to evaluate the phytoplankton communities of three similar dams, situated in the same area in Gorgan, Iran.

Materials and Methods

Physico-chemical parameters of the selected three dams located at Node Khanduz (Dam 1), Seyed Abad (Dam 2) and Marzban (Dam 3) in Azad Shahr, Gorgan, Iran were studied. The dams were having a depth of 6 m and area of 10 ± 2 Ha. The factors were investigated using standard procedures.

Water samples were collected from selected dams for seven months from April to September 2013. Samples were collected periodically every month during morning hours between 9.00 and 11.00 A.M. or 6-8 P.M. according to circumstances by P.V.C tube with 1.5 m length and 6 cm diameter and 2 liters of surface water was collected for further analysis. The collected plankton samples were transferred to polyethylene bottles and preserved with 4% formalin. Quantitative analysis was performed by Sedgwick Rafter Counting Cells. Plankters were studied under microscope and identified with the help of standard references [16,17]. Surface water temperature was recorded on the spot using Centigrade thermometer. The pH of the water samples was measured by using gun pH meter on the spot and transparency were estimated by secchi disk. One-Way ANOVA and Spearman rank correlation were used to analyze data in SPSS 20 and Microsoft Excel 2010 software.

Results and Discussion

During the study period, a total of 8 families and 28 genera were identified, of which 6 genera were belonging to family Bacillariophyceae, 5 genera were from Cyanophyceae, 2 genera from Charophycea, 2 genera from Chrysophyceae, 2 genera from Euglenophyceae, 8 genera from Chlorophyceae, 2 genera from Dinophyceae and 1 genus was belonging to Xanthophyceae.

Maximum phytoplankton density in Dam 1, 2 and 3 was belonging

***Corresponding author:** Mohammad Hasan Gerami, Young Researchers and Elite Club, Shiraz Branch, Islamic Azad University, Shiraz, Iran
E-mail: m.h.gerami@gmail.com

to Chlorophyceae family. Bacillariophyceae and Cyanophyceae family were placed in the next level respectively (Figures 1-3). Although Xanthophyceae was not observed in April and August in dam 1 and 2 yet it was being observed in dam 3 continuously. The results revealed that dam 3 was having better phytoplankton communities. There was a significant difference (P<0.05) in plankton communities and most of the species were different in in all dams, specifically in dam 3.

Change in the phytoplankton communities in the dams may be attributed to the change in habitat across the tropic spectrum [18]. There are different co-variable factors with trophic state including physical environment, carbon problem, resources and energy, herbivory and factor interaction (such as local climate and, hence, latitude, altitude and relative exposure, water clarity and alkalinity). These changes lead to different composition of phytoplankton in different areas or aquatic bodies. These phytoplankton species are also different as they are having different level and limits of tolerance or adaptability to different prevailing conditions such as hazardous environmental setup. Changes in phytoplankton communities are also due to changes in concentration of phosphates and nitrogen as well as light and temperature [19]. Phytoplankton communities establishment is also dependent on the density, wind induced circulation of water and turbulence etc. of that specific dam or aquatic body [20].

Physicochemical factors are greatly influenced by phytoplankton population. In fact, cloudy weather, low transparency and heavy flood caused the decline of phytoplankton density and physico-chemical parameters [7]. The physicochemical parameters of the dams are shown in Table 1.

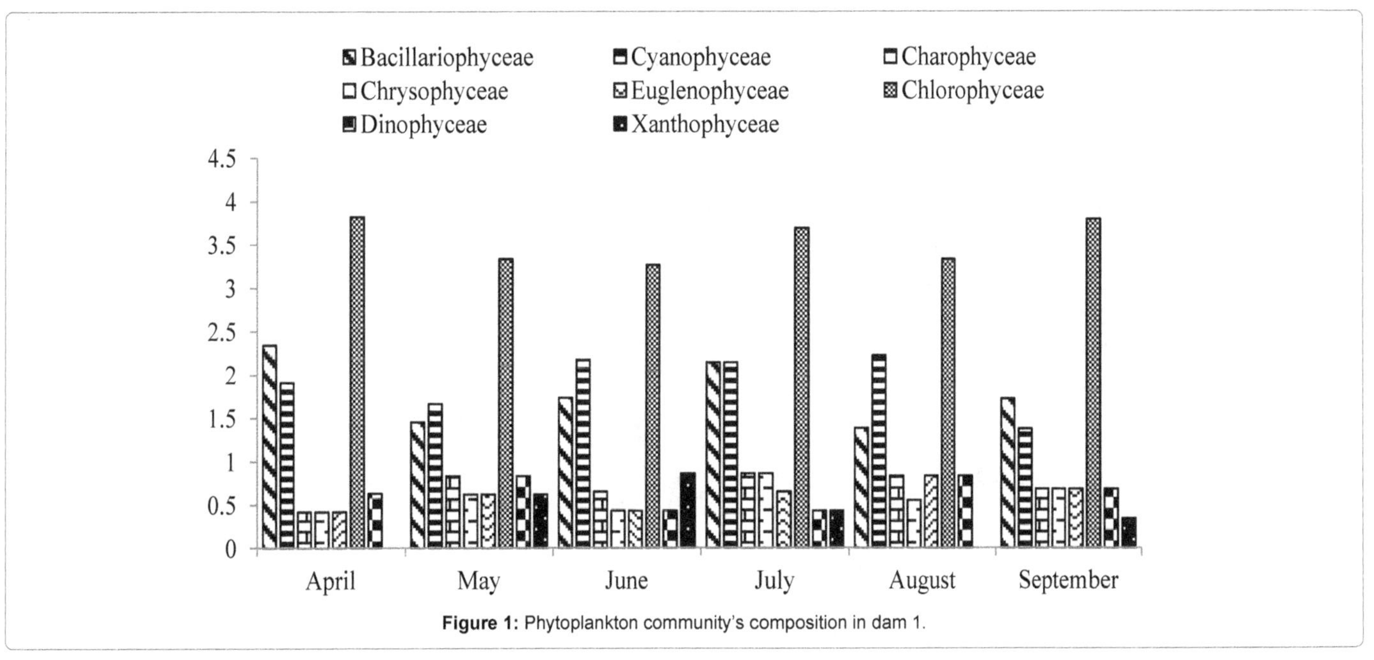

Figure 1: Phytoplankton community's composition in dam 1.

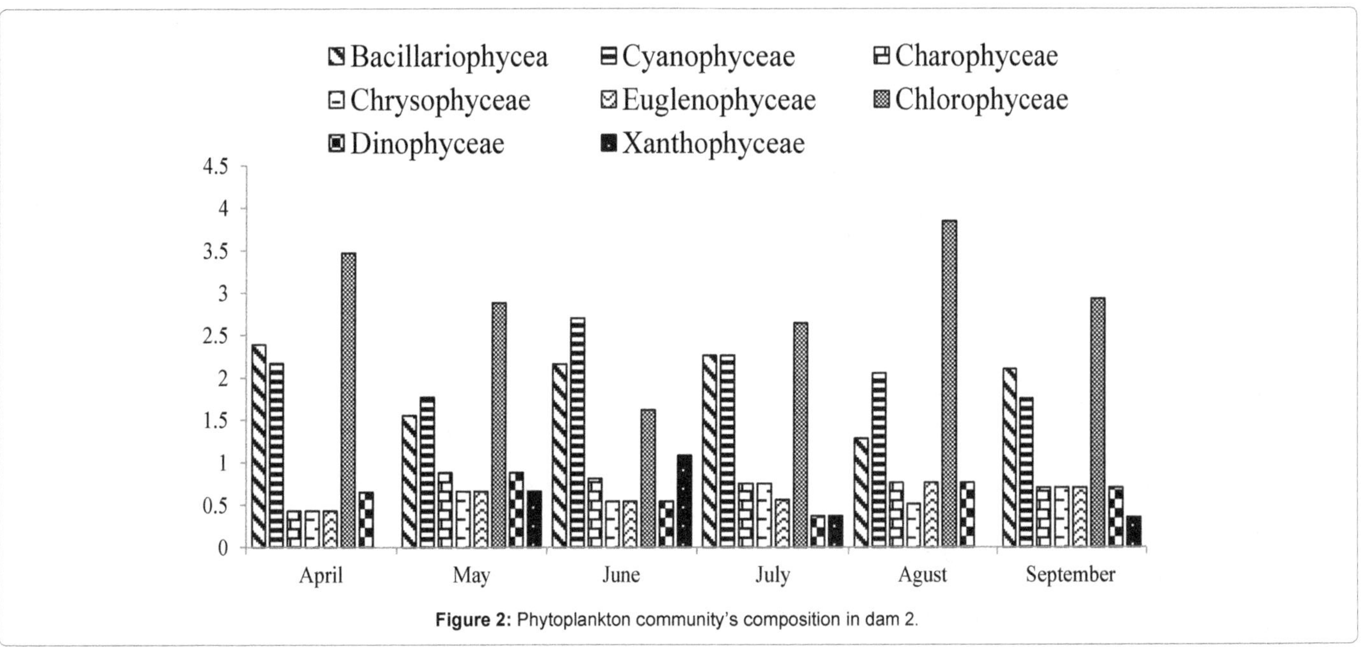

Figure 2: Phytoplankton community's composition in dam 2.

Planktonic communities can describe complex interactions and community structure in lakes habitat [21]. In fact, same habitats have similar species compositions. In this study, all the three dams were not having similar habitat's conditions especially in transparency (Table 2). Results showed that only Euglenophyceae family was having significant correlation with salinity and pH. However, statistical analysis showed that three phytoplankton communities were significantly different in the dams studied.

This study concluded that all the three dams are having same phytoplankton communities on account of being in the same habitat but their compositions, numbers of individuals in each family and relative abundance are significantly different in all the dams. This shows that all the dams are having very well balanced communities of phytoplankton consisting of almost all species. Change in individuals' composition and numbers were significantly varying among the three studied dams which may be due to the dynamic nature of these ecosystems [22]. In order to reduce impacts of pollutants, remedial steps ought to be taken.

It is concluded from the current study that similar habitat was not having similar phytoplankton communities. However, the current study is still scanty on account of its study period and sampling months. Further studies focusing on other aspects such as physico-chemical characterizations and toxicants' presence in the very same habitats are recommended.

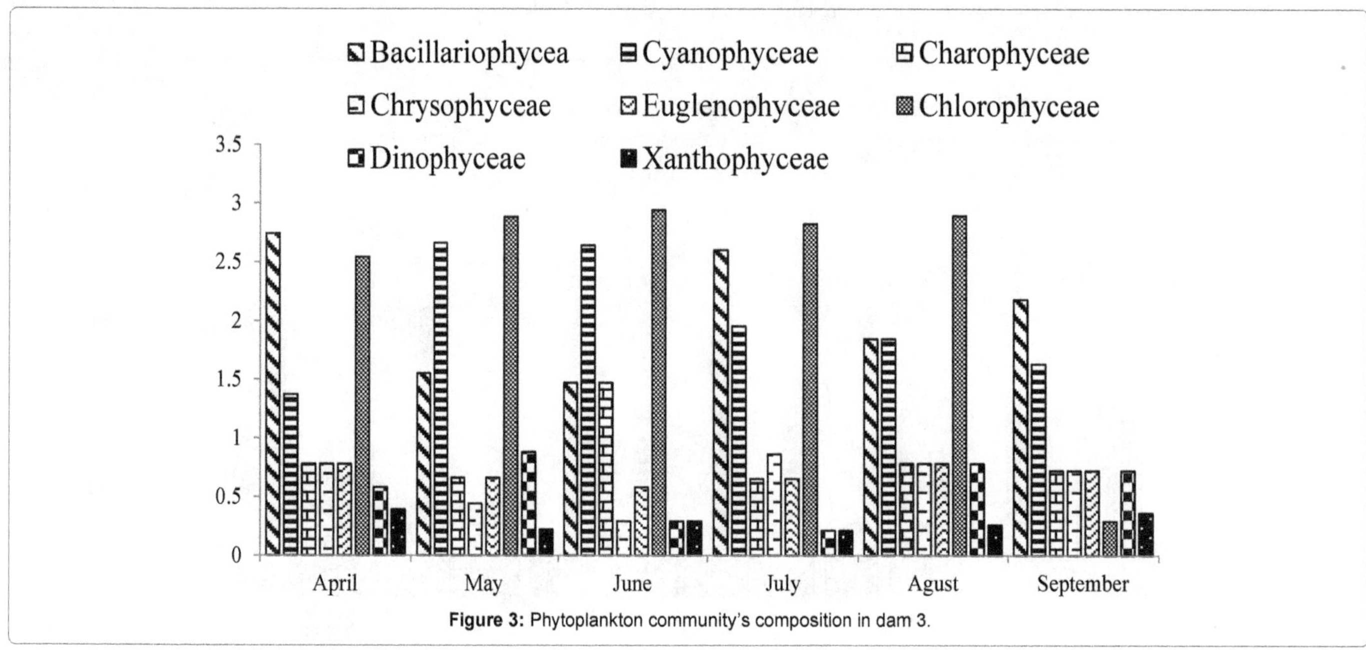

Figure 3: Phytoplankton community's composition in dam 3.

Table 1: Physico–chemical parameters of 3 dams during April to September 2013 (Numbers are monthly average).

Dam	Factors	April	May	June	July	Aug	Sep	Mean ± SD
Dam 1	pH	7.8	8.2	8.5	9	9.3	9.5	8.71 ± 0.6
	Transparency (cm)	25	22	20	20	25	24	22.66 ± 2.13
	Salinty (g L^{-1})	0.18	0.1	0.25	0.28	0.3	0.35	0.24 ± 0.08
	Surface Temperature °C	26	29	33	30	34	32	30.66 ± 2.68
Dam 2	pH	8.2	8.5	8.9	9.1	9.5	9.4	8.93 ± 0.46
	Transparency (cm)	22	25	29	22	20	24	23.66 ± 2.86
	Salinty (g L^{-1})	0.1	0.1	0.1	0.3	0.4	0.45	0.24 ± 0.14
	Surface Temperature °C	28	30	31	32	31	31	30.5 ± 1.25
Dam 3	pH	8.4	8.7	9	9.2	9.4	9.2	8.98 ± 0.33
	Transparency (cm)	25	28	30	25	22	21	25.16 ± 3.13
	Salinty (g L^{-1})	0.3	0.1	0.5	0.45	0.7	0.75	0.46 ± 0.22
	Surface Temperature °C	27	29	31	33	32	30	30.33 ± 1.97

**Shows correlation at the 0.01 level; *shows correlation at the 0.05 level

Table 2: Spearman rank correlation between physicochemical factors and phytoplankton communities.

Factor	Bacillariophyceae	Cyanophyceae	Charophyceae	Chrysophyceae	Euglenophyceae	Chlorophyceae	Dinophyceae	Xanthophyceae
pH	-0.305	-0.103	0.106	0.379	0.620**	0.038	0.059	-0.246
Transparency (cm)	0.028	0.280	0.154	-0.255	-0.053	-0.326	-0.017	-0.090
Salinty (g L^{-1})	0.031	-0.300	0.042	0.381	0.513*	-0.132	-0.227	-0.269
Surface Temperature °C	-0.269	0.227	0.066	0.201	0.210	-0.054	-0.220	-0.006

References

1. Tas S (2014) Phytoplankton composition and abundance in the coastal waters of the Datça and Bozburun Peninsulas, south-eastern Aegean Sea (Turkey). Mediterranean Marine Science 15: 84-94.

2. Polat S, Akiz A, Piner MP (2005) Daily variations of coastal phytoplankton assemblages in summer conditions of the northeastern Mediterranean (Bay of Iskenderun). Pakistan Journal of Botany 37: 715-724.

3. Vareethiah K, Haniffa MA (1998) Phytoplankton pollution indicators of coir retting. J Environ Pollut 3: 117-122.

4. Bianchi F, Acri F, Aubry FB, Berton A, Boldrin A, et al. (2003) Can plankton communities be considered as bioindicators of water quality in the laggon of Venice? Mar Pollut Bull 46: 964-971.

5. Tiwari A, SV (2006) Chauhan Seasonal phytoplanktonic diversity of Kitham lak, Agra. J Environ Biol 27: 35-38.

6. Hoch MP, Dillon KS, Coffin RB, Cifuentes LA (2008) Sensitivity of bacterioplankton nitrogen metabolism to eutrophication in sub-tropical coastal water of Key West. Florida. Mar Pollut Bull 56: 913-926.

7. Rajagopal T, Thangamani A, Archunan G (2010) Comparison of physico-chemical parameters and phytoplankton species diversity of two perennial ponds in Sattur area, Tamil Nadu. Journal of Environmental Biology 31: 787-794.

8. Rees SE (1997) The historical and cultural importance of ponds and small lakes in Wales, UK. Aqu Cons Mar Freshwat Ecosyst 7: 133-139.

9. Narayan R, Saxena KK, Chauhan S (2007) Limnological investigations of Texi temple pond in district Etawah (U.P.). J Environ Biol 28: 155-157.

10. Bishnoi M, Malik R (2008) Ground water quality in environmentally degraded localities of panipat city, India. J Environ Biol 29: 881-886.

11. McCartney MP, Sullivan C, Acreman MC (2000) Ecosystem impacts of large dams. In: Background Paper Nr 2. Prepared for IUCN/UNEP/WCD. Center for Ecology and Hydrology, Wallingford, UK.

12. Li J, Dong S, Liu S, Yang Z, Peng M, et al. (2013) Effects of cascading hydropower dams on the composition, biomass and biological integrity of phytoplankton assemblages in the middle Lancang-Mekong River. Ecological Engineering 60: 316-324.

13. Liu D, Morrison RJ, West RJ (2013) Phytoplankton assemblages of two intermittently open and closed coastal lakes in SE Australia. Estuarine Coastal and Shelf Science 132: 45-55.

14. Ardyna M, Gosselin M, Michel C, Poulin M, Tremblay JE (2011) Environmental forcing of phytoplankton community structure and function in the Canadian High Arctic: contrasting oligotrophic and eutrophic regions. Mar Ecol Prog Ser 442: 37-57.

15. Cardoso SJ, Roland F, Loverde-Oliveira SM, Huszar VLM (2012) Phytoplankton abundance, biomass and diversity within and between Pantanal wetland habitats. Limnologica 42: 235-241.

16. Adoni A, Joshi DG, Gosh K, Chourasia SK, Vaishya AK, et al. (1985) Work book on limnology. Pratibha Publisher, Sagar.

17. Agarker MS, Goswami HK, Kaushik S, Mishra SM, Bajpai AK, et al. (1994) Biology, conservation and management of bhojwtland, Upper lak ecosystem in Bhopal. Bionature 14: 250-273.

18. Reynolds CS (1998) What factors influence the species composition of phytoplankton in lakes of different trophic status? Hydrobiologia 369/370: 11-26.

19. Khan RM (2014) Biodiversity of phytoplankton and zooplanktons of Triveni lake in Amravati district of Maharashtra. International Journal of Innovative and Applied Research 2: 1-4.

20. Bronmark C, Hansoon LA (2005) The biology of lakes and ponds (2ndedn.) Oxford University Press Inc, New York.

21. Lodge DM, Barko JW, Strayer D, Melack JM, Mittelbach GG, et al. (1988) Spatial Heterogeneity and Habitat Interactions in Lake Communities. In: Complex Interactions in Lake Communities. Carpenter SR Springer, New York.

22. Shinde S, Pathan TS, Sonawane DL (2012) Seasonal variations and biodiversity of phytoplankton in Harsool-Savangi dam, Aurangabad, India. Journal of Environmental Biology 33: 643-647.

33

Survey on Phytoplankton Biomass and Water Parameters in the Habitats of Invasive Tigers Shrimps (Penaeus Monodon) in Nigeria

Oketoki TO*

Nigeria Institute for Oceanography and Marine Research (NIOMR), Wilmot point road, Bar beach, Victoria Island Lagos, Nigeria

Abstract

Penaeus monodon is an invasive species found in the coastal waters of Nigeria. Although widely exploited with significant economic importance, investigation into its adaptation and potential ecological impact in the newly found environment is poorly known. This survey provides baseline information on the phytoplankton community and physico-chemical parameters in ten selected stations from five states where they are exploited in Nigeria. These include: Ibeno (Akwa Ibom State), Bonny (Rivers State), Kaa (Rivers State), Brass (Bayelsa State), Aiyetoro (Ondo State), Makoko (Lagos state), Folu (Lagos state), Apapa (Lagos state), Tin Can Island (Lagos state) and Tarkwa Bay (Lagos state). Total of 147 species of phytoplankton from six classes were recorded during the survey with diatoms being the most prevalent (70.4%), green algae (20.4%), Blue-green algae (5.6%), Chrysophyceae (1.9%). Water parameters recorded temperature (range: 27.33 ± 1.53°C-29.00 ± 1.00°C), pH (7.39 ± 0.08-8.13 ± 0.14), dissolved oxygen (5.40 ± 3.22 mgL^{-}1-8.00 ± 1.44 mgL^{-1}), Conductivity (11.22 ± 10.03 μS/cm-39.33 ± 5.87 μS/cm) and salinity (11.02 ± 15.56% -25.98 ± 2.02%). Lowest values for phosphate, nitrate-nitrogen and sulphate were 0.11 ± 0.07mgL^{-1}, 0.10 ± 0.07 mgL^{-1} and 523.67 ± 880.21 mgL^{-1} respectively. Generally, ecological factors in their newly found environment are similar to their native range. However, negative impact as an invasive species most be checked.

Keywords: Phytoplankton; Physico-chemical characteristics; *Penaeus monodon*; Invasive

Introduction

The new millennium has witnessed invasive species which have a renowned and most severe ecological and economic threat globally [1,2]. Generally they have great effects on native biodiversity and cause difficulties in natural ecosystems conservation and management. The Asian tiger shrimp (*Penaeus monodon Fabrizio*) is invasive to the coastal waters of Nigeria [3,4]. The first report of this incidence was about 16 years ago [5]. The Nigerian coastline itself spans approximately 853 km with seven states along the coastal zones namely Lagos, Ondo, Delta, Bayelsa, Rivers, Akwa Ibom and Cross River respectively from South-west to South-South coast. There are at least five families of shrimps contributing to the aquatic resources in Nigeria. These include Penaeidae, Atyidae, Palaemonidae, Alpheidae and Hippolytidae. However Penaeids (Penaeidae) and Macrobrachium species (Palaemonidae) are the predominant species found in Nigeria [6]. Chemonics [7] listed four members of the Penaeids from marine and brackish water namely; *Penaeus (Farfantepenaeus) notialis* (Pink Shrimp), *Penaeus Kerathurus* (striped or zebra shrimp), *Parapenaeopsis atlantica* (Brown Shrimp), *Parapenaeus longirostrics* (Red Shrimp). Not until the sudden emergence of P. monodon, all the aforementioned species, together with *Nematopalaemon hastatus*, formed the basis of the artisanal prawn fishery [8,9].

The black tiger shrimp is a widespread Penaeid shrimp native to the eastern hemisphere from longitude 30˚E to 155˚E and latitude 35˚N to 35˚S (Indo-West Pacific ocean of East Africa, Arabian Peninsula, India, China Japan, the Middle East and North Australia [10,11]. *P. monodon* is now established in many areas presumably due to escapement from aquaculture facilities outside its native range, including West Africa and and South East United States [12-14]. Other regions of invasion include, the Caribbean [15], northern and north-eastern coasts of South America [16-19]. The role of food and feeding habit of this shrimp cannot be overemphasized in its adaptation to novel habitat. Food must be exploited in the new environment and the adaptation for this is related to morphological traits connected to feeding [20]. A major source of food is phytoplankton which comprises complex community of floating micro-algae with size range from about 1 μm to a few millimetres [21,22]. They are microscopic organisms with chlorophyll a, floating on water surfaces or suspended in water column and are dependent on sunshine for photosynthesis [23,24]. Other essential inorganic nutrients dissolved in water are phosphates, nitrates are sulphates. As primary producers, carbon in the form of carbon dioxide is needed in the aquatic environment to initiate the food chain for secondary and tertiary producers [25]. In the native ecosystem of the tiger shrimps, extensive work has been reported on plankton community and physico-chemical parameters. [21] Described the Phytoplankton biomass and Community Structure of Kottakudi and Nari, South East of Tamil Nadu, India. Kannan and Vasantha [26] studied Microphytoplankton of the Pitchavaram Mangals, Southeast Coast of India. Diatoms domination amidst various groups of phytoplankton was reported in the two studies. Also, in the South Eastern coast of India, Uppanar Estuary, Cuddalore, physico-chemical parameters were reported by [27]. In North and North West Australia, Hallegraeff and Jeffrey [28] reported the Tropical phytoplankton species and pigments of the continental shelf waters. The nanoplankton (e.g., amphora species and Navicular species) were documented most abundant species.

***Corresponding author:** Oketoki TO, Nigeria Institute for Oceanography and Marine Research (NIOMR), Wilmot point road, Bar beach, Victoria Island Lagos, Nigeria, E-mail: topeoketoki@gmail.com

Many studies have also examined the phytoplankton biomass of some lagoons and surrounding creeks in Nigeria. [29,30] reported the phytoplankton of Lagos lagoon, eight cyanobacteria species in South Western coast of Nigeria and dinoflagellates list of Lagos lagoon. Diatoms of Olero creek and Lekki lagoon were reported [31]. The Cyanobacteria of a Tropical Lagoon (lekki), Nigeria was also reported by [32]. Many others reports [33-36] attributed seasonal changing hydro environmental characteristics as determinants of the phytoplankton and zooplankton standing crop at anytime. In the Niger delta, [37] studied Warri Forcados estuary phytoplankton and a similar study was earlier reported in the New Calabar river by [38]. Countries in West Africa are generally constituted by many physico-chemical characteristics that make them environmentally sustainable for shrimp farming and coastal areas have water suitable for farming many other aquatic species [39]. However further studies investigating the adaptability of specific shrimp species with climate are essential [40] documented the potentials of Andoni river for production of tiger shrimps in Nigeria haven studied the physico-chemical properties. The conditions were observed to be within satisfactory range limits for warm water fish and shellfish production. Furthermore the physico-chemical qualities of the Andoni River, coupled with the findings of [41] were reported suitable for the production of this alien species.

This survey on phytoplankton and physico-chemical parameters was carried out across selected station in Nigeria as a preliminary investigation to understand the ecology of the invasive tiger shrimp. The information will be useful for aquaculture and wild resource management on the invasive species.

Materials and Methods

Description of study sites

Following a pilot survey on prominent artisanal tiger shrimp fishing areas within the South-West and South-South coastal states of Nigeria, 5 states and 10 stations were selected for the study (between the dry seasons of November, 2013 and January, 2014). Table 1 and Figure 1 show the stations with geographic coordinates and map of South-West and South-South Nigeria.

Lagos State, Nigeria (latitudes 6°23'N and 6°41'N and longitudes 2°42'E and 3°42'E) is located in South-western part of the country on the West Coast of Africa. It is flanked from the north and east by Ogun State, to the west by the Republic of Benin and bounded southward by the Atlantic Ocean (Gulf of Guinea). Lagos Tarkwa-Bay (LT) is a key location that opens the Atlantic ocean as a source of salt water incursion to the Lagos lagoon-Makoko (LM), Apapa (LA) and Tin can island (LC) through the Lagos harbor Lagos-Makoko.

(LM) represents one of the largest fish and shrimp landing sites of the Lagos Lagoon while Folu (LF) is a foremost coastal community along the Atlantic Ocean [42]. Ondo state is bounded southward by the Atlantic Ocean and has many rivers such as R Owena and Oluwa which empties into the sea. Aiyetoro OA in Ilaja community is one of the major coastal settlements in the area [43]. Tarkwa-Bay LT, Folu (LF), Aiyetoro (OA), Brass (BB), Bonny (RB) and Ibeno (AB) are located in proximity to the Atlantic coast of their respective states in Nigeria. Tarkwa-Bay is a major point bounded in the North by the Cowrie creek, in the south by the Atlantic Ocean and the West by the Lagos Harbor. The east end of Tarkwa-Bay opens to the surrounding for water transportation to the rest of the Islands (including Tin can Island) and mainland in Lagos State. The Commodore Channel along the east end is the only significant connection between the Lagos Lagoon the Atlantic Ocean

State	Sampling stations	Coordinates	
		Long	Lat
Lagos	Folu (LF)	4°0'16.128"	6°26'50.438"
Lagos	Makoko (LM)	3°23'8.405"	6°28'35.00"
Lagos	Apapa (LA)	3°21'7.352"	6°27'0.520"
Lagos	Tarkwa Bay (LT)	3°23'42.392"	6°24'10.075"
Lagos	Tin can island (LC)	3°20'30.562"	6°26'8.401"
ONDO	Aiyetoro (OA)	4°40'11.91"	6°11'37.53"
BAYELSA	Brass (BB)	6°15'55.532"	4°17'33.431"
RIVERS	Bori-Kaa Water side (Rk)	7°30'53.352"	4°44'7.9553."
RIVERS	Bonny (RB)	7°9'21.926"	4°39'47.274"
AKWA IBOM	Eket-ibeno (AB)	8°0'30.974"	4°32'31.000"

Table 1: Sampling stations with coordinates.

[42]. Makoko (LM) is located on the west of Lagos lagoon and is one of the major fish landing sites of the lagoon. The commodore channel also connects the Atlantic Ocean to Apapa and environs through the Lagos harbour. Kaa (RK) water front in Rivers is one of the two major landing sites for the obolo (Andoni) fishers that use un-motorized "dugout" canoe for their subsistence occupation.

The Andoni River sand Bonny River are connected with the Niger Delta which is one of the world's largest wetlands covering an area of approximately 70,000 km^2. The Niger Delta is rich in biodiversity with numerous oil exploratory activities [44].

According to [45] he basin of bonny river is enclosed eastward by Andoni basin, westward by the New Calabar basin and then northward is the coastal plain sand. The length of the upper bonny river to the bonny bar is approximately 80 km. Water depth generally decreases up stream. It represents one of the most stressed river system with great economic importance due to the intensity of several oil field, industrial and fishing activities Ibeno (AB) station has the highest number of fishing settlements in Akwa Ibom and represents one of the largest producers of fish in the country. The Southern part of Akwai Ibom is bounded by the Atlantic Ocean at a region commonly called the Bight of Benin [46].

Generally, the climate of the study sites is equatorial in nature with two distinct periods of rainfall: March-July, which peaked in July and Sept-October. The periods of dry season are January-February, August and November-December. The sampling sites are both marine and brackish waters with artisanal fishing being the predominant economic activity within the surrounding settlements.

Collection of water samples and analysis of physico-chemical parameters

Sampling took place between 7 am and 12 pm. For each station, sampling was done 3 times at different locations to obtain their mean values. Water samples for physico-chemical analysis were collected 0.50 m below the water surface, in 1 dm^3 water sampler and stored on ice chest in one litre water bottles. In the laboratory, the water samples were transferred into refrigerator (4°C) and analysed within 24 hr of collection. Surface water temperature was measured in situ using mercury-in-glass thermometers, while pH, conductivity and salinity were analysed in the laboratory using a multi-meter water checker (Horiba U-10). Dissolved oxygen content was determined using Griffin digital meter Dissolve Oxygen (model 40) [47].

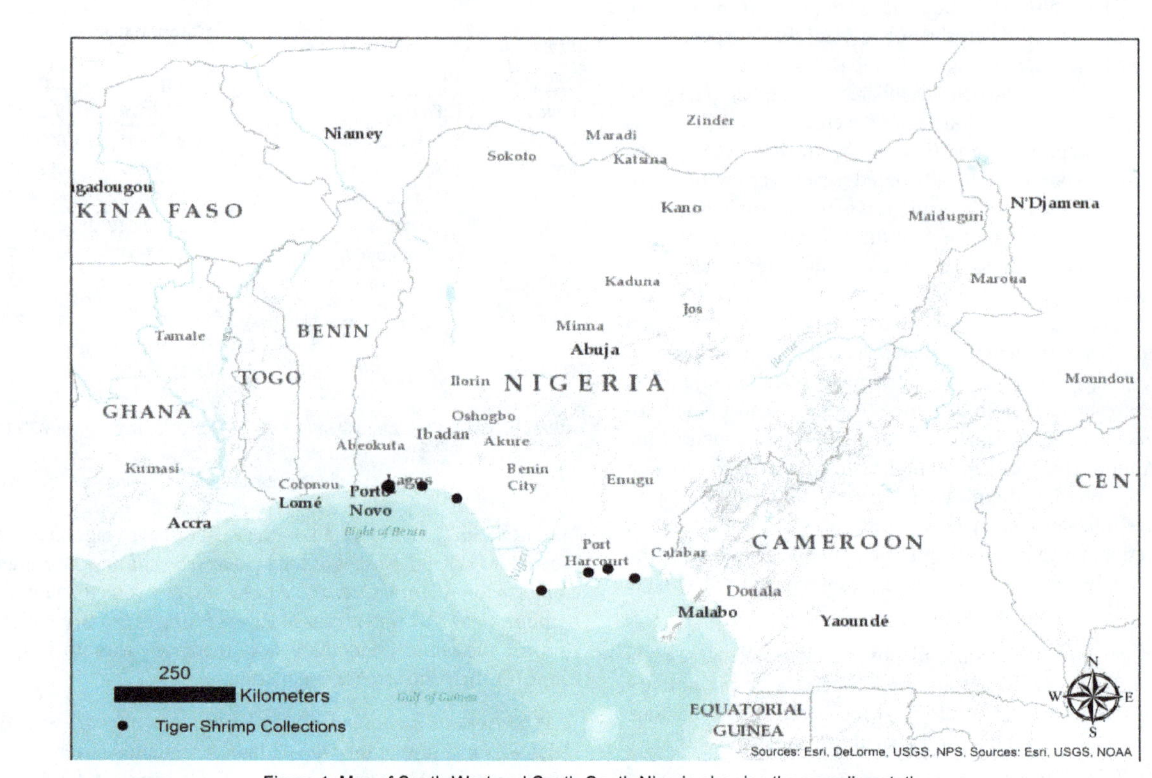

Figure 1: Map of South-West and South-South Nigeria showing the sampling stations.

Phytoplankton biomass

Samples for plankton analysis were collected using 55 μm mesh size standard plankton net and a 10 L plastic container. For each station, plankton samples were collected by filtering 100 L of surface water through the net. Plankton filtrates were then transferred (preserved with 4% formalin) into a well labelled 250 ml plastic container with a screw-cap. Plankton analysis and biomass were determined by counts. Plankton fixation lasted about 48 hr in the lab. The supernatant was decanted leaving behind a concentration of about 40 ml. With the aid of a dropper, two drops (0.2 ml) of each sample was placed on a glass slide with cover slip over the mount. Drop count method was used to analyze. This was done five times for each sampling station. Thorough investigation under light binocular microscope (Olympus BX51) was achieved. Examination, identification, counting and record of mean abundance of was done via varying magnification (x50-x400). A record of total organism was taken and equated per ml [31,32,47]. Authentication of species was confirmed using appropriate text (Hendey 1958, 1964; Patrick and Reimer, 1966, 1975; Wimpenny, 1966; Whitford and Schmacher, 1973; Vanlandingham, 1982; Nwankwo, 1990, 1995, 2004; Bettrons and Castrejon, 1999; Lange-Bertalot, 2001; Witkowski 2000; Siver, 2003; Rosowski, 2003).

Nutrient sampling

Nitrate, Phosphate Sulphate and silicate were measured with LaMotte SMART Spectrophotometer at different wavelengths using their appropriate colour development reagents. The Smart Spectro is an Environmental Protection Agency-Accepted instrument, meets the requirements for instrumentation as found in test procedures that are approved for the National Primary Drinking Water Regulations (NPDWR) or National Pollutant Discharge Elimination System (NPDES) compliance monitoring programs (GCLME-2009, LaMotte Operator's Manual-2012).

Instrumentation: Horiba U-10, LaMotte Smart Spectrophotometer, Centrifuge and Membrane filtration Unit.

Biological indices

- Shannon and Wiener diversity index (H′) according to Ogbeibu [48] was obtained as

$$H = -\sum p_i \ln(p_i)$$

where

p_i=proportion of observations species category

- Species Equitability or Evenness index (j) [49]

Formula for evenness is $J' = H'/H'_{max}$

Where $H'_{max} = \ln(S)$

Simpson Diversity Index (D) obtained with the formula $= \sum_{i=1}^{R} p_i^2$

- Gini Simpson index: (1-D) obtained as the transformation of Simpson index, D since the values of D will always reduce with increasing diversity.

Species Richness Margalef Index (d) assessment of community structures and obtained by

$$d = \frac{S-1}{\ln N}$$

where

d=Species richness index

S=Number of species in subpopulation

N=Sum total of individuals in S species [47,48]

Results

Physico-chemical parameters

The Physico-chemical characteristics of the study areas are presented in Tables 2 and 3. Mean temperature range was between 27.33 ± 1.53°C (station LM) and 29.00 ± 1.00°C (station RB). Mean pH values ranged between 7.39 ± 0.08 (station LA) and 8.13 ± 0.14 (station OA) (buffered ecosystems across sampling station). Range value for dissolved oxygen was 5.40 ± 3.22 mgL^{-1} (station AB)-8.00 ± 1.44 mgL^{-1} (station OA) while conductivity was 11.22 ± 10.03 µS/cm (station LM)-39.33 ± 5.87 µS/cm (station OA). The lowest values for phosphate-phosphorus, nitrate-nitrogen and sulphate were 0.11 ± 0.07mgL^{-1} (station RK), 0.10 ± 0.07 mgL^{-1} (station RK) and 523.67 ± 880.21 mgL^{-1} (station LM) respectively. Salinity recorded the least mean value of 11.02 ± 15.56% (station LC) and highest 25.51 ± 2.02% (station RB.

Phytoplankton biomass

Composition and distribution of phytoplankton across the 10 sampling stations are presented in (Tables 4-6). A total of 147 species of phytoplankton from 6 classes were recorded during the survey. Total number of species recorded per station ranged between 9 (station BB) and 42 (station LT). Furthermore the highest number of individual count was 952 per ml (station AB) and lowest 156 were observed in stations RK and LF.

(Table 7) depicts the phytoplankton assemblage and prevalence recorded in the survey. 6 classes of species vis-à-vis, Bacillariophyceae-Diatoms (70.4%), the Chlorophyceae-Chlorophytes (20.4%), Cyanophyceae-Blue-green algae (5.6%), Chrysophyceae (1.9%) Dinophyceae (1.1%) and Euglenophyceae (0.5%) were recorded.

(Table 8) however tabulates the phytoplankton community's biological indices. Margalef Index (d) Values were from 1.30 (station BB) to 6.14 (station RK), Shannon-Wiener Index (H^1).

were between 1.52 (station BB) and 3.03 (station RK), Equitability (j) values were between 0.52 (station AB) and 0.88 (station RK) and Simpson's Dominance Index ranged between 0.07 (station RK) and 0.32 (AB).

Discussion

In *P. monodon*, temperature is an essential environmental parameter having considerable impact not only on the success of culture, but also on survival and spread in areas of introduction. Temperature is known to influence rate of development, reproductive cycle and timing, migration patterns, growth, metabolism, sensitivity to toxins, susceptibility to parasites and diseases infection [50,51]. Optimum temperature range of 28°C-33°C is essential for survival and growth, but detrimental below 20°C. Although, tolerance between 13°C - 33°C and mortality at <13°C and >33°C were documented, fatal extremes have not been ascertained [52-54]. In this study, the least mean temperature recorded was 27.33°C at Lagos-Makoko. While the highest was 29°C at Rivers (Bonny and Kaa) station, according to Dublin-Green [45] and Komi and Sikoki [40] who reported Physico-chemical Characteristics

STATE	STATIONS	Water Temp (°C)	pH	Cond (µS/cm)	Salinity (ppt)	D.O (mg/l)	Phosphate (mg/l)	Nitrate (mg/l)	Sulphate (mg/l)	Chl a (µg/l)	Chl b (µg/l)	Chl c (µg/l)	Values
AKWA IBOM	AB(1)	27.00	7.78	19.30	26.34	1.80	0.23	0.08	1,980.00	0.0017	0.0019	0.0039	
	AB(2)	29.00	7.73	19.20	11.50	8.00	0.52	0.08	2,400.00	0.0209	0.0003	0.0004	
	AB(3)	28.00	7.49	8.00	30.76	6.40	0.72	0.44	1,960.00	0.0112	0.0055	0.0504	
		28.00	**7.67**	**15.50**	**22.87**	**5.40**	**0.49**	**0.20**	**2,113.33**	**0.01**	**0.00**	**0.02**	mean
		1.00	0.16	6.50	10.09	3.22	0.25	0.21	248.46	0.01	0.00	0.03	S.D
RIVERS	RB(1)	30.00	7.47	4.50	25.00	5.20	0.19	0.12	1,640.00	0.0039	0.0015	0.0024	
	RB(2)	29.00	7.34	22.90	22.02	4.40	0.77	0.15	2,440.00	0.0114	0.0137	0.0470	
	RB(3)	28.00	8.06	45.40	29.50	7.60	1.28	0.13	4,500.00	0.0259	0.0149	0.0421	
		29.00	**7.62**	**24.27**	**25.51**	**5.73**	**0.75**	**0.13**	**2,860.00**	**0.01**	**0.01**	**0.03**	mean
		1.00	0.38	20.48	3.77	1.67	0.55	0.02	1,475.53	0.01	0.01	0.02	S.D
	RK(1)	29.00	7.51	25.40	15.60	4.00	0.06	0.16	3,120.00	0.0187	0.0145	0.0586	
	RK(2)	30.00	7.47	23.89	27.28	5.20	0.19	0.12	1,640.00	0.0039	0.0015	0.0024	
	RK(3)	28.00	7.97	39.20	25.20	7.60	0.07	0.02	3,760.00	0.0295	0.0077	0.0125	
		29.00	**7.65**	**29.50**	**22.69**	**5.60**	**0.11**	**0.10**	**2,840.00**	**0.02**	**0.01**	**0.02**	mean
		1.00	0.28	8.44	6.23	1.83	0.07	0.07	1087.38	0.01	0.01	0.03	S.D
Bayelsa	BB(1)	29.00	7.35	10.40	25.67	8.40	0.16	0.30	3,920.00	0.0504	0.0200	0.0303	
	BB(2)	29.00	8.53	0.81	19.90	8.20	0.19	0.01	31.00	0.0559	0.0095	0.0089	
	BB(3)	28.00	8.01	35.90	22.80	6.79	0.23	0.21	1,960.00	0.0068	0.0078	0.0302	
		28.67	**7.96**	**15.70**	**22.79**	**7.80**	**0.19**	**0.17**	**1,970.33**	**0.04**	**0.01**	**0.02**	mean
		0.58	0.59	18.14	2.89	0.88	0.04	0.15	1,944.52	0.03	0.01	0.01	S.D
Ondo	OA(1)	28.00	8.06	35.70	22.60	9.20	0.98	0.62	3,500.00	0.0068	0.0027	0.0021	
	OA(2)	29.00	8.05	36.20	23.00	8.40	0.34	0.10	3,700.00	0.0060	0.0002	0.0008	
	OA(3)	27.00	8.29	46.10	30.00	6.40	0.01	1.02	440.00	0.0023	0.0007	0.0066	
		28.00	**8.13**	**39.33**	**25.20**	**8.00**	**0.44**	**0.58**	**2,546.67**	**0.01**	**0.00**	**0.00**	mean
		1.00	0.14	5.87	4.16	1.44	0.49	0.46	1,827.17	0.00	0.00	0.00	S.D

Table 2: Physico-Chemical parameters of water samples across stations.

STATE	STATIONS	Water Temp (°C)	pH	Cond (µS/cm)	Salinity (ppt)	D.O (mg/l)	Phosphate (mg/l)	Nitrate (mg/l)	Sulphate (mg/l)	Chl a (µg/l)	Chl b (µg/l)	Chl c (µg/l)	values
	LA(1)	28.00	7.44	1.99	12.00	7.00	0.02	0.03	77.00	0.0057	0.0040	0.0122	
	LA(2)	27.00	7.30	23.55	13.56	6.65	0.67	0.12	1,540.00	0.0029	0.0015	0.0025	
	LA(3)	28.00	7.42	22.08	17.98	6.70	0.01	0.02	9.00	0.0105	0.0062	0.0126	
		27.67	**7.39**	**15.87**	**14.51**	**6.78**	**0.23**	**0.06**	**542.00**	**0.01**	**0.00**	**0.01**	mean
		0.58	0.08	12.05	3.10	0.19	0.38	0.06	864.96	0.00	0.00	0.01	S.D
	LM(1)	29.00	7.36	10.34	19.76	5.30	0.98	0.22	1,540.00	0.0039	0.0015	0.0024	
	LM(2)	26.00	7.95	21.66	0.20	7.00	0.33	0.3	24.00	0.7352	0.3256	0.1834	
	LM(3)	27.00	7.63	1.66	23.56	7.56	0.34	2.03	7.00	0.0117	0.0134	0.0375	
		27.33	**7.65**	**11.22**	**14.51**	**6.62**	**0.55**	**0.85**	**523.67**	**0.25**	**0.11**	**0.07**	mean
		1.53	0.30	10.03	12.53	1.18	0.37	1.02	880.21	0.42	0.18	0.10	S.D
	LT(1)	28.00	8.53	14.50	19.90	8.34	0.19	0.01	31.00	0.0560	0.0065	0.0089	
	LT(2)	28.00	7.22	12.53	12.00	6.34	0.03	0.01	161.00	0.0087	0.0076	0.0185	
Lagos	LT(3)	27.00	7.58	35.40	22.60	5.70	0.01	0.31	1,939.00	1.0029	0.1765	0.4179	
		27.67	**7.78**	**20.81**	**18.17**	**6.79**	**0.08**	**0.11**	**710.33**	**0.36**	**0.06**	**0.15**	mean
		0.58	0.68	12.67	5.51	1.38	0.10	0.17	1,066.04	0.56	0.10	0.23	S.D
	LC(1)	27.00	7.34	22.90	22.02	7.40	0.77	0.15	2,440.00	0.0114	0.0137	0.0470	
	LC (2)	29.00	7.13	24.66	0.00	6.70	0.52	0.33	34.00	0.0409	0.0130	0.0267	
	LC(3)	28.00	7.77	0.31	0.01	6.78	0.01	0.09	7.00	0.0078	0.0088	0.0402	
		28.00	**7.41**	**15.96**	**11.02**	**6.96**	**0.43**	**0.19**	**827.00**	**0.02**	**0.01**	**0.04**	mean
		1.00	0.33	13.58	15.56	0.38	0.39	0.12	1,396.96	0.02	0.00	0.01	S.D
	LF(1)	27.00	8.45	18.44	26.34	7.20	0.23	0.08	1,880.00	0.0017	0.0019	0.0049	
	LF(2)	28.00	7.42	20.99	23.80	7.88	0.18	0.32	11.00	0.0065	0.0045	0.0089	
	LF(3)	28.00	8.06	2.61	27.80	8.55	0.02	0.03	4.00	0.006	0.005	0.009	
		27.67	**7.98**	**14.01**	**25.98**	**7.88**	**0.14**	**0.14**	**631.67**	**0.00**	**0.00**	**0.01**	mean
		0.5774	0.52	9.9575	2.024	0.675	0.109697	0.155	1081.094	0.00264	0.00167	0.0024	S.D

Table 3: Physico-Chemical parameters of water samples across stations.

of the Kaa water station (part of the Andoni River) and its potentials for production of *P. monodon*, temperature varied between 27°C and 31°C. However previous work reported values of 26.2°C to 32.4°C [41]. Generally the temperature range was within satisfactory WHO (2006) and FEPA (1991) limits for warm water fish and shellfish production.

Apart from the temperature of coastal waters, salinity over the years has also been recognized as a key factor influencing the absence, presence and abundance of endemic species [22,55]. In the wild, high salinity in the marine habit may be an important factor in ovarian maturation and egg/embryo development of the tiger shrimp, which may account for the offshore movement of sub-adult in preparation for full maturation and breeding. This study reported the highest mean salinity of 25.98% in Lagos-Folu which is a marine habitat. While the lowest (11.02% and 14.51%) were reported in the brackish waters of Lagos-Tin can Island and Lagos-Makoko stations respectively. Ecologists have connected salinity gradients within lagoon and estuaries to two main factors; influx of floodwater from rivers and nearby creeks of wetlands and tidal seawater inflow [23,56,57]. In Lagos lagoon, it has been reported that rainfall distribution determines salinity gradient and same factor may apply to estuarine system. Least salinity recorded by Nwankwo and Gaya [58] during the dry season was 8.6%. While Onyema [34] reported an average of 18.0% in the months of November and December at Onijeji Lagoon, Lagos. Salinity plays essential role in controlling growth and survival of tiger shrimps. Though euryhaline, *P. monodon* is comfortable at optimum salinity. High salinity is reported to induce slow growth, but promotes high health and resistance to diseases. On the other hand, low salinity may encourage growth but with weak shell and disease susceptibility. In the wild, the need for low salinity may account for movement of larval and protozoa stages to the estuaries where they grow into sub-adults [59,60] observed that *P. monodon* obtained from their native spawning grounds (Indian Ocean) with salinity 33% yielded better maturation and egg fertilization compared with those from Songkhla lake (22-28%). However, lower salinity (15-25%) is optimum to stimulate growth in the grow-out phase [61]. In culture ponds, salinity range of 20-28 % has been demonstrated and survival rate was 87% [62]. In Nigeria, spawning of gravid females from wild invasive species was successfully at 35%. Out of four breeding trials, six were successful [4]. Moreover other authors recommended salinity range of 10-35%). Few reports have observed adaptation of *P. monodon* to freshwater conditions, which may be attributed to its wide range of salinity tolerance [63-65].

H has an important role in metabolism, physiological processes, indicator of presence of metabolites, photosynthetic activity and fertility of aquatic environment. It varies with dead algae, excretory and residual feed (in the case of culture medium). Maximum values for aquatic environment are obtainable at maximum photosynthetic activity. Whereas very high pH is an indication of high fertility and depletion of dissolved oxygen due to plankton bloom, suggestive of eutrophication (nutritive enrichment by nitrogen and phosphorous compounds generated by human activities) [66]. According to [67] pH 6.8-8.7 is optimum for peneaid shrimps. A range of 7.5-8.5, which is in consonance with the range reported in this study, was recommended by Reddy [68]. Alkaline values indicating high amount of CO_2 stored in Carbonate forms in seawater produces a buffering effect [32] A similar inference was reported by [33] for the Lagos lagoon and Onyema and Nwankwo [36] at Iyagbe lagoon.

The species composition of phytoplankton observed in the present study was dominated by the marine phytoplankton. Marine

State	Akwaibom	Rivers		Bayelsa	Ondo	Lagos					
Stations	**AB**	**RB**	**RK**	**BB**	**OA**	**LF**	**LC**	**LT**	**LM**	**LA**	**TOTAL**
CLASS BACILLARIOPHYCEAE											
A. coffeaeformis (Agardh) Kutzing	-	40	-		-	-	-	-	-	-	40
A. holsatica Hustedt	-	1	-	-	-	-	-	-	-	-	1
A. ovalis(Kutzing) Kutzing	-	-	1	-	-	-	-	-	-	-	1
A.veneta Kutzing	-	-	-	-	-	-	-	-	-	3	3
Amphora sp	-	-	-	-	-	-	-	1	-	-	1
Achnanthes exilis Kutzing	-	-	-	-	-	-	-	-	-	1	1
A. parvula Kutzing	-	-	-	-	-	-	-	5	-	-	5
Asterionella formosa Hassal	7	-	-	-	-	-	-	-	-	57	64
A. japonica Cleve	-	-	3	4	-	-	100	54	-	-	161
Bacillaria paxillifer (O.F. Muller) Hendey	-	237	10	-	-	-	-	150	-	-	397
Biddulphia aurita (Lyngbye) Brebisson	-	-	3	7	-	-		1	2	-	13
B. longricruris Greville	-	-	-	-	-	-	65	-	-	-	65
B. rhombus (Ehr.) W. M. Smith	-	-	-	98	-	-	-	-	-	-	98
B. regia Greville	-	-	-	-	-	2	77	9	-	-	88
B. sinensis Greville	-	-	2	125	-	-	70	-	-	-	197
Biddulphia sp	-	-	-	134	-	-	-	-	-	-	134
Caloneis permagna (Bailey) Cleve	-	-	2	-	-	-	-	-	-	-	2
Chaetoceros lorenzianum Grunow	-	-	-	1	-	1	2	-	-	-	4
C. placentula Ehrenberg	1	-	-	-	-	-	-	-	-	-	1
Cocconeis sp	-	1	-	-	-	-	-	-	-	-	1
Coscinodiscus sp	10	2	8	89	-	9	8	11	-	-	137
Cyclotella sp	-	-	5	-	-	-	5	9	-	-	19
*Cymbella amphicephala*Naegeli	-	1	-	-	-	-	-	-	-	-	1
C. ehrenbergii Kutz	-	19	-	-	-	-	-	-	-	-	19
C. prostrata (Berkeley)	-	-	-	-	-	-	-	-	-	3	3
C. silesiaca Bleisch	-	-	-	-	-	-	-	-	-	2	2
Cymbella sp	-	6	8	-	-	-	-	1	-	-	15
Diploneis didyma (Ehr.) Cleve	-	-	1	-	-	-	-	-	-	-	1
Entomoneis costata (Hustedt) Reimer	-	-	1	-	-	-	-	-	-	-	1
E. ornata (Bailey) Reimer	30	-	-	-	-	-	-	-	-	-	30
Entomoneis sp	1	-	1	-	-	-	1	-	-	-	3
Epithemia sp	1	-	-	-	-	-	-	-	-	-	1
Eunotia triodon Ehr.	1	-	-	-	-	-	-	-	-	-	1
Eunotia sp	2	1	-	-	1	-	-	-	-	1	5
Fragilaria capucina Desmarziers	-	-	2	-	-	-	-	-	-	-	2
Fragilaria sp	14	-	11	-	50	-	-	-	-	-	75
Frustulia rhomboides (Ehrenberg) de Toni	7	-	-	-	6	1	-	-	-	-	14
Frustulia sp	-	-	-	-	-	-	-	-	-	1	1

Table 4: Composition and abundance of phytoplankton species across sampling stations.

phytoplankton are mainly composed of microalgae know as diatoms (Bacillariophytes) though other algae (green and blue green algae) can be found in low prevalence as reported in this study. Microalgae are requisite for larval nutrition by direct consumption [69]. In tiger shrimp, larval stage is made up of 6nauplius, 3protozoea, 3mysis and 3-4 megalopa substages. Initially, nauplii utilize yolk granules within their body as food and subsequently feeding on microalgae begins at the protozoea stage [70]. Carbohydrates in microalgae are mostly obtainable as highly digestible starch or as glucose, sugars and other forms of polysaccharides. Protein and fatty acid contents are major factors determining the nutritional value of available microalgae and are essential for zooplankton growth and metamorphosis of larval stages [71]. Phytoplankton are by themselves able to synthesize all amino acids, hence can supply the essential ones to larva and other zooplanktons [72].

Examples of the prevalent microalgae reported are in this survey are: Nitzschia, Navicula, Thalassiothrix, Amphora, Fragilaria, Coscinodiscus, Asterionella, Bacillaria paxillifer, Biddulphia, Melosira, Tabellaria and Surirella species. Some of the green algae include Aulacoseira granulate, Chlorella, Closterium, Chaetoceros and Eudorina elegan. While the blue green algae was dominated by Oscillatoria Sp, Spirulina Sp and Merismopedia Glauca. A good number of the diatoms have been reported by earlier workers especially for the Lagos lagoon and allied tidal creeks [29-32,56]. The presence of Nitzschia, Biddulphia and Thalassiothrix species probably point to their source of recruitment (the marine). According to [56], salinity and floodwater conditions are known to influence the algal composition and abundance in the Lagos lagoon. A similar situation likely exists for the brackish stations under this study. However for the green algae, [29] has already related these species to primarily fresh water conditions in association with the wet

State	Akwaibom	Rivers		Bayelsa	Ondo	Lagos					
Stations	**AB**	**RB**	**RK**	**BB**	**OA**	**LF**	**LC**	**LT**	**LM**	**LA**	**TOTAL**
CLASS BACILLARIOPHYCEAE											
Gyrosigma acuminatum (Kutzing) Rabenhorst	-	-	4	-	-	1	-	-	-	-	5
G. balticum (Ehrenberg) Rabenhorst	-	-	6	-	-	-	1	-	-	-	7
G. obscurum (W. Smith) Griffith & Henfrey	-	-	3	-	-	-	-	-	-	-	3
G. scalproides (Rabh.) Cleve	-	26	4	-	-	-	-	-	-	-	30
G. strigilis (W. Smith) Cleve	-	1	5	-	-	1	-	-	-	-	7
G. peisonis(Grunow) Hustedt	-	-	-	-	-	-	-	4	-	-	4
G. wansbeckii (Donkin) Cleve	-	-	-	-	-	-	-	1	-	-	1
Mastogloia sp	-	-	-	-	-	-	-	-	-	1	1
Melosira sp	95	-	-	-	-	45	1	**125**	4	-	270
Navicula clementis Grunow	-	-	-	-	-	-	-	**1**	-	-	1
N. crucicula (W.Sm.) Donkin	2	-	-	-	-	-	-	1	-	-	3
N. decussis Oestrup	1	-	-	-	-	-	-	-	-	-	1
N. mutica Kutzing	1	-	-	-	-	-	-	-	-	-	1
N. radiosa Kutz	-	1	3	-	-	-	-	46	-	-	50
N. zeta Cleve	-	-	1	-	-	-	-	-	-	-	1
Navicula sp	-	-	5	-	-	-	1	-	-	-	6
Neidium apiculatum Reimer	-	3	-	-	-	-	-	-	-	-	3
*N. binodeformis*Krammer	-	7	-	-	-	-	-	-	-	-	7
N. iridis (Ehreberg) Cleve	6	-	-	-	-	-	-	-	-	-	6
N. ladogensis (Cleve) Foged	-	1	-	-	-	-	-	-	-	-	1
N.septentrionale Cleve-Euler	-	-	-	-	-	1	-	-	-	-	1
N. productum (W. Smith) Cleve	-	1	-	-	-	-	-	-	-	-	1
Neidium sp	12	-	-	-	-	-	1	2	-	-	15
Nitzschia acicularis W.Smith	1	2	2	-	1	18	-	-	1	-	25
N. ignorata Krasske	-	70	2	-	-	2	3	-	-	-	77
N. thermalis Kutzing	-	5	-	-	-	-	-	-	-	-	5
N. sublinearis Hustedt	-	-	-	-	-	-	-	1	-	-	1
*N. subtilis*Grun	-	-	-	-	-	1	-		-	-	1
N. obtusa W. Sm	-	21	-	-	-	-	1	3	-	-	25
N. palea (Kutz) W. Sm	-	2	-	-	-	-	-	-	-	-	2
Nitzschia sp	-	2	-	-	-	5	1	1	-	3	12
Pleurosigma angulatum (Quekett) W. Smith	-	1	35	-	-	-	-	3	-	-	39
P. elongatum W. Smith	-	-	4	-	-	-	-	5	-	-	9
P. salinarum Grunow	-	-	-	-	-	-	10	-	-	-	10
Pinnularia acrosphaeria Brebisson	1	-	-	-	-	-	-	-	-	1	2
P. dactylus Ehrenberg	-	-	-	-	-	-	-	1	-	-	1
P. divergentissima(Grunow) Cleve	1	-	-	-	-	-	-	-	-	-	1
P. gibba Ehrenberg	2	-	-	-	-	-	-	-	-	-	2
P. lundii Hustedt.	-	-	-	-	-	-	-	-	-	1	1
P. macilenta Ehr. Emend. Cleve	-	-	-	-	-	-	-	5	-	-	5
P. maior (Kutzing) Rabenhorst	8	-	-	-	-	-	-	-	1	-	9
P. microstauron (Ehrenberg) Cleve	-	-	-	-	-	-	-	-	2	-	2
P. stomatophora (Grunow) Cleve	2	-	-	-	-	-	-	-	-	-	2
Pinnularia sp	-	-	2	-	-	-	-	-	-	-	2
Stephanodiscus sp	3	-	-	-	-	-	-	-	-	-	3
Surirella elegans Ehr.	-	-	-	-	-	-	-	5	101	-	106
Surirella sp	41	5	-	-	-	-	-	114	18	-	178
Synedra sp	10	-	-	-	2	23	3	45	1	-	84
Tabellaria fenestrata(Lyng) Kutzing	522	2	-	-	4	-	-	9	2	58	597
T. flocculosa (Roth) Kut.	79	-	-	-	-	-	-	-	-	-	79
Thalassiothrix frauenfeldii Grunow	-	-	6	2	-	-	39	8	-	-	55
T. nitzschioides (Grunow) Van Heurck	-	-	-	-	-	-	-	-	-	5	5
Ulnaria ulna (Nitzsch) Ehrenberg	-	1	-	-	-	-	-	-	-	-	1
U. ulna var. *longissima* (W. Sm.) Brun	-	-	-	-	-	-	-	-	-	1	1
Ulnaria sp	-	1	-	-	-	-	-	-	-	-	1
TOTAL INDIVIDAL COUNT	**861**	**460**	**140**	**460**	**64**	**110**	**389**	**621**	**132**	**138**	**3375**

Table 5: Composition and abundance of phytoplankton species across the sampling stations.

State	Akwaibom	Rivers		Bayelsa	Ondo	Lagos					
Stations	AB	RB	RK	BB	OA	LF	LC	LT	LM	LA	TOTAL
CLASS CHLOROPHYCEAE											
Asterococcus sp	-	-	-	-	-	24	-	-	-	-	24
Aulacoseira granulata var. *angustissima* f. *spiralis* (Hust)	-	-	-	-	-	-	-	-	201	-	201
Chlorella sp	-	-	-	-	-	-	-	24	4	-	28
Closterium abruptum (Lynb.) Breb.	-	-	-	-	1	-	-	-	-	-	1
C. moniliferum Ehrenb.	-	-	-	-	-	-	-	-	-	1	1
C. setaceum f. *sigmoideum* Irenee- Marie	-	-	2	-	-	5	-	-	-	-	7
C. peracerosum Gay	-	-	-	-	-	-	-	-	20	-	20
Closterium sp	-	3	-	2	-	-	1	2	-	-	8
Cosmarium binum Nordst	-	-	-	-	-	-	-	1	-	-	1
Cosmarium sp	-	-	-	-	-	-	-	-	3	-	3
Desmidium sp	-	-	5	-	-	-	8	2	-	-	15
Euastrum sp	-	-	-	-	-	-	-	1	-	-	1
Eudorina elegans Ehrenberg	-	-	-	-	150	-	-	32	-	-	182
Gonatozygon sp	1	-	-	-	-	-	-	-	-	-	1
Scenedesmus acuminatus (Lag.) Chodat	-	-	-	-	-	-	-	36	8	-	44
S. armatus var. *bicaudatus* (Gugl. Print) Chodat	-	-	-	-	-	-	-	-	8	-	8
S.bijuga (Turp) Lagerheim	-	-	-	-	-	-	-	-	4	-	4
S. quadricauda (Turp) Brebisson	-	-	-	-	-	-	-	-	2	-	2
Scenedesmus sp	-	-	-	-	-	-	-	8	-	-	8
Selenastrum bibraianum Reinsch	-	-	-	-	-	-	-	-	6	-	6
Spirogyra sp	-	1	-	-	-	-	-	-	-	5	6
Staurastrum americanum (W. and G. S. West) G.M. Smith	2	-	-	-	-	-	-	-	-	-	2
S. cingulum var. *floridense* Scott and Gronblad	-	-	-	-	-	-	-	-	124	-	124
S. tetracerum Ralf	-	-	-	-	-	-	-	-	4	-	4
S. vestitum Ralfs	-	-	-	-	-	-	-	4	-	-	4
Staurastrum sp	1	-	-	-	-	-	-	-	10	-	11
Stigeoclonium sp	-	-	-	-	-	-	-	-	-	6	6
Tetradesmus cumbrucus G.S. West	-	-	-	-	-	-	-	-	16	-	16
Tetraedron gracile Hansgirg	-	-	-	-	-	-	-	-	7	-	7
Pandorina sp	-	-	-	-	-	2	-	-	135	-	137
Pediastrum sp	-	-	-	-	-	-	-	9	7	-	16
Micrasterias sp	-	-	-	-	-	-	-	1	-	-	1
Microspora sp	-	-	-	-	-	10	-	-	-	-	10
Mougeotia sp	-	-	-	-	67	-	-	-	-	-	67
TOTAL INDIVIDUAL COUNT	**4**	**4**	**7**	**2**	**218**	**41**	**9**	**120**	**559**	**12**	**976**
CLASS DINOPHYCEAE											
Ceratium sp	-	-	-	-	-	-	1	-	-	-	1
Gymnodinium sp	50	-	-	-	-	-	-	-	-	-	50
Perinidium cinctum (Muller) Ehrenberg	-	-	-	-	-	-	-	-	-	-	0
Perinidium sp	-	-	-	-	-	-	-	-	-	-	0
TOTAL INDIVIDUAL COUNT	**50**	**0**	**0**	**0**	**0**	**0**	**1**	**0**	**0**	**0**	**51**
CLASS EUGLENOPHYCEAE			-	-						-	
Euglena sp	-	-	-	-	-	-	-		2	-	2

Phacus sp	2	-	-	-	11	-	-	1	9	-	23
TOTAL INDIVIDUAL COUNT	**2**	**0**	**0**	**0**	**11**	**0**	**0**	**1**	**11**	**0**	**25**
CLASS CYANOPHYCEAE				-							
Lyngbya sp	-	2	-	-	-	-	-	-	-	1	3
O. agardhii Gom	20	-	3	-	-	-	-	-	-	-	23
O. amphibia Agardh	-	-	-	-	30	-	-	1	-	8	39
O. limosa (Roth) Ag.	-	1	-	-	-	-	-	-	-	-	1
*O. subuliformis*Gom	-	2	-	-	-	-	-	-	-	-	2
Oscillatoria sp	-	-	-	-	17	-	-	-	-	-	17
S. subsalsa Oersted	-	-	6	-	-	-	-	-	-	-	6
Spirulina sp	-	-	-	-	-	5	-	-	76	3	84
Merismopedia glauca (Ehr) Nageli	-	-	-	-	101	-	-	-	-	-	101
TOTAL INDIVIDUAL COUNT	**20**	**5**	**9**	**0**	**148**	**5**	**0**	**1**	**76**	**12**	**276**
CLASS CHRYSOPHYCEAE											
Dinobryon sp	15	-	-	-	-	-	-	76	-	-	91

Table 6: Composition and abundance of phytoplankton species across the sampling stations.

State	**Akwa ibom**	**Rivers**		**Bayelsa**	**Ondo**	**Lagos**						
Stations	AB	RB	RK	BB	AO	LF	LC	LT	LM	LA	Total	% prevalence
CLASS												
BACILLARIOPHYCEA	861	460	140	460	64	110	389	621	132	138	3375	70.400501
CHLOROPHYCEAE	4	4	7	2	218	41	9	120	559	12	976	20.358782
EUGLENOPHYCEAE	2	0	0	0	11	0	0	1	11	0	25	0.5214852
CYANOPHYCEAE	20	5	9	0	148	5	0	1	76	12	276	5.7571965
CHRYSOPHYCEAE	15	0	0	0	0	0	0	76	0	0	91	1.8982061
DINOPHYCEAE	50	0	0	0	0	0	1	0	0	0	51	1.0638298
											4794	

Table 7: Phytoplankton Class Prevalence Across the 10 stations.

States	**AKWA IBOM**	**RIVERS**		**BAYELSA**	**ONDO**	**LAGOS**				
Stations	**AB**	**RB**	**RK**	**BB**	**AO**	**LF**	**LC**	**LT**	**LM**	**LA**
BIO-INDICES										
Total species diversity (S)	34	32	32	9	13	18	22	42	28	20
Total individual abundance (N)	952	469	156	462	441	156	399	819	778	162
Margalef index (d)	4.81	5.04	6.14	1.30	1.97	3.37	3.42	6.11	4.06	3.37
Shannon-Weiner (H^1)	1.84	1.88	3.03	1.52	1.80	2.21	2.03	2.68	2.26	1.85
Simson Dominance index (D)	0.32	0.29	0.07	0.24	0.21	0.15	0.17	0.10	0.15	0.25
Gini Simson index (1-D)	0.68	0.71	0.93	0.76	0.79	0.85	0.83	0.90	0.85	0.75
Species evenness (J')	0.52	0.54	0.88	0.78	0.65	0.77	0.69	0.72	0.65	0.62

Table 8: Biodiversity indices across the 10 sampling stations.

season and much less salinities. In North and North West Australia (native range of tiger shrimp) prevalence of diatoms and dianoflagellates (*Amphora* species and *Navicula* species) were reported. Large diatoms and blue-green alga Trichodesmium were fairly abundant, with the large dinoflagellates less significant [28]. It was further observed that the large tropical diatoms and dinoflagellates forms were markedly dissimilar from species in subtropical and temperate waters. Possession of large spines, horns, setae and wing-like structures in tropical forms, common symbiotic associations and greater species diversity accounted for the differences. Moreover, [21] also documented diatoms (Thalassiothrix fraunfeldii and Thalassiothrix nitzschioides) and dinoflagellates (*C trichoceros* and *P depressium*) as the most prevalent species composition of phytoplankton observed in waters of South East of Tamil Nadu, India. Conversely, diatoms domination amidst a range of phytoplankton were reported in [73], Pichavaram mangroves, India [74] and Kollidam eastuary, India [75].

Phytoplankton biomass is positively correlated with primary productivity. The upwelling (estuaries, mangroves) and the coastal regions have the highest productivity compared with open sea. One of the major factors responsible for this is nutrient availability in which run-offs from land & sediment disturbance reaches the upwelling region before the coastal and open sea. Consequently higher fish production is found in upwelling region [25,76,77]. Nigeria is a tropical country whose coastal waters, brackish and lagoon systems seem to have favoured the establishment and widespread populations of *P. monodon* in the last 16 years. Hypothetical introductions of *P. monodon* into the Nigerian coastal waters must have been movements or migration (through the trans-Atlantic Guinea current) from established populations in

Gambia, Senegal or Cameroon. These countries have culture facilities for tiger shrimp from which accidental introduction into the Atlantic Ocean must have occurred [77,78]. The same population in Gambia is suggestive of invasion in South East United States through the trans-Atlantic North equatorial current [12,53]. Another possible source is ballast waters (containing a variety of non-native living aquatic organisms). Reports of many decapod crustacean larval stages in viable conditions were documented to be recovered from ballast tanks [12,79]. Regardless of the mode of entry, the aquatic ecosystem is dynamic and for an invader to appear in a system it must first arrive via a transport vector, and then it must be documented [80]. On few accounts, it is likely that the detection of new invaders will be virtually simultaneous with their entry into a system as in the case of premeditated introductions or invasion of large or noticeable species. However, in most cases, time lapse is possible between initial invasion and eventual discovery of the invaders, as there is a strong predisposition for sighting invaders only after they become abundant [81,82]. These lags in detection are critical and could presumably be the case of the tiger shrimp in Nigeria as the invaders were only detected when they were already in abundance. First report of their capture was in [5]. The precise year, exact time and specific source of first introduction into the Atlantic coast of Nigeria is poorly documented. However, in the United States, precise year, time and source of first introduction was detected and reported. About 27 years ago (1988) at Waddell Mariculture Center, South Carolina, a number of tiger shrimps (originally from Hawaii) were inadvertently released into the Atlantic coast. They were initially assumed not to be established. However, two months later close to 300 of the shrimps were recovered in trawl nets off the coasts of South Carolina and two other coastal states. Not until after a time lag of 18 years (September 2006) a single adult male was caught in Mississippi Sound near Dauphin Island, Alabama. Subsequent catches were further reported in increasing amount over the years; (4) 2007, (45) 2009, (32) 2010 and (678) 2011 [12,83,84].

Initial introduction and population explosion have a time lapse (gap observed between an event and the period when its effects are visible) which is critical in ecological studies [80]. Mostly like during this period the invasive species develop adaptation, reproduce and spread throughout the newly colonized region. The time lapse of *P. monodon* invasion in Nigeria is not clear but suggestive of short period. This is not surprising as the invaders themselves have high adaptability to new environment, fecundity (up to 500, 000 eggs per spawn) and fast growth rate [59,85]. Time lag also appears to have coincided with the time, late 1990's when stocks of the most abundant and exploited native marine shrimp, Farfantepenaeus notialis plummeted leading to near collapse of the vibrant shrimp subsector [78]. At the end of 90s and beginning of the 21st century (year 2000), *Penaeus monodon* invasion suddenly became apparent in coastal and creek environments, with populations that surprisingly complemented *P. notialis* in a way significant enough to maintain industrial production. Could there possibly be an interaction between these two events of *P. monodon* explosion and *P. notialis* depletion around the same period [2] reported that invasive species can possess the ability to be well adapted and compete better than native species. The tiger shrimp is a host for the White spot syndrome virus of crustaceans (WSSV) which can be transmitted. *P. notialis* might be at risk of infection and may take some time to develop resistance. Furthermore, [86] reported higher condition factor index in *P. monodon* compared with *P. notialis*. The condition factor is a quantitative parameter of the wellbeing state of a fish reflecting recent feeding condition. Thus delineates weighty fish of a known length are somewhat in better condition. It represents an index of growth and feeding intensity [3,87,88]. Have also compared the weight of the two species observing significant difference stating that higher values obtained for *P. monodon* is due to its ability to grow larger and at a faster rate than other *Peneaids shrimp.*

Conclusion

Invasion of *P. monodon* is known worldwide, stimulating the interest of scientists. This is because the impression that aquatic invasive species rarely possess noticeable ecological impact in the newly found ecosystem was disputed by Carlton [80], highlighting that there is no sufficient experimental research on ecology of known marine crustacean invasions. Ironically, utilization of the tiger shrimp in aquaculture might have declined in the last 10 years due to disease susceptibility and hence low fitness in cages, but they are fast spreading in the wild as they invade new territories across the Pacific, off the coasts of West Africa, South East U.S, Mexican Gulf and the Caribbean. The giant size of this species is suggestive of greater nutritional requirements and higher competitive advantage over native faunas. The high fecundity is suggestive of pressure on limited available plankton and other resources for larval and post larval stages of organisms. If the giant tiger shrimp is known to be a predator, predation on indigenous organism is a concern. In addition, the species is also known to be a host vector of the most deleterious crustacean virus, WSSV that causes White Spot Syndrome, thus the risk of transmitting the disease to other native crustacean species. This and several other ecological studies are grossly inadequately and therefore must be checked.

References

1. Pimentel David S, McNair J, Janecka J, Wightman C, Simmonds, et al. (2001) Economic and environmental threats of alien plant, animal and microbe invasions. Agriculture, Ecosystems & Environment 84: 1-20.

2. Fox MD (1995) Conserving biodiversity: impact and management of exotic organisms. In: Bradstock RA, Auld TD, Keith DA, Kingsford RUT, Lunney D, Sivertsen DP. Conserving Biodiversity: Threats and Solutions. Surrey Beatty & Sons, Chipping Norton, New South Wales 177-183.

3. Yakubu AS, Ansa EF (2007) Length-weight relationships of the pink shrimp Penaeus monodon and giant tiger shrimp P. monodon of Buguma Creek in the Niger Delta Nigeria. The Zool 5: 47-53.

4. Ayinla OA, Anyanwu PE, Solarin BB, Hamzat B, Ebonwu BI, et al. (2009) Collection and maturation of broodstock of black tiger shrimp, Penaeus monodon in Nigeria. Proceedings of the 24th Annual Conference of the Fisheries Society of Nigeria (FISON), Nigeria pp: 91-95.

5. FAO (1999) Report of the four GEF/UNEP/FAO regional work shop on reducing the impact of tropical shrimp trawl fisheries, 15–17 Dec, Lagos, Nigeria. FAO Corporate Document Repository, Fisheries and Aquaculture Department. FAO fisheries Report No-627. pp: 15-17.

6. Dublin-Green CO, Tobor JG (1992) Marine Resources and Activities in Nigeria. Nigerian Institute of Oceanography and Marine Research (NIOMR), Tech Paper No: 84.

7. Chemonics International Incorporated (2002) Subsector Assessment of the Nigerian Shrimp and Prawn Industry. Chemonics International Incorporated. November report pp: 85.

8. Holthuis LB (1980) FAO species catalogue. Shrimps and prawns of the world. An annotated catalogue of species of interest to fisheries. FAO Fisheries Synopsis 125. Food and Agriculture Organization of the United States, Rome.

9. Marioghae IE (1987) An appraisal of the cultivability of Nigerian Palaemonid prawns. Lagos.

10. Lucien-Brun H (1997) Evolution of world shrimp production: fisheries and aquaculture. World Aquaculture 21-33.

11. Motoh H (1985) Biology and Ecology of Penaeus monodon. In Taki Y, Primavera JH, Llobrera JA, Proceeding of the First International Conference on the Culture of Penaeid prawn/shrimp. Aquaculture Department, Southeast Asian Fisheries Development Center, Iloilo, Philippines 27-36.

12. Fuller Pam L, David Knott M, Peter R, Kingsley-Smith, James Morris A, et al. (2014) Invasion of Asian tiger shrimp, Penaeus monodon Fabricius, 1798, in the western north Atlantic and Gulf of Mexico. Aquatic Invasions 9: 12.

13. Anyanwu PE, Ayinla OA, Ebonwu BI, Ayaobu-Cookey IK, Hamzat MB, et al. (2011) Culture possibilities of Penaeus monodon. Nigerian Fisheries and Aquatic Sciences 6: 499-505.

14. Global Biodiversity Information Facility (2013) Biodiversity occurrence data published by: Senckenberg: Collection Crustacea-ZMB.

15. Gómez-Lemos LA, Campos NH (2008) Presencia de Penaeus monodon Fabricius (Crustacea: Decapoda: Penaeidae) en aguas de la G.

16. Coelho PA, Santos MCF, Ramos-Porto M (2001) Ocorrência de Penaeus monodon Fabricius, 1798 no litoral dos estados de Pernambuco e Alagoas (Crustacea, Decapoda, Penaeidae Boletim Técnico Cienifico CEPENE, Tamandaré 9: 149-153.

17. Silva KCA, Ramos-Porto M, Cintra IHA (2002) Registro de Penaeus monodon Fabricius, 1798, na plataforma continental do estado do Amapá (Crustacea, Decapoda, Penaeidae Boletim Técnico Cientifico Centro de Pesquisa e Gestão de Recursos Pesqueiros do Litoral Norte (CEPNOR), Belém 2: 75-80.

18. Aguado NG, Sayegh J (2007) Presencia del camarón tigre gigante Penaeus monodon (Crustacea, Penaeidae) en la costas del Estado Anzoátegui, Venezuela. Boletim do Instituto Oceanográfico, Venezuela 46: 107-111.

19. Cintra IHA, Paiva KS, Botelho MN, Silva KCA (2011) Presence of Penaeus monodon in the continental shelf of the State of Para, northern Brazil (Crustacea, Decapoda, Penaeidae Revista de Ciencias Agrarias 54: 314-317.

20. Baskar S, Narasimhan N, Swamidass Daniel G, Ravichelvan R, Sukumaran M, et al. (2013) Food and Feeding Habits of Penaeusmonodon (Fabricius) from Mallipattinam Coast in Thanjavur Dist, Tamil Nadu, India. International Journal of Research in Biological Sciences 3: 1-4.

21. Thirunavukkarasu K, Soundarapandian P, Varadharajan D, Gunalan B (2013) Phytoplankton Composition and Community Structure of Kottakudi and Nari Backwaters, South East of Tamil Nadu. Journal of aquaculture research development 5: 1-9.

22. Nwankwo DI (2004) A Practical Guide to the study of algae. JAS Publishers, Lagos. Nigeria 84.

23. Onyema IC (2007) The phytoplankton composition, abundance and temperature variation of a polluted estuarine creek. Turk. J. Fish. Aqua. Sci 7: 89-96.

24. Verlencar XN, Desai S (2004) Phytoplankton Identification Manual. National Institute of Oceanography. Dona paula, Goa India 33.

25. Rabalais NN (2002) Nitrogen in Aquatic Ecosystems. BioOne 31: 102-112.

26. Kannan L, Vasantha K (1992) Microphytoplankton of The Pitchavaram Mangals, Southeast Coast of India. Species composition and population density. Journal of Hydrobiology 247: 76-86.

27. Soundarapandian P, Premkumar T, Dinakaran GK (2009) Studies on the Physico-chemical Characteristic and Nutrients in the Uppanar Estuary of Cuddalore, South East Coast of India. Current Research Journal of Biological Sciences 1: 102-105.

28.

29. Hallegraeff GM, Jeffrey SW (1984) Tropical phytoplankton species and pigments of continental shelf waters of North and North-West Australia. Marine Ecology-Progress Series 20: 59-74.

30. Nwankwo DI (1988) A preliminary checklist of planktonic algae in Lagos lagoon Nigeria. Nigeria. Journal of Botanical Applied Sciences 2: 73-85.

31. Nwankwo DI (1997) A first list of dinoflagellates (Pyrrophyta) from Nigerian coastal waters (creeks, estuaries, lagoons) Pol. Arch. Hydrobiol. 44: 317-321.

32. Adesalu TA, Nwankwo DI (2005) Studies on the phytoplankton of Olero creek and parts of Benin river, Nigeria. The Ekologia 3: 21-30.

33. Adesalu TA, Nwankwo DI (2010) Cyanobacteria of tropical lagoon, Nigeria. Nature and Science 8: 77-82.

34. Onyema IC, Otudeko OG, Nwankwo DI (2003) The distribution and composition of plankton around sewage disposal site at Iddo, Nigeria. Journal Science Research Development 7: 11-24.

35. Onyema IC (2013) The Physico-Chemical Characteristics and Phytoplankton of the Onijedi Lagoon, Lagos. Nature and Science 11: 1-9.

36. Emmanuel BE, Onyema IC (2007) The plankton and fishes of a tropical creek in South-western Nigeria. Turkish Journal of Fisheries 7: 105-113.

37. Onyema IC, Nwankwo DI (2009) Chlorophyll a dynamics and environmental factors in a tropical estuarine lagoon. Academia Arena 1: 18-30.

38. Opute FI (1992) Contribution to the knowledge of algae of Nigeria 1: Desmids from the Warri/Forcados estuaries. The genera Euastrum and Micrasterias. Archiwum Hydrobiologii. Supplement 93: 73-92.

39. Nwadiaro CS, Ezefill EO (1986) Preliminary Checklist of the Phytoplankton of New Calabar River, Lower Niger Delta. Nig. Hydrobiol. Bull 19: 133-138.

40. Sahel and West Africa Club/OECD (2006) Exploring Economic Opportunities in Sustainable Shrimp Farming in West Africa: Focus on South-South Cooperation.

41. Komi GW, Sikoki FD (2013) Physico-chemical Characteristics of the Andoni River and its potentials for production of the Giant Tiger Prawn (Penaeus monodon) in Nigeria. Journal of Natural Sciences Research 3: 83-89.

42. Ansa EJ, Sikoki FD, Francis A, Allison ME (2007) Seasonal variation in interstitial fluid quality of the Andoni flats, Niger Delta Nigeria. J. Appl. Sci. Environ. Manage 11: 123-127.

43. DEEP (2011) Bathymetric, object detection, hydrodynamic and sediment concentration survey-Oando Jetty project, Lagos.

44. Fabiyi, Oluseyi O, Gabriel kinbola A, Joseph Oloukoi, Funmilayo Thonteh, et al. (2012) Integrative approach of indigenous knowledge and scientific methods for flood risk analyses, responses and adaptation in rural coastal communities in Nigeria. START GRANT REPORT.

45. Francis A (2003) Studies on the Icthyofauna of the Andoni River System in the Niger Delta of Nigeria. Ph. D thesis, University of Port-Harcourt, Nigeria 281.

46. Dublin-Green (1990) Seasonal variation in some physico-chemical parameters of the bonnyestuary Niger delta. Nigeria institute for Oceanography and marine research technical paper 59: 1-24.

47. Ekpo, Imaobong Emmanuel, Mandu Asikpo Essien-Ibok (2013) Development, Prospects and Challenges of Artisanal Fisheries in Akwa Ibom State, Nigeria. International Journal of Environmental Science, Management and Engineering Research 2: 69-86.

48. Onyema IC, Harris-Sanni MO (2014) Changes in Water Chemistry, Chlorophyll a Concentration (Phytoplankton Biomass) and Zooplankton Characteristics at a Mariculture Site in the Lagos Lagoon International Journal of Environmental Sciences 3: 51-59.

49. Ogbeibu AE (2005) Biostatistics: A practical approach to research and data handling. Mindex Publishing Company limited, Benin city, Nigeria 264.

50. Boyce, Richard L (2005) Life under your feet: Measuring soil invertebrate diversity. Teaching Issues and Experiments in Ecology 3, Ecological Society of America 4-5.

51. Abowei JFN (2010) Salinity, Dissolved Oxygen, pH and Surface Water Temperature Conditions in Nkoro River, Niger Delta, Nigeria. Advance Journal of Food Science and Technology 2: 36-40.

52. Kelly Addy, Linda Green (1997) Dissolved oxygen and temperature. Natural Resources Facts, University Rhode Island fact sheet No 96.

53. Jintoni B (2003) Water quality requirements for Penaeus monodon culture in Malaysia. Department of Fisheries, Sabah, Malaysia.

54. Kingsley-smith PR, David Knott, Pam Fuller, Amy Benson, Matt Cannister, et al. (2012) The Asian tiger shrimp, Penaeus monodon: updates on a recent invader to the coastal marine and estuarine waters of the southeastern United States. USGS Invasive Species Interest Group.

55. Knott DM, Fuller PL, Benson AJ, Neilson ME (2015) Penaeus monodon. USGS Nonindigenous Aquatic Species Database, Gainesville, FL.

56. Onyema IC (2008) A checklist of phytoplankton species of the Iyagbe lagoon, Lagos. Journal of Fisheries and Aquatic Sciences 3: 167-175.

57. Nwankwo D (1986) I Phytoplankton of a sewage disposal site in Lagos lagoon, Nigeria. The Journal of Biological Sciences 1: 89-96.

58. Nkwoji JA, Onyema IC, Igbo JK (2010) Wet Season spatial occurrence of Phytoplankton and Zooplankton in Lagos lagoon. Science world Journal 2: 214-218.

59. Nwankwo DI, Gaya EA (1996) The Algae of an Estuarine Mari-culture site in

South-western Nigeria. Trop. Freshwater Biol 5: 1-11.

60. Hoa ND (2009) Domestication of black tiger shrimp (Penaeus monodon) in recirculation systems in Vietnam. PhD thesis, Ghent University, Belgium.

61. Ruangpanit N, Maneewongsa S, Pechmanee T, Tanan T, Kraisingdeja P (1984) Induced ovaries maturation and rematuration by eyestalk ablation of Penaeus monodon Collected from Indian Ocean and Songkhla lake. First Intl. Conference on the culture of Penaeids prawns/shrimps, Illoilo city, Phillipines pp: 4-7.6.

62. Yano I (2000) Cultivation of Broodstock in Closed Recirculation System in specific Pathogen free (SPF) Penaeid Shrimp. Suisanzoshoku 48: 249-257.

63. Pushparajan N, P Soundarapandian (2009) Recent Farming of Marine Black Tiger Shrimp, Penaeus Monodon (Fabricius) in South India. African Journal of Basic & Applied Sciences 2: 33-36.

64. Muthu MS (1980) Site selection and type of farms for coastal aquaculture of prawns. Proceedings of the Symposium on shrimp farming, Bombay, Marine Products Export Development Authority 97-106.

65. Karthikean J (1994) Aquaculture (Shrimp farming) its influence on environment. Technical paper submitted to the seminar 'Our Environment-Its challenges to development projects'. American Society of Civil Engineers, Culcutta, India.

66. Chen HC (1985) Water quality criteria for farming the grass shrimp, Penaeus monodon in: Proceedings of the first International conference on culture of Penaid prawns/shrimps p: 165.

67. Langland M, Cronin T (2003) A Summary Report of Sediment Processes in Chesapeake Bay and Watershed. In Water-Resources Investigations Report pp: 03-4123. New Cumberland, PA: US Geological Survey.

68. Ramanathan NP, Padmavathy T, Francis S, Athithian, Selvaranjitham N (2005) Manual on polyculture of tiger shrimp and crabs in freshwater. Tamil Nadu Veterinary and Animal Sciences University, Fisheries College and Research Institute, Thothukudi 1-161.

69. Reddy R (2000) Culture of the tiger shrimp Penaeus monodon (Fabricius) in low saline waters. M.Sc. Thesis. Annamalai University, Chidambaram, Tamil Nadu, India.

70. Helbling EW, Villafane V, Holm-Hansen O (1994) Effects of Ultraviolet Radiation on Antarctic Marine Phytoplankton Photosynthesis with Particular Attention to the Influence of Mixing. In Ultraviolet Radiation in Antarctica: Measurements and Biological Effects. Antarctic Research Series 62.

71. Primavera JH (1982) Studies on broodstock of sugpo Penaeus monodon Fabricius and other penaeids at the SEAFDEC Aquaculture Department. Marine Biological Association of India, Proceedings of the Symposium on Coastal Aquaculture, Cochin, India pp: 28-36.

72. Nichols DS (2003) Prokaryotes and the input of polyunsaturated fatty acids to the marine food web. FEMS Microbiology Letter 219: 1-7.

73. Guil-Guerrero JL, Navarro-Juárez R, López-Martínez JC, Campra-Madrid P, Rebolloso-Fuentes MM (2004) Functionnal properties of the biomass of three microalgal species. Journal of Food Engineering 65: 511-517.

74. Ignatiades L, Vassilion A, Karydis M (1985) A Comparison of Phytoplankton Biomass Parameter and Their Inter-Relation with Nutrients in Saronicos Gulf Greece Hydrobiol 128: 201.

75. Mani P (1994) Phytoplankton in Pichavaram Mangroves, East Coast of India. Indian J Mar Sci 23: 22-26.

76. Edward, Patterson JK, Ayyakkannu K (1991) Studies on the Ecology of Plankton Community of Kollidam Estuary, Southeast Coast of India. I Phytoplankton. Magasagar-Bull Natl Inst Oceanogr. 24: 89-97.

77. Hairston G, Nelson and Hairston G, Nelson (1993) Cause-effect relationship in energy flow, trophic structure and interspecific interactions. The American Naturalist. 142: 379-411.

78. Global change (2006) World Fisheries: Declines, Potential and Human Reliance.

79. Zabbey N, Erondu ES, Hart AI (2010) Nigeria and the prospect of shrimp farming: critical issues. Livestock Research for Rural Development 22.

80. Carlton JT (2011) The global dispersal of marine and estuarine crustaceans. In: Galil BS, Clark PF, Carlton JT, In the wrong place-alien marine crustaceans: distribution, biology and impacts, invading nature. Springer Series in Invasion Ecology 6: 3-23.

81. Crooks, Jeffrey A (2005) Lag times and exotic species: The ecology and management of biological invasions in slow-motion. Ecoscience 12: 316-329.

82. Lewin R (1987) Ecological invasions offer opportunities. Science 238: 752-753.

83. Crooks JA, Soulé ME (1999) Lag times in population explosions of invasive species: Causes and implications: 103-125. In Sandlund OT, Schei PJ, Viken A. Invasive Species and Biodiversity Management. Kluwer Academic Press, Dordrecht.

84. FAO (2005) Introductions and movement of two penaeid shrimp species in Asia and the Pacific. FAO fisheries technical paper No.476: 1-78.

85. McCann JA, LN Arkin, JD Williams (1996) Nonindigenous aquatic and selected terrestrial species of Florida. Report to US Fish and Wildlife Service, Washington, DC.

86. CAB International (2004) Prevention and Management of Alien Invasive Species: Forging Cooperation throughout West Africa. In: Proceedings of a workshop held in Accra, Ghana. CAB International, Nairobi, Kenya.

87. Ajani E, Gloria, Bello O, Beatrice, Osowo Olufem (2013) Comparative condition factor of two Penaeid shrimps, Peneaus notialis (Pink shrimp) and Peneaus monodon (Tiger shrimp) in a coastal state, Lagos, South West Nigeria. Nat. Sci. J 11: 1-3.

88. Fagade SO (1980) The structure of the otoliths of Tilapia guineensis (Dumeril) and their use in age determination Hydrobiologia 69: 169-173.

89. Bagenal TB, FW Tesch (1978) Age and growth. In: Bagenal T, Methods of assessment of fish production in Fresh Waters. Oxford Blackwell Scientific Publication: London, UK 101-136.

Permissions

All chapters in this book were first published in FAJ, by OMICS International; hereby published with permission under the Creative Commons Attribution License or equivalent. Every chapter published in this book has been scrutinized by our experts. Their significance has been extensively debated. The topics covered herein carry significant findings which will fuel the growth of the discipline. They may even be implemented as practical applications or may be referred to as a beginning point for another development.

The contributors of this book come from diverse backgrounds, making this book a truly international effort. This book will bring forth new frontiers with its revolutionizing research information and detailed analysis of the nascent developments around the world.

We would like to thank all the contributing authors for lending their expertise to make the book truly unique. They have played a crucial role in the development of this book. Without their invaluable contributions this book wouldn't have been possible. They have made vital efforts to compile up to date information on the varied aspects of this subject to make this book a valuable addition to the collection of many professionals and students.

This book was conceptualized with the vision of imparting up-to-date information and advanced data in this field. To ensure the same, a matchless editorial board was set up. Every individual on the board went through rigorous rounds of assessment to prove their worth. After which they invested a large part of their time researching and compiling the most relevant data for our readers.

The editorial board has been involved in producing this book since its inception. They have spent rigorous hours researching and exploring the diverse topics which have resulted in the successful publishing of this book. They have passed on their knowledge of decades through this book. To expedite this challenging task, the publisher supported the team at every step. A small team of assistant editors was also appointed to further simplify the editing procedure and attain best results for the readers.

Apart from the editorial board, the designing team has also invested a significant amount of their time in understanding the subject and creating the most relevant covers. They scrutinized every image to scout for the most suitable representation of the subject and create an appropriate cover for the book.

The publishing team has been an ardent support to the editorial, designing and production team. Their endless efforts to recruit the best for this project, has resulted in the accomplishment of this book. They are a veteran in the field of academics and their pool of knowledge is as vast as their experience in printing. Their expertise and guidance has proved useful at every step. Their uncompromising quality standards have made this book an exceptional effort. Their encouragement from time to time has been an inspiration for everyone.

The publisher and the editorial board hope that this book will prove to be a valuable piece of knowledge for researchers, students, practitioners and scholars across the globe.

List of Contributors

Kazumi Sakuramoto
Tokyo University of Marine Science and Technology, Konan, Minato-ku, Tokyo, Japan

Nwipie GN, Erondu ES and Zabbey N
Department of Fisheries, Faculty of Agriculture, University of Port Harcourt, East-West Road, PMB 5323, Choba, Rivers State, Nigeria

Masami Fujiwara
Department of Wildlife and Fisheries Sciences, Texas A&M University College Station, TX 77843-2258, USA

Michael KG and Sogbesan OA
Department of Fisheries, ModibboAdama University of Technology, Yola, Nigeria

Mudasir Rashid, Masood ul Hassan Balkhi and Gulzar Ah.Naiko
Faculty of Fisheries, Sher-e-Kashmir University of Agricultural Sciences &Technology of Kashmir, Rangil, Ganderbal, India

Tamim Ahamad
Department of Fisheries, J&K, India

Jahangir Sarker Md, Indrani Kanungo, Mehedi Hasan Tanmay and Shamsul Alam Patwary Md
Department of Fisheries and Marine Science, Noakhali Science and Technology University, Noakhali-3814, Bangladesh

Kyuji Watanabe
Hokkaido National Fisheries Research Institute, Japan Fisheries Research and Education Agency, 2-2 Nakanoshima, Toyohira-ku, Japan

Gulzar Naik, Mudasir Rashid, Balkhi MH and Bhat FA
Sher-e-Kashmir University of Agricultural Sciences and Technology of Kashmir, Rangil, Ganderbal, Jammu and Kashmir, India

Naser Ahmed Bhouiyan, Mohammad Abdul Baki, Anirban Sarker and Md. Muzammel Hossain
Department of Zoology, Jagannath University, Dhaka-1100, Bangladesh

Chandasudha Goswami
Department of Zoology, Govt. Institute of Science and Humanities, Amravati-444604, India

Zade VS
Govt. Institute of Science and Humanities, VMV Road, Amravati-444604, India

Atul K.Singh
Exotic Fish Germplasm Section of Fish Health Management Division of National Bureau of Fish Genetic Resources, Canal Ring Road, P.O. Dilkusha, Lucknow-226002 (Uttar Pradesh), India

Srivastava PP
Central Institute of Fishery Education, Mumbai-400061

Sven M Bergmann, Michael Cieslak and Dieter Fichtner
FLI Insel Riems, Südufer 10, 17493 Greifswald-Insel Riems, Germany

Juliane Dabels
University of Rostock, Aquaculture and Sea Ranching, Justus-von-Liebig-Weg 6, Rostock 18059, Germany

Sean J Monaghan
Aquatic Vaccine Unit, Institute of Aquaculture, School of Natural Sciences, University of Stirling, Stirling, FK9 4LA, UK

Qing Wang and Weiwei Zeng
Pearl-River Fisheries Research Institute, Xo. 1 Xingyu Reoad, Liwan District, Guangzhou 510380, P. R. of China

Jolanta Kempter
West Pomeranian Technical University, Aquaculture, K. Królewicza 4, 71-550, Szczecin, Poland

Carlos Rosas, Maite Mascaró, Richard Mena and Claudia Caamal-Monsreal
Unit Multidisciplinary Teaching and Research, Faculty of Sciences UNAM, Puerto de Abrigo s/n Sisal Yucatán, México

Pedro Domingues
Spanish Institute of Oceanography, Oceanographic Centre of Vigo, Stay out, Canido, 36390 VIGO, Spain

Asadujjaman M and Hossain MA
Department of Fisheries, University of Rajshahi, Rajshahi, Bangladesh

Shahangir Biswas, Matiar Rahman and Islam MA
Department of Biochemistry and Molecular Biology, University of Rajshahi, Rajshahi, Bangladesh

Manirujjaman M
Department of Biochemistry, Gonoshasthaya Samajvittik Medical College and Hospital, Savar, Dhaka 1344, Bangladesh

Abdel Moneim Yones M and Atallah Metwalli A
National Institute of Oceanography and Fisheries (NIOF), Shakshouk Aquatic Research Station, El Fayoum, Egypt

J Selvam, D Varadharajan, A Babu and T Balasubramanian
Faculty of Marine Sciences, Centre of Advanced Study in Marine Biology, Annamalai University, Parangipettai, Tamil Nadu, India

Hans Ulrik Riisgård and Florian Lüskow
Marine Biological Research Centre, University of Southern Denmark, Hindsholmvej 11, DK-5300 Kerteminde, Denmark

Josephine Goldstein and Kim Lundgreen
Marine Biological Research Centre, University of Southern Denmark, Hindsholmvej 11, DK-5300 Kerteminde, Denmark

Max-Planck Odense Center on the Biodemography of Aging and Department of Biology, Campusvej 55, DK-5230 Odense M, Denmark

Marianna Vaz Rodrigues and Jacqueline Kazue Kurissio
Department of Microbiology and Immunology, Biosciences Institute, Univ. Estadual Paulista (UNESP), Distrito de Rubião Júnior s/n, Botucatu, São Paulo, Brazil

Agar Costa Alexandrino de Pérez and Thaís Moron Machado
Reference Unit Laboratory Technology of Seafood-Instituto de Pesca, Agência Paulista de Tecnologia do Agronegócio, Secretaria da Agricultura e Abastecimento, Av Bartolomeu de Gusmão 192, Santos, São Paulo, Brazil

Fátima Maria Orisaka
Freelance veterinary Distrito de Rubião Júnior s/n, Botucatu, São Paulo, Brazil

Andrea Lafisca
Veterinary, In-lingua scientific translations and linguistic services, Distrito de Rubião Júnior s/n, Botucatu, São Paulo, Brazil

Comfort Adetutu Adeniji, Ajani Murano Rasheed and Rasheed Bolaji
Department of Fisheries, Lagos State University, Lagos, Nigeria

Pius Abimbola Okiki
Department of Biological Sciences, College of Sciences, Afe Babalola University, Ado Ekiti, Nigeria

Guoqiang Wang and Iain J McGaw
Department of Ocean Sciences, 0 Marine Lab Road, Memorial University, St John's, NL A1C 5S7, Canada

Tombi Jeannette, Akoumba John Francis, Mieguim Ngninpogni Dominique and Bilong Bilong Charles Felix
Department of Animal Biology and Physiology, University of Yaounde I, Yaounde, Cameroon

Katherine R Brown, Michael E Barnes and Timothy M Parker
McNenny State Fish Hatchery, South Dakota Department of Game, Fish and Parks, 19619 Trout Loop, Spearfish, South Dakota-57783, USA

Brian Fletcher
Cleghorn Springs State Fish Hatchery, South Dakota Department of Game, Fish and Parks 4725 Jackson Boulevard, Rapid City, South Dakota-57702, USA

Monjurul Hasan Md, Bhakta Supratim Sarker, Mahabubur Rahman, Shamsul Alam Patwary Md and Jahangir Sarker Md
Department of Fisheries and Marine Science, Noakhali Science and Technology University, Sonapur, Noakhali, Bangladesh

Shahriar Nazrul KM
Department of Fisheries, Ministry of Fisheries and Livestock, Bangladesh

Mohammed Rashed Parvej
Bangladesh Fisheries Research Institute, Mymensingh, Bangladesh

Mahmoud Abdelghagffar Emam
Histology and Cytology Department, Faculty of Veterinary Medicine, Benha University, Egypt

Badia Abughrien
Histology and Anatomy Department, Faculty of Veterinary Medicine, Tripoli University, Libya

Rohit Kumar, Lalit Singh, Lata Sharma, Neha Saxena, Dimpal Thakuria, Atul K Singh and Prabhati K Sahoo
ICAR-Directorate of Coldwater Fisheries Research, Bhimtal, Nainital, Uttarakhand, India

Veena Pande
Department of Biotechnology, Kumaun University, Bhimtal, Nainital, Uttarakhand, India

Jahangir Sarker Md, Borhan Uddin AMM, Shamsul Alam Patwary Md, Mehedi Hasan Tanmay, Farhana Rahman and Moshiur Rahman
Department of Fisheries and Marine Science, Noakhali Science and Technology University, Sonapur, Noakhali-3814, Bangladesh

Hanaa MM El-Khayat, Hoda Abdel-Hamid, Kadria MA Mahmoud and Hassan E Flefel
Department of Environmental Researches and Medical Malacology, Theodor Bilharz Research Institute, Imbaba, PO Box-30, Giza, Egypt

Hanan S Gaber
National Institute of Oceanography and Fisheries, Cairo, Egypt

Ahmed M Al-Beak
General Authority for Fish Resources Development (GAFRD), Egypt

Ghoneim, SI and Salem M
Suez Canal University, Egypt

El-Dakar AY
Suez University, Egypt

Jahangir Sarker Md, Shamsul Alam Patwary Md, Borhan Uddin AMM, Monjurul Hasan Md, Mehedi Hasan Tanmay and Indrani Kanungo
Department of Fisheries and Marine Science, Noakhali Science and Technology University, Noakhali, Bangladesh

Mohammed Rashed Parvej
Bangladesh Fisheries Research Institute, Mymensingh, Bangladesh

Elaine Espino-Barr, Marcos Puente-Gómez and Arturo Garcia-Boa
Instituto Nacional de Pesca, Playa Ventanas s/n, Manzanillo, Colima, México

Manuel Gallardo-Cabello
Instituto de Ciencias del Mar y Limnología, Universidad Nacional Autónoma de México, México

Fayaz Ahmad, Tanveer A Sofi, Bashir A Sheikh
Department of Zoology, University of Kashmir, Srinagar, India

Khalid M Fazili, Ajaz A Waza and Rabiya Rashid
Department of Biotechnology, University of Kashmir, Srinagar, India

Tantry Tariq Gani
Sheri Kashmir Institute of Medical Science-Soura, Srinagar, India

Narges Rostamian and Sirvan Azizpour
Department of Fisheries, University of Environment, Karaj, Iran

Ebrahim Masoudi
Department of Fisheries, Gorgan University of Agricultural and Natural Resources, Gorgan, Iran

Mohammad Hasan Gerami
Young Researchers and Elite Club, Shiraz Branch, Islamic Azad University, Shiraz, Iran

Sana Ullah
Fisheries and Aquaculture Lab, Department of Animal Sciences, Quaid-i-Azam University Islamabad, Pakistan

Oketoki TO
Nigeria Institute for Oceanography and Marine Research (NIOMR), Wilmot point road, Bar beach, Victoria Island Lagos, Nigeria

Index